AF775578

Additive Manufacturing with Novel Materials

Scrivener Publishing
100 Cummings Center, Suite 541J
Beverly, MA 01915-6106

Publishers at Scrivener
Martin Scrivener (martin@scrivenerpublishing.com)
Phillip Carmical (pcarmical@scrivenerpublishing.com)

Additive Manufacturing with Novel Materials

Processes, Properties and Applications

Edited by

R. Rajasekar

Department of Mechanical Engineering, Kongu Engineering College, Tamil Nadu, India

C. Moganapriya

Department of Mining Engineering, Indian Institute of Technology, Kharagpur, West Bengal, India

and

P. Sathish Kumar

The Sirindhorn Thai-German Graduate School of Engineering, King Mongkut's University of Technology, Bangkok, Thailand

Scrivener Publishing

WILEY

This edition first published 2024 by John Wiley & Sons, Inc., 111 River Street, Hoboken, NJ 07030, USA and Scrivener Publishing LLC, 100 Cummings Center, Suite 541J, Beverly, MA 01915, USA
© 2024 Scrivener Publishing LLC
For more information about Scrivener publications please visit www.scrivenerpublishing.com.

All rights reserved. No part of this publication may be reproduced, stored in a retrieval system, or transmitted, in any form or by any means, electronic, mechanical, photocopying, recording, or otherwise, except as permitted by law. Advice on how to obtain permission to reuse material from this title is available at http://www.wiley.com/go/permissions.

Wiley Global Headquarters
111 River Street, Hoboken, NJ 07030, USA

For details of our global editorial offices, customer services, and more information about Wiley products visit us at www.wiley.com.

Limit of Liability/Disclaimer of Warranty
While the publisher and authors have used their best efforts in preparing this work, they make no representations or warranties with respect to the accuracy or completeness of the contents of this work and specifically disclaim all warranties, including without limitation any implied warranties of merchantability or fitness for a particular purpose. No warranty may be created or extended by sales representatives, written sales materials, or promotional statements for this work. The fact that an organization, website, or product is referred to in this work as a citation and/or potential source of further information does not mean that the publisher and authors endorse the information or services the organization, website, or product may provide or recommendations it may make. This work is sold with the understanding that the publisher is not engaged in rendering professional services. The advice and strategies contained herein may not be suitable for your situation. You should consult with a specialist where appropriate. Neither the publisher nor authors shall be liable for any loss of profit or any other commercial damages, including but not limited to special, incidental, consequential, or other damages. Further, readers should be aware that websites listed in this work may have changed or disappeared between when this work was written and when it is read.

Library of Congress Cataloging-in-Publication Data

ISBN 978-1-394-19791-0

Cover image: Pixabay.Com
Cover design by Russell Richardson

Set in size of 11pt and Minion Pro by Manila Typesetting Company, Makati, Philippines

Printed in the USA

10 9 8 7 6 5 4 3 2 1

Contents

Preface

It is well-known that technology is rapidly moving towards additive manufacturing (AM) for the design and manufacture of high-performance materials for aerospace, medical, energy, automotive applications, and so on. AM is being utilized everywhere in the manufacturing process of parts to avoid difficulties and produce parts with high quality at a high rate. It is expected that the AM processes can bring another industrial revolution. AM is becoming more and more popular for producing components with intricate geometries, customizing products, and generating small quantities and prototypes.

This book provides insights into various AM processes, novel materials and their properties, as well as basic knowledge of AM techniques and their applications. It also explores the fundamental concepts and recent advancements in the development of novel materials for several applications, with special emphasis on platforms like AM techniques for polymers, metallic materials, composites, nanomaterials, hydrogels, etc. Special topics like environmental aspects of 3D printing and advanced 4D printing are also introduced. The technological aspects of AM are discussed in a very clear and understandable way with the help of self-illustrating artworks.

In this context, the book provides insight for all researchers, academicians, post-graduate or senior undergraduate students, and industry professionals working in this important area. We thank all the authors for their valuable research and input, and we offer our sincere thanks to the Scrivener and Wiley publishing teams for their help with this book.

R. Rajasekar
C. Moganapriya
P. Sathish Kumar
November 2023

1

Brief Glimpses of Additive Manufacturing Techniques

V. Bhuvaneswari

Department of Mechanical Engineering, KPR Institute of Engineering and Technology, Coimbatore, Tamil Nadu, India

Abstract

Manufacturing has undergone various changes since the beginning of wheel manufacturing in the stone age. Subtractive manufacturing was predominantly used for many years. Additive manufacturing started in the 1980s, moving manufacturing towards a different path, which led to freedom to produce sophisticated components very easily. This chapter focuses on the current trends in 3D printing techniques along with envisaging its scope through the patent landscape. At the end of this chapter, the researcher gets a clear picture of the current state of additive manufacturing techniques and the future directions in this field.

Keywords: Additive manufacturing, 3D printing, polymer, metal, ceramics

1.1 Introduction

Layer-by-layer fabrication, which has been used to create 3D objects since the 1980s [1, 2], paves the path for intricate structures through carefully tailored characteristics and functions. Rapid prototyping was the name given to this novel technique of producing physical prototypes. Raw materials could either be solid, liquid, or powder in this developing technology [3]. Additive manufacturing (AM) processes have found broad use in industry and medicine because of their versatility in producing complex and precisely crafted items with preferentially constructed characteristics [4]. As a result, AM techniques have accelerated product growth by less

Email: bhuvanashankar82@gmail.com

R. Rajasekar, C. Moganapriya and P. Sathish Kumar (eds.) Additive Manufacturing with Novel Materials: Processes, Properties and Applications, (1–22) © 2024 Scrivener Publishing LLC

expensive, shorter manual interaction and product cycle, and fewer constraints. Binder jetting (BJ), direct ink writing, sheet lamination, and digital light processing are just some of the many techniques and processes that have advanced in AM processes [5]. The collection of AM methods is often referred to as "3D printing (3DP)." Rather than being confined to the creation of static, single-function 3D objects, 3DP techniques can be expanded to include the creation of multifunctional, ever-changing shapes for the duration of an object's lifespan. Four-dimensional printing (4DP) allows for greater customization and variety in the manufacturing process. The same methods as AM are used in 4DP, with the addition of a gradual alteration to the material's properties [6]. Time is considered a fourth dimension, with smart materials having the ability to adapt to changing conditions over time. As a result, these substances can be used in a variety of biomedical contexts, including sensors, robots, complex components, scaffold implants, and several other uses [7–10].

Computer-aided design (CAD) modeling software is used to define the shape and structure of smart materials with the desired characteristics and the material is then sliced into layers of the appropriate thickness for 3DP or 4DP. Using external stimuli to modify the material's specific item and, by extension, the product's shape, would be the next step in the evolution of products made from some other sort of smart materials. The pH, temperature, electromagnetic field, light, moisture, and/or moisture content are all examples of these kinds of externally applied stimuli [11–13]. Whether it is a phase shift, stress relaxation, deformation, or anything else, these factors have the potential to alter the material's chemical and physical properties. Materials are further defined by their mechanical properties, which include their response to external stimuli in terms of flexing, inflammation, contortion, and size reduction [14, 15]. There must be extensive research into the development of smart materials in order to produce a product with the desired qualities for a given set of applications. Concerns remain about the ability to fabricate new resources using AM procedure and to trigger their characteristics using exterior factors.

Several smart materials fabricated using various AM techniques are described in the literature [16–22], each with their own set of applications and associated constraints like toxic effects, degradation rate, bioactivity, and so on. However, research into remotely triggered materials, such as polymers nanocomposites, is lacking in the existing literature. Functional materials and novel designs, such as cardio stents [23] and porous bone used in the biological field [24], are essential for smart systems and devices. In this paper, we present the results of a systematic analysis focused on the glimpse into3D printing and the identification of polymeric composites that can be activated by outside factors and have the tailored architecture necessary for use in various fields of application. This paper provides a comprehensive examination of smart nanocomposites constructed from functional polymers, focusing on their design and the most appropriate AM method for their various applications in the medical field. The paper begins with a brief introduction to various AM approaches that are predicated on the fed material, then moves on to discuss operational polymeric material with varying stimuli-reaction behaviors and distinctive design, before finally wrapping up with potential future applications and fields of study. This comprehensive overview will offer the investigators information on the most recent polymeric nanocomposite materials and their potential applications. Additive manufacturing is a novel technique that utilizes modeling software and CAD files to create real-world 3D objects at full scale [25]. Time and money spent on manufacturing have been drastically reduced due to its use [26]. This is especially true for products with multiple materials and intricate shapes. In contrast to traditional methods, 3D manufacturing allows for the fabrication of complex designs by means of the accurate implantation of thin layers, thus, eliminating the need for assembly protocol and resulting in a lightweight, one-piece geometry. As a result, it has simplified issues associated with stocking and fabricating components. Although traditional methods are gradually being phased out for mass production, they are still occasionally considered for use on a smaller scale [27]. Because of its benefits in targeted therapy and as a suitable therapeutic, AM has gained widespread acceptance in a number of biomedical applications [28, 29], despite the fact that 3D printing has one major drawback: a slow printing rate. New solutions inspired by robust needs have been made possible by the development of AM methods. Polymers and their nanocomposites, for example, have led to the development of a number of distinct protocols [30–33]. This chapter evaluates some of the most recent and cutting-edge resources.

1.2 Polymer-Based AM

As a result of the synergistic combination of regular polymers and the electrochemical characteristics of metals or semiconductors, research into electrical conductor polymer composites has attracted significant attention. The increased design freedom, more complex shapes, and faster manufacturing times made possible by AM are all reasons for optimism in the field of conductive polymer composites. Materials can be 3D printed using a variety of AM techniques, such as injection molding, vat polymerization, material jetting, powder bed fusion, and laminated object manufacturing. In this chapter, we provide a glimpse into recent developments in the sector of conductive polymeric materials created for AM, speed up the configuration and advancement of 3D printable electrical devices. A close look is taken at the advantages and disadvantages of several AM techniques, resource necessity and recent developments in 3D printing of conductive polymeric materials are discussed, and some of the most exciting electronic applications, including wearable electronics, power storage, power electronics, and more are highlighted (Figure 1.1) [34]. Recent advances in 3DP technologies and potential uses for electrically powered CPCs are discussed. 3D printing processes that can be used to produce these materials are also presented. In this chapter, a deep dive into the world of conductive materials for 3D printing is taken. This motivates the discussion of few design constraints and dynamic applications of 3DP in bio-electronics. There has been a lot of work done over the past few years to develop this new field of study. Still, it has a way to go before reaching its full potential, as several obstacles must be overcome. When creating CPCs for AM, a number of factors must be taken into account. There is a need for so many polymeric systems which can be utilized in AM processes and have high electrical conductivity. The small number of CPCs currently amenable to 3DP may be a result of the challenge of integrating semiconducting substances within polymer matrices. This necessitates a synergistic combination of materials to improve dispersibility while maintaining high

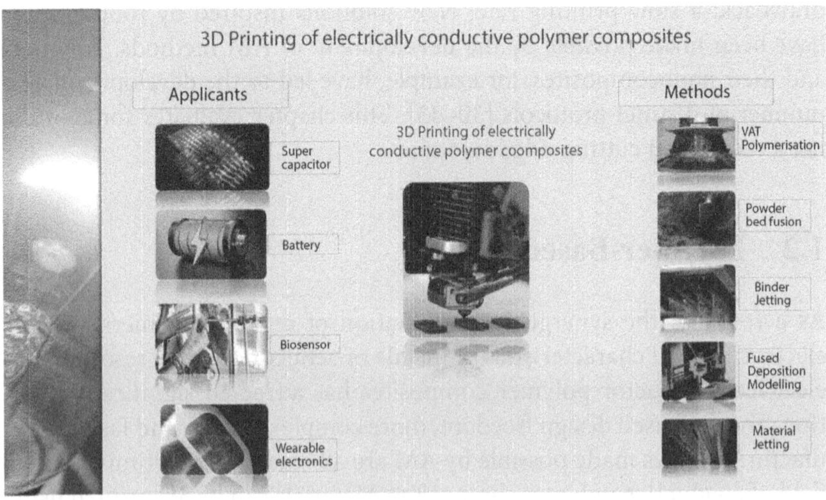

Figure 1.1 Polymer composite printing methods and applications [34].

electrical conductivity. The use of functional fillers in 3DP of CPCs has emerged as an exciting zone with numerous potential uses. The relationship among fabrication techniques, framework, and the final appearance of CPCs for AM needs to be studied further. This allows for a more nuanced investigation into the effects of the printing method and environmental factors on the electrochemical characteristics of printed specimen fabricated from various conductive specimens. Given that the effectiveness of a printable product may be affected by factors that differ across AM processes, it is important to develop electronic test procedures that are method specific. Despite its many advantages over traditional manufacturing methods, AM is limited by factors like minimal print velocities, pixel size, and indirect control over the crystal structure and features of the finished item. Therefore, it is critical to create cutting-edge 3DP methods suitable for CPCs of varying densities and characteristics (light transmission) without compromising overall printing speed. Improvements in printing resolution necessitate a closer look at matrices and fillers and the matching of their properties. Since only some parts of a 3D object need to be processed at a high resolution, research into 3DP methods that permit adaptable alteration of printing resolution is warranted such that large volume generation rates can be achieved. The evolution of printing technology and the introduction of new materials have enabled it to print structures out of responsive components, which can change their shape or properties over time. By invoking a transformation in form or properties in 3D-printed objects, a temporal dimension into the printing process is introduced and paves the way for the advent of 4DP. Changes in environmental conditions (heat, illumination, magnetic forces, electrical pulses, etc.) can cause shape memory resources to return to their natural programmed dimensions [35–40]. The added flexibility and adaptability are a result of these features. Recent years have seen extensive research into developing semiconducting composite materials for 4DP of shape patterns [41, 42]. For the purpose of creating temperature-sensitive electrical devices for use in atmospheric regulation, Chan *et al.* [43] used post-printing electroplating conductive coating method to instil a high conductivity Cu outer skin on 3D-printed shape memory composites. The use of such adaptable structures could improve the performance of new sensors, portable electronic devices, design field of robotics, and non-invasive medical equipment.

It is also important to think about things like digital file encryption and the security of the printing process. The steps involved in AM are complex and varied. If the CAD folders are altered, the safety of 3D-printed functional parts can be compromised, including components of critical safety mechanisms in automotive and aerospace systems. Due to these factors,

AM processes are an enticing objective for security-based attacks, so protecting the design configuration document from corruption is essential. Yampolskiy *et al.* [44]'s research offers an additional thorough initiation and evaluation. More and more people have access to 3D printers and the raw resources require to create their own prints as 3D printing systems become more widely available. One viable option is to host 3D digital printing designs of commonly used components on shared digital repositories; this would allow users to access the files, print the 3D goods, and fix any flaws in the model using their own printing equipment.

1.3 Surgical Planning

Recent decades have seen a meteoric rise in medical innovation, driven largely by the introduction of cutting-edge surgical techniques and their accompanying high-tech equipment. This has helped doctors and surgeons pinpoint the onset of a disease, track its progression over time, prepare for surgery, perform surgery, and develop postoperative protocols. Technological developments in minimally invasive surgery, virtual reality, and robotics have allowed surgeons to better plan for the preoperative, intraoperative, and postoperative phases of a procedure even when they have only a limited field of vision. Recently, surgeons have been able to develop and formulate cutting-edge pre- and post-operative planning thanks to advancements in AM. By using AM, a physical representation of a body part can be created in three dimensions, which can help surgeons visualize, comprehend, and evaluate the part's physical condition before actually operating on it. Thus, these models are employed in preoperative preparation for actual surgical procedures. Postoperative rehabilitation is another area where AM has proven useful. The structure and dimensions of the patient's body part can be measured with the help of a 3D-printed model prototype. The conclusion of the operation and any potential changes that may occur after can be evaluated more precisely with the help of this data. Medical fields that can benefit from 3D-printed prototypes and prosthetics after surgery include orthopedics, ceramic and polymer blends scaffolds for implants, ocular prostheses, and crystal images for facial deformities, to name a few. In addition to requiring less time to produce than their conventional counterparts, these devices can also provide postoperative patients with ergonomic comfort and support. It goes without saying that AM-made prototypes are essential for providing efficient assistance to the health industry in the event of both pre- and post-surgical strategic plan in aspects of improved precision and repetitions for simulated surgical procedure, reduced surgical duration, accelerated fabrication time, enhanced product quality, and a wide variety of individualized product options [45].

1.4 Titanium Alloy

Applying a pulsed magnetic field (PMF) to a metal during its solidification process has been shown to reduce grain size and boost the material's mechanical properties. Laser additive manufacturing (LAM) has shown promise, but fewer works have been reported in this area, and the method of grain growth following the PMF is still unknown. To investigate the thermal-fluid characteristics of the Ti-alloy melt pool produced by the laser scanning process under the influence of a merged direct current (DC) electric field and PMF, numerical models were developed in the present work. Microstructure change after LAM was explained by discussing the temperature distribution, melt pool evolution, and magneto-oscillation impact in the weld zone. According to the findings, electric-magnetic fields applied during the LAM process can significantly alter the melt pool's temperature distribution, shape, and dimensions, as well as the melt flow characteristics. The augmented fluid-fluid motion and constant stimulus force in the melt, as well as the rising temperature slope, crystallisation rate of growth, and cooling rate at the liquid-solid interface, are all likely to play a role in producing the desired sophisticated crystal structure [46].

1.5 Thickness Control Using Machine Learning

To get high-resolution printing in a variety of materials, non-contact AM via aerosol jet printing (AJP) is a hopeful new option. Thinness control is a major issue that has prevented AJP from gaining wider acceptance in the materials science community. In this chapter, we propose, choose, and measure the optimum model to deliberate and optimize the AJP production procedure by using a model-based design of experiments (MBDoE) structure which incorporates physics-informed designs, regression models, and information criteria. Four applicant physics-informed designs are proposed and trained using data that is derived from commissioning of system (for example, previous single variable sensitivity experiment). To verify these forecasting analytics with quantitative ambiguity, MBDoE finds a single extra ideal experiment, utilized to find the optimal experimental status to regulate the width of the printed film. Nonparametric

Gaussian process regression (GPR) model, which does not include original data, is applied to the same dataset as a comparison benchmark. Our results show that the nonlinear physics-informed parametric model can be calibrated by just five experiments using MBDoE principles, and that it can outperform the black-box machine learning GPR model by means of this small amount of data. This important finding exemplifies a current pattern in data science: using physical information in predictive models can often greatly reduce the amount of data needed. As a result of implementing MBDoE, data efficiency was improved even further. The suggested data science structure is generic by design, making it amenable to adaptation for use with systems other than AJP that employ exploratory and AM techniques [47].

For the first time, advanced tools like model-based design of experiments (MBDoE) are used to enable fine-grained regulation of printed film thickness in AJP procedures. Figure 1.2 shows the schematic of layer thickness controlled by ML algorithms. As a starting point, a dataset with 22 experiments is used from the commissioning of AJP apparatus and other projects that is typically ignored. These preliminary data are used to propose a library of eight applicant physics-informed models, which are then narrowed down to four models using Akaike Information Criteria (AIC). To enhance the reliability of these models' predictions, MBDoE zeroes in on a single experiment that will yield the best results. Four more experiments

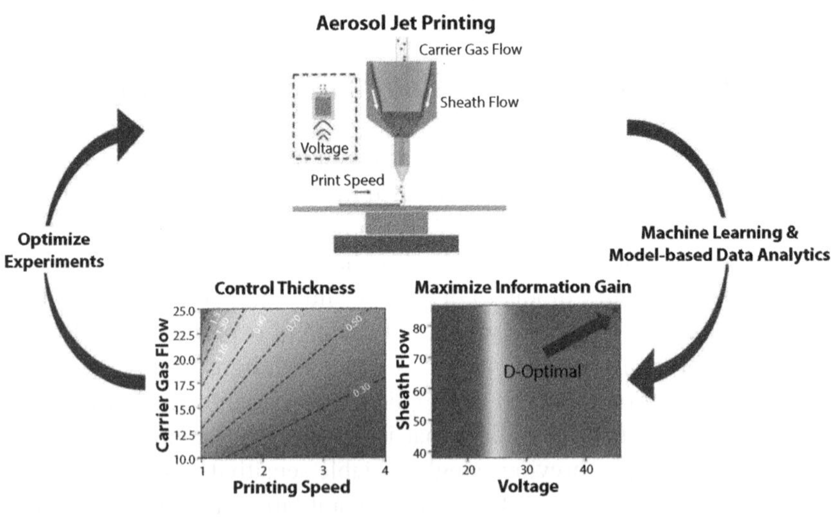

Figure 1.2 Layer thickness controlled by machine learning [47].

are carried out, and the results confirm that MBDoE-informed experimental conditions are much more insightful than those chosen predicated purely on specialist instinct. Finally, the cross-validated best model is used to devise strategies for adjusting the power output, carrier fuel water flow rate, layer water flow rate, and scan speed to attain the preferred layer depth with negligible prototype estimation uncertainty. Additionally, with the help of Fisher information matrix for retroactive analysis, we find five optimal investigations are required to treat the recognized physics-informed multiple linear regression model. Based on that, it appears that, a small number of experiments is required to adapt the planned work order for use with AJP in another substance scheme. The method's efficacy is measured against that of a black box nonparametric model trained on the repeated information. Researchers observed that the GPR model is most effective in areas with a large amount of available data. On the contrary, the nonlinear parametric model that is grounded in physical principles performs better than GPR in sparsely sampled regions. When looking at just the five best experiments in the dataset, this trend becomes especially apparent. In light of current trends in the ML literature informed by physics, these findings are not surprising. To reduce the amount of data needed for training, mathematical models should integrate science and experiential information whenever possible. This chapter provides a second example to back up this claim. In the future, we hope to apply the established correlation between output power, carrier gas water flow rate, circulation in the sheath, swiftness in printing, and depth to other AJP material systems for thermoacoustic metals and gadgets, such as AgSe [48] or BiSbTe [49]. Between five and ten optimum tests may be required for initial modeling when making changes to new material systems, with two important adjustments: (1) the limits of factors must be up to date to take into account shifts inside the toner viscosity, and (2) the linear regression is likely to differ for every scheme. The improvement of the model's accuracy close to the operational bounds is another area worthy of further study. If the most promising model is taken as a starting point, it can be seen, for instance, that the linear and quadratic assumptions might not seize close to the functioning limits. Scientists speculate that slight model-form error accounts for the improved presentation of black-box GPR when dealing with a large amount of data. A hybrid model [50, 51] could be utilized to account for this minor logical error in thickness predictions in the future. In addition, the ink composition for AJP and other AM processes can be optimized with the help of the recognized physics-informed designs and the suggested data scientist's structure.

1.6 Carbon Fiber-Based AM

Today, AM is the backbone of the engineering industry's ability to discover and develop new and useful materials and tools. However, little task was done in the resource modeling work of each stage, and numerous experiments have demonstrated inconsistent results and emphasised various deficiencies in the structure, calling into question the implementation of the AM technique in real engineering applications. A micromechanical analysis using the sample quantity aspect has been considered, to take into account the consequences of the accumulated defects, and a state-field mathematical model was used as a prototype to show the impact of inter-fiber crack propagation. While satisfying geometric and material periodicity, MLA were used to acquire fiber dimensions and content for use in creating 3D models of reflective control volumes in the Abaqus environment. Furthermore, in both the cross line and in-plane shear test cases, periodic limit requirements were assumed for all representative volume elements (RVE). The model essentials and description of Abaqus were used in the analysis rather than a custom-built UDE because the analysis was performed using an open-source UMAT subprogram that related the phase-field equilibrium equations to the built-in heat transfer equation. Experimental outcomes for circumferential tension and compression for the filled carbon composites were used to calibrate the interface properties, and three main quantity component dimensions were used to verify the model. Results of the experiment and calculated results procured to use a phase-field fracture method were found to be consistent with one another for the circumferential tension and compression and shear actions of synergistically fabricated continuous-fiber-reinforced composites [52]. The remainder showed reliance on the type of reflective bulk aspect for certain pressure cases. Here, we present the arithmetic operations of the micromechanical behaviour of AM CFRP materials in tensile stress and in-plane shear, grounded in infinitesimal assessment and the macroscopic response are presented. The methods used in producing materials and getting them ready for microscopic inspections are covered, along with the benefits and drawbacks of this approach. Many of the advantages of CT over microscopy are presented, and the RVE design, which is derived from simple geometric hypotheses and assisted by the micro structural metrics is also discussed. Because of its support for the built-in features of Abaqus, the micro structural analysis presented here allows for the progress and assessment of recurrent RVEs with the help of the phase-field rupture hypothesis. It is possible because of the analogy proposed among phase-field rupture, the heat transfer differential calculations. This method made it easier to visualize the acquired result and to divide the domains depending on the different resources and functionality. In addition, differences in AM composite materials can be traced back to the deposit's contact zones, as evidenced by both data dispersion and DIC images. The proposed RVEs were too small to accommodate the magnitude of these deficiencies, so they were integrated into the system alongside the fiber/matrix impact at the interface of the constituents. The limitations of this method were highlighted, and a multiscale strategy utilizing interface properties obtained experimentally was proposed for future research. Overall, the results of this study confirmed that the transverse tensile behavior of AM CFRP composites is consistent with the numerical simulations that use the phase-field fracture approach. When compared to the experimental scatter, the mathematical outcomes were consistent regardless of the RVE used. However, when comparing the in-plane shear behavior of the tested materials, RVE-1 and RVE-2 showed discrepancies with both experimental results and outcomes procured for the RVE-3, which leads to underestimating the shear strength and modulus. In the case of in-plane shear, however, RVE-3 overestimated the mechanical properties. Although this phase-field modeling approach was shown to characterize the transverse micromechanical behavior well, it was also shown that in-plane shear cases do not obey the exploratory outcomes after 1% of shear strain. Further, the model predicts a complete specimen failure even after surpassing the shear strength measured based on ASTM instructions, which does not occur in the experiments. Therefore, future research should take into account a variety of methods to experimentally identify and design shear behavior in AM-fiber-reinforced composites [52].

1.7 Ceramics-Based AM

Complex ceramic materials that emit light and scintillate using stereolithography 3DP are of interest for use as specialized phosphors and detectors. It is important to investigate the potential impact of the powders' luminescent characteristics on the printing process because the starting substances for that kind of ceramic materials may acquire UV absorption bands. Complex garnet oxides, such as Y3Al5O12 and Gd3Al2Ga3O12, are discussed in this chapter. These compounds are well-known hosts for luminescent components. Slurries made from luminescent powders made via different chemical routes have their photopolymerization rates and printable regimes investigated. Compared to non-doped slurries, those containing Ce-doped powders that have a broad UV absorption band have much slower photopolymerization rates, and a high Ce doping virtually prevents being able to print with layers thicker than 25-50 m. In addition, it demonstrated how printing is affected by the powder synthesis method of choice. Layer thicknesses between 25 and 100 m are printable using slurries containing Tb-doped powder because of the powder's high photopolymerization activity and similarity to that of the undoped powder. Using digital light processing 3DP, Gadolinium Aluminum Gallium Garnet (GAGG) was used to make net-like objects: wall thicknesses of 200-400 m and round holes of 200-700 m are examples of the sub millimeter features that can be achieved with Ce and GAGG:Tb ceramic. The luminescent properties of the starting powders, and more specifically the interconnection of the photoluminescence lattice vibrations band(s) with the 3D printer's source of light emission, were shown to be a big negative in curing dynamics, thereby limiting the printing dictators that can be used. For this reason, at equivalent dopant concentrations, thicker layers could be printed using Yttrium Aluminium Garnet: Cerium (YAG:Ce) than GAGG:Ce. The photoluminescence vibrational band density and the required illumination dose for printing a given layer thickness were both lower for powder particles acquired by uniform precipitate than for powders synthesised by co-precipitation. If required to print luminescent ceramics with a DLP printer, options are limited unless you either use powders with undeveloped luminescent properties or select a printer whose light source does not overlap with the powder's absorption bands [53].

1.8 Wire Polymer-Based AM

Fused filament fabrication (FFF) was used to 3D print a CWPC thermal gradient sensor, with the integrated wires acting as impedance detecting components. Thermoset polyurethane (TPU) and Cu wire, polylactic acid (PLA) and Ni wire, and PLA and Cu wire were all investigated as potential sensor combinations. Samples were 3D printed with varying numbers of layers for each composition to examine how sensor thickness affected performance. Distances between elements, electrical conductivity of polymers, and temperature coefficients of resistance of wires were measured. The 3D-printed CWPC performed admirably as a heat flux sensor in performance tests, proving its viability as a heat flux sensor in low temperature and low heat flux applications. Errors in measurements were below 10% for 2-layer specimens at the high thermal fluxes and below 17% for all other configurations. To accurately estimate heat loss from an insulated system, a 3D-printed flexible CWPC thermal gradient sensor was used in a study.

Given its adaptability to a wide variety of polymers and sensing elements, the approach to sensor fabrication that is described here has broad potential for use. Gradient-type heat flux sensors with integrated wires that acted as impedance heat sensors were 3D-printed using continuous wire polymer composites (CWPC). Two-, three-, and four-layer structures containing the combinations PLA + Cu, PLA + Ni, and TPU + Cu were created. The measurement of heat flux was characterized quantitatively with respect to key design parameters that govern performance and accuracy. The sensors were tested by having an external heat flux superimposed and the two values compared. There was a good deal of concordance between the two methods, with the calculated heat flux within the margin of experimental error. For all heat fluxes, the sensors made of PLA + Cu and TPU + Cu had error rates of less than 17%, while those made of PLA + Ni had error rates of less than 12%. Uncertainty analysis revealed that a thicker sensor reduced the margin of error for the evaluated thermal gradient, however, a higher resistance is expended. When compared to other wires, the Ni wire used in the sensor was more sensitive because of its higher resistance. In general, the sensors functioned adequately for low-temperature, low-heat-flux applications, and their main benefit is that they can be modified to fit a specific geometric setting. To test the feasibility of measuring heat loss from an insulated pipe, a flexible TPU + Cu detector was manufactured and wrapped around a shielded heated rod. Good agreement was found between this study's predictions and the conceptual thermal failure [54].

1.9 Nanomaterial-Based AM

There is a growing trend of combining the relatively new technologies of 3DP and nanotech to make all sorts of exciting and reasonably priced new items. In fact, 3DP technology has quickly come to dominate a number of industries and the medical field due to its remarkable advantages that were either unavailable before or prohibitively expensive when compared to other production techniques in both fields. As an industry, 3DP is built on the shoulders of two pillars: the advanced resources used in the fabrication, and the technical details required for printing them. Since 3DP materials can be put to use in so many different contexts, their development is crucial to the growth of the industry as a whole. This chapter takes a look at the many different nanomaterials that can be used in 3DP, as well as their advantages and how they affect the properties of the final products. The chapter also discusses the gaps in risk assessment related to 3DP and nanotechnology, which are necessary given the need to evaluate idea at the nano-scale to confront any possible hazards to both people and the environment. Furthermore, the manufacturing effect of this promising technology is discussed by examining the various implementations and nanomaterials now incorporated into 3DP technology at a commercial level [55].

1.10 Direct Ink Writing (DIW)

The most practical approach to 3DP hydrogels is likely to be direct ink writing (DIW) technology. Research into the joint creation of photosensitive hydrogel and DIW printing machinery for targeted application is ongoing. The advancement of a soluble photo catalyst that can be initiated under light waves is of significant importance for 3DP hydrogel innovation as it allows for the prevention of UV-induced skin damage. By choosing a surfactant with stacking ability, the water-based photoinitiator (Aqu-PI) can be formed via complex molecular arrangement with TPO without solvents emitting any volatile organic compounds (VOCs). Aqu-PI has a particle size of about 15 nm and an absorption peak in the visible region of the UV-visible spectrum. By combining Aqu-PI with monofunctional monomer and bifunctional monomer (BM), a photosensitive hydrogel can be made (MBA). The photosensitive hydrogel's viscosity is increased after the addition of the tackifier, resulting in hydrogels with improved mechanical properties. The DIW printer's extrusion device uses a 405 nm LED laser to photocure the photosensitive hydrogel, creating the interlayer connectivity structure and a 3D printout hydrogel that matches the digital production model's structure [56].

1.11 Hull of Soy

The lack of material diversity and subpar robotic, heat, and rheology properties severely restrict the FFF technique's potential uses. Most FFF filaments are made from either acrylonitrile butadiene styrene (ABS) or polylactic acid (PLA). Some fiberglass fiber components are used in the FFF process alongside pure thermoplastics. To obtain a diverse set of material properties to meet a variety of application needs, it is crucial to investigate materials for the FFF filaments. Polylactic acid serves as the matrix while soy hulls and soy protein isolate (SPI) are used as the fillers to create a bio-based and biodegradable fiberglass fiber material. The printability of the biocomposite filaments developed is evaluated using the

FFF process, and a variety of soy hull-to-PLA blends are tested. As for the FFF filament, the amounts of soy hulls studied are 5 wt%, 7.5 wt%, and 10 wt%. In this study, we use experimental analysis and statistical methods to look into the effects of different soy hulls proportions on tensile, melt flow index (MFI), glass transition temperature (Tg), and storage modulus. Experiment results indicate that, of the three percentages tested, soy hulls at 5 wt% are the most effective. Furthermore, two fillers, SPI (at 2.5 wt%) and soy hulls (at 7.5 wt%), are mixed with PLA to examine the effects of each filler on these four properties. In this study, soy hulls and SPI-based bioproducts are successfully used to create FFF filaments. This research demonstrated the viability of 3DP with PLA, soy hulls, and SPI mixed biomaterials [57].

1.12 Laser Powder Bed Fusion

The NiTi-centered super elastic composite materials decorated with TiC nanoparticles were made using the laser powder bed fusion (LPBF) method. The high energy density of the laser beam prompted strong dispersion actions of carbon molecules from the TiC nano particles, resulting in the precipitation of TiCx and the associated Ni-rich even Ti-rich intermetallics, in comparison to the lack of reaction between TiC and NiTi observed in standard techniques. These nano precipitates were dispersed along the intricate connectivity of interconnected deformations, where they established an innovative cell membrane reinforcement framework on a submicron scale and helped to bring about a hierarchical diversity in the micro-structural. Submicron cellular fiber reinforced structure formation and TiC nano particle migration and distribution were then discussed in detail. According to the results, the rate at which the laser was scanned had a major impact on how the TiC nano particles were dispersed, the size of the cells, and the orientation of the matrix grains. Nano-TiC/NiTi composites fabricated using LPBF at the optimal parameter [58] exhibited a weak direction surface along the building direction, a finer cellular structure, and a significant steady reclaimable strain of 2.3% at the highest stress load of 300 MPa.

1.13 Future Challenges

Recent developments in 3DP and its derivative, 4DP, have paved the way for the creation of numerous novel materials with potential in fields as

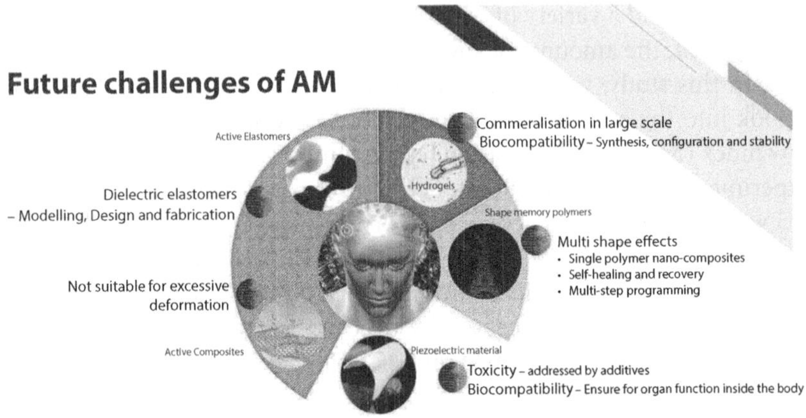

Figure 1.3 Future challenges of AM [Redrawn from reference 59].

Table 1.1 Additive manufacturing patent count based on country, year, and IPC code [60].

S. no.	Countries	Count	IPC code	Count	Year	Count
1	USA	32,535	B33Y	16,575	2014	1685
2	PCT	21,788	B29C	13,064	2015	2513
3	China	7187	B22F	8,380	2016	3496
4	Europe	2836	H01L	3,759	2017	4855
5	UK	1199	B23K	3,601	2018	6178
6	Canada	696	G06F	3,174	2019	7368
7	India	582	A61B	2,550	2020	8529
8	Japan	264	A61F	2,529	2021	8875
9	Australia	255	F01D	2,498	2022	8995
10	Korea	251	C22C	1,937	2023	502

Patent landscape analysis is carried out for the keyword "additive manufacturing" in the "English All" category; the total count is 68018 (as of 29th Jan 2023).

diverse as biomedicine, soft robots, detectors and effectors, automotive and aerospace. Table 1.1 shows the number of patents filed on AM technologies. Because of advancements in 4DP technology, materials like polymeric composites can now be used to fabricate intricate and reactive structures. These are created using fabrication processes like extrusion-related polymerization, photo polymerization, and powder-based AM techniques. Artificial muscles, bone scaffolds, and cardiovascular stents are all examples of biomedical applications that benefit from the unique properties bestowed upon them by the interplay between the polymerization of advanced materials, the construction method adopted, and post-curing processes. There are possible concepts for future studies in various classes of materials [59]. Figure 1.3 reveals the future challenges of AM.

1.14 Future Scope of AM

Future scope is analyzed with the aid of a patent database which leads to understanding the scope of this domain in the near future. The reason for choosing patent landscape is it is updated every minute. Thus, it is one of the effective databases to predict the future of AM.

1.15 Conclusion

A glimpse into AM is summarized in this chapter. Polymer-based AM for electrical devices, surgical planning using AM, printing of titanium alloy using pulsed magnetic field, thickness control using machine learning, carbon fiber-based AM, ceramics-based AM, wire polymer-based AM, nanomaterial-based AM, direct ink writing of hydrogels, AM of hull of soy, NiTi using laser powder bed fusion, and future challenges for AM are discussed. Finally, a route map of the future scope of AM is assessed with the aid of patent landscape. Thus, the researchers in this domain get a clear vision of AM techniques.

References

1. Sharma, A. and Singholi, A.K., Effect of nanosilica on shape memory and mechanical characterization of polylactic acid wood composite. *Polym. Compos.*, *42*, 5, 2502–2510, 2021.
2. Al Rashid, A., Khan, S.A., Al-Ghamdi, S.G., Koc, M., Additive manufacturing: Technology, applications, markets, and opportunities for the built environment. *Autom. Constr.*, *118*, 103268, 2020.
3. Al Rashid, A., Khan, S.A., Al-Ghamdi, S.G., Koc, M., Additive manufacturing of polymer nanocomposites: Needs and challenges in materials, processes, and applications. *J. Mater. Res. Technol.*, *14*, 910–941, 2021.
4. Moganapriya, C., Rajasekar, R., Kumar, P.S., Pal., S.K., Investigation of TiAlN, AlTiN, and TiAlSiN coated inserts on the machining performance of AISI 420 steel and multi-objective optimization of process parameters. *J. Inst. Eng. (India): Ser. D*, 103, 2, 563–573, 2022.
5. Klippstein, H., Diaz De Cerio Sanchez, A., Hassanin, H., Zweiri, Y., Seneviratne, L., Fused deposition modeling for unmanned aerial vehicles (UAVs): A review. *Adv. Eng. Mater.*, *20*, 2, 1700552, 2018.
6. Shie, M.Y., Shen, Y.F., Astuti, S.D., Lee, A.K.X., Lin, S.H., Dwijaksara, N.L.B., Chen, Y.W., Review of polymeric materials in 4D printing biomedical applications. *Polymers*, *11*, 11, 1864, 2019.
7. Valentine, A.D., Busbee, T.A., Boley, J.W., Raney, J.R., Chortos, A., Kotikian, A., Berrigan, J.D., Durstock, M.F., Lewis, J.A., Hybrid 3D printing of soft electronics. *Adv. Mater.*, *29*, 40, 1703817, 2017.
8. Zolfagharian, A., Kaynak, A., Khoo, S.Y., Kouzani, A.Z., Polyelectrolyte soft actuators: 3D printed chitosan and cast gelatin. *3D Print. Addit. Manuf.*, *5*, 2, 138–150, 2018.
9. Dey, A. and Yodo, N., A systematic survey of FDM process parameter optimization and their influence on part characteristics. *J. Manuf. Mater. Process.*, 3, 3, 64, 2019.
10. Brunello, G., Sivolella, S., Meneghello, R., Ferroni, L., Gardin, C., Piattelli, A., Zavan, B., Bressan, E., Powder-based 3D printing for bone tissue engineering. *Biotechnol. Adv.*, *34*, 5, 740–753, 2016.
11. Ulag, S., Kalkandelen, C., Bedir, T., Erdemir, G., Kuruca, S.E., Dumludag, F., Ustundag, C.B., Rayaman, E., Ekren, N., Kilic, B., Gunduz, O., Fabrication of three-dimensional PCL/BiFeO3 scaffolds for biomedical applications. *Mater. Sci. Eng.: B*, *261*, 114660, 2020.
12. Jain, S. and Parashar, V., Critical review on the impact of EDM process on biomedical materials. *Mater. Manuf. Processes*, *36*, 15, 1701–1724, 2021.
13. Holman, H., Kavarana, M.N., Rajab, T.K., Smart materials in cardiovascular implants: Shape memory alloys and shape memory polymers. *Artif. Organs*, *45*, 5, 454–463, 2021.
14. Shin, D.G., Kim, T.H., Kim, D.E., Review of 4D printing materials and their properties. *Int. J. Precis. Eng. Manuf.-Green Technol.*, 4, 349–357, 2017.

15. Chaidas, D. and Kechagias, J.D., An investigation of PLA/W parts quality fabricated by FFF. *Mater. Manuf. Processes*, *37*, 5, 582–590, 2022.
16. Yang, K., Hai, L., Wang, Z., Li, H., Yi, W., Luo, Y., Li, J., Deng, L., He, D., A biofilm microenvironment-responsive one-for-all bactericidal nanoplatform for photothermal-augmented multimodal synergistic therapy of pathogenic bacterial biofilm infection. *J. Mater. Chem. B*, *10*, 38, 7744–7759, 2022.
17. Haidari, H., Kopecki, Z., Sutton, A.T., Garg, S., Cowin, A.J., Vasilev, K., pH-responsive "smart" hydrogel for controlled delivery of silver nanoparticles to infected wounds. *Antibiotics*, *10*, 1, 49, 2021.
18. Zhang, R., Huang, Z., Xue, M., Yang, J., Tan, T., Detailed characterization of an injectable hyaluronic acid-polyaspartylhydrazide hydrogel for protein delivery. *Carbohydr. Polym.*, *85*, 4, 717–725, 2011.
19. Narupai, B., Smith, P.T., Nelson, A., 4D printing of multi-stimuli responsive protein-based hydrogels for autonomous shape transformations. *Adv. Funct. Mater.*, *31*, 23, 2011012, 2021.
20. Uribe-Lam, E., Treviño-Quintanilla, C.D., Cuan-Urquizo, E., Olvera-Silva, O., Use of additive manufacturing for the fabrication of cellular and lattice materials: A review. *Mater. Manuf. Processes*, *36*, 3, 257–280, 2021.
21. Chinnasamy, M., Rathanasamy, R., Kannayiram, P., Palaniappan, S.K., Pal, S.K., Somasundaram, M., Kaliyannan, G.V., Experimental investigation on the effect of multilayer TiCN/TiAlN/WC-C coating on the tribological behaviour of tool inserts for machining applications. *Int. J. Mater. Eng. Innovation*, 10, 3, 186–202, 2019.
22. Ullah, M.S., Yazıcı, N., Wis, A.A., Özkoç, G., Kodal, M., A review on polyhedral oligomericsilsesquioxanes as a new multipurpose nanohybrid additive for poly (lactic acid) and poly (lactic acid) hybrid composites. *Polym. Compos.*, *43*, 3, 1252–1281, 2022.
23. Amin, F., Ali, M.N., Ansari, U., Mir, M., Minhas, M.A., Shahid, W., Auxetic coronary stent endoprosthesis: Fabrication and structural analysis. *J. Appl. Biomater. Funct. Mater.*, *13*, 2, 127–135, 2015.
24. Polley, C., Distler, T., Detsch, R., Lund, H., Springer, A., Boccaccini, A.R., Seitz, H., 3D printing of piezoelectric barium titanate-hydroxyapatite scaffolds with interconnected porosity for bone tissue engineering. *Materials*, *13*, 7, 1773, 2020.
25. Wong, K.V. and Hernandez, A., A review of additive manufacturing. *Int. Sch. Res. Notices*, *2012*, 10 pages, 2012.
26. Parandoush, P., Li, X., Chang, B., Sorensen, C.M., Shi, J., Liu, Y., Lin, D., Additive manufacturing of continuous carbon fiber reinforced epoxy composite with graphene enhanced interlayer bond toward ultra-high mechanical properties. *Polym. Compos.*, *43*, 2, 934–945, 2022.
27. Falahati, M., Ahmadvand, P., Safaee, S., Chang, Y.C., Lyu, Z., Chen, R., Li, L., Lin, Y., Smart polymers and nanocomposites for 3D and 4D printing. *Mater. Today*, *40*, 215–245, 2020.

28. Mokhtari, F., Azimi, B., Salehi, M., Hashemikia, S., Danti, S., Recent advances of polymer-based piezoelectric composites for biomedical applications. *J. Mech. Behav. Biomed. Mater.*, *122*, 104669, 2021.
29. Feng, H., Zhao, C., Tan, P., Liu, R., Chen, X., Li, Z., Nanogenerator for biomedical applications. *Adv. Healthcare Mater.*, *7*, 10, 1701298, 2018.
30. Osswald, T.A., Jack, D., Thompson, M.S., Polymer composites: Additive manufacturing of composites. *Polym. Compos.*, *43*, 6, 3496–3497, 2022.
31. Monticeli, F.M., Neves, R.M., Ornaghi Jr., H.L., Almeida Jr., J.H.S., A systematic review on high-performance fiber-reinforced 3D printed thermoset composites. *Polym. Compos.*, *42*, 8, 3702–3715, 2021.
32. Aktitiz, İ., Delibaş, H., Topcu, A., Aydın, K., Morphological, mechanical, magnetic, and thermal properties of 3D printed functional polymeric structures modified with Fe2O3 nanoparticles. *Polym. Compos.*, *42*, 12, 6839–6846, 2021.
33. Chen, X.G., Li, A., Zhang, Y.F., Temperature effect on thermal, mechanical, and electrical properties of Ag nanoparticle filled thermoplastic polyurethane composites. *Polym. Compos.*, *42*, 11, 5952–5961, 2021.
34. Yan, Y., Jiang, Y., Ng, E.L.L., Zhang, Y., Owh, C., Wang, F., Song, Q., Feng, T., Zhang, B., Li, P., Loh, X.J., Progress and opportunities in additive manufacturing of electrically conductive polymer composites. *Mater. Today Adv.*, *17*, 100333, 2023.
35. Chan, B.Q.Y., Low, Z.W.K., Heng, S.J.W., Chan, S.Y., Owh, C., Loh, X.J., Recent advances in shape memory soft materials for biomedical applications. *ACS Appl. Mater. Interfaces*, *8*, 16, 10070–10087, 2016.
36. Kai, D., Prabhakaran, M.P., Chan, B.Q.Y., Liow, S.S., Ramakrishna, S., Xu, F., Loh, X.J., Elastic poly (ε-caprolactone)-polydimethylsiloxane copolymer fibers with shape memory effect for bone tissue engineering. *Biomed. Mater.*, *11*, 1, 015007, 2016.
37. Kai, D., Tan, M.J., Prabhakaran, M.P., Chan, B.Q.Y., Liow, S.S., Ramakrishna, S., Loh, X.J., Biocompatible electrically conductive nanofibers from inorganic-organic shape memory polymers. *Colloids Surf. B: Biointerfaces*, *148*, 557–565, 2016.
38. Chan, B.Q.Y., Liow, S.S., Loh, X.J., Organic–inorganic shape memory thermoplastic polyurethane based on polycaprolactone and polydimethylsiloxane. *RSC Adv.*, *6*, 41, 34946–34954, 2016.
39. Chan, B.Q.Y., Heng, S.J.W., Liow, S.S., Zhang, K., Loh, X.J., Dual-responsive hybrid thermoplastic shape memory polyurethane. *Mater. Chem. Front.*, *1*, 4, 767–779, 2017.
40. Ang, J.Y., Chan, B.Q.Y., Kai, D., Loh, X.J., Engineering porous water-responsive poly (PEG/PCL/PDMS urethane) shape memory polymers. *Macromol. Mater. Eng.*, *302*, 9, 1700174, 2017.
41. Zarek, M., Layani, M., Cooperstein, I., Sachyani, E., Cohn, D., Magdassi, S., 3D printing of shape memory polymers for flexible electronic devices. *Adv. Mater.*, *28*, 22, 4449–4454, 2016.

42. Jiang, Y., Leng, Q.Y., Yan, Y., Ng, E.L.L., Chee, H.L., Wang, F., Chan, S.Y., Loh, X.J., Wang, J., Chan, B.Q.Y., 4D printing of single-network shape memory polyurethanes with two-way actuation properties. *ACS Appl. Polym. Mater.*, *4*, 11, 8574–8583, 2022.
43. Chan, B.Q.Y., Chong, Y.T., Wang, S., Lee, C.J.J., Owh, C., Wang, F., Wang, F., Synergistic combination of 4D printing and electroless metallic plating for the fabrication of a highly conductive electrical device. *Chem. Eng. J.*, *430*, 132513, 2022.
44. Yampolskiy, M., King, W.E., Gatlin, J., Belikovetsky, S., Brown, A., Skjellum, A., Elovici, Y., Security of additive manufacturing: Attack taxonomy and survey. *Addit. Manuf.*, *21*, 431–457, 2018.
45. Kashyap, S., Additive manufacturing for presurgical and postsurgical planning: A state-of-the-art review, in: *Additive Manufacturing Processes in Biomedical Engineering*, pp. 117–142, 2023.
46. Priyadharshini, M., Balaji, D., Bhuvaneswari, V., Rajeshkumar, L., Sanjay, M.R., Siengchin S., Fiber reinforced composite manufacturing with the aid of artificial intelligence–A state-of-the-art review. *Arch. Comput. Methods Eng.*, 29, 7, 5511–5524, 2022.
47. Wang, K., Zeng, M., Wang, J., Shang, W., Zhang, Y., Luo, T., Dowling, A.W., When physics-informed data analytics outperforms black-box machine learning: A case study in thickness control for additive manufacturing. *Digital Chem. Eng.*, *6*, 100076, 2023.
48. Turan, N., Saeidi-Javash, M., Chen, J., Zeng, M., Zhang, Y., Go, D.B., Atmospheric pressure and ambient temperature plasma jet sintering of aerosol jet printed silver nanoparticles. *ACS Appl. Mater. Interfaces*, *13*, 39, 47244–47251, 2021.
49. Hollar, C., Lin, Z., Kongara, M., Varghese, T., Karthik, C., Schimpf, J., Eixenberger, J., Davis, P.H., Wu, Y., Duan, X., Zhang, Y., High-performance flexible bismuth telluride thin film from solution processed colloidal nanoplates. *Adv. Mater. Technol.*, *5*, 11, 2000600, 2020.
50. Eugene, E.A., Gao, X., Dowling, A.W., Learning and optimization with Bayesian hybrid models, in: *2020 American Control Conference (ACC)*, 2020, July, IEEE, pp. 3997–4002.
51. Wang, J., Eugene, E.A., Dowling, A.W., Scalable stochastic programming with Bayesian hybrid models, in: *Computer Aided Chemical Engineering*, vol. 49, pp. 1309–1314, Elsevier, UK, 2022.
52. Gljušćić, M., Lanc, D., Franulović, M., Žerovnik, A., Microstructural analysis of the transverse and shear behavior of additively manufactured CFRP composite RVEs based on the phase-field fracture theory. *J. Compos. Sci.*, *7*, 1, 38, 2023.
53. Ermakova, L.V., Dubov, V.V., Saifutyarov, R.R., Kuznetsova, D.E., Malozovskaya, M.S., Karpyuk, P.V., Dosovitskiy, G.A., Sokolov, P.S., Influence of luminescent properties of powders on the fabrication of scintillation ceramics by stereolithography 3D printing. *Ceramics*, *6*, 1, 43–57, 2023.

54. Saleh, M.A., Elkholy, A., Melenka, G.W., Kempers, R., Heat flux measurement using 3D-printed continuous wire polymer composite sensors. *Case Stud. Therm. Eng.*, 42, 102739, 2023.
55. Ahmed, W., Zaneldin, E., Al Hassan, A., Al-Marzouqi, A.H., Nanomaterial used in 3D printing technology, in: *Smart 3D Nanoprinting*, pp. 165–182, CRC Press, USA, 2023.
56. Shi, X., Gu, X., Zhang, Y., Dong, L., Meng, Q., Preparation of an aqueous photoinitiator through supramolecular assembly oriented towards 3D printing hydrogel without VOC emission. *J. Sol-Gel Sci. Technol.*, 105, 7, 577–586, 2023.
57. Dey, A., Rahman, M.M., Yodo, N., Grewell, D., Development of biocomposite filament for fused filament fabrication from soy hulls and soy protein isolate. *Mater. Today Commun.*, 34, 105316, 2023.
58. Ma, C., Ge, Q., Yuan, L., Gu, D., Dai, D., Setchi, R., Wu, M., Liu, Y., Li, D., Ma, S., Peng, X., The development of laser powder bed fused nano-TiC/NiTi superelastic composites with hierarchically heterogeneous microstructure and considerable tensile recoverable strain. *Compos. Part B: Eng., 250*, 110457, 2023.
59. Ikram, H., Al Rashid, A., Koç, M., Additive manufacturing of smart polymeric composites: Literature review and future perspectives. *Polym. Compos.*, *43*, 9, 6355–6380, 2022.
60. https://patentscope.wipo.int/search/en/result.jsf?_vid=P20-LDGURD-48623 – Accessed on 29th Jan 2023.

2

Recent Developments in Additive Manufacturing Equipment's and Its Processes

Ganesh Angappan[1]*, Santhosh Sivaraj[2], Madhan Mohankumar[2], Elango Vaidyanathan[2] and Arun Joseph[2]

[1]Department of Mechanical Engineering, Kongu Engineering College, Perundurai, Erode, India
[2]Department of Robotics and Automation, Easwari Engineering College, Chennai, India

Abstract

Additive manufacturing (AM) processes are done through the layer-by-layer deposition of precursor materials in the form of solid wire, sheet, and liquid resin. Soon after the application of the layered deposition of raw materials of a finished product, a certain time period is given to achieve better curing of the deposited material. The improper curing of deposited layer causes internal cracks or trapping of blow holes, which leads to the increase in the brittle nature of the fabricated product. In case of complex curvatures in the required design, there is a need to employ the proper structural supports for withstanding the deposited layers initially. After obtaining the fabricated product, the supports are carefully removed from the main structure. The base of the additive-manufactured product is separated from the base of the 3D printer by applying wax, a strong plastic sheet, and a stainless-steel remover. There are various AM techniques to fabricate the desired product, but the most advanced technique is found to be four-dimensional (4D) printing technique, which considers the time as well as external stimuli factors involved in the fabrication process.

***Keywords*:** Additive manufacturing, stereolithography, 3D printing, fused deposition modeling, 4D printing

**Corresponding author*: ganesh.mech@kongu.edu

R. Rajasekar, C. Moganapriya and P. Sathish Kumar (eds.) Additive Manufacturing with Novel Materials: Processes, Properties and Applications, (23–48) © 2024 Scrivener Publishing LLC

2.1 Introduction

Rapid prototyping (RP) creates three-dimensional (3D) solid items from computer-aided drawings, and was developed initially using 3D printing process [1]. Researchers at the University of Texas developed the initial additive manufacturing (AM) process in 1986 as a result of the advancement of RP approaches, which included the ability to successfully produce complex parts. Many novel AM techniques have been created over the past 30 years. These procedures exhibit substantial benefits, such as flexible geometric capabilities, absence of human engagement, and shortened design cycle times [2]. Several industries have successfully adopted AM. Automobile [3], space [4, 5] and aviation engine and airframe [6, 7] components have all been manufactured using functional AM components that have complicated geometries. The basic procedures for producing materials made of polymers, metals, and ceramics via AM are described in this chapter. A review is given of the most recent AM tools and procedures. On the effectiveness and quality of the goods of AM, the effects of fabrication settings and processing factors are discussed. Also, the advanced materials for AM methods are discussed, such as multimaterials and biological tissues.

The AM process is broadly classified based on the medium of precursor material, type of base material used, and methodology of fabrication technique. Depending on the type of base material, some AM processes are categorized as solid wire based [fused deposition modeling (FDM), wire and arc AM, electron beam free foam manufacturing (EBF3), etc.], powder based [electron beam melting (EBM), selective laser sintering, laser metal deposition, and selective laser melting], and liquid based (stereolithography, direct light processing, polyjet printing, etc.). Also, the AM process is further divided into three types: laser beam, thermal based, and ultraviolet (UV) based. Sheet lamination (SL), energy deposition, powder bed fusion (PBF), material extrusion, binder jetting (BJ), and material jetting were all technically included in the AM process along with different production techniques. Several improvements to the 3D printing process were found to be more promising in various fields such as medical, aerospace, automobile, electronics, robotics, dental implants, etc. Through AM process, the prototypes, which hold equivalent functionality, can be achieved in a short period of time with more accuracy. Prototypes are being used to understand the exact problem and also give a better solution to the problem in a minimum amount of time and material expenses. Rapid prototyping is a process that helps beginners understand the intensity of an acquired problem and predict solutions in an appropriate manner. Nowadays, there have been attempts to use RP to replace defective organs to minimize death rates.

Medical devices and healthcare are key sectors that are using AM. In addition to producing medical devices faster and more effectively, AM can also help establish the feasibility of such innovations outside of the operating room and convince other stakeholders about their benefits [8]. Additive manufacturing methods employed nowadays in medical applications can be divided into portions composed of polymers and those manufactured with metals depending on the raw material [9]. A number of implantable devices are made possible by AM technologies such as heart valves, rib cages, and bones. In dentistry, AM has been applied to implant-supported dentures as well as other prosthetic devices. Anatomical models can also be created easily using AM, which can be used in teaching and planning surgery [10]. The chin and jaw can be fractured if patients are traumatized. The infected tissue must be removed in patients where cancer has spread to the jaw. An entire jaw or part of the jaw can be printed in three dimensions [11]. Partially and completely removable dentures have been made using AM. A method for making metal dentures is called DMLS (direct metal laser sintering). These dentures are made from an alloy of cobalt, chrome, and molybdenum [12]. Before doing the procedure, the surgeon must plan the operation, and the jaw models, teeth models, and other organ models must be already developed. Preparatory surgery ensures precision and reduced errors in the operation. The development of superior materials facilitates the development of improved models [13].

In scaffold-based tissue engineering, cells are deposited on a 3D scaffold, and scaffold-less tissue engineering allows cells to multiply without a scaffold [14]. Researchers have concentrated on improving AM technology in this area to create products with more resemblance to natural tissues. However, the mechanical stability of the products made by inkjet-based printing systems is not satisfactory, even though the print resolution can reach 20–100 μm. Microfilament-based extrusion machines are capable of up to 400 μm resolution, but the high resolution causes cells in bio inks to be damaged. Both inkjet and extrusion printing have the same resolution issue, specifically the nozzles clogging above 50 μm [15]. For drug delivery and medical products, fusion deposition method and inkjet printing are the most preferred printing methods. In the medical industry, four-dimensional (4D) printing is a fairly new innovation and it has been used in a limited number of applications, like direct ink writing for intravenous drug delivery systems [15].

Because of the differences in pharmacokinetic traits between patients, the conventional healthcare method applies a cookie-cutter approach that provides inconsistent results. "Personalized medicine" has come a long way with the advancement in pharmacokinetic testing [16]. This biodegradable polymer is synthetic polyvinyl alcohol. Despite its colorlessness, it is a highly soluble organic solvent that is partly soluble in water but mostly nonsoluble in ethanol. Polyvinyl acetate after its acetate group is removed is hydrolyzed to make it. All devices for drug delivery made of this polymer are biocompatible, especially FDM. As aqueous solutions may cause undesirable swelling if used to load up to 10% of drugs into polyvinyl alcohol (PVA), it is usually done in an ethanolic solution [17]. The use of thermoplastic, very durable, recyclable polymer polylactic acid has not been connected to any toxicological or cost issues. Chemically, it is insoluble in organic solvents and has a high melting point. For the synthesis of this polymer, two different methods are used: ring opening polymerization (ROP) and direct polymerization (DP) [18].

A prospective application of geopolymer cements made from fly ash is for the AM of on-site construction elements. Geopolymer mortar, a green infrastructure material, was printed directly from a 3D computer-aided design (CAD) model using a layer-by-layer technique. Geopolymer, a sustainable building material, might be utilized in AM to help promote environmental sustainability [19]. In a sense, there are three processes for AM, and the differences are in the applications and materials for which they are developed, which result in distinct advantages for each. In a similar way to Z-Corp's 3D printing process, D-Shape uses powder deposition to form objects that are selectively hardened with a binder [20]. Through the building up of successive layers of cement-based adhesive, the contour crafting process creates a smooth finish by extruding paste against a trowel. At the moment, the material is being deposited quickly to build a full structural wall with the least amount of material [21]. The third method is concrete printing process, which is similar to the conventional method of building solid geometry. Machine instructions are created from this geometry, the product is printed, and the subsequent postprocessing of the component, including the elimination of support structures, as well as any finishing required [22].

The building information modeling (BIM) system that connects virtual and actual buildings will be essential to transforming the construction industry into a 4.0 revolution. A construction project's core functions

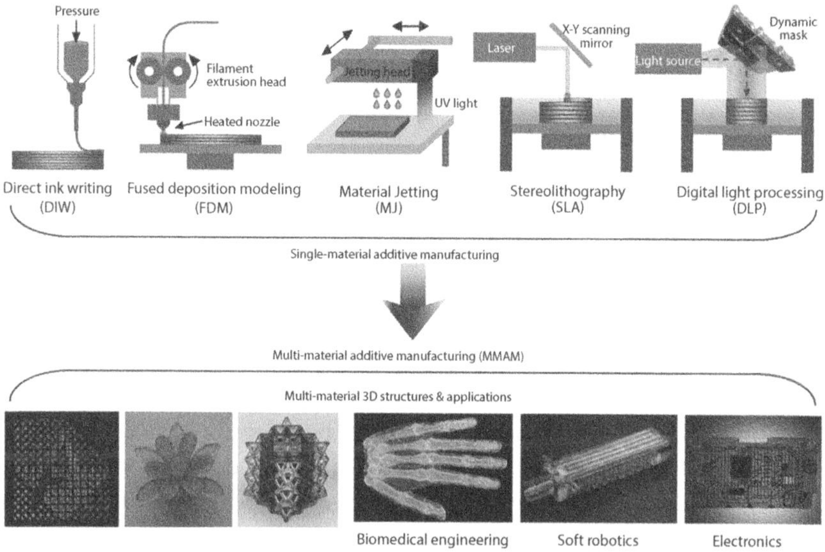

Figure 2.1 Various advancements in additive manufacturing process [25].

include building survey and advanced planning. Figure 2.1 shows various advancements in additive manufacturing process. Building structures, pipes, and other objects, like water pipes, can be digitized using laser 3D technology, which can then be transmitted to a planning software for use by BIM teams [23]. Building information modeling technology is often used to facilitate data sharing between stakeholders, so stakeholders from different geographical locations can work together on a project utilizing data taken from the Cloud. By allowing seamless access to cloud-based BIM data from mobile devices, such as smartphones and tablets, decision-making is improved while meeting delivery targets, thus improving and enhancing efficiency [24]. This chapter discusses the various types of AM processes and their importance. The broad classifications of AM process in various sectors are explained individually. In the beginning stage, AM process only involves single material fabrication. Later on, due to the complexity and enhanced demand in the market, AM technique grabs huge attention in the manufacturing sector. Multimaterial-based AM process is especially found to be more useful than the single material. The advantages, advancements, and applications of various AM processes are described in reference to recent trends in detail.

2.2 Equipment and Procedures for Polymer Additive Manufacturing

2.2.1 Stereolithography

Stereolithography is a type of AM technique that uses a liquid photopolymer resin and ultraviolet laser to fabricate 3D objects layer by layer. The CAD model is subdivided into thin layers and uploaded to a stereolithography machine. In the stereolithography machine, a platform is lowered into a vat of liquid photopolymer resin. A UV laser then traces the first layer of the object, solidifying the resin in the desired shape. The platform is then lowered slightly, and the laser traces the next layer on top of the previous one. This process is repeated layer by layer until the entire object is created. Once the object is complete, it is removed from the machine and rinsed in a solvent to remove any excess resin. The object may also be cured with UV light to strengthen it and improve its mechanical properties. Stereolithography is used in a variety of applications, including rapid prototyping, product design, and manufacturing. It is particularly useful for creating complex geometries [26]. They discovered that the exterior quality, construction time, and the support framework were the three main standards for evaluating an selective laser annealing (SLA) part. There was a suggestion for a software program that might automatically look for the best orientation. The throughput of SLA in relation to 50 processing parameters will be examined by an experimental design described by the authors [27]. They came to the conclusion that making thicker layered parts would cut runtime and costs in half but would compromise surface quality. The best SLA parameters have been proposed by researchers [28] utilizing an evolutionary approach. The fabrication of an "H" formed part at ideal conditions was contrasted with parts under nominal conditions. Researchers [29] investigated the two main elements that influence the process of curing were once believed to be temperature and light intensity. We evaluated a double-step reaction process that uses cycloepoxy groups and vinyl ether monomers. Also, they refined an equation previously published by the author [30], which evaluated the rate of monomer consumption, by taking the intensity of the light into account [31]. High-resolution projected micro stereolithography, which was made using an electronic micromirror system as a dynamic mask. This system allows for the parallel fabrication of intricate microstructures like micro-electro mechanical systems (MEMS). The behavior of photon-induced curing was investigated using a process model. Parameters can alter the features of an item made using SLA. Researchers built a process model using part strength

[32]. Layer width, post-curing duration, and direction were the three main model parameters. Using these three variables, this process model can forecast the printed part's strength. Recent research [33] established an SLA process model using the Taguchi approach. Nine parameters were taken into consideration when conducting their research. These parameter values were used to carry out an orthogonal array, an experiment design. This model can determine the part's quality and strength. A better SLA technology called continuous liquid interface generation was created by Tumbleston *et al.* [34] to shorten the development time of continuous liquid interface production (CLIP) polymer printing. Whereas the CLIP produces the piece continuously, the standard SLA does not. Simultaneous printing, curing the element only where necessary, is made possible by an oxygen permeable screen placed under the plane of the light projection.

2.2.2 Ink Jetting

Another method for printing functional materials at the micro- and nanoscale using polymers is ink jetting. The procedure was created on the foundation of traditional two-dimensional (2D) inkjet printing. Ink-like liquid-phased materials were deposited layer by layer for attaining the desired finished product. An ink-jetting equipment, with a minimal dimension of 1 μm, must be used for making nanodots [35]. An array of carbon nanotubes with a pattern was manufactured using catalyst ink made of transition-metal nanoparticles. The most significant difficulties for inkjet printing are material limitations. Droplet generation is the main obstacle to polymer high velocity formation. A description of acoustic resonance jetting was given [36]. By creating high-pressure gradients close to the nozzle opening with the help of ultrasonic waves, ink is jetted repeatedly. Nevertheless, viscous friction or vitrification inside the nozzle has a limit on how much the ink-jetting performance may be improved. It has been investigated [37] how the orientation of the polymer jetted portion affects its mechanical properties. According to results of tensile and hardness tests, the printing orientation may have an impact on how much light energy is absorbed, which can lead to anisotropic physical characteristics.

2.3 Equipment and Procedures for Metal Additive Manufacturing

A significant topic of discussion right now is AM of metallic materials. Due to the numerous industrial uses for metal parts, their utilization is transitioning from research to commercialization. Metals require more intricate AM processing than other materials frequently do. Metal production cannot be done using vat polymerization, and some of the other methods, including material extrusion [38] and material jetting [39], are still under research. There are four different types of metal additive manufacturing techniques:

- Powder bed fusion
- Directed energy deposition
- Binder jetting
- Sheet lamination

The definitions in the ASTM standard are used to categorize the systems. The processing information is also supplied, including the laser focus diameter and layer thickness range. The most widely used method of PBF utilizes lasers, and depending on the manufacturer, the laser strength ranges from 100 to 1000 W. Since every build layer in the laser PBF process can have thicknesses as little as 20 μm, it offers an edge over other approaches in terms of precision. The only manufacturer of PBF based on EBM at the moment is Arcam (bought by GE in 2017). E-beam has a significantly greater strength than lasers, and each scan can create a layer that is thicker. Additional metal AM procedures include laser PBF and powder-fed directed energy deposition (DED), both offered by Trumpf simultaneously. Another is BJ and SL technologies appropriate for metal AM. Producers of these technologies include ExOne and Fabrisonic.

2.3.1 Powder Bed Fusion

Powder bed fusion selectively melts or sinters a metallic powder collectively over the bed. It can be classified as laser melting or sintering and EBM. To contain the powder and build upon it, a platform is required for both processes. Although the fundamentals of the two processes are identical, they differ greatly in their processing steps. Figure 2.2 depicts how the laser melting arrangement works. In the laser powder bed method, the light from the laser passes through a set of lenses while being redirected by a reflecting surface onto the platform's face.

The movement of the laser beam spot along the intended paths in planar directions is controlled by mirrors. The laser scanning technique is repeated once a thin film of powder has been selectively melted, then the platform descends. A laser AM machine's build chamber is often argon-filled to prevent oxidation of the metal as it dissolves and resolidifies.

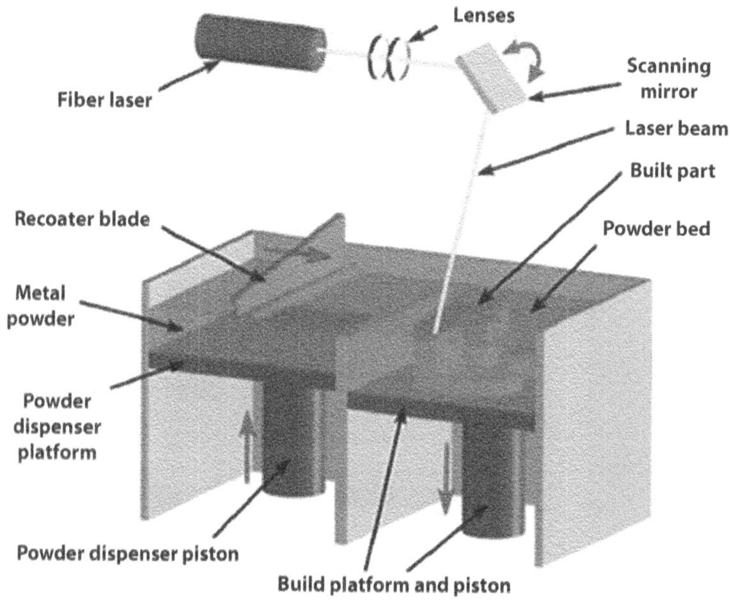

Figure 2.2 Selective laser melting [40].

The scanning electron microscope (SEM) method served as the foundation for the development of the EBM process [42]. To selectively melt the powder, it uses an electron beam with significantly greater power. The EBM process needs low pressure, close to vacuum conditions. According to Figure 2.3 [43], the powder bed is placed on top of the electron beam source. The motion of the electron beam is directly controlled by a lens mechanism.

2.3.2 Directed Energy Deposition

Directed energy deposition is a production technology utilizing metal in powder or wire form as feedstock combined with high energy laser or electron beam to melt and deposit the material on top of the substrate. Based on feedstock, it is classified as the first category, which consists of techniques that have been improved upon from conventional welding, and utilizes metal wire as a feedstock. The second technique, known as "laser engineered net shaping (LENS)," was created by Sandia National Laboratories, employing flow of metal powder as a feedstock [44]. Figure 2.4 displays the LENS machine's operation. A build chamber uses a lens system to focus a neodymium-doped yttrium aluminium garnet (Nd:YAG) laser beam

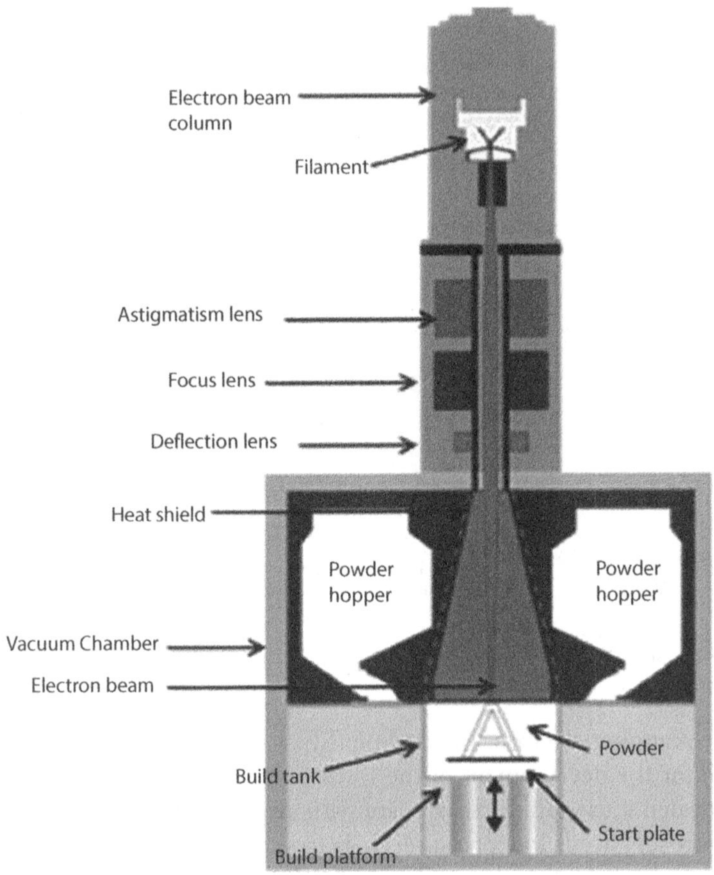

Figure 2.3 Electron beam melting [41].

on the construction base of desired product while simultaneously injecting metal powder through a powder nozzle into the laser spot. Following the construction of one layer, the laser lens and powder nozzle are moved higher, and the laser heating-powder injection process is repeated for the subsequent layer [45]. Thinner layers and slower deposition rates are possible using an electron beam, leading to improved surface finishing and higher precision. NASA developed the electron beam freeform fabrication (EBF3) technique [46]. Majority of its applications are space based. Instead of using a powder injection system, EBF3 uses a metal wire filament. The metal wire's front end is melted and then carefully sprayed with a material to create a new material layer over substrate using a laser or electron beam.

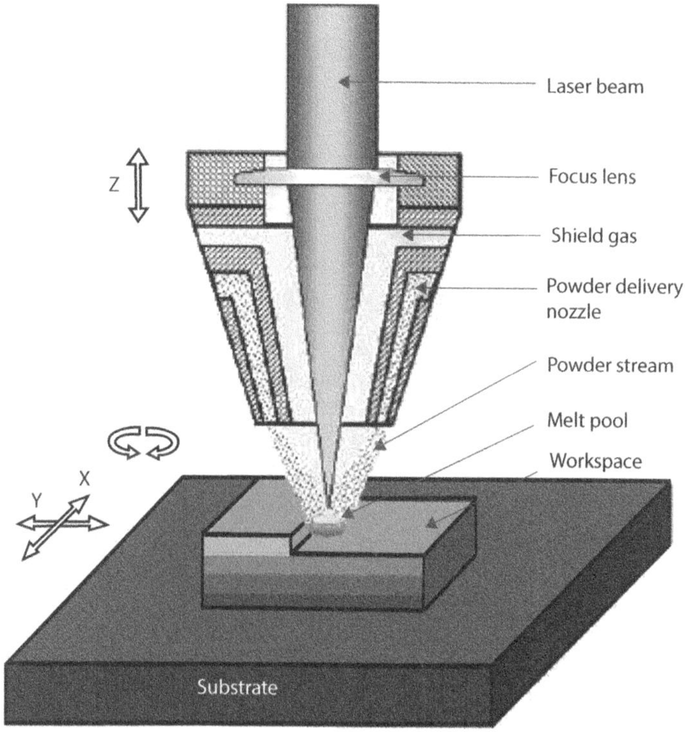

Figure 2.4 Schematic of laser engineered net shaping process [47].

2.3.3 Binder Jetting

Inkjet head and powder bed 3D printing are also known as BJ. In 1993, the invention was created and copyrighted [48]. The concept behind it is to add a third dimension to conventional 2D printing. A liquid binder is injected through one or more nozzles onto a powder bed to bind the powder together. Once a tiny layer of powder is added, the nozzle continues to move along the intended route. When the layers are built up, a 3D object is created. Figure 2.5 depicts a schematic of a BJ apparatus. The system is made up of a powder tank and a build platform. Binder jetting is a 3D printing technology that uses a powder bed and a liquid binding agent to create 3D objects layer by layer. The process begins with the creation of a 3D model using CAD software. The model is then uploaded to a BJ machine.

The object is sintered or infiltrated with a secondary material to improve its mechanical properties. Binder jetting is used in a variety of applications,

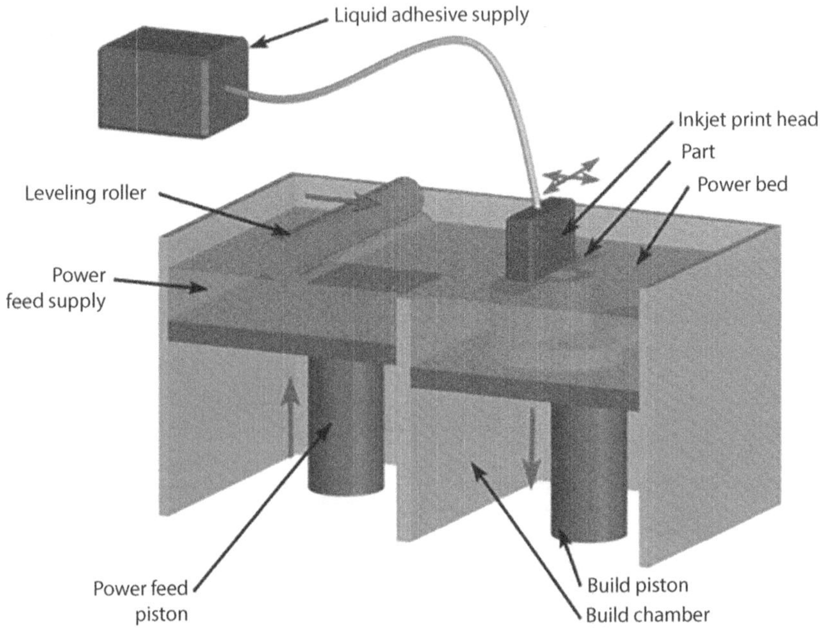

Figure 2.5 Additive manufacturing — binder jetting [49].

including rapid prototyping, product design, and manufacturing. It is particularly useful for creating large, complex objects with fine details, and for producing metal parts with high accuracy and surface quality. Advancements in BJ technology include improved printing speeds, higher resolution, and the ability to print with a wider range of materials, including metals, ceramics, and composites. Binder jetting is becoming increasingly popular in industries such as aerospace, automotive, and healthcare, where it is used to create functional parts and components with complex geometries. Binder-jetted items' post processing is frequently more difficult than other methods of additive manufacturing, particularly for metallic materials.

Green body must be cured at about 6 to 12 h before being heated for 24–36 h at temperatures above 1000°C to create a solid metallic component with the desired mechanical strength. In addition to sintering, consolidation, and occasionally infiltration and burning of the binder, heat treatment can also result in improved mechanical strength. A metal with a smaller melting point, like bronze, can occasionally be used to penetrate a metal with a higher melting point. The flexibility of the BJ material can be improved by doing this [50].

2.3.4 Sheet Lamination

Metal sheets are used as feedstocks in the SL or laminated object manufacturing (LOM) process [51]. Ultrasonic additive manufacturing (UAM), which was created and licensed [52, 53], is the most widely used manufacturing method. The interfaces of layered sheets are connected by diffusion rather than melting when ultrasonic wave and mechanical stress are applied to them at room temperature. Without the aid of any frictional force acting as a heat source, the stacked sheets are joined one by one to create a 3D object. Conventional polishing is performed after fabrication of the desired product. In Figure 2.6, the UAM equipment's operational procedure is depicted. On a base plate are arranged and layered metallic sheets. For ultrasonic vibration and pressure, a sonotrode that is digitally controlled rolls in the rolling direction. Because of the high-frequency sonotrode vibration, a fresh metallic sheet is consequently bound to the earlier-built structures. Frictional heat generated at the bonded surfaces during this process raises the temperature of the consolidated region. There is a brief cooling period in between each layer's manufacture to prevent residual thermal tensions. The product is ejected from the substrate once all the layers have been constructed, and the surface finish is then polished to make it better.

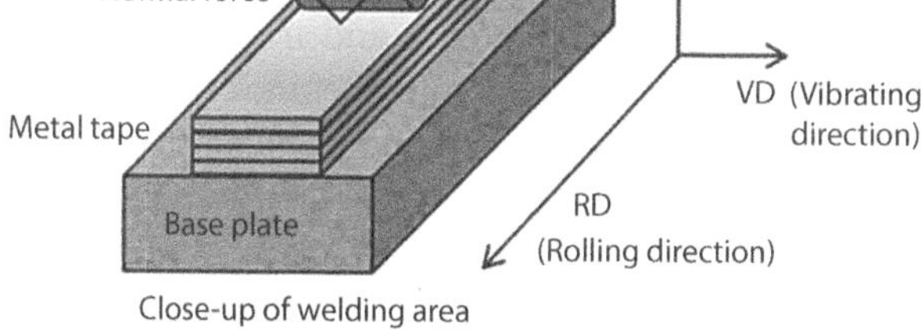

Figure 2.6 Ultrasonic additive manufacturing process [54].

2.4 Equipment and Procedures for Ceramics Additive Manufacturing

2.4.1 Binder Jetting

Binder jetting is a 3D printing technology that uses a powder bed and a liquid binding agent to create 3D objects layer by layer. The model is then uploaded to a BJ machine [48]. In the BJ machine, a thin layer of powder is spread onto a build platform. A print head then deposits a liquid binding agent onto the powder layer, which binds the particles together in the desired shape. The platform is then lowered slightly, and the process is repeated layer by layer until the entire object is complete [55]. Surface quality is also another crucial factor that influences how well the finished product performs. Printing is done using bimodal ceramic powders [56]. Once the object is printed, it is removed from the machine and cleaned to remove any excess powder. The object may also be sintered or infiltrated with a secondary material to improve its mechanical properties [57]. Binder jetting is used in a variety of applications, including rapid prototyping, product design, and manufacturing. It is particularly useful for creating large, complex objects with fine details, and for producing metal parts with high accuracy and surface quality.

Advancements in BJ technology include improved printing speeds, higher resolution, and the ability to print with different materials. Binder jetting is becoming increasingly popular in industries such as aerospace, automotive, and healthcare, where it is used to create functional parts and components with complex geometries.

2.4.2 Ink Jetting

The ceramic powder is combined with binders in the ink-jetting apparatus to create printable filaments or suspension. Little droplets of the filament are discharged from the printer nozzle and imprinted on the substrate. Ceramic ink jetting is mostly used in microceramic devices like micropillar arrays. It was investigated how ceramic micropillar arrays were inked-jetted [58]. Lead Zirconate Titanate (PZT) and Titanium dioxide (TiO2) were used in their research to print micropillar arrays. With the right printing settings and binder materials, a dense micropillar array can be produced for a particular micropillar array configuration. The formulation of the filament and droplet disintegration of many inks with various compositions were examined [59]. The Rayleigh number, thread minimum radius, droplet breakup time, and filament length were specifically acquired. The

Taguchi optimization and ANOVA techniques were used by Rahul *et al.* to identify the best process variables for printing dense ceramic coatings over glass substrates [60]. According to their findings, layer thickness will increase with a greater nanopowder concentration; however, the shape of the printed layer will have flaws. Thus, they outlined the ideal parameters for obtaining thick layers while also managing surface flaws.

2.4.3 Stereolithography

Stereolithography can be used to create ceramic objects using a process called ceramic stereolithography. In this process, a ceramic-filled photopolymer resin is used as the printing material. The ceramic particles are mixed with a liquid resin turns into solid when exposed to UV light. The process of ceramic stereolithography is similar to traditional stereolithography, with the difference being that ceramic-filled resin is used instead of a standard photopolymer resin. The process involves creating a digital 3D model using modeling software, which is then sliced into thin layers. The ceramic-filled resin is spread onto a build platform, and a UV laser is used to selectively solidify the resin in the desired shape. After printing, the object is removed from the printer and cleaned to remove any excess resin. The object is then fired in a kiln to burn off the organic binder and sinter the ceramic particles together to form a dense, solid ceramic object.

Ceramic stereolithography has several advantages over traditional ceramic manufacturing techniques, including the ability to create complex geometries with high accuracy and resolution. Ceramic stereolithography has applications in industries such as aerospace, biomedicine, and electronics, where ceramic parts with high precision and complex geometries are required.

2.5 FDM 3D Printing Technique

Fused deposition modeling (FDM) is the one that is believed to be the least expensive and most widely used. This method is based on the melting or softening of the raw materials first, then depositing several layers of thermoplastic components on top of them other [61]. Fused deposition modeling printers are now increasingly available for private use in houses. These 3D printers are sometimes known as desktop 3D printers since they operate similarly to a conventional 2D printer in terms of ease of use. You may create your own FDM 3D printer as well. The Fab@Home and RepRap became the first two freely available do-it-yourself (DIY) 3D printers.

With the introduction of these DIY 3D printers, the technology of 3D printing became more accessible and affordable. The Prusa i3 is one of the most well-known versions of the RepRap 3D printer. A FDM 3D printer can operate with a wide variety of materials. The materials might be thermoplastics [like polylactic acid (PLA) and acrylonitrile butadiene styrene (ABS)] as well as engineering materials (such polyethylene terephthalate glycol (PETG), thermoplastic polyurethane (TPU), and poly(ether-ether-ketone) (PEEK) and polyether imide (PEI)). The two thermoplastics that are most frequently used are PLA and ABS. A form of plastic, ABS, has been used to create LEGO, electronic housings, and automotive bumper pieces. It is mostly used in the injection molding sector. In contrast, PLA is derived from renewable resources like sugarcane or maize starch and is biodegradable. It is inferior to ABS in terms of heat resistance, elongation, and flexural strength. As a result, ABS is frequently chosen because it offers superior mechanical qualities. Once PLA gets close to 60°C, it can easily lose its structural integrity and start to droop and distort, especially if it is under strain. With the right circumstances, PLA, a bioplastic, can decompose. In normal air conditions, PLA can break down in 50 days, whereas in water, it takes 48 months. While recyclable, ABS cannot be degraded by nature. The strength, ductility, machinability, and thermal stability of ABS make it ideal for usage in a variety of applications. When aesthetics matter, PLA is appropriate [62]. For FDM 3D-printed objects, postprocessing is typically necessary because the print layers of the final product are frequently visible. To give it a shiny sheen, ABS is frequently polished using acetone. After printing, ABS can also be simply machined or sanded. Fused deposition modeling is typically the most economical method for making prototypes and customized thermoplastic parts. Because of the technology's wide availability, FDM has quick lead times. For FDM printing, a variety of thermoplastic materials are available. These materials consist of those that can be used for both functioning non-commercial applications and prototyping.

2.6 Hot Melt Extrusion Method—Polymers

Hot melt extrusion (HME) is a manufacturing process that involves melting a solid material and extruding it through a die to form a continuous profile with predefined shape and size. It is widely used in the plastics industry to produce a variety of products, including films, tubes, and sheets [63]. A schematic illustration of the hot melt extrusion procedure is shown in Figure 2.7.

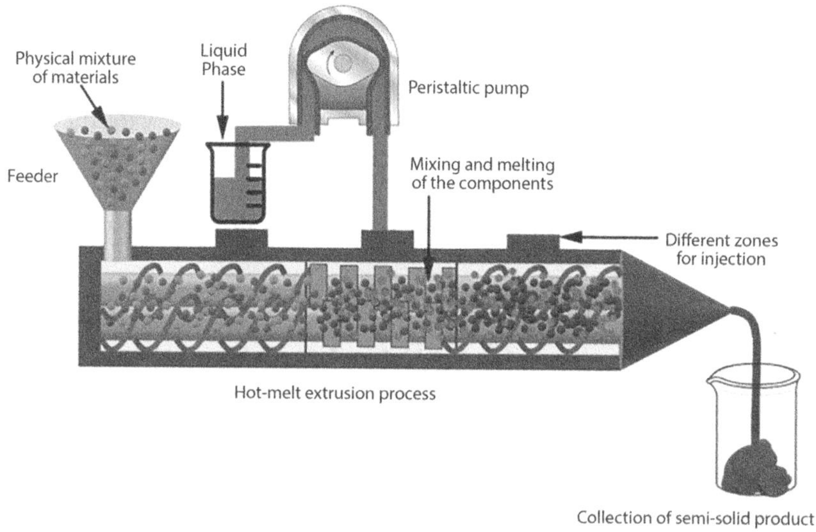

Figure 2.7 Hot melt extrusion process [64].

More recently, HME has been used in the medical and healthcare fields, where it is employed in the production of medical equipment and the blending of active pharmaceutical ingredients (APIs). This is due to HME's capacity to increase the bioavailability and solubility of APIs. Twin screw extruders are preferable when creating products composed from several formulation ingredients because they offer superior mixing to create a homogenous solid with evenly dispersed API particles. The primary method for producing the filaments required for FDM 3D printing is also owned by HME. Due to the growing acceptance of 3D printers as an alternative to conventional manufacturing techniques, there has been an increase in research interest in the production of various types of filaments from various materials. The HME process starts with feeding solid material, typically a thermoplastic polymer, into the extruder barrel. The material is heated to a molten state and is then forced through a die to form a continuous profile. The die determines the shape and size of the extruded product [65]. The HME process can be modified by adding other materials to the polymer melt, such as fillers, pigments, or APIs. This allows for the production of composites, color-coded products, or pharmaceutical dosage forms.

Thus, it is now feasible to combine FDM 3D printing with HME in an effort to expand the number of polymers that may be modified using FDM [66]. Hot melt extrusion has several advantages over traditional manufacturing processes, including the ability to produce complex shapes and sizes with consistent quality, high productivity, and low cost. It is also a continuous process, which allows for better control of the final product's properties, and is widely used in the pharmaceutical industry for the manufacture of solid oral dosage forms, such as tablets and capsules. It is also used in the food industry to produce products such as snacks and breakfast cereals, and in the chemical industry to produce adhesives, coatings, and sealants.

2.7 Advancements in Additive Manufacturing

Even though AM was first developed in the middle of the 1980s, its development for biomedical equipment has lagged behind that of AM for structural applications. This is mostly caused by the demanding performance standards and worries about reproducibility and product quality [67]. Laser assisted three dimensional bioprinting (LAB) is particularly useful for the production of complex biological structures, such as tissue scaffolds or organoids. It is also being used to create bio inks for use in extrusion-based bio printing. LAB is a technology in which efforts to optimize the printing process and improve the quality and functionality of the printed structures were undertaken. LAB has the potential to revolutionize the field of tissue engineering and regenerative medicine by enabling the creation of functional tissues and organs for transplantation and drug testing [68] created a 3D plotting method with controlled pore size gradients and distribution to get around the problem of low cell seeding effectiveness in material-extruded scaffolds. LAB is a 3D printing method utilizing laser beams to deposit living cells layer by layer to create complex biological structures. It enables high-precision printing with high cell viability and low shear stress [69]. In LAB, a laser beam is focused onto a solution containing living cells and biomaterials, such as hydrogels or extracellular matrix materials. The laser energy causes a small amount of the solution to vaporize, creating a tiny bubble that pushes the surrounding solution out of the way and forms a droplet that contains living cells and biomaterials [70, 71]. These droplets are then deposited onto a substrate to form a 3D structure, layer by layer. The advantage of LAB is that it allows for the precise deposition of cells and biomaterials in a way that does not damage the cells or cause them to clog up the printer nozzle. The laser energy can also be controlled to vary the size and spacing of the droplets, which allows for the creation of structures with varying densities and properties. Figure 2.8 shows the various parameters influencing 4D printing process.

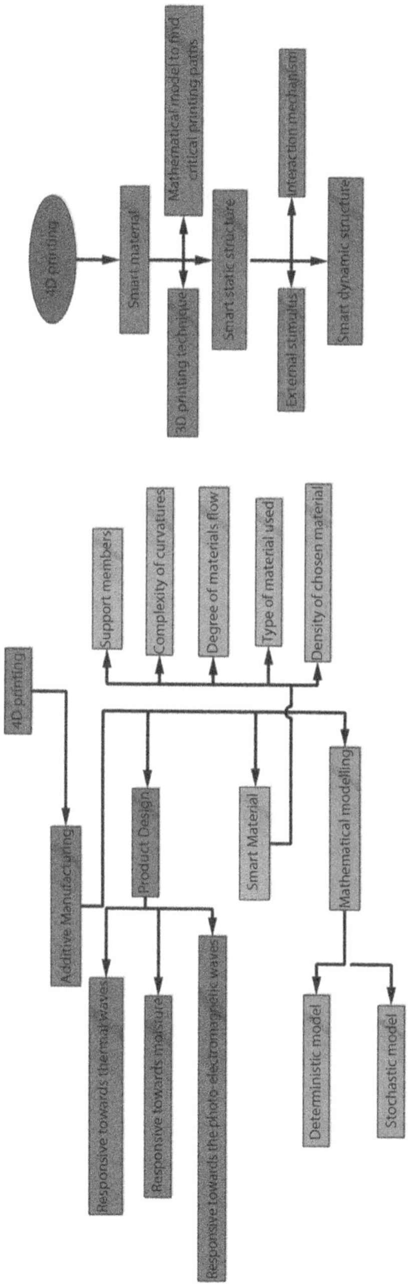

Figure 2.8 Parameters influencing the 4D printing process [76].

For inkjet printers, a multimaterial printing approach has been created [72]. They fabricate the product with minimal cost and better functionality [73]. They proposed that improving system productivity, building volume, and layer thickness would significantly improve the efficiency of the inkjet printing process. Massive data availability, modern manufacturing techniques, and digitalization have all increased the need for an interest in merging machine learning (ML) and optimization strategies to improve product quality. The use of ML-integrated manufacturing techniques improves the acceptability of novel ideas, conserves time, energy, and resources, and prevents waste. Machine learning-integrated assembly procedures contribute to the development of "smart manufacturing," in which faults are automatically corrected in real-time to avoid spillage. The manufacturing industry employs a variety of computing methods and tools, but more contemporary approaches, such as ML and data mining, are crucial for addressing difficult industrial and academic challenges [74]. The Precision Medicine Initiative, a project to reform healthcare delivery, was introduced by the White House in 2015. Advanced manufacturing, which includes the creation of tailored medicines, drug delivery systems, and drug screening platforms, will be crucial in bringing the idea of precision medicine from the lab to the clinic. These goods need manufacturing methods that are both reliable and adaptable because they are highly personalized. Over the past 5 years, the field of advanced manufacturing has significantly advanced. A quick examination of processing materials and their properties will be given after the introduction of the items made for precision medicine in this state-of-the-art review. A discussion of several manufacturing techniques that apply to the aforementioned products is given [75]. A very interesting field of engineering study is the use of multiple (3D) printing and AM to create composite materials that are smart and intelligent and have several functions.

It is described how to integrate different ML applications at different subprocesses. It has been determined what the difficulties are and some viable solutions for standardizing such combined procedures [77]. The choice of materials as well as the integration of disparate material processing are the key challenges for metal materials. It is vital to employ the choice of optimum materials to address these issues. The effectiveness of deposition is strongly impacted by a number of wire arc additive manufacturing (WAAM) process approaches reviewed [78]. Several efforts have been made in recent years to find new materials that can work with 3D printing techniques. The advancement of 4D printing from the initial 3D printing technique was made possible by the intense research interest in smart materials. In 2013, a research team from MIT made the first formal presentation of the concept of 4D printing.

The process of 4D printing involves creating a 3D structure using materials that have been engineered to have specific properties. These materials can be programmed to change their shape, size, or stiffness when exposed to a particular stimulus. For example, a printed structure made from a shape-memory polymer can be designed to change its shape in response to temperature changes. One of the key advantages of 4D printing is its ability to create structures that can adapt and transform in response to changing environmental conditions. This has applications in various fields, including architecture, robotics, and medicine. For example, 4D printing can be used to create self-assembling structures, such as furniture that can be flat-packed for easy transportation and then self-assembled into the final product when exposed to heat or moisture. In the medical field, 4D printing has the potential to revolutionize the design of medical devices and implants. For example, 4D-printed stents can be designed to expand and contract in response to changes in blood pressure, improving their functionality and reducing the risk of complications. Although 4D printing is still in the early stages of development, it has the potential to transform the way we design and manufacture products. It offers new opportunities for creating self-assembling structures, adaptive materials, and responsive devices that can change and adapt over time. The basis for this technology's operation is the 3D printing of smart materials [79]. Therefore, 3D objects that can change in a predetermined way in response to outside stimuli can be printed via 4D printing. This implies that the 3D-printed object can alter shape and function in response to stimulation [80]. For 4D printing, there are often two prerequisites. For printed items to be programmable, materials must first be responsive to outside stimuli. Second, there must be outside factors that can act as catalysts for the conversion to happen, such as heat, pH, light exposure, or water [81]. The 4D-printed device can only be activated to gain full functionality when these two conditions are satisfied [62]. One concept that influenced the development of 4D printing is the art of origami. In order to offer some creative ways for cramming enormous objects into a tiny volume of space, researchers have successfully investigated this sort of traditional art [80]. As a result, one of the main benefits of 4D printing is that it allows for the simplification of difficult or massive structure designs during printing. Once triggered and configured, the printed object can change into its ultimate complicated form. A more cost-effective manufacturing process will be achieved by reducing the object's complexity during printing. This will also help to reduce the printing duration and waste of materials [79].

2.8 Conclusion

Additive manufacturing process paved a new way to fabricate desired 3D products with very little material wastage along with greater accuracy. Microprocessor chips were also fabricated by using multimaterial AM process, resulting in better flexibility and strength. The soft grippers of robots for handling materials in a smooth manner and the flexible grippers were prepared using AM technique. The brickless building with strength equivalent to conventional buildings were fabricated in very minimal time through AM technique and which reduced the overall cost. Biomedical implants such as tooth structure, knee joints, hip implants, jaw (upper and lower), and prosthetic limbs were not possible without the impact of 3D printing. With the consideration of external stimuli affecting the performance of the product, the corresponding AM process became known as 4D printing. Thus, in the future, all supplementary parts of machineries, body parts, and parts of space vehicles can be replicated through the AM process.

References

1. Wong, K.V. and Hernandez, A., A review of additive manufacturing. *Int. Sch. Res. Notices*, 2012, 208760, 1–12, 2012.
2. Ashley, S., Rapid prototyping is coming of age. *Mech. Eng.*, 117, 7, 62, 1995.
3. Bak, D., Rapid prototyping or rapid production? 3D printing processes move industry towards the latter. *Assem. Autom.*, 23, 4, 340–345, 2003.
4. Bartolotta, P.A. and Krause, D.L., Titanium aluminide applications in the high speed civil transport. *Technical Memorandum*, NASA/TM-1999-209071, 1–12, 1999.
5. Wu, T. *et al.*, A framework for optimizing the design of injection molds with conformal cooling for additive manufacturing. *Proc. Manuf.*, 1, 404–415, 2015.
6. Campbell, T., *Could 3D printing change the world?: Technologies, potential, and implications of additive manufacturing*, Atlantic Council - JSTOR, Washington, 2011.
7. Marks, P., *3D printing takes off with the world's first printed plane*, Elsevier, London, 2011.
8. Gibson, I. and Srinath, A., Simplifying medical additive manufacturing: Making the surgeon the designer. *Proc. Technol.*, 20, 237–242, 2015.
9. Calignano, F. *et al.*, Design of additively manufactured structures for biomedical applications: A review of the additive manufacturing processes applied to the biomedical sector. *J. Healthcare Eng.*, 2019, 9748212, 1–7, 2019.
10. Bhargav, A. *et al.*, Applications of additive manufacturing in dentistry: A review. *J. Biomed. Mater. Res. Part B: Appl. Biomater.*, 106, 5, 2058–2064, 2018.

11. Liu, J. *et al.*, Direct 3D-printing of cell-laden constructs in microfluidic architectures. *Lab. Chip*, 16, 8, 1430–1438, 2016.
12. Carrel, J.P. *et al.*, A 3D printed TCP/HA structure as a new osteoconductive scaffold for vertical bone augmentation. *Clin. Oral. Implants Res.*, 27, 1, 55–62, 2016.
13. Krishnan, S. *et al.*, A review of rapid prototyped surgical guides for patient-specific total knee replacement. *J. Bone Joint Surg. Br. Vol.*, 94, 11, 1457–1461, 2012.
14. Alharbi, N., Osman, R.B., Wismeijer, D., Factors influencing the dimensional accuracy of 3D-printed full-coverage dental restorations using stereolithography technology. *Int. J. Prosthodont.*, 29, 5, 503–510, 2016.
15. Xie, Z. *et al.*, 3D bioprinting in tissue engineering for medical applications: The classic and the hybrid. *Polymers*, 12, 8, 1717, 2020.
16. Reddy, R.D.P. and Sharma, V., Additive manufacturing in drug delivery applications: A review. *Int. J. Pharm.*, 589, 119820, 2020.
17. Qamar, N. *et al.*, Personalized 3D printed ciprofloxacin impregnated meshes for the management of hernia. *J. Drug Delivery Sci. Technol.*, 53, 101164, 2019.
18. Sin, L.T., Rahmat, A., Rahman, W., Synthesis and production of poly (lactic acid), in: *Polylactic Acid*, vol. 1, pp. 71–107, 2013.
19. Panda, B. *et al.*, Additive manufacturing of geopolymer for sustainable built environment. *J. Cleaner Prod.*, 167, 281–288, 2017.
20. Sanjayan, J.G. and Nematollahi, B., 3D concrete printing for construction applications, in: *3D Concrete Printing Technology*, pp. 1–11, Elsevier, Massachusetts, United States, 2019.
21. Khoshnevis, B. *et al.*, Experimental investigation of contour crafting using ceramics materials. *Rapid Prototyping J.*, 7, 1, 32–42, 2001.
22. Khoshnevis, B., Automated construction by contour crafting—Related robotics and information technologies. *Autom. Constr.*, 13, 1, 5–19, 2004.
23. Xu, J., Ding, L., Love, P.E., Digital reproduction of historical building ornamental components: From 3D scanning to 3D printing. *Autom. Constr.*, 76, 85–96, 2017.
24. Yazdani-Chamzini, A. *et al.*, Developing a fuzzy model based on subtractive clustering for road header performance prediction. *Autom. Constr.*, 35, 111–120, 2013.
25. Han, D. and Lee, H., Recent advances in multi-material additive manufacturing: Methods and applications. *Curr. Opin. Chem. Eng.*, 28, 158–166, 2020.
26. Lan, P.-T. *et al.*, Determining fabrication orientations for rapid prototyping with stereolithography apparatus. *Comput.-Aided Des.*, 29, 1, 53–62, 1997.
27. Schaub, D.A., Chu, K.-R., Montgomery, D.C., Optimizing stereolithography throughput. *J. Manuf. Syst.*, 16, 4, 290–303, 1997.
28. Cho, H. *et al.*, Determining optimal parameters for stereolithography processes via genetic algorithm. *J. Manuf. Syst.*, 19, 1, 18–27, 2000.

29. Esposito Corcione, C. and Maffezzoli, A., Photopolymerization kinetics of an epoxy-based resin for stereolithography. *J. Appl. Polym. Sci.*, 92, 6, 3484–3491, 2004.
30. Nelson, E. *et al.*, Photo-differential scanning calorimetry studies of cationic polymerizations of divinyl ethers. *Polymer*, 36, 24, 4651–4656, 1995.
31. Sun, C. *et al.*, Projection micro-stereolithography using digital micro-mirror dynamic mask. *Sens. Actuators A: Phys.*, 121, 1, 113–120, 2005.
32. Chockalingam, K. *et al.*, Establishment of process model for part strength in stereolithography. *J. Mater. Process. Technol.*, 208, 1-3, 348–365, 2008.
33. Raju, B. *et al.*, Establishment of process model for rapid prototyping technique (stereolithography) to enhance the part quality by Taguchi method. *Proc. Technol.*, 14, 380–389, 2014.
34. Tumbleston, J.R. *et al.*, Continuous liquid interface production of 3D objects. *Science*, 347, 6228, 1349–1352, 2015.
35. Murata, K., Super-fine ink-jet printing for nanotechnology, in: *Proceedings International Conference on MEMS, NANO and Smart Systems*, IEEE, 2003.
36. Margolin, L., *Ultrasonic droplet generation jetting technology for additive manufacturing: An initial investigation*, Georgia Institute of Technology, 2006.
37. Kęsy, A. and Kotliński, J., Mechanical properties of parts produced by using polymer jetting technology. *Arch. Civ. Mech. Eng.*, 10, 3, 37–50, 2010.
38. Ladd, C. *et al.*, 3D printing of free standing liquid metal microstructures. *Adv. Mater.*, 25, 36, 5081–5085, 2013.
39. Visser, C.W. *et al.*, Toward 3D printing of pure metals by laser-induced forward transfer. *Adv. Mater.*, 27, 27, 4087–4092, 2015.
40. Santecchia, E., Spigarelli, S., Cabibbo, M., Material reuse in laser powder bed fusion: Side effects of the laser—Metal powder interaction. *Metals*, 10, 3, 341, 2020.
41. Azam, F.I. *et al.*, An in-depth review on direct additive manufacturing of metals, in: *IOP Conference Series: Materials Science and Engineering*, IOP Publishing, 2018.
42. Sames, W.J. *et al.*, The metallurgy and processing science of metal additive manufacturing. *Int. Mater. Rev.*, 61, 5, 315–360, 2016.
43. Jamshidinia, M., Kong, F., Kovacevic, R., Numerical modeling of heat distribution in the electron beam melting® of Ti-6Al-4V. *J. Manuf. Sci. Eng.*, 135, 061010, 1–14, 2013.
44. Griffith, M.L. *et al.*, *Laser engineered net shaping (LENS) for the fabrication of metallic components*, Sandia National Lab. (SNL-NM), Albuquerque, NM (United States), 1996.
45. Mudge, R.P. and Wald, N.R., Laser engineered net shaping advances additive manufacturing and repair. *Weld. J.-New York-*, 86, 1, 44, 2007.
46. Domack, M.S., Taminger, K.M., Begley, M., Metallurgical mechanisms controlling mechanical properties of aluminium alloy 2219 produced by electron

beam freeform fabrication, in: *Materials Science Forum*, Trans Tech Publ., 2006.
47. Popoola, P. *et al.*, Laser engineering net shaping method in the area of development of functionally graded materials (FGMs) for aero engine applications-A review, in: *Fiber Laser*, vol. 2, p. 64, 2016.
48. Sachs, E.M. *et al.*, Three-dimensional printing techniques. Google Patents, 1993.
49. Moritz, T. and Maleksaeedi, S., Additive manufacturing of ceramic components, in: *Additive Manufacturing*, pp. 105–161, Elsevier, Massachusetts (United States), 2018.
50. Sodhi, D.S., Nonsimultaneous crushing during edge indentation of freshwater ice sheets. *Cold Reg. Sci. Technol.*, 27, 3, 179–195, 1998.
51. Obikawa, T., Yoshino, M., Shinozuka, J., Sheet steel lamination for rapid manufacturing. *J. Mater. Process. Technol.*, 89, 171–176, 1999.
52. White, D. Ultrasonic object consolidation. Google Patents, 2003.
53. White, D.R., Ultrasonic consolidation of aluminum tooling. *Adv. Mater. Processes*, 161, 1, 64–66, 2003.
54. Shimizu, S. *et al.*, Mechanism of weld formation during very-high-power ultrasonic additive manufacturing of Al alloy 6061. *Acta Mater.*, 74, 234–243, 2014.
55. Moon, J. *et al.*, Ink-jet printing of binders for ceramic components. *J. Am. Ceram. Soc.*, 85, 4, 755–762, 2002.
56. Lanzetta, M. and Sachs, E., Improved surface finish in 3D printing using bimodal powder distribution. *Rapid Prototyping J.*, 9, 3, 157–166, 2003.
57. Sun, C. *et al.*, Effect of particle size gradation on the performance of glass-ceramic 3D printing process. *Ceram. Int.*, 43, 1, 578–584, 2017.
58. Lejeune, M. *et al.*, Ink-jet printing of ceramic micro-pillar arrays. *J. Eur. Ceram. Soc.*, 29, 5, 905–911, 2009.
59. Bienia, M. *et al.*, Inkjet printing of ceramic colloidal suspensions: Filament growth and breakup. *Chem. Eng. Sci.*, 149, 1–13, 2016.
60. Rahul, S., Balasubramanian, K., Venkatesh, S., Optimizing inkjet printing process to fabricate thick ceramic coatings. *Ceram. Int.*, 43, 5, 4513–4519, 2017.
61. Melocchi, A. *et al.*, 3D printing by fused deposition modeling of single- and multi-compartment hollow systems for oral delivery–A review. *Int. J. Pharm.*, 579, 119155, 2020.
62. Tan, D., Nokhodchi, A., Maniruzzaman, M., 3d and 4d printing technologies: Innovative process engineering and smart additive manufacturing, in: *3D and 4D Printing in Biomedical Applications: Process Engineering and Additive Manufacturing*, pp. 25–52, 2019.
63. Maniruzzaman, M. *et al.*, A review of hot-melt extrusion: Process technology to pharmaceutical products. *Int. Sch. Res. Notices*, 2012, 436763, 1–10, 2012.
64. Tambe, S. *et al.*, Hot-melt extrusion: Highlighting recent advances in pharmaceutical applications. *J. Drug Delivery Sci. Technol.*, 63, 102452, 2021.

65. Melocchi, A. *et al.*, Hot-melt extruded filaments based on pharmaceutical grade polymers for 3D printing by fused deposition modeling. *Int. J. Pharm.*, 509, 1-2, 255–263, 2016.
66. Pietrzak, K., Isreb, A., Alhnan, M.A., A flexible-dose dispenser for immediate and extended release 3D printed tablets. *Eur. J. Pharm. Biopharm.*, 96, 380–387, 2015.
67. Bose, S., Vahabzadeh, S., Bandyopadhyay, A., Bone tissue engineering using 3D printing. *Mater. Today*, 16, 12, 496–504, 2013.
68. Sobral, J.M. *et al.*, Three-dimensional plotted scaffolds with controlled pore size gradients: Effect of scaffold geometry on mechanical performance and cell seeding efficiency. *Acta biomaterialia*, 7, 3, 1009–1018, 2011.
69. Guillotin, B. *et al.*, Laser assisted bioprinting of engineered tissue with high cell density and microscale organization. *Biomaterials*, 31, 28, 7250–7256, 2010.
70. Williams, J.M. *et al.*, Bone tissue engineering using polycaprolactone scaffolds fabricated via selective laser sintering. *Biomaterials*, 26, 23, 4817–4827, 2005.
71. Kalita, S.J. *et al.*, Development of controlled porosity polymer-ceramic composite scaffolds via fused deposition modeling. *Mater. Sci. Eng.: C*, 23, 5, 611–620, 2003.
72. Sitthi-Amorn, P. *et al.*, MultiFab: A machine vision assisted platform for multi-material 3D printing. *ACM Trans. Graphics (Tog)*, 34, 4, 1–11, 2015.
73. Zhang, Y. *et al.*, Additive manufacturing of metallic materials: A review. *J. Mater. Eng. Perform.*, 27, 1–13, 2018.
74. Kumar, S. *et al.*, Machine learning techniques in additive manufacturing: A state of the art review on design, processes and production control. *J. Intell. Manuf.*, 34, 1, 21–55, 2023.
75. Awad, A. *et al.*, A review of state-of-the-art on enabling additive manufacturing processes for precision medicine. *J. Manuf. Sci. Eng.*, 145, 1, 010802, 2023.
76. Quanjin, M. *et al.*, Recent 3D and 4D intelligent printing technologies: A comparative review and future perspective. *Proc. Comput. Sci.*, 167, 1210–1219, 2020.
77. Babu, S.S. *et al.*, Recent developments in the application of machine-learning towards accelerated predictive multiscale design and additive manufacturing. *Virtual Phys. Prototyping*, 18, 1, e2141653, 2023.
78. Huang, L. *et al.*, A review of challenges for wire and arc additive manufacturing (WAAM). *Trans. Indian Inst. Met.*, 1–17, 2023.
79. Momeni, F., Liu, X., Ni, J., A review of 4D printing. *Mater. Des.*, 122, 42–79, 2017.
80. Khoo, Z.X. *et al.*, 3D printing of smart materials: A review on recent progresses in 4D printing. *Virtual Phys. Prototyping*, 10, 3, 103–122, 2015.
81. Lee, J.-Y., An, J., Chua, C.K., Fundamentals and applications of 3D printing for novel materials. *Appl. Mater. Today*, 7, 120–133, 2017.

3

Computational Modeling of Additive Manufacturing—Overview, Principles, and Simulations in Different Scales

B. K. Sivaraj, R. Nitheesh Kumar and V. Karthik*

Department of Metallurgical and Materials Engineering, National Institute of Technology, Tiruchirappalli, Tamil Nadu, India

Abstract

Computational modeling of materials has become an essential tool for scientists and engineers in the materials science field. This chapter provides an overview of the multiple scales of computational materials modeling, specifically focusing on simulation of additive manufacturing. The chapter first highlights the importance of computational modeling in understanding the behavior of materials at different length scales. A few methods in each scale are explored in detail. It is followed by four sections that describe separate scales and domains of materials modeling—atomic scale, mesoscale, macroscale, and incorporation of machine learning in additive manufacturing simulation. The constitutive relations are discussed, along with the important aspects of relevant literature in each scale. The post-processing tools available in each method are highlighted along with a sample simulation. The chapter concludes with a brief outlook on developments, multiscale modeling, and future scope.

Keywords: Additive manufacturing, molecular dynamics, phase-field modeling, finite element method, crystal plasticity, machine learning

**Corresponding author*: karthikv@nitt.edu

R. Rajasekar, C. Moganapriya and P. Sathish Kumar (eds.) Additive Manufacturing with Novel Materials: Processes, Properties and Applications, (49–118) © 2024 Scrivener Publishing LLC

3.1 Introduction

Computational modeling has emerged as a powerful tool for understanding and predicting the behavior of materials. It allows researchers to investigate the properties of materials at various length scales, from the atomic level to the continuum, which is essential to the design and development of new materials and manufacturing processes. The history of materials modeling can be traced back to the early 20th century, with the development of quantum mechanics and the study of atomic structure. Over the years, as computing power has increased and new theoretical frameworks have been developed, materials modeling has become increasingly sophisticated and accurate. Today, it is an interdisciplinary field that draws on physics, chemistry, materials science, and computer science. The heart of materials science and engineering is the ability to modify the structure of a material at various levels. Any framework that simulates the structural behavior of material under the applied conditions is extremely useful, since it can be utilized to study the effect of conditions on the evolution of structures without experiments, break down different stages in the temporal and spatial evolution and analyze them, and predict and optimize the properties obtained under certain processing conditions. In recent years, materials modeling has had a significant impact on additive manufacturing (AM), which is a rapidly growing field with the potential to revolutionize the current realm of manufacturing processes. By using materials modeling to understand the behavior of materials during AM, optimizing printing parameters and design materials has become faster, more efficient, and cost-effective.

Additive manufacturing being a relatively new manufacturing approach, has a promising future but requires lot of prerequisites to economically produce components on an industrial scale. This poses a difficult challenge to the conventional experimental techniques to obtain the optimal parameters. The process involves a combination of phenomena at different scales and there are many variables that impact the product, which makes it nearly impossible and uneconomic to understand the process by conventional techniques. Additionally, the defects encountered in the process and interplay of parameters must be studied. These can be achieved by computational modeling and optimization of the process. Approaches to different scales offer unique perspectives of the AM process, which are instrumental for the rapid pace of process integration. Materials modeling has been well established at different length scales, depending on the level of detail and phenomena of interest. At the atomic scale, molecular dynamics (MD) and density functional theory (DFT) calculations can be used to predict the structural properties of materials. At the mesoscale, phase-field modeling and crystal plasticity can be used to simulate the behavior of materials under various conditions. At the macroscale, continuum models such as finite element analysis (FEA) can be used to predict the behavior of materials in complex systems. Each scale highlights a particular aspect of the AM process and has a unique set of features and limitations. Additional insights and predictions can be developed by incorporating machine learning (ML) during material modeling.

The following methods of materials modeling in AM are discussed in the upcoming sections:

- Atomistic simulation – molecular dynamics (MD)
- Mesoscale simulation – phase-field modeling, crystal plasticity
- Macroscale simulation – finite element analysis (FEA)
- ML in computational modeling

3.2 Atomistic Simulation

3.2.1 Introduction

Atomic simulations emphasize the phenomena occurring at the scale of nanometers (10^{-9} m) with a time span in the order of picoseconds (10^{-12} s), such as crystallization, diffusion and non-equilibrium processes. Such simulations have many advantages, such as:

- Phenomena at extremely short time and length scales can be clearly observed
- Non-equilibrium processes can be simulated, observed, and controlled easily
- Study domains where conventional techniques are too difficult to carry out

Molecular Dynamics is a type of atomic scale modeling method. It considers atoms as fundamental particles and disregards electronic and sub-atomic structures. The complex quantum effects are neglected, and classical mechanics can be used for describing and predicting atomic movements with time [1]. Initially in the 1950s only a handful of molecules were simulated (hence called "molecular dynamics"), but with the growth in computation power, a large number of particles and various types of materials including bulk solids, liquid, gasses, powders, polymers, and nanostructures were simulated. Prominent software used for such simulations include LAMMPS, ATOMSK, NAMD and Materials Studio. For visualization of the results and to carry out further analysis, software like OVITO, AtomEye and VMD are used.

In any given system, the atoms are considered as hard spheres that interact classically. The system evolves with time from an initial state by successively solving Newton's equations of motion for each atom at every time step. The forces arising due to their interaction are calculated from the interatomic potential energy function (U). This function also accounts for the chemical nature of the atoms. But the electronic and magnetic properties do not come into the picture. Several types of potentials are used, which are determined based on the type of interaction simulated.

In embedded atom method (EAM) style, the total energy of a system is given by the function,

$$E_{total} = \sum_i \left[F_i(\rho_i) + \frac{1}{2} \sum_{j(\neq i)} \Phi_{i,j}(R_{ij}) \right] \tag{3.1}$$

where $F_i(\rho_i)$ is the embedding energy of the atom i[th] caused by its neighbors, $\Phi_{i,j}(R_{ij})$ is the electrostatic potential between atoms i and j separated by a distance R_{ij} [2].

Tersoff potential relates potential energy 'E' with bonding energy V_{ij} between atoms i and j as

$$E = \frac{1}{2} \sum_i \sum_{j(\neq i)} V_{ij} \tag{3.2}$$

Lennard-Jones (L-J) potential is frequently used in multi-component systems, defined as:

$$U(r) = 4\varepsilon \left[\left(\frac{\sigma}{r} \right)^{12} - \left(\frac{\sigma}{r} \right)^{6} \right] \tag{3.3}$$

where ε is the depth of the potential well, and σ is the distance of separation at which the interatomic potential is zero, and r is the distance of separation of the atoms.

The required parameters are computed by relating the derivative of interatomic potential with Newton's second law of motion. A certain order of computation is carried out for every atom during each time step, as illustrated by Equation 3.4:

$$-\frac{\partial U}{\partial r} = F = ma = m\frac{dv}{dt} = m\frac{d^2r}{dt^2} \tag{3.4}$$

1. By differentiating interatomic potential (U) by position, force (F) on each atom is calculated.
2. Force obtained is used to calculate the new velocity (v) and position (r) by integration.
3. This is used as input parameters for the next time step, and so on.

As the number of particles increases, the accuracy and precision of the simulation results increase. This iteration is carried out at least till reaching equilibrium, such that the system properties do not change with time. However, remember that the timescales are very small, and the system cannot attain equilibrium in such a short duration of time. Typically, thousands of timesteps are required to equilibrate the system, translating to a few nanoseconds (10^{-9} s) in real life. Figure 3.1 illustrates the process flow during an MD simulation—initialization, integration and output. Each stage is explained further in the subsequent sections.

3.2.1.1 Initialization

During the initialization, some important inputs are given. Usually, the most time-consuming part of such simulations is the computation of forces from potentials, since the effect of all surrounding atoms needs to be calculated. However, as potential falls off at long distances and becomes close to zero, we can disregard the force calculation for atoms at distances beyond a certain cutoff distance. It is usually around 5 Å. This drastically reduces the computation, at a tiny cost to accuracy. When simulating bulk behavior using a few thousands of atoms, the fraction of atoms exposed to the surface is very large, which causes the system to behave differently.

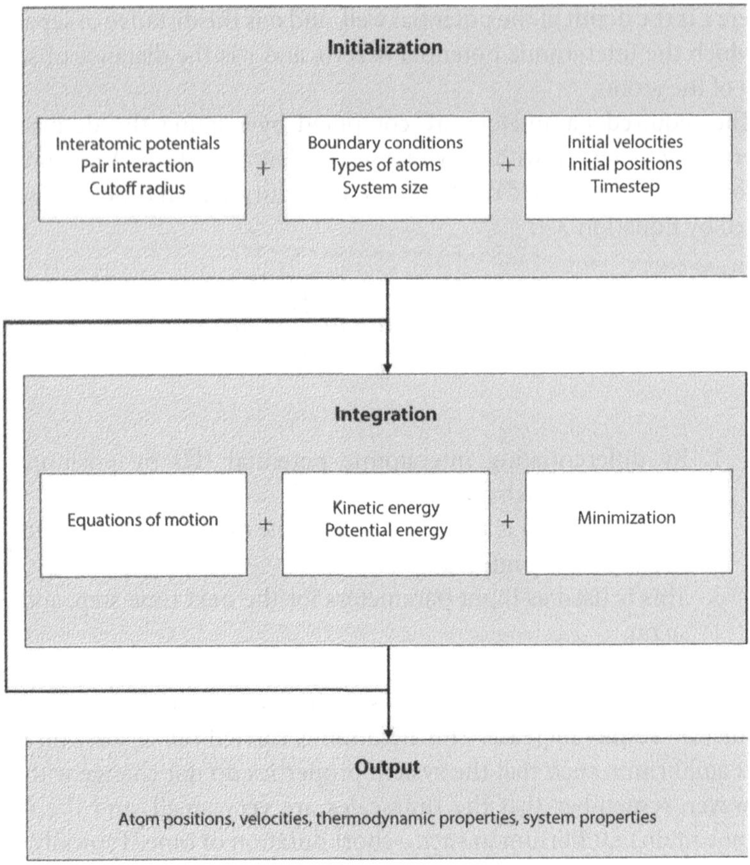

Figure 3.1 Schematic of process flow in molecular dynamics.

To overcome this, we impose periodic boundary condition, which means that the system is surrounded by a number of its copies all around, hence behaves like a bulk material. Then, the number of atoms, their types, structure (hence initial positions) are given. To initialize the velocity, the molecular velocity at a temperature with a type of distribution (Gaussian, Boltzmann, etc.) is specified. Next are the crucial parameters—timestep and number of steps. Smaller timesteps increase computational load but improve accuracy. Based on the level of accuracy and computational power, we need to choose the timestep, usually a few picoseconds (10^{-12} s). To provide sufficient time for simulation, a sufficient number of steps (in thousands) needs to be given. Finally, to equilibrate, a suitable set of conditions, or ensemble, is provided.

3.2.1.2 *Simulation*

The actual experimentation begins only after equilibration. Using special commands like fixes, the desired conditions are imposed on the system, for the required time period. Often, several nanoseconds of simulation time are required to observe a phenomenon proceeding. It also depends on the size of the system and the number of particles involved. For obtaining the results, various data of each atom and the total system are stored, including atomic positions, momentum, energy, force, pressure, and temperature. This raw data is stored in several formats for post-processing to extract meaningful information and validate the simulation.

3.2.1.3 *Output and Post-Processing*

A complete picture of the simulation is achieved only after proper post-processing of the calculated data. Several properties of the system can be obtained directly or by using statistical analysis from the stored data, such as mechanical properties, thermal properties, phase diagram, free energy differences between phases, grain boundary structure, and defect structure. Some of the properties can be directly processed using variables during the simulation itself. For example, one can observe the specific density of a group of atoms at required time intervals by declaring a variable to explicitly calculate and store the value in output files. The values of variables can be tabulated and plotted to observe the variation graphically. To extract more information and visualize the simulation, tools like OVITO, VMD, and ATOMEYE are used. OVITO provides a slew of options for inferring different conclusions. From simple visualization, additional modifiers can be toggled such as atomic strain, elastic strain calculation, bond structures, cluster analysis, dislocation analysis, defect analysis, and centrosymmetry parameters. Pizza.py is a python toolkit which can be used to analyze, plot, and visualize for LAMMPS simulations. It can be invoked in software like Matlab and Gnuplot. It is a handy tool for converting file formats, processing output files, creating plots, and animating simulation snapshots. Other post-processing tools and commands include colvars, eff, replica, smd, spin, and xmgrace.

3.2.2 Molecular Dynamics Simulation—Demonstration

In order to demonstrate some of the capabilities of MD, the authors have carried out a simple simulation of melting and rapid cooling a block of Ni in LAMMPS [3]. A cuboidal block of dimensions, 10 Å x10 Å x100 Å, was defined with an Face Centered Cubic (FCC) structure of lattice parameter

3.52 Å, filled with 1026 Ni atoms based on EAM interatomic potential. From room temperature, the system was heated to 2000 K using NPT fix and equilibrated for a few picoseconds for homogenous and complete melting. Using another NPT ensemble, the block was rapidly quenched to room temperature with an effective cooling rate of 10^{10} K/s. During the process, several properties of the entire system and atom-specific values are stored. Upon post-processing the results using OVITO, and observing the log file from the output, several observations are noted. A few of them have been illustrated in Figure 3.2.

Figure 3.2.a represents the Ni block at 2000 K, while Figure 3.2.b illustrates the block after quenching to room temperature. The block has contracted and atoms are significantly ordered after rapid cooling. The bond types can be visualized with additional modifiers (Figures 3.2.c and 3.2.d), such as CNA, which is discussed in subsequent sections. The highly disordered atoms (blue) after complete melting are clearly visible in Figure 3.2.c,

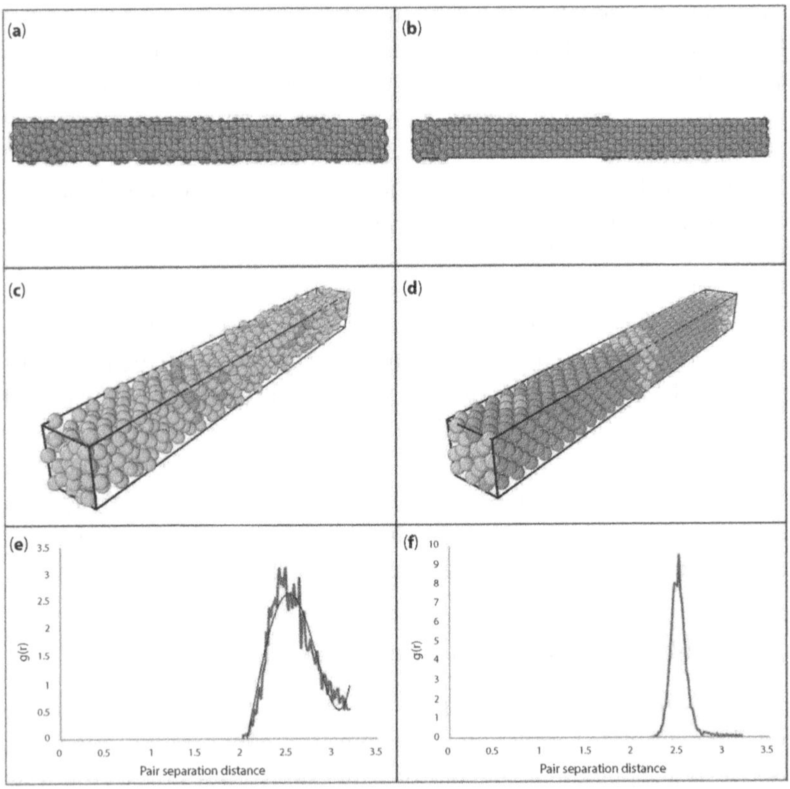

Figure 3.2 Some of the postprocessed results from rapid cooling simulation of Ni block.

as more than 97% of the FCC bonds were dissociated. However, as seen in Figure 3.2.d, upon rapid cooling, most of the system crystallizes into FCC structures (green atoms), indicating poor glass-forming ability as expected. Finally, Figures 3.2.e and 3.2.f present the plots of radial distribution function (RDF) for both the cases.

3.2.3 Atomistic Simulation of Additive Manufacturing

Simulation of AM using MD is effective to understand various phenomena occurring at atomic scale during the process. The associated non-equilibrium processes can also be studied. Metal AM is of greatest interest when compared to other types of AM, due to the versatile applications of metal-based parts. The primary objective is to obtain a defined region of atoms, which can either be of a single type or a mixture of elements (which is essentially a spherical particle of an alloy). Then the process is simulated by directing energy and movement of the energy source through the region. Subsequent processing of data obtained from the simulation can be used to study several aspects of the AM process such as structure of the atomic bonding, sintering characteristics, and the effect of process parameters. In the following sections, these phenomena are discussed in further detail along with previous studies conducted.

3.2.3.1 Sintering

Sintering is an important phenomenon associated with processes involving nanoparticles, metals, and metal AM technologies. The sintering mechanisms and effect of process parameters has a profound effect on the final product. Molecular simulations offer an easy way to observe and study such phenomena to gain root level insights. For this, several parameters are calculated and their values are studied [2].

Dihedral angle: To describe the grain boundary structure when two particles come in contact, this angular measure is used. From dihedral angle, neck width can be calculated as

$$X = Gsin\left(\frac{\theta}{2}\right) \tag{3.5}$$

where θ is the dihedral angle, G is particle size and X is the neck width of the sintered clump.

Radial Distribution Function (RDF): It is used to obtain the structural properties and the variation of density of the system as a function of distance. Several properties of an N-atom system can be calculated by using RDF, as given by:

$$g(r) = \frac{V}{N^2} \frac{\Sigma_i \; n_i(r)}{4\pi r^2 \Delta r} \tag{3.6}$$

Here r is the radial distance, $n_i(r)$ is the coordination number and Δr is the interval in which i^{th} atom is separated by distance r and V is total volume of the simulated region.

Shrinkage ratio: It is calculated as

$$f = \frac{\Delta L}{L_0} \tag{3.7}$$

where L_0 is the initial length of the cluster and ΔL is the change in length of the cluster. The extent of diffusion and neck width is inversely proportional to the rate of change of f.

Mean Square Displacement (MSD): To obtain the diffusion coefficient, slope of the MSD versus time plot is used. For a N particle system, MSD is calculated as

$$MSD = [r^2(t)] = \frac{1}{N} \sum_{i=0}^{N} [r_i(t) - r_i(0)]^2 \tag{3.8}$$

where $r_i(t) - r_i(0)$ is the distance travelled by the particle in time t.

Radius of Gyration (ROG): It is used for studying the agglomeration of the particles due to a net reduction in surface energy. It is calculated as

$$R_g^2 = \frac{1}{M} \sum_i m_i (r_i - r_{cm})^2 \tag{3.9}$$

where M is the total mass of the cluster and r_i is center of mass position of the group.

The driving force of sintering is related to the radius of curvature and surface energy as

$$\sigma = \gamma k = \gamma \left(\frac{1}{R_1} + \frac{1}{R_2} \right) \tag{3.10}$$

where k is the radius of curvature of the surface of the particle, γ is the surface energy, R_1 and R_2 are the radii of the particles. To study the sintering mechanism of AlSi10Mg system, Nandy *et al.* [2] simulated three arrangements of particles: two equal sized particles, two dissimilar particles, and a combination of particles with different sizes. It was observed that during the initial stages of sintering, due to higher surface diffusion under heating, the equal-sized particle system shrank faster than the two dissimilar particles. Mean square displacement and ROG decrease as the size of the second nanoparticle decreases. In a similar way, two arrangements, Ni nanoparticles (two-particle and three-particle systems), were used by Sestito *et al.* [4] to investigate the role of grain boundary misorientation on the sintering behavior, densification rates and pore shrinkage. Gao *et al.* [5] studied the sintering of Cr_2O_3 (chromia) coated cementite powders using MD-ReaxFF reactive force field, in which interaction of different kinds of atoms and their impact on necking was analyzed. Similar simulations found in literature are tabulated in Table 3.1.

These studies emphasize the impact and influence of sintering phenomena during metal AM. Mathematical descriptions as discussed above provide the ability to understand and control a particular aspect of the process to obtain the required outcome. This is true for a range of materials such as AlSi10Mg, stainless steel, and similar industrially significant alloys. The mechanisms of sintering play a major role in the structure and properties of the product. Kinetics and thermomechanical behavior can be studied by using such simulations. It is seen that most of the simulations of such phenomena are carried out using LAMMPS, and OVITO is the preferred visualization tool, which has additional modifications and dedicated studies.

3.2.3.2 Crystallization

The inherent characteristics of AM produce different regions of scan track to have different types of thermal histories, which in turn produces different crystallites and bond structures. This is elevated to a more complex level when multi-component systems like metallic glasses (MGs) and high entropy alloys (HEAs) are involved. The mechanism of glass forming is complex and depends on many parameters. Crystallization leads to

Table 3.1 Various aspects of sintering simulations from literature.

Material	Tools used	Aspect studied	Reference
AlSi10Mg	LAMMPS	Sintering behavior and dimensional changes	Nandy *et al.* [2]
Ni	LAMMPS, OVITO	Sintering kinetics and densification with regard to GB misorientation	Sestito *et al.* [4]
Cr_2O_3 and Fe_3C	ReaxFF	Sintering behavior during AM of Cr_2O_3 coated Stainless Steel	Gao *et al.* [5]
AlCoCrFeNi (MPEA)	LAMMPS, OVITO	Mechanisms during solid state sintering	Sreeramagiri *et al.* [6]
AlSi10Mg	LAMMPS, OVITO	Shape dependency of nanoparticles sintering	Nandy *et al.* [7]
Ni	LAMMPS, OVITO	Effect of heating rate on mechanical properties	Zhang *et al.* [8]
AlSi10Mg	LAMMPS, OVITO	Coalescence behavior and neck growth	Nandy *et al.* [9]
Ti	ATOMSK, LAMMPS, OVITO	Mechanical properties of Ti nanoparticle chains	Jeon *et al.* [10]
Hollow Pd	LAMMPS	Thermomechanical properties and optimal process parameters	Lai *et al.* [11]

deterioration of properties and even failure during fabrication of MGs. Hence, the suppression of crystallization during AM of MGs is an important aspect in such systems [12]. In HEAs, the variety of possible phase transformations and formation of intermetallic phases are the areas that require extensive study. Crystallization is inevitable during AM, especially in the regions of Heat affected zone (HAZ) and overlapping zones. Additionally, the previous amorphous layer may crystallize due to the cyclic heating and cooling during laser scanning [13]. The complex mechanisms in the formation and crystallization of multicomponent systems are difficult to observe and understand using conventional experimental techniques. The analysis of crystalline fraction in the system with respect to solidification, cooling rates and alloying elements is quite easy to carry out using molecular simulations. Usual steps followed for such systems are:

1. Defining a region, with number of atoms, types and structure
2. Replacing the required fractions of atoms with that of another type
3. Initializing with all the necessary parameters
4. Equilibrating beyond its melting point by NPT ensemble
5. Quenching by very high cooling rates
6. Carrying out the process of interest
7. Analysis (crystalline nature, structure, etc.)

There are two approaches for studying the atomic arrangements and crystallization.

Common Neighbor Analysis (CNA)

It is a method used to identify amorphous and crystal bond types present in the system. Four integers, denoted like an index as i j k l, represents the type of atomic arrangement. Crystalline bond types are mainly of three classes - FCC, Hexagonal Close Packed (HCP), and Body Centered Cubic (BCC). The 1421 and 1422 bond pairs signify FCC and HCP phases, while 1441 and 1661 bond pairs indicate BCC phase. Amorphous and liquid bond types are mainly of two classes based on icosahedral structure. The 1431 and 1541 bond pairs signify defective icosahedral structure, and 1551 indicates a perfect icosahedron. The fraction of crystalline phase in the system is determined by the fraction of crystal bond pairs.

Voronoi Analysis (VA)

This technique, also called Voronoi tessellation, provides structural information about atoms clusters, i.e., atom accumulation and aggregation. The system into polyhedrons is called Voronoi polyhedral (VP). A VP is denoted by the indices <n3, n4, n5, n6> where ni denotes the number of i-edged faces present around a central atom. Each polyhedron contains one atom. It is used to describe the structure of the atom cluster. From the polyhedron structure, information about the bond network of the cluster can be studied. Such analysis tools provide a direct way to understand the crystallization mechanisms and associated phenomena during the simulation of multi-component systems. Yang *et al.* [14] analyzed crystallization during AM of Fe-Cu-Ni metallic glass. After producing the MG system, AM process was simulated using powder particles, for which spheres were cut out from the MG block. Then, the laser was mimicked using the "heat"

Table 3.2 Various aspects of crystallization simulations from literature.

Material	Tools used	Aspect studied	Reference
$Fe_{50}Cu_{50-x}Ni_x$ (x = 0, 5, 10, 15, 20, 30)	LAMMPS, OVITO	Crystallization and atomic rearrangement	Yang *et al.* [14]
$Ni_{75}Al_{25}$	LAMMPS	Crystallization in the HAZ of the HEA	Guo *et al.* [15]
$Fe_{50}Ni_{50}$	LAMMPS, OVITO	Crystallization behavior and structure	Jiang *et al.* [16]
Cu	LAMMPS, OVITO	Stress-induced crystallization	Zhao *et al.* [17]

command, which adds kinetic energy to a defined area in the powder layer. Subsequently, scan speed and directional aspects were incorporated. It was seen that as Ni content increases, there is increased inhibition of nucleation and growth of crystals, which increases the amorphous phase fraction. The microstructural evolution of Ni-Al metallic glass during cyclic heating process was analyzed using MD simulation by Guo *et al.* [15]. A $Ni_{75}Al_{25}$ block was formed, melted by equilibrating at high temperature, and cooled to 300 K at high cooling rates to form metallic glass. The ratio of crystal bond types during the cyclic heating process was calculated. It was observed that the system with a higher variation rate of temperature will have lesser time for atoms to rearrange, therefore, lesser crystallization. Even though sufficient driving force is present, the relaxation time during the cyclic heating process is the main control parameter that affects the extent of crystallization. Similar studies are summarized in Table 3.2.

These studies indicate the scope of MD simulations to understand crystallization during AM of multi-component systems like MGs and HEAs. Using visualization tools like OVITO, extended analysis on bond structure, arrangement, and atomic modifications can be carried out. This enables huge savings in material development and provides several starting points for further processing and fabrication.

3.2.3.3 Structural Evolution

The local thermal history during the AM process is central to many phenomena. Cycles of rapid heating and cooling, presence of layers of material, difference in thermal history add to the complexity of understanding and controlling the structure. During the evolution of microstructure, the temperature distribution in and around melt pool is a critical aspect that controls it. Simplified relations can be used to study the thermal aspects.

The radial temperature distribution in melt pool is given by Kurian and Mirzaeifar [18] as

$$T(r) = T_{max} - mR \tag{3.11}$$

where T_{max} is the temperature melt pool center (which is the maximum), R is the distance of the atom from the center and m is a constant.

A linear variation in temperature is used for simplifying calculations for temperature distribution with time. During laser melting, the temporal distribution of temperature is given as:

$$T(r, t) = T_{min} + [T(r) - T_{min}]t \tag{3.12}$$

whereas during solidification, the distribution has an exponential function,

$$T(r, t) = T_{min} + [T(r) - T_{min}]e^{-ct} \tag{3.13}$$

Here, T_{min} is 293 K, c is a constant, and t is the fraction of time elapsed to the total time.

Huang *et al.* [19] investigated the effect of process parameters on melt pool and consequently on the structural evolution selective laser melting (SLM) of Ni-based superalloy. Zhang *et al.* [20] studied the structural evolution and mechanisms in melt pool and heat-affected zone. Similar studies are summarized in Table 3.3.

Previous studies elucidate the importance of modeling in atomistic scale to understand the melting and solidification behavior and structural evolution. The effect of process parameters is relatively easier to study using such simulations. The difference in materials and system only requires a mere change in the potential used in the model, which enables a less demanding development route for any material. This is especially relevant for AM.

3.2.3.4 *Associated Phenomena in Additive Manufacturing*

Apart from the aspects discussed above, a wide spectrum of phenomena related to AM can be simulated and studied using MD. Processes like aerosol deposition for AM of ceramics [24], printing debonding sintering [25], binder jetting [26] and aspects like elastic behavior, defect structures,

Table 3.3 Various aspects of structural evolution simulations from literature.

Material	Tools used	Aspect studied	Reference
Al	LAMMPS, OVITO	Melting and solidification behavior	Kurian *et al.* [18]
IN718	LAMMPS, OVITO	Influence of process parameters on structure	Huang *et al.* [19]
$Cu_{50}Zr_{50}$	LAMMPS	Structural evolution in the molten pool and HAZ	Zhang *et al.* [20]
Ti-48Al-2Nb-2Cr	LAMMPS, OVITO	Microstructural changes, deformation mechanisms	Wang *et al.* [21]
Al bronze (7.5% Al)	LAMMPS, OVITO	Grain growth and deformation	Nikonov *et al.* [22]
Cu/Al and Ti/Al	ATOMSK, LAMMPS	Melting and solidification behavior	Rahmani *et al.* [23]

and atomic packing are studied using MD. Defect structures, atomic , and stresses are also simulated in this scale. As molecular motion governs metallurgical phenomena, such simulations are a powerful tool to understand and engineer them. Some examples are indicated in Table 3.4.

From these studies, it is clear that atomistic simulation provides vital insights into AM, and helps in understanding the associated phenomena during the process for almost all the industrially significant materials. Most of the simulations are carried out using LAMMPS, due to its versatile features and logical operation commands. It seems that computational modeling of non-equilibrium processes and atomistic phenomena surpassed the experimental information.

3.2.4 Limitations

Simulation of AM using MD is effective to understand various phenomena occurring at atomic level during the process. However, there are several limitations for this scale of simulation. The large scaling factor between the continuum and atomic levels may introduce mismatch in such simulations. Also, simulation of a limited system size may not give the complete picture of the phenomena occurring at larger scales. Difficulty in obtaining

Table 3.4 Various aspects of mixed type of atomistic simulations from literature.

Material	Tools used	Aspect studied	Reference
Anatase (TiO_2)	LAMMPS, OVITO	Plasticity in aerosol deposition (AD) for ceramic AM	Daneshian *et al.* [24]
AISI 316L stainless steel	LAMMPS, OVITO	Sintering in printing debinding sintering (PDS)	Wang *et al.* [25]
HA/NBCA	Material Studio	Effect of binder molecules on mechanical properties	Wang *et al.* [26]
Ti-6Al-4V	VASP	Elastic behavior of α' martensitic phase by AM	Dumontet *et al.* [27]
Ti	LAMMPS, OVITO	Capabilities and limitations of interatomic potentials for AM	Etesami *et al.* [28]
Cu	LAMMPS, OVITO	Defect structures during metal AM	Singh *et al.* [29]
FeNiCrCoCu (HEA)	LAMMPS, OVITO	Effect of varying laser configurations in PBF	Jamil *et al.* [30]
Ti-6Al-4V	LAMMPS, OVITO	Electron beam scanning	Shugurov *et al.* [31]
AISI 316L stainless steel	ReaxFF Simulation	Structure of binder species	Gao *et al.* [32]
AISI 316L stainless steel	ReaxFF Simulation	Atomistic level mechanism	Gao *et al.* [33]
TiAl	Unspecified	Atomic level stresses and packing structures	Xiangyang *et al.* [34]
FeNiCr (MEA)	LAMMPS, OVITO	Effect of process parameters on properties	Jamil *et al.* [35]
Ag	LAMMPS, OVITO	Thermodynamic properties of nanoclusters	Vo *et al.* [36]

accurate experimental data for non-equilibrium processes, including rapid solidification, makes it cumbersome to correlate and predict the required data in some situations. Some of the physical phenomena might get neglected in order to simplify the interactions. Lack of sophisticated potentials limits the composition of the system to be simulated. Sometimes the potentials might be too simple and fail to incorporate certain interactions. It is not applicable in case of any chemical reactions occurring in the system. There is limited access for high computation power to simulate sufficiently large and complex systems. Albeit such limitations, atomic scale simulation is an important method in understanding such advanced manufacturing processes, especially involving non-equilibrium phenomena. It remains the widely used way to study and predict phenomena at nanoscale, and will be crucial for quick integration of disruptive technologies to the existing workflow, or as a mainstream process.

3.3 Mesoscale Modeling

3.3.1 Introduction

Mesoscale modeling seems to be the broadest branch of computational materials science. It deals with phenomena occurring in length scale of several microns (10^{-6} m) and in time span of several microseconds (10^{-6} s). The emphasis of this scale is mainly on microstructural evolution, which involves non-equilibrium processes and governed by kinetics. This means that there will be a plethora of defective structures and a range of interactions occurring simultaneously. Such a simulation needs to encompass relatively large parts of a material, at least several microns. This is achieved by using the relevant constitutive relations, rate equations and functions. Therefore, there is no unique or fixed methodology to simulate mesoscale phenomena. A large variety of mesoscale simulation methods are used, such as dislocation dynamics, kinetic Monte Carlo, phase-field modeling, cellular automata, crystal plasticity, and kinetic Potts model. Two prominent methods relevant to mesoscale modeling of AM—phase-field modeling and crystal plasticity—are discussed in the following sections.

3.3.2 Phase-Field Modeling

Developed from the ideas of Van der Waals, Cahn-Hilliard, and Ginzburg-Landau, phase-field modeling is a computational approach that uses partial differential equations to simulate the evolution of microstructures and phases in materials. The basic idea is to represent the microstructure and phases as a continuous field using functions, called "phase fields", which evolves over time according to certain rules. It has fixed values in each phase, and but has a continuous and smooth variation at the interfaces. It describes the dynamics of an evolving interface using a variable, from which the properties of the system at any point in space and time can be calculated by solving the differential equations. The physical reality of structural evolution seems to have a sharp interface with negligible thickness that evolves with time, but such a behavior is very difficult to model mathematically with variables and equations. This is because the exact position of the interface is necessary for obtaining the solutions, which is difficult since the interface itself varies with time. To overcome this, the interfaces are considered to be diffuse, which has smooth variation of a parameter, called "order parameter." If we specify how it varies with space as a function, then there is no need to explicitly track the interface [37]. The phase-field variable, or order parameter, ϕ, represents a state that a system can evolve towards, for example: phase, spin, crystal lattice, and composition. The parameter varies as $0 \leq \phi \leq 1$, where the fixed ends correspond to a particular state (e.g., phases α and β). Figure 3.3 illustrates the difference between sharp interface and diffuse interface in the perspective of phase-field variable.

The phenomena at the interfaces are described by coupling an equation of motion with transport equations for heat or chemical components. As the method involves solving differential equations coupled with defined

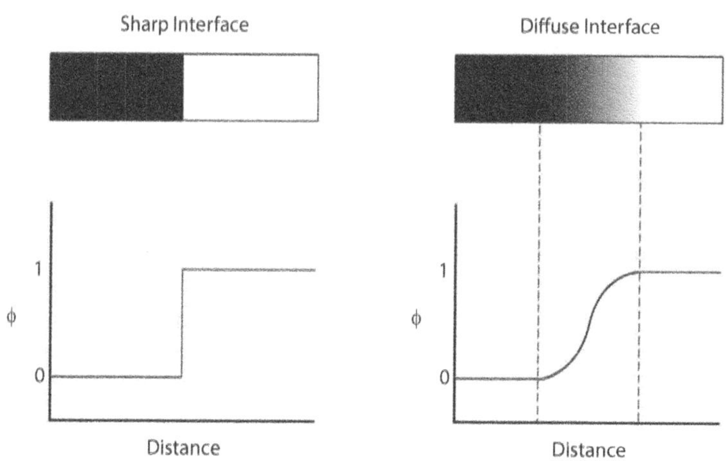

Figure 3.3 Concept of sharp and diffuse interface using order parameter.

functions, many types of numerical methods can be used for performing simulations. Some of them are summarized here.

MOOSE: Multiphysics object-oriented simulation environment (MOOSE) is an open-source software package developed by Idaho National Laboratory. It allows users to perform phase-field simulations and solve complex multiphysics problems in a flexible and efficient manner.

FEniCS: FEniCS is a popular finite element software package that enables users to solve partial differential equations, including phase-field equations, in a flexible and efficient manner. It is open source and has a user-friendly interface.

COMSOL Multiphysics: COMSOL Multiphysics is a commercial software package that allows users to simulate multiphysics problems, including phase-field simulations. It provides a graphical user interface and a variety of solvers to simulate complex phenomena.

Abaqus: Through user-defined subroutines, simulation of phase transformations and microstructural evolution in complex systems can be carried out in Abaqus. It also offers preprocessing and postprocessing tools for phase-field simulations, making it a versatile tool for materials modeling.

PRISMS-PF: It is an open-source phase-field simulation package that enables users to model microstructural evolution in materials under complex thermodynamic and mechanical conditions. The modular architecture and extensive customization make PRISMS-PF a great choice.

PFHub: Phase field hub (PFHub) is a web-based platform that provides access to various open-source phase-field software packages, including MOOSE and FEniCS. It allows users to share and access code, data, and best practices in phase-field modeling.

3.3.2.1 Fundamental Relations

Several fundamental relations form the basis for phase-field modeling. However, based on the application requirements, new relations and equations can be derived with suitable boundary conditions, and can be integrated with fundamental relations. This makes it possible to use phase-field modeling for any type of microstructural simulation, which is perhaps the biggest advantage of the technique. The free energy f_0 is expressed as a polynomial of the order parameter ϕ and the temperature T. The number of order parameters and the equilibrium values reflect the symmetry relations between the phases. The general format of the function is given by

$$f_0(\phi,T) = f_0(T,\phi = 0) + \sum_{n=2} \frac{a_n(T)}{n}(\phi)^n \tag{3.14}$$

The coefficients of the function are dependent on temperature as well as on other thermodynamic variables. This free energy function will be modified to the required phase transformation phenomena. Some of the examples are discussed below:

Binary alloy with a symmetric phase diagram: Here, the free energy is identical on both sides of the spinodal concentration at $\phi = \frac{1}{2}$. Only even powers in the expansion need to be retained.

Gas-liquid phase transition: The even powers in the expansion are retained. The free energy variation can be accounted for by retaining at least one third order term.

Two co-existing phases merging into one: By adding a third-degree term with negative sign to the free energy function, such phenomena can be modeled.

The double-well potential function is the simplest function that describes the free energy density of two phases. It has two minima at $\phi = \pm 1$, which represents the two stable phases of the system, and a maximum at $\phi = 0$, representing the unstable interface between the two phases.

This can be extended to a general multiphase system with p coexisting phases. It will then be described by using p phase fields ϕ_k (k=1, 2, .., p). Then the total free energy of a system defined by a concentration field x_B (at constant temperature, pressure and molar volume, and neglecting elastic, magnetic or electric fields) is given by

$$\begin{aligned} F(x_B,\phi_k) &= \int_V f(x_B,\phi_k,\nabla x_B,\nabla \phi_k)dr \\ &= \int_V \left[f_0(x_B,\phi_k) + \frac{\epsilon}{2}(\nabla x_B)^2 + \sum_k \frac{\kappa_k}{2}(\nabla \phi_k)^2 \right] dr \end{aligned} \tag{3.15}$$

Here, the function $f(x_B, \phi_k, \nabla x_B, \nabla \phi_k)$ represents the local free energy density of the system. $f_0(x_B, \phi_k)$ is the free energy of a homogeneous system.

The gradient terms $\frac{\epsilon}{2}(\nabla x_B)^2$ and $\frac{\kappa_k}{2}(\nabla\phi_k)^2$ are the cause of diffuse nature of the interfaces, through which anisotropy in interfacial energy is introduced. ϵ and κ_k are called gradient energy coefficients [38]. The non-conserved variables govern a time-dependent Ginzburg-Landau equation, also known as the Allen-Cahn equation, given by

$$\frac{\partial \phi_k}{\partial t} = -L_k\left[\frac{\partial f_0}{\partial \phi_k} - \nabla .\kappa_k \nabla \phi_k\right] \tag{3.16}$$

Here, L_k is a positive kinetic parameter, which is related to the interfacial mobility. The equation has two terms: a Laplacian term that smooths out the order parameter profile and a free energy term that drives the phase separation. The Allen-Cahn equation is the basis for phenomena like solidification (liquid-solid), recrystallization (deformed grains-new grains), phase separation (mixture-separate phases), phase transformation (austenite-martensite), grain growth, and pattern formation on surfaces.

Conservative phase-field variables obey the Cahn-Hilliard equation, which is

$$\frac{\partial x_B}{\partial t} = -\nabla .J_B = \nabla .M\nabla\left[\frac{\partial f_0}{\partial x_B} - \nabla .\epsilon \nabla x_B\right] \tag{3.17}$$

The parameter M is related to the interdiffusion coefficient. This equation incorporates the effects of surface tension and diffusion, and describes the evolution of the concentration field. It is applied for phenomena like solidification, spinodal decomposition, pattern formation, surface segregation, and evolution of grain boundaries. In addition to these, established relations like heat conduction equations, Gibbs-Thomson equation, and Navier-Stokes equation, are added using non-equilibrium thermodynamics to construct models that include plastic deformation, electric and magnetic effects, viscosity and so on.

3.3.3 Phase Field Modeling of Additive Manufacturing

To give an idea on the versatile applications of phase-field modeling, some broad classes of simulations relevant to additive manufacturing are described in this section.

Solidification and crystal growth: Aspects like dendritic growth, effects of anisotropic surface energy, polycrystal growth, multi-component alloys, columnar growth, eutectic growth, crystal growth under forced flow, directional solidification, interface morphology, and kinetics are the focus in such simulations.

Microstructure evolution: Studying phenomena like solidification, phase transformation, plastic deformation, recrystallization, grain growth, sintering, additive manufacturing, welding, and fatigue require simulation and analysis of how the microstructure evolves and responds to conditions.

Multiscale simulations: Bridging different length scales, such as the atomistic and continuum scales, phase-field models can incorporate information from higher-resolution simulations. By coupling physics-based phenomena, multi-physics simulation can also be performed with certain approaches.

As discussed in previous sections, there is a spectrum of phase-field models for performing simulations, each based on certain dominant physical phenomena. Some prominent models used include cellular automata (CA), Monte Carlo (MC), finite element (FE), lattice Boltzmann (LB), level set, and phase field dislocation dynamics (PFDD). Different models imply that they have different types of output data, hence there is no common set of post-processing tool. In the subsequent sections, a few important models and the available literature relevant to AM are discussed.

3.3.3.1 Cellular Automata-Based Models

Cellular automata are a discrete mathematical model used to simulate the behavior of complex systems by breaking down the system into a group of cells. It predicts the evolution of the system in time and space from the specified rules governing the state of each cell based on the state of its neighboring cells. It can simulate the microstructure evolution of materials during solidification, melting, and phase transformations. It can also capture complex microstructural patterns with minimal computational cost, making it a popular choice for simulating large-scale systems. Based on the required level of complexity, it can be two dimensional or three dimensional. These models are also highly customizable, allowing users to define their own rules for the state of each cell based on the dominant physics involved.

In phase-field modeling, the CA model is often coupled with other models such as the MC method and the FE method to capture the broad range of phenomena occurring in the system. Some of the previous studies using CA relevant to AM are highlighted in Table 3.5.

Table 3.5 Various aspects of simulations based on cellular automata.

Material	Models used	Aspect studied	Reference
IN718	2D CA-LB model	Microstructure evolution during AM process	Rai *et al.* [39]
Ti-6Al-4V	CA	Microstructure evolution during SLM process	Yang *et al.* [40]
Ti-6Al-4V	3D CA-FD model	Kinetic, statistical, and topological aspects of microstructure evolution	Zinovieva *et al.* [41]
SS 304	3D CA	Effect of nucleation mechanisms on grain structure	Li & Tan [42]
Ti-6Al-4V	2D CA-PF model	Microstructure and solutal evolution of Al and V	Liu & Shin [43]
IN718	3D CA	Effect of process parameters on grain structure	Koepf *et al.* [44]
IN718	2D CA-LB model	Microstructure evolution during SEBM process	Rai *et al.* [45]
Ti-6Al-4V	2D CA	Microstructure evolution	Akram *et al.* [46]

3.3.3.2 *Monte Carlo-Based Models*

In MC simulations, a large number of random moves are made in a defined system, and the energy of the system is calculated at each step. These random moves are based on the probability of the system's energy changing due to the movement of atoms or molecules. The final result is a statistical representation of the behavior of the system, which can be used to analyze the system's properties. In phase-field modeling, the MC method is used to study solid-state transformations, including nucleation, growth, and coarsening of phases. Monte Carlo simulations are often coupled with other methods, such as the phase-field method or kinetic MC simulations. This allows for the simulation of more complex systems, such as multi-phase systems, and the study of the kinetics of phase transformations. Some of the previous studies using MC models relevant to AM are highlighted in Table 3.6.

Table 3.6 Various aspects of simulations based on Monte Carlo model.

Material	Models used	Aspect studied	Reference
Ti-6Al-4V	Potts kinetic Monte Carlo (KMC)	Microstructure evolution	Rodgers *et al.* [47]
SS 304L	FDMC model	Microstructure evolution during AM process	Rodgers *et al.* [48]
Inconel 625	MC Potts model and EVP-FFT	Effect of hatch spacing and layer thickness on microstructure	J Pauza [49]
AA6061-T6	MC-Kampmann-Wagner Numerical (KWN) model	Integrated modeling of friction stir AM	Zhang *et al.* [50]

Table 3.7 Various aspects of simulations based on finite element-coupled models.

Material	Models used	Aspect studied	Reference
Al	Enhanced Discrete Element Method (EDEM)	Effect of particles on the porosities and densities of SLS products	Zhang *et al.* [51]
Ti-6Al-4V	Finite volume method	Evolution of dendritic microstructure during AM process	Zhang *et al.* [52]
IN718	FEM and stochastic analysis	Microstructure evolution during AM process	Nie *et al.* [53]
Ni-Nb	FE-PF model	Morphology and cellular segregation structure	Karayagiz *et al.* [54]
Ti-6Al-4V	FE-PF model	Microstructure evolution during AM process	Yang *et al.* [55]
AlSi10Mg	FE and statistical analysis	Prediction of material response and reliability of AM products	Simon & Frantziskonis [56]
Ti-6Al-4V	FE-PF model	Microstructure evolution	Salsi *et al.* [57]

3.3.3.3 *Finite Element-Based Models*

Finite element modeling (FEM) is a widely used technique for simulating and analyzing various physical phenomena. In the context of phase-field modeling, it is employed to solve the governing equations of phase-field models. Finite element modeling involves discretizing the domain of interest into finite elements and representing the field variables within each element by polynomial functions. The solution of the equations is then obtained by solving the algebraic equations derived from the element equations. It can be used to handle complex geometries, incorporate material anisotropy, and model a wide range of physical phenomena. The capabilities of FEM in phase-field modeling are vast, with the ability to simulate various phenomena such as solidification, fracture, and deformation. Some of the previous studies using finite element models relevant to AM are highlighted in Table 3.7.

3.3.3.4 *Other Phase-Field Models*

Apart from the above-discussed models, a wide spectrum of phenomena can be simulated and studied using various other phase-field models. A few miscellaneous examples, with focus on AM, are summarized in Table 3.8.

Table 3.8 Various aspects of phase-field modeling from literature.

Material	Models used	Aspect studied	Reference
Ti-6Al-4V	PF model	Effect of process parameters on melt pool and porosity	Lu *et al.* [58]
IN718 and IN738LC	Microstructure-based creep model	Cavity formation kinetics and creep deformation	Wu *et al.* [59]
Ti	PF-entropy approach	Thermal fluid phenomena and elastic structure response	Li & Fan *et al.* [60]
IN718	Multi-phase PF model	Microstructure evolution, tip velocity, solute undercooling	Kundin *et al.* [61]

(*Continued*)

Table 3.8 Various aspects of phase-field modeling from literature. (*Continued*)

Material	Models used	Aspect studied	Reference
SS 316L	PF-Rosenthal model	Microstructure evolution during AM process	Chadwick & Voorhees [62]
Ti-Nb	PF-CALPHAD-multiphysics	Solidification kinetics and microstructure evolution	Berry *et al.* [63]
Ti-6Al-4V	PF-FORTRAN	Microstructure evolution during EBAM process	Sahoo & Chou [64]
IN718 and Ti-6Al-4V	Computational welding mechanics	Microstructure evolution during direct energy deposition (DED) process	Lindgren *et al.* [65]
Ti-6Al-4V	CA, PF, KMC and DNN models	Microstructure evolution during AM process (review of models)	Li *et al.* [66]
SS 316L	Thermal-fluid flow model	Grain evolution of TiB2 during SLM process	Yang *et al.* [67]
IN718	3D PF-LB model	Porosity and microstructure evolution during AM process	Laskowski *et al.* [68]
Ti-6Al-4V	Multiscale PF models	Multiscale simulation of microstructure evolution	Ji *et al.* [69]
Ni-based alloys	2D MPF-CALPHAD model	Microstructure evolution during Wire Arc Additive Manufacturing (WAAM) process	Nomoto *et al.* [70]
General	FE, CFD, KMC, CA, PF	Microstructure evolution during AM process (review of models)	Tan *et al.* [71]
Ni-based superalloys	Transient thermal analysis	Modifying the AM solidification texture	Tabei *et al.* [72]
Al-4.5Cu	PF model	Local dendrite growth in fused coating AM	Geng *et al.* [73]

From previous studies, it is evident that almost all types of industrially significant materials can be studied to understand their microstructural evolution and associated phenomena using phase-field modeling. The major advantage of the method is that any type of mesoscale phenomena can be easily coupled together to simulate the exact set of conditions that is required. Some of the most used models integrated for simulation of AM under phase-field modeling seems to be two-dimensional (2D) CA, MC, and FEM. Phenomena ranging from solidification, growth of phases, morphology, defect structure behavior, and effect of AM process parameters are simulated using this method.

3.3.4 Limitations of Phase-Field Modeling

Most of the phase-field simulations are computationally expensive and require a significant amount of time and computational resources to run, especially when modeling large-scale systems or long time frames. The accuracy and reliability of phase-field simulations can be highly dependent on the choice of parameters used in the model. Small variations in the parameters can lead to significant changes in the simulation results. The spatial resolution of a phase-field simulation is limited by the mesh size used in the finite element method, which can lead to errors and inaccuracies in the results, especially at interfaces. Certain simplifications and idealizations, such as homogeneity and isotropy, may not hold true in some real systems, limiting the applicability of a particular model. The phase-field model can become more complex and difficult to handle when dealing with multi-phase systems and highly dynamic systems.

3.3.5 Crystal Plasticity

Crystals are mechanically anisotropic owing to the different orientations of atomic arrangements. In metallic systems, the activation of crystallographic deformation mechanisms like dislocations and twins depends on the local crystallographic orientations. Crystallographic textures are commonly used to describe this anisotropy in orientations in polycrystalline materials. With the evolution of the crystallographic texture during any mechanical deformation, the strength, strain hardening, damage, and deformation-induced surface roughening also vary consequently. Plastic deformation in materials is a result of shearing between the orderly arranged atoms in the crystals. Plastic shear creates both shape and orientation changes. Any theory describing the mechanical properties of crystals must include the crystallographic and anisotropic nature of those mechanisms that create shear, and the orientation of the crystals studied relative to the applied boundary conditions. Crystal plasticity FEM was developed as an improvement over many initial works involving simplifying assumptions of strain or stress homogeneity to cope with the intricate interactions within a polycrystal. It is a multiscale modeling approach that extends between the meso and continuum scales. Crystal plasticity simulations are based on phenomenological models. These models are built around the deformation gradient that differentiates the deformed and undeformed structures. The deformation gradient is often decomposed into elastic (F^e), plastic (F^p), and thermal (F^{th}) gradients. The generalized Hooke's law can be used to relate the stress state to the elastic deformation gradient. In the plastic regime, plastic velocity gradient (L^p) comes into picture. This plastic velocity gradient considers the shear rate calculated using the flow rule with hardening and yielding. Mathematically the phenomenological models can be presented as follows. The deformation gradient can be represented as,

$$F = F^e.\ F^p.\ F^{th} \tag{3.18}$$

The elastic deformation gradient (F^e) is related to the elastic strain as,

$$E^e = \frac{1}{2}(F^{e^T} F^e - I) \tag{3.19}$$

With this equation and Hooke's law, the elastic stresses can be related to the deformation gradient.

The plastic deformation gradient is related to the plastic velocity gradient as,

$$L^p = \sum_n \dot{\gamma}^n (b^n \otimes a^n) \tag{3.20}$$

where $\dot{\gamma}$ is the shear rate, b is the unit vector along shear direction, a is the unit vector along shear planed normal and n is the number of slips systems. The shear rate is calculated as function dependent on the resolved shear stress (τ^α) and the resistance to flow (R^α) of the slip system.

A generalized expression for the evolution of shear rate (given by the flow rule) and the resistance to flow (hardening) are given by,

$$\dot{\gamma}^{\alpha} = f(\tau^{\alpha}, R^{\alpha}) \tag{3.21}$$

$$\dot{R}^{\alpha} = f\left(\dot{\gamma}^{\beta}\right) h^{\alpha\beta} \tag{3.22}$$

where $\dot{\gamma}^{\alpha}$ is the shear rate of slip system α, $\dot{\gamma}^{\beta}$ is the shear rate of slip system β, τ^{α} is the resolved shear stress in slip system α, R^{α} is the resistance to flow offered by the slip system α, and $h^{\alpha\beta}$ represents the dependency matrix between the α and β slips systems.

These generalized expressions take different forms with different simulation tools. The flow rule and Voce hardening model in a tool called FEPX can be given by mathematical expressions as follows,

Flow rule:

$$\dot{\gamma}^{\alpha} = \dot{\gamma}_0 \left(\frac{|\tau^{\alpha}|}{g^{\alpha}} \right)^{\frac{1}{m}} sgn(\tau^{\alpha}) \tag{3.23}$$

where, $\dot{\gamma}_0$ is the fixed-state strain rate scaling coefficient, τ^{α} is the resolved shear stress of the αth slip system, m is the rate sensitivity coefficient and g^{α} is the current slip system strength or resistance to shear. This is common for most of the other tools like WARP3D and DAMASK.

Hardening rule and yield condition:

$$g^{\alpha} = h_0 \left(\frac{-g^{\alpha}}{g_s(\dot{\gamma}) - g_0} \right)^{n} g_0 \tag{3.24}$$

where, h_0 is the fixed-state hardening rate scaling coefficient, g_0 is the initial slip system strength (which is considered equal for all slip systems), n is the non-linear Voce hardening exponent, and $g_s(\dot{\gamma})$ describes the slip systems saturation strength and is given by a power law. Through crystal plasticity, localized stress-strain behavior of materials can be analyzed. Factors influencing the microstructural behavior are given as an input, and the mechanical response is analyzed. With this, the impact of the orientation, grain size distribution, phase fractions, and other microstructural features on the mechanical properties of the materials are analyzed. Some popular tools used for CPFEM are Abaqus, DAMASK, Neper and FEPX.

3.3.6 Crystal Plasticity Simulation—Demonstration

For any AM procedure, after the sample is built through characterization techniques, the grain size distribution of phases present and textures can be obtained. By feeding the information to polycrystal generation tools, the as-built sample can be replicated as tessellations in the simulation domain. Finite element analysis is then done on the tessellations and analysis on the impact of crystal orientations, grain size distribution, and grain morphology on the mechanical properties of the AM samples. Not just in terms of AM, but even in many other aspects of research in crystal plasticity, Neper and FEPX tools are yet to be explored deeply. Here is some evidence of crystal plasticity modeling using Neper and FEPX to demonstrate their capabilities. Neper is a polycrystal generation and meshing tool. It has four modules, tessellation (T), meshing (M), simulation (S), and visualization (V). The T module takes input as grain size distribution, crystal orientation distribution, grain morphology, and domain size and shapes, which can be found through certain characterization techniques. The M module meshes the polycrystal generated while taking the order of meshing, the relative size of each mesh (or the relative characteristic length), and partitioning. The S module is used in postprocessing the results of the FEPX FE solver. The V module is used to visualize the tessellations generated, post-mesh tessellations, and the results obtained after the use of the S module. Figure 3.4 shows a few capabilities in terms of grain morphologies that can be created in Neper [74]. Tessellations of Voronoi, grain growth, and elongated grain morphologies are shown.

FEPX [74] is a finite element plasticity solver that models the global and local mechanical behavior of large polycrystalline solids. By segregating each grain into finite elements, the local mechanical properties of every polycrystal can be studied, hence the influence of every microstructural feature and their evolution can be analyzed. For the FE simulation, a configuration file is necessary along with the tessellation and mesh files. This configuration file contains information on the material parameters and the process parameters. In the case of uniaxial tension, information on the axis of load application, target strain value, and the number of timesteps are required for the simulation to run. Nodal and elemental outputs can be printed out in terms of nodal coordinates, nodal displacement, critical resolved shear stress (CRSS), deformation rate, elemental volume, crystallographic orientations, slip system shear, slip system shear rate, plastic spin rate, total strain, elastic strain, plastic strain, stress, equivalent stress, velocity gradient, work rate, plastic work rate, and surface forces. Outputs can be obtained in terms of tensors with values for every active slip system and equivalents of the same can also be made available after the simulations. Postprocessing of the FE results is done using the S module of Neper. The results are segregated as nodal and elemental outputs and put into separate files. This facilitates proper visualization of the results. Here is a simple simulation of the uniaxial tension of a polycrystal. Tensile loading across the y-axis is done for a copper sample with a strain rate of 10^{-3} to a target strain of 1%. The representative volume element presented in Figure 3.4a is used in this simulation. The localized distribution of stresses and strains can be obtained. A visualization of local stress distribution in terms of the stress tensor components (σ_{11}, τ_{12}, τ_{13}, σ_{22}, τ_{23}, and σ_{33}) is provided in Figure 3.5. Every output obtained can be visualized in a similar way. The normal stress in the loading direction (σ_{22}) appears to have a major influence on the simulation sample generated than the other stress components. Neper and FEPX coexist as a duo of tools to generate polycrystals, mechanical analysis of the generated crystals, and postprocessing of the data. Interestingly, the polycrystals generated in Neper can be exported out in other formats and the FE simulations and the postprocessing can be done using other tools.

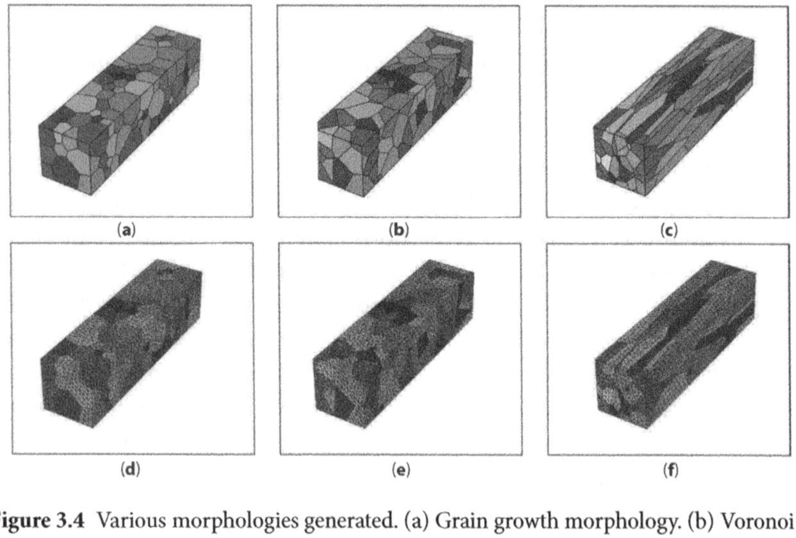

Figure 3.4 Various morphologies generated. (a) Grain growth morphology. (b) Voronoi morphology. (c) Elongated morphology. (d), (e) and (f) are the corresponding meshed systems.

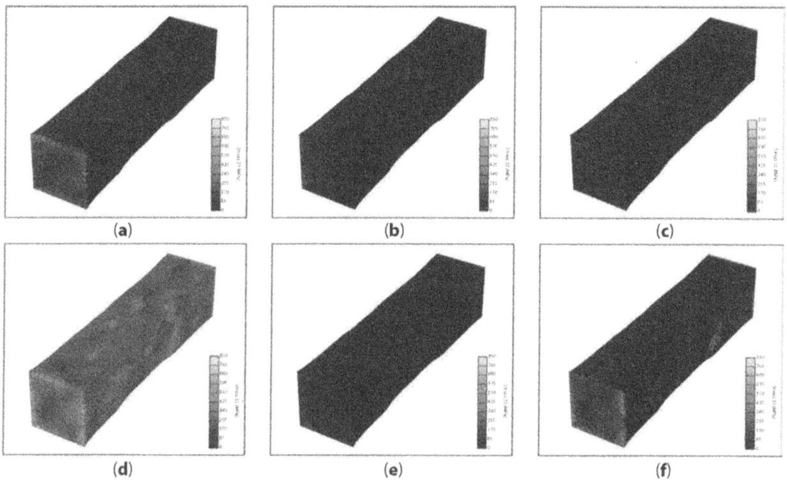

Figure 3.5 Stress tensor components. (a) σ11. (b) τ12. (c) τ13. (d) σ22. (e) τ23 and (f) σ33.

3.3.7 CPFEM Simulation of Additive Manufacturing

Crystal plasticity simulations using Abaqus to model the anisotropic microstructures and the study of effect of different grain orientations on additively manufactured AA6061 aluminum alloy [75] are done by Acar *et al.* Grilli *et al.* developed a crystal plasticity finite element model [76] to evaluate residual stresses and the correlation between plastic deformation, depth, and residual stresses of SLM-manufactured specimens with the use of DAMASK tool. Ameen *et al.* [77] discussed the design, FEA, and fabrication of Ti-6Al-4V alloy using electron beam-based AM for cranial implants. Pal *et al.* [78] used a dislocation density-based crystal plasticity finite element modeling (CPFEM) to predict subgrain formation during ultrasonic AM of Al 3003 H-18 tempered foils. With the capability of phase-to-phase analysis in CPFEM, Kim *et al.* [79] evaluated the stress-strain relationship of constituting phases in AlSi10Mg alloy produced by SLM. High-cycle fatigue life time prediction of additively manufactured AlSi10Mg [80] was discussed in Zhang *et al.*, exploring the versatility of CPFEM using DAMASK tool. Geng *et al.* [81] investigated the CPFEM of functionally graded Ti6Al4V fabricated by AM. Efforts in integrating CPFEM with phase-field modeling [82] were done by Liu *et al.* to predict the PSP (process-structure-property) relationships in additively manufactured metallic materials.

Previous works on CPFEM modeling are presented in Table 3.9. In terms of tools, Abaqus and MOOSE were extensively used. DAMASK, Neper

Table 3.9 Applications of mesoscale FEM in additive manufacturing.

Material	Tools used	Models used	Aspects studied	Reference
AA6016 Al alloy	Abaqus (UMAT sub-routine)	Phenomenological model including elastic and plastic deformation gradients (Kroner-Lee assumption) Peirce and Asaro's self-hardening law	Analysis of elongation of grains and orientation alignment in AM components	Acar *et al.* [75]
316L SS	MOOSE – CPFEM tool	Phenomenological model including elastic, plastic and thermal deformation gradients Power law flow rule (Schmid law) Exponential model to predict Critical resolved shear stress Green-Lagrange strain tensor	Simulation of Residual stresses and its correlation with plastic deformation and grain orientation in SLM	Grilli *et al.* [76]
Al-Si alloy	Abaqus (UMAT sub-routine)	Phenomenological model including elastic and plastic deformation gradient Jaumann rate of Kirchhoff stress. Power law-based flow rule (Schmid law) Anisotropic hardening rule with self-hardening and latent-hardening terms	Evaluation of stress-strain relationship between Al and Si constituents	Kim *et al.* [79]

(*Continued*)

Table 3.9 Applications of mesoscale FEM in additive manufacturing. (*Continued*)

Material	Tools used	Models used	Aspects studied	Reference
SLM $AlSi_{10}Mg$	DAMASK – CPFEM tool	Phenomenological model including elastic and plastic deformation gradient Power law-based flow rule (Schmid law) Anisotropic hardening rule with self-hardening and latent-hardening terms	Analysis of high cycle and very high cycle fatigue	Zhang *et al.* [80]
Ti6Al4V	Rhinoceros with grasshopper plugin to generate Voronoi tessellations. Abaqus tool as CP solver	Phenomenological model including thermo-elastic and plastic deformation gradient Power law-based flow rule (Schmid law) Anisotropic hardening rule with self-hardening and latent-hardening terms	Dual-phase crystal plasticity modeling to investigate effects of secondary α_s phase and primary α' phase in functionally graded	Geng & Harrison [81]
LPBF processed 316L SS	MOOSE-CPFEM tool	Phenomenological model including thermos-elastic and plastic deformation gradient Plastic slip rates are modelled by Orowan's law Taylor hardening equation with bow-out effect to model CRSS Hybrid Galerkin CCD model	To understand the origin of dislocation density in AM metals and how it is affected by thermal fields and stresses, loading conditions and position in a grain through continuum dislocation dynamics models	Hu *et al.* [83]

and FEPX are not used to a large extent. The governing models behind every work are unique and chosen based on the objectives and complexity. Through the table, it could be inferred that crystal plasticity analysis is useful with analysis on residual stresses.

3.3.8 Limitations of CPFEM

Crystal plasticity simulations are based on the microstructural features of the manufactured samples. In the case of AM, the thermal history plays a crucial role in determining the nature of the microstructure and phase distribution. Although the microstructural features are sufficient details, a thermal model to bring the thermal history into picture might give more accurate results. To understand the impact of orientations, pole figures can be used effectively. Although these tools have the capability to visualize the data as pole figures, the capability is not yet explored. In many polycrystal generation tools like Neper and DAMASK, the actual length scale units of the domain are not clearly known.

3.4 Macroscale Modeling

3.4.1 Introduction

Continuum is the only scale that is easy to perceive for human beings, which ranges from several millimeters to meters. Still, the behavior of materials in this scale is too complex to compute by hand. Finite element modeling through FEA and FEM is an efficient tool to understand and quantify the physical, mechanical, and thermal behavior of entire components manufactured by any method. To understand how it works, let us consider a component. Every physical phenomenon associated with the above-mentioned behavior of the component can be and is indeed quantified accurately in terms of mathematical expressions (such as partial differential equations and tensors). With the change of dominant physics at different parts and domains of a component, the nature of these expressions changes at these different parts. By discretizing the component to multiple finite elements and solving the mathematical expression specific to those elements, the change of dominant physics can be captured effectively. Finite element analysis is the method of solving these mathematical expressions and FEM is the method to simulate these results for systematically visualizing the results and drawing correlations. Extensive research has already recorded the FEM of AM at multiple scales, exploring the idea behind the material science tetrahedron.

The modern-day revolution in AM started with the introduction of the laser. Although laser-based AM offers many advantages, the large temperature gradient in related processes results in large residual stresses. Understanding the modeling of the heat source and heat distribution on the substrate is essential to model the thermal field. Thermal history dictates the mechanical response of the material components and hence the stress and strain states. Macroscale modeling in AM is mainly used in thermomechanical studies, including temperature profiles and the resulting thermal and residual stresses. With additively manufactured components playing a vital role to fabricate complex components in crucial applications, the above-mentioned analysis is very important.

3.4.2 Thermal Modeling

Thermal modeling is done to take into account the thermal transport phenomena happening during the AM process. During rapid solidification processes like some laser- and electron beam-based AM and welding, although all three modes of heat transfer happen, convection and radiation phenomenon are neglected. This is because major heat transfer happens on the surface and into the substrate volume through conduction. Thus, a mathematical equation that considers this conduction heat transfer can be used as the governing equation to obtain the temperature distribution in terms of thermal fields. Since only conduction heat transfer is considered, the generalized Fourier heat conduction equation is widely used in building the governing equation. It is framed by an energy balance approach for a small control volume of the whole body.

$$\frac{\partial T}{\partial t}+\left(\vec{U}\cdot\vec{\nabla}\right)T=\alpha\nabla^{2}T+\frac{q}{\rho c_p} \tag{3.25}$$

Equation 3.25 is the widely accepted form of the governing equation used for thermal modeling, where U is the relative speed of scanning to the substrate, T is the temperature as a function of spatial coordinates x, y, and z, t is the time, α is the thermal diffusivity, ρ is the density of the substrate, c_p is the specific heat and q is the heat energy from the source. Here, the gradient of temperature with time $\left(\frac{\partial T}{\partial t}\right)$ represents the transient heat transfer aspect, the term with the relative speed and gradient of temperature with spatial coordinates ($\left(\vec{U}\cdot\vec{\nabla}\right)T$) represents advective heat transfer, the term with thermal diffusivity ($\alpha \nabla^2 T$) represents the diffusive mode of heat transfer and the term $\frac{q}{\rho c_p}$ represents the volumetric heat generated in the control volume considered. Based on the initial assumptions, any of these terms can be neglected or simplified to a different form.

For instance, if thermal modeling for AM is to be done, to reduce complexity of the equation, the heat transfer will be assumed to be of steady state, the scanning speed is assumed constant (along a straight line along the x-axis negative direction), and the substrate material is assumed isotropic and homogenous with constant thermophysical properties throughout. These assumptions simplify Equation 3.25 to Equation 3.26 as the final governing equation.

$$\frac{\partial^2 T}{\partial x^2} + \frac{\partial^2 T}{\partial y^2} + \frac{\partial T^2}{\partial z^2} + \frac{U}{\alpha}\frac{\partial T}{\partial x} = 0 \tag{3.26}$$

Through heat source modeling, heat flux distribution from a heat source can be quantified to mathematical expressions. By assuming some boundary conditions, based on the heat source modeling (which can be Gaussian, hemispherical, ellipsoidal, and Goldak's double ellipsoidal models) the governing equation can be solved to obtain a mathematical expression for the thermal field. These expressions will have temperature as a function of spatial coordinates x,y and z.

Based on this methodology, many analytical models for the thermal field have been built. An example is Rosenthal's solutions [120] for temperature field for thick plate welding assuming a point heat source.

$$T(x,y,z) = T_0 + \frac{q}{2\pi k r} exp\left(-\frac{U}{2\alpha}(r+x)\right) \tag{3.27}$$

where, k is the thermal conductivity of the plate, $r = \sqrt{x^2 + y^2 + z^2}$ and T_0 is the preheat temperature. The thermal field equation can be solved further based on the requirements to extract more information just using process parameters like scanning speed and heat energy from the source and thermophysical properties of the material.

Heat source modeling

The heat from the heat source does not always reach the substrate completely. The energy reaching the substrate is also not completely used in melting.

With different AM techniques and welding involving different heat sources, the distribution of input heat fluxes on the surface and volume of the substrate varies. It is thus important to model the heat flux distribution from the source. It depends on many parameters like the nature of heat source, symmetry, path, and scanning speed. Some derived models of heat flux distributions are as follows,

- Pavelic's model assumes a Gaussian heat flux distribution at the surface in a circular profile [84]. Heat flux is maximum at the center and it decreases exponentially as moved outwards radially. It can be quantified as follows,

$$q(r) = q(0) \cdot exp\,(-Cr^2) \tag{3.28}$$

  $q(r)$ = surface flux at radius r, $q(0)$ = maximum flux at the center of the heat source, C = distribution width coefficient and r = radial distance from the center of the heat source.
- Hemispherical model was built to consider the volumetric distribution of heat flux into the substrate [84]. However, it was suitable for only spherically symmetrical pools. It can be quantified by the equation,

$$q(x, y, \varepsilon) = \frac{6\sqrt{3}Q}{c^3\pi\sqrt{\pi}} e^{\frac{-3x^2}{c^2}} e^{\frac{-3y^2}{c^2}} e^{\frac{-3\epsilon^2}{c^2}} \tag{3.29}$$

  where, $\varepsilon = z + v(\tau - t)$ represents the movement in the z-axis, with velocity v and a lag factor τ.
- An ellipsoidal model was an improvement of the hemispherical model [84]. It is quantified by the equation,

$$q(x, y, \varepsilon) = q_m e^{-Ax^2 - By^2 - C\varepsilon^2} \tag{3.30}$$

  where A, B and C are distribution parameters with $A = \frac{3}{a^2}, B = \frac{3}{b^2}, C = \frac{3}{c^2}$ and a, b, and c are the semi-axis lengths of the elliptical cross-section on the surface of the substrate along the x, y and z axes respectively.

- Goldak developed a double-ellipsoidal model to take into consideration the trailing effect of the heat source in materials with low thermal conductivity [85]. The heat flux distribution can be represented in the form of two ellipsoidal distributions having different minor axis lengths along the direction of movement of heat source. The distribution is quantified with separate equations for the front and rear ends. Front end:

$$q_f(x,y,z,t) = \frac{6\sqrt{3} f_f Q}{abc\pi\sqrt{\pi}} e^{\frac{-3x^2}{a_f^2}} e^{\frac{-3y^2}{b^2}} e^{\frac{-3[z+v(\tau-t)]}{c^2}} \tag{3.31}$$

Rear end:

$$q_r(x,y,z,t) = \frac{6\sqrt{3} f_f Q}{abc\pi\sqrt{\pi}} e^{\frac{-3x^2}{a_r^2}} e^{\frac{-3y^2}{b^2}} e^{\frac{-3[z+v(\tau-t)]}{c^2}} \tag{3.32}$$

Here, f_f and f_r are the fraction of total heat distributed in the front and rear regions. They are considered as the distribution is non-symmetric. They are related to the semi-axis lengths in the direction of motion of heat source a_f and a_r as,

$$\frac{f_f}{a_f} = \frac{f_r}{a_r} \tag{3.33}$$

Goldak's double ellipsoidal model is widely used in many simulations. With the heat flux distributions known, thermal modeling can be done for further analysis.

Initial and boundary conditions

The governing equation involved in the thermal modeling is a partial differential equation, which requires known conditions at well-defined boundaries to obtain a solution. The initial condition is most often the preheat temperature of the substrate. Boundary conditions represent the effect of phenomenon that are not directly modeled with the governing equation. The finite element solutions to governing equations should also satisfy the imposed boundary conditions.

These boundary conditions can be of three different forms—Dirichlet, Neumann, and mixed. The Dirichlet boundary conditions are direct relationships with the field variable. Neumann boundary conditions are derivatives of the field variables. Mixed boundary conditions are a combination of Dirichlet and Neumann types. In thermal modeling, since the governing equations used are mostly conduction based, convection and radiation phenomenon are considered as boundary conditions.

Convection boundary conditions are represented by the Newton's law of cooling,

$$Q_{conv} = hA(T_s - T_\infty) \tag{3.34}$$

where Q_{conv} is the convective heat transfer, h is the convective heat transfer coefficient, A is the surface area, T_s is the surface temperature and T_∞ is the temperature of fluid in operation or simply the temperature of the environment of operation.

Radiation boundary conditions are represented by the Stefan-Boltzmann law,

$$Q_{rad} = \varepsilon\sigma(T_s^{\,4} - T_\infty^{\,4}) \tag{3.35}$$

where Q_{rad} is the radiative heat transfer, ε is the surface emissivity and σ is the Stefan-Boltzmann constant.

3.4.3 Mechanical Modeling

Modeling the stress and strain response of the material and for a particular geometry of components is essential for effectively manufacturing the components. Modeling of stress and strain gives information about the distribution of stress and strain on the component, residual stress distribution, and many other useful information to analyze the mechanical response of the material towards different loading conditions. Often the mechanical modeling is done followed by thermal modeling. The thermal history and gradients from the thermal modeling results can be taken as input in form of thermal load for modeling. The modeling procedure involves mathematical expressions for stress and strain at a particular state, relationship between stress and strains in elastic and plastic regimes, and governing equations for how stress and strain evolve with every load step with or without strain hardening. The stress state is often quantified based on the static force equilibrium. It is expressed in equations as

$\Delta\sigma = 0$ and $\sigma_{ij} + P = 0$, where σ and P indicate stress and forces, respectively. Strain state can be expressed as the sum of elastic, plastic, thermal, transformation induces plasticity, and volumetric strain components. It is given by the expression, $\varepsilon_{tot} = \varepsilon_E + \varepsilon_P + \varepsilon_T + \varepsilon_{Trp} + \varepsilon_{\Delta V}$. Stress–strain relationships can be given by Hooke's law, Ludvik-Holloman expression, or Johnson-Cook expression. Hooke's law is the best expression of the elastic behavior. Modeling plasticity requires information on at what state of stress yielding has occurred, a flow rule to model how materials flow after initial yielding, and a hardening rule to model the evolution of yield surface with plastic strain. Plasticity finite element simulations are modeled using constitutive laws which are categorized as rate-independent and rate-dependent plasticity laws. Rate-independent models are used in case of low temperatures and low strain rates and rate-dependent models are used in case of high temperatures and high strain rates. Stress and strain analysis begins with the determination of the effective stress and its comparison with the yield stress of the material at each time step or load step. The yield stresses are calculated using yield criterion based on the stress states. The comparison would give the information whether elastic or plastic yielding has occurred or not. Based on that, the deformation can be decomposed to plastic and elastic parts. The next load is applied incrementally in terms of incremental strain and the comparison is done. In case of elastic deformation, the stress–strain relationship can be easily established by Hooke's law using the total stress-strain formulation. However, with initiation of plastic yielding, the relationship between stress and plastic strain is modeled using plastic flow rules. Prandtl-Reuss equations based on the von Mises criterion is widely used to relate the incremental strains with the deviatoric stress. For comparisons to indicate yielding, mathematical expressions representing yield criterion are proposed. Most commonly, von Mises' and Tresca's yield criteria are used. The equivalent stress acting for the principal stresses, acting on three axes (σ_1, σ_2 and σ_3), can be given by the following equations,

von Mises:

$$\sqrt{\frac{1}{2}\left[(\sigma_1-\sigma_2)^2+(\sigma_2-\sigma_3)^2+(\sigma_3-\sigma_1)^2\right]} \tag{3.36}$$

$$\text{Tresca: } max|\sigma_1-\sigma_2|, |\sigma_2-\sigma_3|, |\sigma_3-\sigma_1| \tag{3.37}$$

On loading a material until plastic deformation, unloading it and then reloading it, due to strain hardening the resistance to plastic flow is increased and hence the yielding is delayed. This factor has to be taken into consideration while modeling plastic deformation. Two approaches used commonly are isotropic and kinematic hardening laws.

Isotropic hardening describes the increase in material strength due to plastic strength. Isotropic hardening law models the yield surface to increase in size without changing its shape, as a result of plastic straining. It is, however, not applicable for cycle loading conditions. The mathematical expression for isotropic hardening is framed between the yield stress (Y) and plastic strain (ε^p). The relationship can be of three forms, perfectly plastic, linear strain hardening, and power law hardening. The mathematical expressions are as follows,

Perfectly plastic:

$$Y = constant \tag{3.38}$$

Linear strain hardening:

$$Y = Y_0 + h\varepsilon^p \tag{3.39}$$

Power-law hardening:

$$Y = Y_0 + h(\varepsilon^p)^{\frac{1}{m}} \tag{3.40}$$

where Y_0, h and m are material properties [86].

Kinematic hardening law models the yield surface to translate without changing its shape. The kinematic hardening laws are of two types: linear kinematic hardening and Armstrong-Frederick hardening law. The mathematical expressions representing each are given by,

Linear kinematic hardening:

$$d\alpha_{ij} = \frac{2}{3} c\, d\varepsilon_{ij}^p \tag{3.41}$$

Armstrong-Frederick hardening:

$$d\alpha_{ij} = \frac{2}{3} c\, d\varepsilon_{ij}^p - \gamma\alpha_{ij}\, d\varepsilon_{eq}{}^p \tag{3.42}$$

where, α_{ij} = back stress, ε_{ij}^p = Plastic strain, γ = kinematic hardening parameter and ε_{eq}^p = equivalent plastic strain [119].

The flow rule relationships for isotropic hardening and linear kinematic hardening to relate incremental strain to deviatoric stresses are given by the following equation

$$d\varepsilon_{ij}^{p} = d\underline{\varepsilon}^{p}\frac{3}{2}\frac{S_{ij}}{Y} \tag{3.43}$$

3.4.4 Thermomechanical Modeling—Demonstration

In sequentially coupled thermal and mechanical analysis, the temperature field is predefined and stored as a function of time. Stress analysis depends on temperature distribution. The flowchart in Figure 3.6 indicates the processing steps involved in thermomechanical modeling. The thermal history of the manufactured sample is decided by the process parameters used in the additive manufacturing procedure. The distribution of heat flux into the substrate is given by the heat source model and the thermal modeling gives the governing equation describing the heat transfer. Based on these the thermal history and temperature distribution are expressed in terms of temperature contours. Since these decide the microstructure and hence the properties of the manufactured samples, these are important inputs for mechanical analysis. During mechanical analysis, the mechanical governing equations for elastic and plastic regimes are fed, along with boundary conditions.

On creating a simulation domain based on the geometry of the component, it is discretized to finite elements and nodes through meshing. The governing equations are applied to every element of the mesh. Boundary conditions and initial conditions are fed to the domain. Solving the governing equations would give the necessary outputs for every node and element. Postprocessing involves visualizing the results graphically and facilitates effective numerical and geometric interpretation.

A demonstration of an Laser Bed Fusion (LBF)-manufactured cylindrical geometry is presented by the authors using Simufact Additive software. The heat source used is Goldak's model considering transient Fourier's heat conduction equation as the governing equation. The possible outcomes that could be obtained are the stress tensor components (6 components including the normal and shear stresses), temperature field, equivalent strain (plastic and elastic), residual stresses, and displacements of the finite elements (distortion). The temperature field could be visualized in terms of temperature contours based on nodal outputs. Thermal strain, thermal

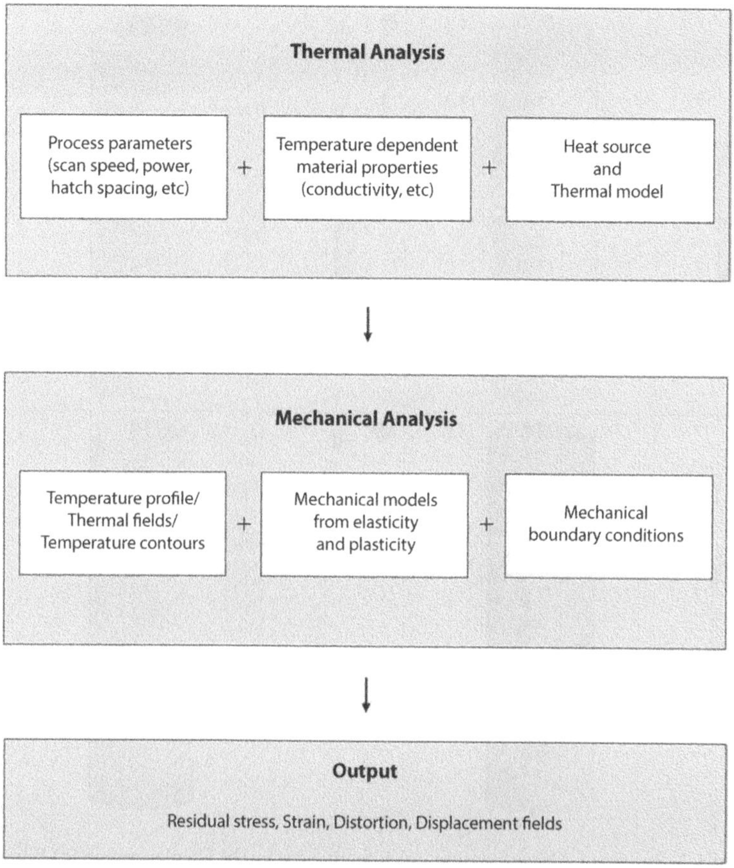

Figure 3.6 Schematic of process flow in thermomechanical analysis.

stress, and heat flow can also be visualized. Stress and strain maps as elemental outputs can be obtained to assess the mechanical properties. The nodal displacements would effectively be used to understand the geometrical variations and hence be used in visualizing distortion.

A visualization of the distortion, displacement, effective stress and effective strain of the built component is presented in Figure 3.7. A hole is made at a point on the cylinder to understand its impact on the mechanical properties. The tangential regions close to the holes appear to experience maximum stress and strain. The distortion analysis and the AM-layer z displacement reveal that for the region around the hole, the displacement is more at the part where the distortion is minimal. The bottom regions of the cylinder also indicate a similar trend.

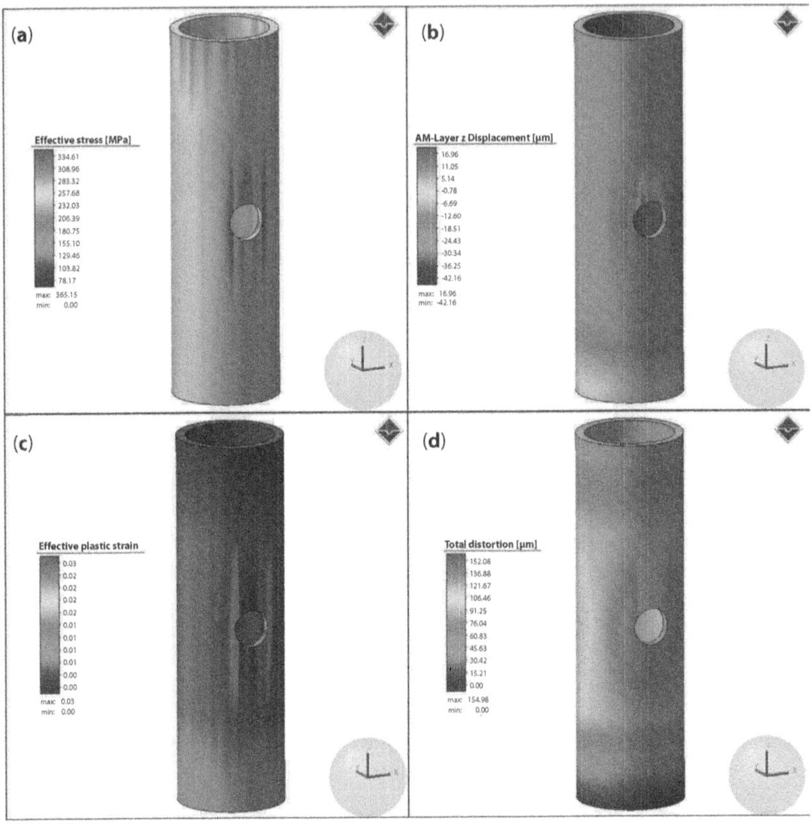

Figure 3.7 Results of AM simulation. (a) Effective stress; (b) Displacement; (c) Effective plastic strain; (d) Total distortion.

3.4.5 FEM Simulation of Additive Manufacturing

Thermomechanical analysis of laser-based and electron beam-based additively manufactured components using powders and wire feed are done extensively to study the stress and strain distribution in the components, the residual stresses, fatigue life, and distortion analysis of the components. Zhao *et al.* [87] analyzed the residual stress fields of laser additive-manufactured, functionally gradient material LAM TC4/Inconel 718, using FEM through Abaqus tool. Wang *et al.* [88] proposed a method to measure the residual stresses distribution using neutron diffraction between two Inconel 625 walls fabricated by laser-based DED, with the use of thermomechanical modeling to validate the results. Promoppatum *et al.* [89] examined the effects of material plasticity models on thermomechanical predictions in the laser bed fusion process of Ti-6Al-4V using Abaqus tool. Vastola *et al.* [90] performed FE method calculations to understand residual stresses in electron beam-based additively manufactured Ti-6Al-4V, and how it is related to process parameters like beam size, beam power, scan speed, and bed preheating temperature. Chen *et al.* [91] presented a numerical finite element model for SLM-manufactured ceramic materials at scale of beads during unitary and successive multi-pass deposition. Montevecchi *et al.* [92] proposed a method to effectively mesh for FEM, for distortion prediction in wire arc AM technology, and also an innovative technique to optimize the idle time between layer passes. Ding *et al.* [93] developed an efficient modeling approach to analyze thermomechanical behavior in terms of residual stresses and distortion of wire and arc AM technology with the help of a transient thermomechanical finite element model. Yang *et al.* proposed a volumetric heat source model based on semi-analytical thermal approach and by thermal modeling using the model, the volumetric heat source was validated [94].

Previous studies on FE thermal, mechanical and thermomechanical modeling are presented in a tabular form in Table 3.10. The thermal model used in most of these cases is the Fourier's heat conduction equation in transient state. For analytical models used in making predictions, the Rosenthal's solution is used as thermal model. Goldak's double ellipsoidal model was widely used as the heat source model followed by a Gaussian flux distribution. For mechanical modeling, the static force equilibrium is widely used. Strain is modeled as a sum of elastic, plastic, and thermal strains. Phase transformation-induced and volumetric strains are neglected in most cases. Elasticity modeling in every case is done using the Hooke's law, however with variations based on the nature of the material. Different plasticity models are used in different cases based on the nature of the material and the objective of the study. Yield stress calculations are done using von Mises yield criterion. Apart from these, FEM has been used to model defects in AM components. In some AM techniques like SLM, porosity and the presence of unmelted powders may lead to deterioration of properties. Modeling the defects and proper understanding of their effects is necessary to optimize the process parameters and minimize them. Teng *et al.* [110] has reviewed the models based on defects for laser-based materials processing and optimization methods for process parameters.

Table 3.10 Applications of macroscale FEM in additive manufacturing.

Material	Tools used	Models used	Aspect studied	Reference
Functionally gradient laser additive-manufactured TC4/In718	DFLUX and FILM subroutines in Abaqus software	Heat source: double ellipsoidal (Goldak's model) Mechanical: von Mises yield criterion	Residual stress field	Zhao *et al.* [87]
Laser-based DED fabricated In625	Project Pan (Autodesk Inc., State College, PA) finite element solver	Heat source: double ellipsoidal (Goldak's model) Mechanical model: static force equilibrium, sum of plastic, elastic and thermal strains, and Hooke's law	Residual stress mapping	Wang *et al.* [88]
Laser powder bed fusion of Ti6Al4V	Abaqus	Heat source: double ellipsoidal (Goldak's model) Mechanical model: static force equilibrium and sum of plastic, elastic and thermal strains Elastic: Hooke's law Plastic: mechanical threshold stress (MTS) and Johnsen-Cook (JC) models	Effect of material plasticity models on thermomechanical simulations and predictions	Promoppatum and D. Rollett [89]

(Continued)

Table 3.10 Applications of macroscale FEM in additive manufacturing. (*Continued*)

Material	Tools used	Models used	Aspect studied	Reference
Powder-based AM in Ti6Al4V	Abaqus	Heat source: moving Gaussian heat source, tailored for electron-beam technique Mechanical model: static force equilibrium and sum of plastic, elastic and thermal strains Elastic: Hooke's law Plastic: Isotropic hardening	Influence of electron beam melting process parameters on the residual stresses	Vastola *et al.* [90]
WAAM on S235JR structural steel using ER70S-6 wire	LS-DYNA (a Crank-Nicolson time integration scheme)	Heat source: double ellipsoidal (Goldak's model) model tailored for Gas Metal Arc Welding (GMAW)	Monitoring the thermal behavior through interpass temperature of each layer	Montevecchi *et al.* [92]
SLM-processed SS 316L and Ti6Al4V	-	Heat source model: half-ellipsoidal volumetric heat source	A volumetric heat source model based on semi-analytical thermal approach for SLM AM process	Yang *et al.* [94]
Laser powder bed fusion of Ti6Al4V	Fenic's code	Thermal modeling: Rosenthal's solution for semi-infinite solid under a moving point heat source	Superposition-based finite element approach	Moran *et al.* [95]

(*Continued*)

Table 3.10 Applications of macroscale FEM in additive manufacturing. (*Continued*)

Material	Tools used	Models used	Aspect studied	Reference
Laser solid forming of Ti6Al4V	Comet [International Center for Numerical Methods in Engineering (CIMNE)]	Heat source: summation of all volume elements in the melt pool Mechanical model: static force equilibrium and sum of plastic, elastic and thermal strains Elastic: Hooke's law	Sensitivity analysis of the mechanical properties to find the optimal parameters for the numerical simulation	Lu *et al.* [96]
Laser beam powder bed fusion of Ti6Al4V	Abaqus	Heat source: power distribution based on process build parameters Mechanical model: static force equilibrium, sum of plastic, elastic and thermal strains, Elastic: Hooke's law Plastic: elastic-perfectly plastic model with von Mises yield criterion	Interdependency of resolution, energy and time	Zhang *et al.* [97]
DED of Ti6Al4V	-	Thermal model: analytical Rosenthal's solutions for moving heat source	Predictions on width, height and deposition area of the layers	Ertay *et al.* [98]

(*Continued*)

Table 3.10 Applications of macroscale FEM in additive manufacturing. (*Continued*)

Material	Tools used	Models used	Aspect studied	Reference
Ti6Al4V	CUBIC by Pan Computing LLC	Heat source model: Goldak's double ellipsoidal model tailored for electron-beam Mechanical model: Static force equilibrium and Sum of plastic, elastic and thermal strains Elastic: Hooke's law Plastic: perfectly plastic with negligible hardening	Predict distortion accumulation	Denlinger *et al.* [99]
Wire arc additive manufactured Ti6Al4V	Simufact Welding	Viscoplastic model: Norton and Bailey creep model	Viscoplastic creep model to analyze residual stresses and distortion	Springer *et al.* [100]
Ti-6Al-4V L-PBF component	GSAS software	Mechanical model: residual strains calculated through lattice parameters and generalized Hooke's law for isotropic materials Maxwell type viscoplasticity model	Residual stresses and performance	Strantza *et al.* [101]

(*Continued*)

Table 3.10 Applications of macroscale FEM in additive manufacturing. (*Continued*)

Material	Tools used	Models used	Aspect studied	Reference
Microplasma transferred arc additive manufactured Ti6Al4V	Ansys 17.2 and SOLID 70	Heat source: Gaussian distribution	Dilution modeling	Sawant *et al.* [102]
Laser metal additive manufactured Ti6Al4V	ANSYS	Heat source: volumetric moving point heat source Mechanical modeling: static force equilibrium and sum of plastic, elastic and thermal strains Elastic: Hooke's law Plastic: Swift law	Elasto-plastic thermomechanical modeling	Mirkoohi *et al.* [103]
LPBF AlSi10Mg	Abaqus	Heat source: Gaussian distribution Mechanical modeling: Static force equilibrium and sum of plastic, elastic and thermal strains. von Mises yield criterion	High-efficient FE thermomechanical model	Cao *et al.* [104]
SnTe thermoelectric material	ANSYS	Heat source: Gaussian distribution	Simulation of temperature and stress fields	Luo *et al.* [105]

(*Continued*)

Table 3.10 Applications of macroscale FEM in additive manufacturing. (*Continued*)

Material	**Tools used**	**Models used**	**Aspect studied**	**Reference**
Laser engineered net shaping (LENS) processed Ti-6Al-4V	-	Heat source: double ellipsoidal volumetric source Mechanical model: static force equilibrium Elastic: Hooke's law Plastic: J2 von Mises plasticity law	FEM and validation of thermomechanical behavior	Yang *et al.* [106]
WAAM on S235JR structural steel using ER70S-6 wire	LS-DYNA (a Crank-Nicolson time integration scheme)	Thermal model: Rosenthal's analytical solutions Heat source: double ellipsoidal (Goldak's model) model tailored for GMAW	Distortion prediction	Montevecchi *et al.* [107]
LENS-processed 304L SS	SIERRA	Heat source: spherical volumetric with Gaussian distribution Mechanical model: von Mises yield criterion	Workflow to model DED AM	Stender *et al.* [108]
LBAM polycarbonate, 316L SS, titanium, pure iron and Ti6Al4V	ANSYS	Heat source: Gaussian distribution	Presentation and validation of a macroscopic thermal FEA model	Stathatos *et al.* [109]

3.4.6 Limitations

The governing equations behind the modeling are developed based on assumptions that neglect certain physical phenomena. Simpler models that do not take into consideration complex physical phenomena would not require more data and the simulation time is also minimal. However, the results obtained are not accurate. To obtain more accurate results, a more complex physical phenomenon involved with any manufacturing process has to be brought into the picture. In such a case, the simulations become computationally intensive, demanding more computer memory and time. Because computers work with only a limited number of digits, the units of working and rounding off may result in some inaccurate optimization.

3.5 Machine Learning

3.5.1 Introduction

Machine learning has become inevitable in every field. With plenty of data available with every operation in various domains, ML facilitates the effective use of this data. By analyzing large amounts of data and creating and training algorithms through statistical models, ML algorithms can identify patterns to make accurate classifications and effective predictions can be drawn from the data. These help with uncovering important and useful insights that are too complex or time consuming for humans to understand on their own. Machine learning has the capability to improve the efficiency, quality, and reliability of AM. This would result in greater adoption of AM in a wide range of industries. With the evolution of AM techniques, in terms of the technique and materials, the efficiency of the process is dependent on a lot of data. Optimization of process parameters based on the properties of the material to be manufactured is important. The performance of the material after manufacturing should be predicted. Indicating defects in manufactured products and potential problems in technique is also important. Machine learning can be majorly categorized into three types based on the way in which the model is trained: supervised learning, unsupervised learning, and reinforcement learning. In supervised learning, the machine is trained on a labeled dataset, where the output variable (or the target) is already known for each input data point. The goal is to learn a mapping function from the input variables to the output variables. The algorithm learns by minimizing the error between predicted output and actual output using various techniques like regression or classification.

Some examples of supervised learning algorithms are linear regression, logistic regression, decision trees, neural networks, and support vector machines. In unsupervised learning, the machine is trained on an unlabeled dataset, where there are no predefined output variables or targets. The goal is to find hidden patterns or structures in the data. The algorithm learns by exploring the underlying relationships in the data using clustering, dimensionality reduction, or association rule learning techniques. Some examples of unsupervised learning algorithms are k-means clustering, hierarchical clustering, principal component analysis (PCA), and the Apriori algorithm. In reinforcement learning, the machine learns by interacting with its environment and receiving feedback in the form of rewards or penalties. The goal is to learn a policy that maximizes the cumulative reward over time. The algorithm learns by trial and error, where it takes an action in a given state and receives feedback from the environment. The feedback is used to update the policy to choose better actions in the future. Some examples of reinforcement learning algorithms are Q-learning, policy gradient methods, and deep reinforcement learning. Research on ML applications in AM is focused on the design for AM (DfAM), in situ monitoring and detection of defects, material analytics, and property prediction. The tools used in research are discussed briefly.

3.5.2 Regression and Optimization

Regression is the process of identifying the relationship between the dependent variable and one or more independent variables. The main goal of regression is to build a model that can accurately predict the values of the output variable for new input data. In regression, optimization algorithms are used to find the best set of parameters for the model that minimize the error between the predicted output and the actual output. The most commonly used optimization algorithms in regression include gradient descent, stochastic gradient descent, and Newton's method. Optimization is the process of finding the optimal value of any quantity that dictates a process such that the process is the most efficient. The goal of optimization is to train a ML model that can accurately predict outcomes on new, unseen data. It is most useful when a mathematical expression exists between the quantities to be optimized and the efficiency of the process. In mathematically modeling any phenomenon, the constants of proportionality can be effectively found using optimization procedures. Regression is done in two ways: linear regression, and logistic regression. In linear regression, the dependent and independent variables are related linearly. The dependent variable is continuous. While in logistic regression, they are related nonlinearly with a categorical dependent variable. The choice between linear regression and logistic regression depends on the nature of the problem being solved and the type of data being analyzed. Mathematically, they can be expressed as follows,

$$y = w_0 a + w_1 b + c \tag{3.44}$$

$$y = w_0 a^2 + w_1 b^2 + c \tag{3.45}$$

where y is the dependent variable and a and b are independent variables. In terms of AM, y can be called a function representing the AM process, a, and b are process parameters, w_0 and w_1 are weight factors to represent the influence of these parameters on the working of the process, and c is a constant. Optimization begins with framing the loss or cost function that has to be minimized or the objective function that has to be maximized, such that the desired efficiency is obtained. The loss or cost function is most often the mean square error for linear regression and logarithmic mean square error for logarithmic regression. These quantities are chosen since they give a good measure of the model performance.

$$MSE = \frac{1}{n}\sum_{i=1}^{n}(\hat{y}_i - y_i)^2 \tag{3.46}$$

$$LMSE \quad \frac{1}{n}\sum_{i\;1}^{n}\left(log\left(\frac{\hat{y}_i}{y_i}\right)\right)^2 \tag{3.47}$$

The cost function is presented by the above equations, where $\hat{y}_i$ is the measured quantity, y_i is the predicted quantity, and n is the sample size. In order to optimize a model, the parameters are adjusted iteratively through an optimization algorithm, such as gradient descent, that updates the parameters in the direction of the steepest descent of the loss function. The optimization algorithm continues to update the parameters until it reaches a minimum of the loss function, indicating that the model has converged and can no longer improve significantly. Gradient descent is an iterative optimization algorithm used to minimize a function by iteratively adjusting the parameters. The algorithm works by calculating the gradient of the function at a given point and then adjusting the parameters in the direction of the negative gradient. This process is repeated until the algorithm converges to a minimum of the function.

The gradient descent algorithm starts by computing the gradient of the function with respect to the parameters. This gradient represents the direction of the steepest ascent of the function at the current parameter values. The algorithm then updates the parameters by subtracting a fraction of the gradient from the current values. This fraction, known as the learning rate, controls the step size of the parameter updates and is typically a small positive number. The process of computing the gradient and updating the parameters is repeated until a stopping criterion is met. Common stopping criteria include reaching a maximum number of iterations, reaching a specified tolerance level for the gradient, or reaching a minimum value of the function. Through regression and optimization, models can be built to make effective predictions and are evaluated.

3.5.3 K-Means Clustering

K-means clustering is an unsupervised ML algorithm, that tries to partition a set of data points into K clusters, where each data point belongs to the cluster whose center is closest to it. The value of K is determined before applying the algorithm and represents the number of clusters to be formed. The K-means algorithm works by first selecting K initial centroids (center points) randomly. Each data point is then assigned to the nearest centroid based on the Euclidean distance. After all data points have been assigned to a centroid, the centroid is moved to the mean of all the data points assigned to it. This process is repeated until convergence, that is, until the centroids no longer move.

The algorithm aims to minimize the sum of squared distances between the data points and their assigned centroids, which is also known as the objective function. K-means is a computationally efficient algorithm, and it works well on a large dataset with a small number of clusters. However, it is sensitive to the initial random selection of centroids, and it may converge to a local minimum rather than a global minimum. Therefore, it is often necessary to run the algorithm several times with different initial centroids to obtain the optimal clustering.

3.5.4 Random Forest Algorithm

Random forest is an ensemble learning algorithm used for classification, regression, and other tasks. The algorithm creates a large number of decision trees and then aggregates their predictions to arrive at a final output. The name "random forest" comes from the fact that each tree is constructed using a random subset of the training data and a random subset of the features. It is built by selecting a random subset of the training data and of the features. A decision tree is constructed using the selected data and features. Similarly, a large number of trees are created. Predictions of all the trees are aggregated to arrive at a final output.

3.5.5 Self-Organizing Maps

A self-organizing map (SOM), also known as a Kohonen map, is a type of artificial neural network that is used for unsupervised learning. The main goal of a SOM is to reduce the dimensionality of large datasets and to visualize the relationships between the data points. In a SOM, a set of input data is represented as a 2D array of nodes or "neurons". Each neuron is associated with a weight vector, and the SOM algorithm works by adjusting these weight vectors so that the neurons are arranged in a way that reflects the underlying structure of the input data. The adjustment process is done iteratively, based on the similarities between the input data and the weights. During the training process, the input data is presented to the SOM one by one, and the SOM calculates the distance between each input data point and the weight vectors of all the neurons. The neuron with the closest weight vector is called the "winning neuron", and its weight vector is adjusted to be more similar to the input data. The weight vectors of the neighboring neurons are also adjusted to a lesser degree, in order to smooth the transition between different regions of the input space. Once the training process is complete, the SOM can be used to visualize the relationships between the input data points. Each neuron represents a different region of the input space, and the colors or labels associated with the neurons can be used to represent different classes or clusters within the data. By visualizing the SOM, it is possible to identify patterns and relationships that might be difficult to discern from the raw input data.

3.5.6 Application of ML in Additive Manufacturing

As a part of the DfAM, ML is used in part design, design optimization, topology optimization, and shape deviation. Defect detection through ML is based on image-based and sensor-based approaches. It also facilitates process and property modeling in AM. For every application, a wide range of ML techniques are used and investigations on usage of various algorithms are extensively studied in previous research.

Previous works involving applications of ML in AM are presented in Table 3.11. It could be inferred that the aspects studied are not subjected to any one particular application. ML has a role to play in every possible corner of the additive manufacturing procedure. With the diversity in application, the models used are also very diverse not confining to only a few popular ones. Although, ML being a very effective tool in modeling AM, it is necessary to understand that it is something that involves just the numbers and not any governing physics as in other modeling approaches. It is thus necessary to understand the physical significance of the problem

Table 3.11 Review of ML-based studies in additive manufacturing.

Reference	Aspect studied	Algorithm used
Zhu *et al.* [111]	Shape deviation & tolerancing	Transformation perspective and a Gaussian process multitask Learning
Yaseer *et al.* [112]	Layer roughness	Random forest algorithm and MSE evaluation of the model
Li *et al.* [113]	Defect detection	Bagging of trees, gradient boosting, random forest, KNN, and Linear SVM
Zhu *et al.* [114]	Prediction of melt pool shape (SS304)	SVR, XGBoost, and BPNN
Gan *et al.* [115]	Microstructure and microhardness design (Inconel)	Self-organizing maps
Akbari *et al.* [116]	Melt pool characteristics predictions	Optimization and development of MeltpoolNet_resource package for ML use in AM
Jiang *et al.* [117]	Optimizing better connection between deposited lines	Neural networks
Gor *et al.* [118]	Density prediction	ANN, KNN, and SVM algorithms MAPE, MAE, and MAD used for evaluation

encountered before handling it through ML, results obtained from the algorithm and proper validation using experimental or physics based models to ensure accurate modeling.

3.6 Conclusion

Computational modeling of materials has become an indispensable tool in the modern era of materials science and engineering. Its potential for predicting the properties and performance of materials at various scales, from the atomic to the macroscopic level, has opened new avenues for the design and development of novel materials with unprecedented functionalities. In fact, it is probably one of the reasons why there is rapid progress in the domain of AM. The ability to predict the outcome and study various phenomena, which are conventionally difficult to perceive provides huge savings in the resources associated with material development and process integration. The potential of AM to be an industrially significant manufacturing process is well-established with the integration of computation modeling during the development of process and materials. However, there are still many challenges that need to be addressed to fully realize its potential.

The outlook for computational modeling of materials is promising, especially in the domain of AM. With the rapid growth of computing power and the emergence of new modeling techniques and algorithms, the accuracy and predictive power of these methods are expected to improve significantly. Coupling of different scales of modeling will pave the way for more holistic understanding of material behavior in complex processes like AM. The integration of ML and artificial intelligence techniques into computational modeling is expected to revolutionize the way materials are designed and developed. This is particularly relevant for AM, where the ability to predict and control the microstructure and properties of the printed parts is essential for achieving optimal performance, even achieving the first print to be perfect components, which was deemed impossible using conventional methods of materials and process engineering.

References

1. Lee, J.G., *Computational materials science - An introduction*, Second Edition, CRC Press, Taylor & Francis, Boca Raton, 2016, doi: https://doi.org/10.1201/9781315368429.

2. Nandy, J., Yedla, N., Gupta, P., Sarangi, H., Sahoo, S., Sintering of AlSi10Mg particles in direct metal laser sintering process: A molecular dynamics simulation study. *Mater. Chem. Phys.*, 236, 121803, Oct. 2019, doi: 10.1016/j.matchemphys.2019.121803.
3. Thompson, A.P. *et al.*, LAMMPS - a flexible simulation tool for particle-based materials modeling at the atomic, meso, and continuum scales. *Comput. Phys. Commun.*, 271, 108171, Feb. 2022, doi: 10.1016/j.cpc.2021.108171.
4. Sestito, J.M., Abdeljawad, F., Harris, T.A.L., Wang, Y., Roach, A., An atomistic simulation study of nanoscale sintering: The role of grain boundary misorientation. *Comput. Mater. Sci.*, 165, 180–189, Jul. 2019, doi: 10.1016/j.commatsci.2019.04.015.
5. Gao, Y., Clares Pastrana, A.P., Manogharan, G., van Duin, A.C.T., Molecular dynamics study of melting, diffusion, and sintering of cementite chromia core–shell particles. *Comput. Mater. Sci.*, 199, 110721, Nov. 2021, doi: 10.1016/j.commatsci.2021.110721.
6. Sreeramagiri, P., Sharma, P., Das, C., Balasubramanian, G., Examining solid-state sintering of AlCoCrFeNi multi-principal element alloy by molecular simulations. *Comput. Mater. Sci.*, 216, 111875, Jan. 2023, doi: 10.1016/j.commatsci.2022.111875.
7. Nandy, J., Sahoo, S., Sarangi, H., Study on shape dependency of Al-alloy nanoparticles during coalescence in direct metal laser sintering: A molecular dynamics approach. *Mater. Today: Proc.*, 41, 347–351, 2021, doi: 10.1016/j.matpr.2020.09.557.
8. Zhang, Y. and Zhang, J., Sintering phenomena and mechanical strength of nickel based materials in direct metal laser sintering process - A molecular dynamics study. *J. Mater. Res.*, 31, 15, 2233–2243, Aug. 2016, doi: 10.1557/jmr.2016.230.
9. Nandy, J., Sahoo, S., Yedla, N., Sarangi, H., Molecular dynamics simulation of coalescence kinetics and neck growth in laser additive manufacturing of aluminum alloy nanoparticles. *J. Mol. Model.*, 26, 6, 1–14, Jun. 2020, doi: 10.1007/s00894-020-04395-4.
10. Jeon, J., Jiang, S., Rahmani, F., Nouranian, S., Molecular dynamics study of temperature and heating rate–dependent sintering of titanium nanoparticles and its influence on the sequent tension tests of the formed particle-chain products. *J. Nanopart. Res.*, 22, 1, 1–12, Jan. 2020, doi: 10.1007/s11051-019-4747-3.
11. Lai, L.F., Lu, D.M., Lu, J.M., Molecular dynamics simulation of thermomechanical properties of hollow palladium nanoparticle pairs during additive manufacturing. *Sens. Mater.*, 34, 10, 3911–3921, 2022, doi: 10.18494/SAM4030.
12. Liu, H., Jiang, Q., Huo, J., Zhang, Y., Yang, W., Li, X., Crystallization in additive manufacturing of metallic glasses: A review. *Addit. Manuf.*, 36, 101568, Elsevier B.V., Dec. 01, 2020, doi: 10.1016/j.addma.2020.101568.

13. Li, X., Additive manufacturing of advanced multi-component alloys: Bulk metallic glasses and high entropy alloys. *Adv. Eng. Mater.*, 20, 5, 1700874, Wiley-VCH Verlag, May 01, 2018, doi: 10.1002/adem.201700874.
14. Yang, D., Liu, H., Jiang, Q., Jiang, Y., Wang, X., Yang, W., Atomic-level understanding of weakening crystallization in additive manufactured ternary Fe-based metallic glasses with Ni addition. *J. Non. Cryst. Solids*, 582, 121435, Apr. 2022, doi: 10.1016/j.jnoncrysol.2022.121435.
15. Guo, S., Wang, M., Zhao, Z., Zhang, Y.Y., Lin, X., Huang, W.D., Molecular dynamics simulation on the micro-structural evolution in heat-affected zone during the preparation of bulk metallic glasses with selective laser melting. *J. Alloys Compd.*, 697, 443–449, 2017, doi: 10.1016/j.jallcom.2016.11.393.
16. Jiang, Q., Liu, H., Li, J., Yang, D., Zhang, Y., Yang, W., Atomic-level understanding of crystallization in the selective laser melting of Fe50Ni50 amorphous alloy. *Addit. Manuf.*, 34, 101369, Aug. 2020, doi: 10.1016/j.addma.2020.101369.
17. Zhao, Z., Liu, J., Soh, A.K., Tang, C., Temperature-mediated fabrication, stress-induced crystallization and transformation: Atomistic simulations of additively manufactured amorphous Cu pillars. *Model. Simul. Mater. Sci. Eng.*, 27, 7, 075012, Aug. 2019, doi: 10.1088/1361-651X/ab3758.
18. Kurian, S. and Mirzaeifar, R., Selective laser melting of aluminum nanopowder particles, a molecular dynamics study. *Addit. Manuf.*, 35, 101272, Oct. 2020, doi: 10.1016/j.addma.2020.101272.
19. Huang, X., Chen, H., Liu, B., Mohammadzadeh, R., Li, J., Fang, Q., Thermal behavior and microstructural evolution of additively manufactured Ni-based superalloys via multi-scale simulation. *Optik (Stuttg)*, 243, 167456, Oct. 2021, doi: 10.1016/j.ijleo.2021.167456.
20. Zhang, Y. *et al.*, Atomic-scale structural evolution in selective laser melting of Cu50Zr50 metallic glass. *Comput. Mater. Sci.*, 150, 62–69, Jul. 2018, doi: 10.1016/j.commatsci.2018.03.072.
21. Wang, J., Luo, Q., Wang, H., Wu, Y., Cheng, X., Tang, H., Microstructure characteristics and failure mechanisms of Ti-48Al-2Nb-2Cr titanium aluminide intermetallic alloy fabricated by directed energy deposition technique. *Addit. Manuf.*, 32, 101007, Mar. 2020, doi: 10.1016/j.addma.2019.101007.
22. Nikonov, A.Y., Lychagin, D.V., Bibko, A.A., Novitskaya, O.S., Growth and deformation simulation of aluminum bronze grains produced by electron beam additive manufacturing. *Metals (Basel)*, 12, 1, 114, Jan. 2022, doi: 10.3390/met12010114.
23. Rahmani, F., Jeon, J., Jiang, S., Nouranian, S., Melting and solidification behavior of Cu/Al and Ti/Al bimetallic core/shell nanoparticles during additive manufacturing by molecular dynamics simulation. *J. Nanopart. Res.*, 20, 5, 1–11, May 2018, doi: 10.1007/s11051-018-4237-z.
24. Daneshian, B., Gärtner, F., Assadi, H., Vidaller, M.V., Höche, D., Klassen, T., Features of ceramic nanoparticle deformation in aerosol deposition explored by molecular dynamics simulation. *Surf. Coat. Technol.*, 429, 127886, Jan. 2022, doi: 10.1016/j.surfcoat.2021.127886.

25. Wang, F., You, S., Jiang, D., Ning, F., Study on sintering mechanism for extrusion-based additive manufacturing of stainless steel through molecular dynamics simulation. *Addit. Manuf.*, 58, 102991, Oct. 2022, doi: 10.1016/j.addma.2022.102991.
26. Wang, Y., Wei, Q., Pan, F., Yang, M., Wei, S., Molecular dynamics simulations for the examination of mechanical properties of hydroxyapatite/poly α-n-butyl cyanoacrylate under additive manufacturing. *Bio-Med. Mater. Eng.*, 24, 1, 825–833, 2014, doi: 10.3233/BME-130874.
27. Dumontet, N., Connétable, D., Malard, B., Viguier, B., Elastic properties of the α' martensitic phase in the Ti-6Al-4V alloy obtained by additive manufacturing. *Scr. Mater.*, 167, 115–119, Jul. 2019, doi: 10.1016/j.scriptamat.2019.03.042.
28. Etesami, S.A., Laradji, M., Asadi, E., Reliability of molecular dynamics interatomic potentials for modeling of titanium in additive manufacturing processes. *Comput. Mater. Sci.*, 184, 109883, Nov. 2020, doi: 10.1016/j.commatsci.2020.109883.
29. Singh, G., Waas, A.M., Sundararaghavan, V., Understanding defect structures in nanoscale metal additive manufacturing via molecular dynamics. *Comput. Mater. Sci.*, 200, 110807, Dec. 2021, doi: 10.1016/j.commatsci.2021.110807.
30. Jamil, I.R., Mustaquim, A.M., Islam, M., Hasan, M.N., Molecular dynamics perspective of the effects of laser thermal configurations on the dislocation and mechanical characteristics of FeNiCrCoCu HEA through powder bed fusion process. *Mater. Today Commun.*, 33, 104998, Dec. 2022, doi: 10.1016/j.mtcomm.2022.104998.
31. Shugurov, A.R., Nikonov, A.Y., Dmitriev, A.I., The effect of electron-beam treatment on the deformation behavior of the EBAM TI-6AL-4V under scratching. *Facta Univ. Ser. Mech. Eng.*, 20, 2, 307–319, 2022, doi: 10.22190/FUME211110001S.
32. Gao, Y., Shin, Y.K., Martinez, D., Manogharan, G., Van Duin, A.C.T., A ReaxFF molecular dynamics study of molecular-level interactions during binder jetting 3D-printing. *Phys. Chem. Chem. Phys.*, 21, 38, 21517–21529, 2019, doi: 10.1039/c9cp03585k.
33. Gao, Y., Clares, A.P., Manogharan, G., van Duin, A.C.T., A reactive molecular dynamics study of bi-modal particle size distribution in binder-jetting additive manufacturing using stainless-steel powders. *Phys. Chem. Chem. Phys.*, 24, 19, 11603–11615, Apr. 2022, doi: 10.1039/d2cp00630h.
34. Xiangyang, Y., Qi, Y., Junjun, L., Lin, Z., Atomic simulations of melting behaviours for TiAl alloynanoparticles during heating. *Bull. Mater. Sci.*, 43, 241, 2020.
35. Jamil, I.R., Mustaquim, A.M., Islam, M., Thakur, M.S.H., Hasan, M.N., Effect of powder bed fusion process parameters on microstructural and mechanical properties of FeCrNi MEA: An atomistic study. *Met. Mater. Int.*, 29, 659–673, 2022, doi: 10.1007/s12540-022-01263-z.

36. Vo, T.Q. and Kim, B.H., Molecular dynamics study of thermodynamic properties of nanoclusters for additive manufacturing. *Int. J. Precis. Eng. Manuf. - Green Technol.*, 4, 3, 301–306, Jul. 2017, doi: 10.1007/s40684-017-0036-8.
37. Faras Al Balushi, T.D., https://personal.ems.psu.edu/~fkd/courses/EGEE520/2019Deliverables/phase.pdf.
38. Nele Moelans, B.B.P.W., An introduction to phase-field modeling of microstructure evolution. *Calphad*, 32, 2, 268–294, 2008.
39. Rai, A., Markl, M., Körner, C., A coupled cellular automaton–lattice Boltzmann model for grain structure simulation during additive manufacturing. *Comput. Mater. Sci.*, 124, 37–48, Nov. 2016, doi: 10.1016/j.commatsci.2016.07.005.
40. Yang, J., Yu, H., Yang, H., Li, F., Wang, Z., Zeng, X., Prediction of microstructure in selective laser melted Ti–6Al–4V alloy by cellular automaton. *J. Alloys Compd.*, 748, 281–290, Jun. 2018, doi: 10.1016/j.jallcom.2018.03.116.
41. Zinovieva, O., Zinoviev, A., Ploshikhin, V., Three-dimensional modeling of the microstructure evolution during metal additive manufacturing. *Comput. Mater. Sci.*, 141, 207–220, Jan. 2018, doi: 10.1016/j.commatsci.2017.09.018.
42. Li, X. and Tan, W., Numerical investigation of effects of nucleation mechanisms on grain structure in metal additive manufacturing. *Comput. Mater. Sci.*, 153, 159–169, Oct. 2018, doi: 10.1016/j.commatsci.2018.06.019.
43. Liu, S. and Shin, Y.C., Integrated 2D cellular automata-phase field modeling of solidification and microstructure evolution during additive manufacturing of Ti6Al4V. *Comput. Mater. Sci.*, 183, 109889, Oct. 2020, doi: 10.1016/j.commatsci.2020.109889.
44. Koepf, J.A., Gotterbarm, M.R., Markl, M., Körner, C., 3D multi-layer grain structure simulation of powder bed fusion additive manufacturing. *Acta Mater.*, 152, 119–126, Jun. 2018, doi: 10.1016/j.actamat.2018.04.030.
45. Rai, A., Helmer, H., Körner, C., Simulation of grain structure evolution during powder bed based additive manufacturing. *Addit. Manuf.*, 13, 124–134, Jan. 2017, doi: 10.1016/j.addma.2016.10.007.
46. Akram, J., Chalavadi, P., Pal, D., Stucker, B., Understanding grain evolution in additive manufacturing through modeling. *Addit. Manuf.*, 21, 255–268, May 2018, doi: 10.1016/j.addma.2018.03.021.
47. Rodgers, T.M., Madison, J.D., Tikare, V., Simulation of metal additive manufacturing microstructures using kinetic Monte Carlo. *Comput. Mater. Sci.*, 135, 78–89, Jul. 2017, doi: 10.1016/j.commatsci.2017.03.053.
48. Rodgers, T.M. *et al.*, Simulation of powder bed metal additive manufacturing microstructures with coupled finite difference-Monte Carlo method. *Addit. Manuf.*, 41, 101953, May 2021, doi: 10.1016/j.addma.2021.101953.
49. Joseph Pauza, A.R., Simulation study of hatch spacing and layer thickness effects on microstructure in laser powder bed fusion additive manufacturing using a texture-aware solidification potts model. *J. Mater. Eng. Perform.*, 30, 9, 7007–7018, 2021, doi: 10.1007/s11665-021-06110-7.

50. Zhang, Z., Tan, Z.J., Li, J.Y., Zu, Y.F., Sha, J.J., Integrated modeling of process–microstructure–property relations in friction stir additive manufacturing. *Acta Metall. Sin. (Engl. Lett.*, 33, 1, 75–87, Jan. 2020, doi: 10.1007/s40195-019-00945-9.
51. Zhang, Z., Yao, X.X., Ge, P., Phase-field-model-based analysis of the effects of powder particle on porosities and densities in selective laser sintering additive manufacturing. *Int. J. Mech. Sci.*, 166, 105230, Jan. 2020, doi: 10.1016/j.ijmecsci.2019.105230.
52. Zhang, J., Wu, L., Zhang, Y., Meng, L., Phase field simulation of dendritic microstructure in additively manufactured titanium alloy. *Met. Powder Rep.*, 74, 1, 20–24, Jan. 2019, doi: 10.1016/j.mprp.2018.11.001.
53. Nie, P., Ojo, O.A., Li, Z., Numerical modeling of microstructure evolution during laser additive manufacturing of a nickel-based superalloy. *Acta Mater.*, 77, 85–95, Sep. 2014, doi: 10.1016/j.actamat.2014.05.039.
54. Karayagiz, K. *et al.*, Finite interface dissipation phase field modeling of Ni–Nb under additive manufacturing conditions. *Acta Mater.*, 185, 320–339, Feb. 2020, doi: 10.1016/j.actamat.2019.11.057.
55. Yang, X., Barrett, R.A., Tong, M., Harrison, N.M., Leen, S.B., Towards a process-structure model for Ti-6Al-4V during additive manufacturing. *J. Manuf. Process.*, 61, 428–439, Jan. 2021, doi: 10.1016/j.jmapro.2020.11.033.
56. Snider-Simon, B. and Frantziskonis, G., Reliability of metal additive manufactured materials from modeling the microstructure at different length scales. *Addit. Manuf.*, 51, 102629, Mar. 2022, doi: 10.1016/j.addma.2022.102629.
57. Salsi, E., Chiumenti, M., Cervera, M., Modeling of microstructure evolution of Ti6Al4V for additive manufacturing. *Metals (Basel)*, 8, 8, 633, Aug. 2018, doi: 10.3390/met8080633.
58. Lu, L.X., Sridhar, N., Zhang, Y.W., Phase field simulation of powder bed-based additive manufacturing. *Acta Mater.*, 144, 801–809, Feb. 2018, doi: 10.1016/j.actamat.2017.11.033.
59. Wu, S. *et al.*, A microstructure-based creep model for additively manufactured nickel-based superalloys. *Acta Mater.*, 224, 117528, Feb. 2022, doi: 10.1016/j.actamat.2021.117528.
60. Li, J.Q. and Fan, T.H., Phase-field modeling of metallic powder-substrate interaction in laser melting process. *Int. J. Heat Mass Transf.*, 133, 872–884, Apr. 2019, doi: 10.1016/j.ijheatmasstransfer.2018.12.158.
61. Kundin, J., Mushongera, L., Emmerich, H., Phase-field modeling of microstructure formation during rapid solidification in Inconel 718 superalloy. *Acta Mater.*, 95, 343–356, Jun. 2015, doi: 10.1016/j.actamat.2015.05.052.
62. Chadwick, A.F. and Voorhees, P.W., The development of grain structure during additive manufacturing. *Acta Mater.*, 211, 116862, Jun. 2021, doi: 10.1016/j.actamat.2021.116862.

63. Berry, J. *et al.*, Toward multiscale simulations of tailored microstructure formation in metal additive manufacturing. *Mater. Today*, 51, 65–86, Dec. 2021, doi: 10.1016/j.mattod.2021.09.024.
64. Sahoo, S. and Chou, K., Phase-field simulation of microstructure evolution of Ti-6Al-4V in electron beam additive manufacturing process. *Addit. Manuf.*, 9, 14–24, Jan. 2016, doi: 10.1016/j.addma.2015.12.005.
65. Lindgren, L.E., Lundbäck, A., Fisk, M., Pederson, R., Andersson, J., Simulation of additive manufacturing using coupled constitutive and microstructure models. *Addit. Manuf.*, 12, 144–158, Oct. 2016, doi: 10.1016/j.addma.2016.05.005.
66. Li, J., Zhou, X., Brochu, M., Provatas, N., Zhao, Y.F., Solidification microstructure simulation of Ti-6Al-4V in metal additive manufacturing: A review. *Addit. Manuf.*, 31. Elsevier B.V, 100989, Jan. 01, 2020, doi: 10.1016/j.addma.2019.100989.
67. Yang, M., Wang, L., Yan, W., Phase-field modeling of grain evolution in additive manufacturing with addition of reinforcing particles. *Addit. Manuf.*, 47, 102286, Nov. 2021, doi: 10.1016/j.addma.2021.102286.
68. Laskowski, R. *et al.*, Concurrent modeling of porosity and microstructure in multilayer three-dimensional simulations of powder-bed fusion additive manufacturing of INCONEL 718. *Addit. Manuf.*, 60, 103266, Dec. 2022, doi: 10.1016/j.addma.2022.103266.
69. Ji, Y., Chen, L., Chen, L.Q., Understanding microstructure evolution during additive manufacturing of metallic alloys using phase-field modeling, in: *Thermo-Mechanical Modeling of Additive Manufacturing*, pp. 93–116, Elsevier, Butterworth-Heinemann, 2017, doi: 10.1016/B978-0-12-811820-7.00008-2.
70. Nomoto, S., Kusano, M., Kitano, H., Watanabe, M., Multi-phase field method for solidification microstructure evolution for a Nni-based alloy in wire arc additive manufacturing. *Metals (Basel)*, 12, 10, 1720, Oct. 2022, doi: 10.3390/met12101720.
71. Tan, J.H.K., Sing, S.L., Yeong, W.Y., Microstructure modelling for metallic additive manufacturing: A review. *Virtual Phys. Prototyp.*, 15, 1. Taylor and Francis Ltd., 87–105, Jan. 02, 2020, doi: 10.1080/17452759.2019.1677345.
72. Tabei, A., Mirkoohi, E., Garmestani, H., Liang, S., Modeling of texture development in additive manufacturing of Ni-based superalloys. *Int. J. Adv. Manuf. Technol.*, 103, 1–4, 1057–1066, Jul. 2019, doi: 10.1007/s00170-019-03555-y.
73. Geng, R.W., Du, J., Wei, Z.Y., Zhao, G.X., Simulation of microstructure evolution in fused-coating additive manufacturing based on phase field approach. *China Foundry*, 14, 5, 346–352, Sep. 2017, doi: 10.1007/s41230-017-7124-9.
74. Quey, R. and Kasemer, M., The Neper/FEPX project: Free/open-source polycrystal generation, deformation simulation, and post-processing. *IOP Conf. Ser. Mater. Sci. Eng.*, 1249, 1, 012021, Jul. 2022, doi: 10.1088/1757-899x/1249/1/012021.

75. Sadik Sefa Acar, T.Y. and Bulut, O., Crystal plasticity modeling of additively manufactured metallic microstructures. *Proc. Struct. Integr.*, 35, 219–227, 2022, doi: 10.1016/j.prostr.2021.12.068.
76. Nicolò Grilli, W.Y., Hu, D., Yushu, D., Chen, F., Crystal plasticity model of residual stress in additive manufacturing using the element elimination and reactivation method. *Comput. Mech.*, 69, 825–845, 2022, doi: 10.1007/s00466-021-02116-z.
77. Ameen, W., Al-Ahmari, A., Mohammed, M.K., Abdulhameed, O., Umer, U., Moiduddin, K., Design, finite element analysis (FEA), and fabrication of custom titanium alloy cranial implant using electron beam melting additive manufacturing. *Adv. Prod. Eng. Manage.*, 13, 3, 267–278, Sep. 2018, doi: 10.14743/apem2018.3.289.
78. Pal, D. and Stucker, B., A study of subgrain formation in Al 3003 H-18 foils undergoing ultrasonic additive manufacturing using a dislocation density based crystal plasticity finite element framework. *J. Appl. Phys.*, 113, 2013, doi: 10.1063/1.4807831.
79. Kim, D.-K., Hwang, J.-H., Kim, E.-Y., Heo, Yoon-Uk, S.-H.C., Woo, W., Evaluation of the stress-strain relationship of constituent phases in AlSi10Mg alloy produced by selective laser melting using crystal plasticity FEM. *J. Alloys Compd.*, 714, 687–697, 2017, doi: 10.1016/j.jallcom.2017.04.264.
80. Zhang, J., Li, J., Wu, S., Zhang, W., Sun, J., Qian, G., High-cycle and very-high-cycle fatigue lifetime prediction of additively manufactured AlSi10Mg via crystal plasticity finite element method. *Int. J. Fatigue*, 155, 106577, 2022, doi: 10.1016/j.ijfatigue.2021.106577.
81. Geng, Y. and Harrison, N., Functionally graded bimodal Ti6Al4V fabricated by powder bed fusion additive manufacturing: Crystal plasticity finite element modelling. *Mater. Sci. Eng. A*, 773, 138736, 2020, doi: 10.1016/j.msea.2019.138736.
82. Liu, P.W. *et al.*, Integration of phase-field model and crystal plasticity for the prediction of process-structure-property relation of additively manufactured metallic materials. *Int. J. Plast.*, 128, 102670, 2020, doi: 10.1016/j.ijplas.2020.102670.
83. Hu, D., Grilli, N., Yan, W., Dislocation structures formation induced by thermal stress in additive manufacturing: Multiscale crystal plasticity modeling of dislocation transport. *J. Mech. Phys. Solids*, 173, 105235, Apr. 2023, doi: 10.1016/j.jmps.2023.105235.
84. Kik, T., Heat source models in numerical simulations of laser welding. *Mater. (Basel)*, 13, 11, 2653, Jun. 2020, doi: 10.3390/ma13112653.
85. Kiran, A., Li, Y., Hodek, J., Brázda, M., Urbánek, M., Džugan, J., Heat source modeling and residual stress analysis for metal directed energy deposition additive manufacturing. *Mater. (Basel)*, 15, 7, 2545, Apr. 2022, doi: 10.3390/ma15072545.

86. Liu, H. and Liou, F., Residual stress modeling and deformation measurement in laser metal deposition process, in: *New Challenges in Residual Stress Measurements and Evaluation*, IntechOpen, Rijeka, 2020, doi: 10.5772/intechopen.90539.
87. Zhao, H., Liu, Z., Yu, C., Liu, C., Zhan, Y., Finite element analysis for residual stress of TC4/Inconel718 functionally gradient materials produced by laser additive manufacturing. *Opt. Laser Technol.*, 152, 108146, Aug. 2022, doi: 10.1016/j.optlastec.2022.108146.
88. Wang, Z., Denlinger, E., Michaleris, P., Stoica, A.D., Ma, D., Beese, A.M., Residual stress mapping in Inconel 625 fabricated through additive manufacturing: Method for neutron diffraction measurements to validate thermomechanical model predictions. *Mater. Des.*, 113, 169–177, Jan. 2017, doi: 10.1016/j.matdes.2016.10.003.
89. Promoppatum, P. and Rollett, A.D., Physics-based and phenomenological plasticity models for thermomechanical simulation in laser powder bed fusion additive manufacturing: A comprehensive numerical comparison. *Mater. Des.*, 204, 109658, Jun. 2021, doi: 10.1016/j.matdes.2021.109658.
90. Vastola, G., Zhang, G., Pei, Q.X., Zhang, Y.W., Controlling of residual stress in additive manufacturing of Ti6Al4V by finite element modeling. *Addit. Manuf.*, 12, 231–239, Oct. 2016, doi: 10.1016/j.addma.2016.05.010.
91. Chen, Q., Guillemot, G., Gandin, C.A., Bellet, M., Three-dimensional finite element thermomechanical modeling of additive manufacturing by selective laser melting for ceramic materials. *Addit. Manuf.*, 16, 124–137, Aug. 2017, doi: 10.1016/j.addma.2017.02.005.
92. Montevecchi, F., Venturini, G., Grossi, N., Scippa, A., Campatelli, G., Idle time selection for wire-arc additive manufacturing: A finite element-based technique. *Addit. Manuf.*, 21, 479–486, May 2018, doi: 10.1016/j.addma.2018.01.007.
93. Ding, J., Colegrove, P., Mehnen, J., Williams, S., Wang, F., Almeida, P.S., A computationally efficient finite element model of wire and arc additive manufacture. *Int. J. Adv. Manuf. Technol.*, 70, 1–4, 227–236, Jan. 2014, doi: 10.1007/s00170-013-5261-x.
94. Yang, Y. and Zhou, X., A volumetric heat source model for thermal modeling of additive manufacturing of metals. *Metals (Basel)*, 1–17, 2020, doi: 10.3390/met10111406.
95. Moran, T.P., Li, P., Warner, D.H., Phan, N., Utility of superposition-based finite element approach for part-scale thermal simulation in additive manufacturing. *Addit. Manuf.*, 21, 215–219, May 2018, doi: 10.1016/j.addma.2018.02.015.
96. Lu, X. *et al.*, Finite element analysis and experimental validation of the thermomechanical behavior in laser solid forming of Ti-6Al-4V. *Addit. Manuf.*, 21, 30–40, May 2018, doi: 10.1016/j.addma.2018.02.003.
97. Zhang, W., Tong, M., Harrison, N.M., Resolution, energy and time dependency on layer scaling in finite element modelling of laser beam powder bed

fusion additive manufacturing. *Addit. Manuf.*, 28, 610–620, Aug. 2019, doi: 10.1016/j.addma.2019.05.002.

98. Ertay, D.S., Vlasea, M., Erkorkmaz, K., Thermomechanical and geometry model for directed energy deposition with 2D/3D toolpaths. *Addit. Manuf.*, 35, 101294, Oct. 2020, doi: 10.1016/j.addma.2020.101294.
99. Denlinger, E.R., Irwin, J., Michaleris, P., Thermomechanical modeling of additive manufacturing large parts. *J. Manuf. Sci. Eng. Trans. ASME*, 136, 6, 061007, Dec. 2014, doi: 10.1115/1.4028669.
100. Springer, S. *et al.*, Implementation of a viscoplastic substrate creep model in the thermomechanical simulation of the WAAM process. *Weld. World*, 66, 3, 441–453, Mar. 2022, doi: 10.1007/s40194-021-01232-x.
101. Strantza, M. *et al.*, Coupled experimental and computational study of residual stresses in additively manufactured Ti-6Al-4V components. *Mater. Lett.*, 231, 221–224, Nov. 2018, doi: 10.1016/j.matlet.2018.07.141.
102. Sawant, M.S., Jain, N.K., Nikam, S.H., Theoretical modeling and finite element simulation of dilution in micro-plasma transferred arc additive manufacturing of metallic materials. *Int. J. Mech. Sci.*, 164, 105166, Dec. 2019, doi: 10.1016/j.ijmecsci.2019.105166.
103. Mirkoohi, E., Sievers, D.E., Garmestani, H., Liang, S.Y., Thermo-mechanical modeling of thermal stress in metal additive manufacturing considering elastoplastic hardening. *CIRP J. Manuf. Sci. Technol.*, 28, 52–67, Jan. 2020, doi: 10.1016/j.cirpj.2020.01.002.
104. Cao, Y. *et al.*, A novel high-efficient finite element analysis method of powder bed fusion additive manufacturing. *Addit. Manuf.*, 46, 102187, Oct. 2021, doi: 10.1016/j.addma.2021.102187.
105. Luo, C., Qiu, J., Yan, Y., Yang, J., Uher, C., Tang, X., Finite element analysis of temperature and stress fields during the selective laser melting process of thermoelectric SnTe. *J. Mater. Process. Technol.*, 261, 74–85, Nov. 2018, doi: 10.1016/j.jmatprotec.2018.06.001.
106. Yang, Q., Zhang, P., Cheng, L., Min, Z., Chyu, M., To, A.C., Finite element modeling and validation of thermomechanical behavior of Ti-6Al-4V in directed energy deposition additive manufacturing. *Addit. Manuf.*, 12, 169–177, Oct. 2016, doi: 10.1016/j.addma.2016.06.012.
107. Montevecchi, F., Venturini, G., Grossi, N., Scippa, A., Campatelli, G., Heat accumulation prevention in wire-arc-additive-manufacturing using air jet impingement. *Manuf. Lett.*, 17, 14–18, Aug. 2018, doi: 10.1016/j.mfglet.2018.06.004.
108. Stender, M.E. *et al.*, A thermal-mechanical finite element workflow for directed energy deposition additive manufacturing process modeling. *Addit. Manuf.*, 21, 556–566, May 2018, doi: 10.1016/j.addma.2018.04.012.
109. Stathatos, E. and Vosniakos, G.C., A computationally efficient universal platform for thermal numerical modeling of laser-based additive manufacturing. *Proc. Inst. Mech. Eng. Part C J. Mech. Eng. Sci.*, 232, 13, 2317–2333, Jul. 2018, doi: 10.1177/0954406217720230.

110. Teng, C. *et al.*, A review of defect modeling in laser material processing. *Addit. Manuf.*, 14, 137–147, 2017, doi: 10.1016/j.addma.2016.10.009.
111. Zhu, Z., Anwer, N., Huang, Q., Mathieu, L., Machine learning in tolerancing for additive manufacturing. *CIRP Ann.*, 67, 1, 157–160, Jan. 2018, doi: 10.1016/j.cirp.2018.04.119.
112. Yaseer, A. and Chen, H., Machine learning based layer roughness modeling in robotic additive manufacturing. *J. Manuf. Process.*, 70, 543–552, Oct. 2021, doi: 10.1016/j.jmapro.2021.08.056.
113. Li, R., Jin, M., Paquit, V.C., Geometrical defect detection for additive manufacturing with machine learning models. *Mater. Des.*, 206, 109726, Aug. 2021, doi: 10.1016/j.matdes.2021.109726.
114. Zhu, X., Jiang, F., Guo, C., Wang, Z., Dong, T., Li, H., Prediction of melt pool shape in additive manufacturing based on machine learning methods. *Opt. Laser Technol.*, 159, 108964, Apr. 2023, doi: 10.1016/j.optlastec.2022.108964.
115. Gan, Z. *et al.*, Data-driven microstructure and microhardness design in additive manufacturing using a self-organizing map. *Engineering*, 5, 4, 730–735, Aug. 2019, doi: 10.1016/j.eng.2019.03.014.
116. Akbari, P. *et al.*, MeltpoolNet: Melt pool characteristic prediction in metal additive manufacturing using machine learning. *Addit. Manuf.*, 55, 102817, Jul. 2022, doi: 10.1016/j.addma.2022.102817.
117. Jiang, J., Yu, C., Xu, X., Ma, Y., Liu, J., Achieving better connections between deposited lines in additive manufacturing via machine learning. *Math. Biosci. Eng.*, 17, 4, 3382–3394, Apr. 2020, doi: 10.3934/MBE.2020191.
118. Gor, M. *et al.*, Density prediction in powder bed fusion additive manufacturing: Machine learning-based techniques. *Appl. Sci.*, 12, 14, 7271, Jul. 2022, doi: 10.3390/app12147271.
119. Roostaei, A.A. and Jahed, H., Fundamentals of cyclic plasticity models, in: *Cyclic Plasticity of Metals*, 23–51, Elsevier, 2022, doi: 10.1016/B978-0-12-819293-1.00011-5.
120. Wood, G. Calibrated expressions for welding and their application to isotherm width in a thick plate. Soldagem & Inspeção, 19, 212–220, 2014, doi: 10.1590/0104-9224/SI1903.03

4

Characterization Methodologies for Additive Manufacturing: From Feedstock to the Final Component

Koduru Venkatesh, A. Muthuchamy and V. Karthik*

Department of Metallurgical and Materials Engineering, National Institute of Technology, Tiruchirappalli, Tamil Nadu, India

Abstract
Additive manufacturing (AM) processes provide new possibilities in the fabrication of complex structures with reduced material costs and defects, and they may become the primary manufacturing route in the near future. It has been reported that the quality of feedstock used in AM influences the quality of the manufactured parts. Therefore, knowing the quality of feedstock and fabricated part helps us understand the optimum conditions required for AM techniques. This chapter's primary goal is to group together all the characterization methods used in AM processes, from the feedstock materials to finished components. The methods used to characterize the feedstock materials and additively manufactured parts are explained in detail in different sections in this chapter.

Keywords: Additive manufacturing, feedstock materials, additively manufactured parts, materials characterization

List of Abbreviations

AM	Additive manufacturing
FDM	Fused deposition modelling
PBF	Powder bed fusion
EBM	Electron beam melting

**Corresponding author*: karthikv@nitt.edu

R. Rajasekar, C. Moganapriya and P. Sathish Kumar (eds.) Additive Manufacturing with Novel Materials: Processes, Properties and Applications, (119–162) © 2024 Scrivener Publishing LLC

SLM	Selective laser melting
DED	Direct energy deposition
LENS	Laser-engineered net shaping
DLP	Digital light processing
SL	Stereolithography
AD	Apparent Density
ASTM	American Society for Testing and Materials
MPIF	Metal Powder Industries Federation
TD	Tap Density
ISO	International Organization for Standardization
SEM	Scanning electron microscope
XRD	X-ray diffraction
EDS	Energy-Dispersive X-Ray Spectroscopy
XPS	X-Ray Photoelectron Spectroscopy
FT-IR	Fourier transforms infrared spectroscopy
IGC	Inverse gas chromatography
DSC	Differential scanning calorimetry
EBAM	electron beam additive manufacturing
LBDED	Laser-based direct energy deposition
BHN	Brinell hardness number
VHN	Vickers hardness number
KN	Knoop hardness number
IPF	Inverse pole figure
TEM	Transmission electron microscope
XRF	X-Ray Fluorescence Spectroscopy

4.1 Introduction

In the near future, additive manufacturing (AM) processes may become the primary manufacturing route for the fabrication of engineering components owing to its capability for the fabrication of complex shapes with reduced defects and material costs. Additive manufacturing can be used to fabricate engineering components that are made of either metals or ceramics or polymers or composite materials. Different types of AM processes have been developed for different types of feedstock materials. The AM process can be classified according to the state of the feedstock material as processes which uses the solid feedstock materials, and processes which utilizes the liquid feedstock materials. The solid feedstock materials are used in different forms such as filament, sheet, and powder during the additive manufacturing process. The few examples are;

- Filament form
  - Fused deposition modeling (FDM)
- Sheet form
  - Laminated object manufacturing
- Powder form
  - Binder jetting
  - Powder bed fusion (PBF)
    - Electron beam melting (EBM)
    - Selective laser melting (SLM)
  - Direct energy deposition (DED)
    - Laser-engineered net shaping (LENS)

The AM of polymeric materials mostly involves liquid feed stock materials. Some of the AM processes which utilizes liquid feed stock materials are:

- Vat polymerization processes
  - Digital light processing (DLP)
  - Stereolithography (SL)
- Material jetting processes
  - Inkjet printing

Additive manufacturing processes are going to revolutionize the manufacturing sector. It is known fact that the quality of the fabricated part is significantly affected by the quality of the feedstock material. Thus, it is necessary to have a thorough understanding of the properties of the feedstock, or raw materials, to make parts with consistent and predictable qualities. Additionally, every AM component or part must have characteristics that are specified for the finished product. As a result, material characterization is crucial in AM. The common properties and characterization techniques used in the AM sector are explained in detail in the ensuing sections.

4.2 Characterization of Solid Feedstock Materials

The important properties of powder form feedstock materials and their measurement techniques are explained in detail in this section. The common characteristics which must be studied before AM of solid-form feedstock are density, flowability, particle size, particle morphology, particle size distribution, compressibility, crystal structure, elemental composition, etc., and their measurement techniques are discussed in this section. For sheet and filament/wire form feedstock materials, dimensional accuracy is the important property along with composition and crystal structure. Let us start with the important properties of powder feedstock materials.

4.2.1 Density

There are three types of densities of powders that must be controlled to effectively minimize the defects in the additively manufactured parts. They are apparent, tap, and skeletal density.

4.2.1.1 Apparent Density (AD)

As per the American Society for Testing and Materials (ASTM) B243 standard [1], apparent density is defined as "the mass of a unit volume of powder, usually expressed as grams per cubic centimeter, determined by a specified method". The AD can be affected by many variables like particle size distribution, particle mass, particle shape, particle resilience, antiparticle friction, mode of filling, container dimensions, vibration during packing, etc. The common methods used for measuring the apparent density are Hall flowmeter funnel method, Carney funnel method, Scott volumeter method, and Arnold meter method. The measurement principles of each method are as follows.

4.2.1.1.1 Hall Flowmeter Funnel Method

The AD of metal powders that flow freely through a specific funnel can be measured using this technique. In this method, first measure the weight of the density cup before placing it concentrically below the Hall flowmeter funnel. Then load the funnel with powder and let it flow through the orifice to fill the density cup. Rotate the flowmeter funnel 90° in the horizontal plane after filling the density cup to allow any remaining powder to fall out from it. The density cup's top can be levelled with powder by using a non-magnetic spatula. To settle the powder and avoid spillage while transferring it, lightly tap the density cup. Weigh the density cup filled with powder to determine the mass of the powder. The apparatus used to measure the apparent density of metal powders by using hall flowmeter funnel method is shown in Figure 4.1. The formula to find the apparent density of powder using this method is given by,

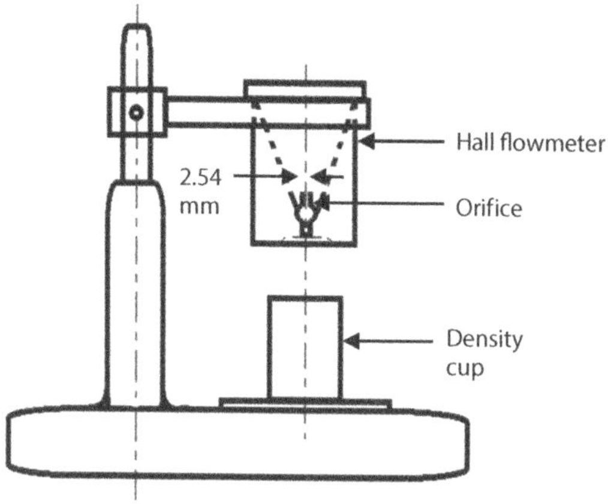

Figure 4.1 Hall flowmeter funnel method apparatus.

$$AD_H = \frac{M}{V} \tag{4.1}$$

Where, M is the mass of powder in density cup in grams (g), V is the volume of the density cup in cubic centimeters (cm^3), and AD_H is the AD of Hall flowmeter funnel method in g/cm^3.

The detailed methodology to measure the AD by using the Hall flowmeter funnel method for free-flowing metal powders can be found in ASTM B212 [2]. Zhai *et al.* [3] used this method to calculate the AD of four stainless steel powders namely SS 316L, ball-milled SS 316L, ball-milled SS 316L-0.3% Y_2O_3, and ball-milled SS 316L-1% Y_2O_3, and calculated the apparent densities as 4.386 g/cm^3, 4.168 g/cm^3, 4.202 g/cm^3,

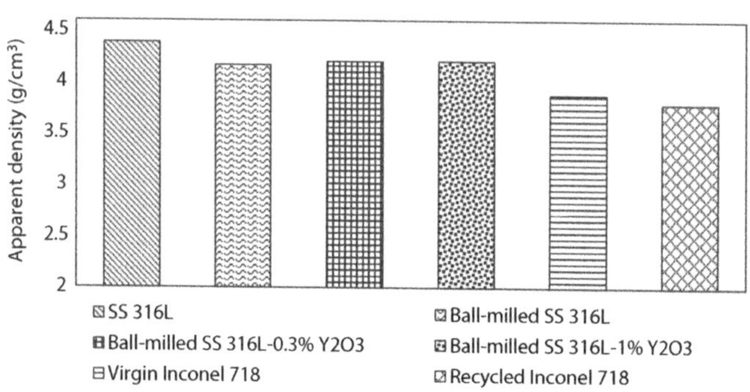

Figure 4.2 Apparent densities of different powders measured by Hall flowmeter funnel method.

and 4.206 g/cm^3, respectively. Nguyen *et al.* [4] also used this method to determine the AD of virgin Inconel 718 as 3.8780 g/cm3 and recycled Inconel 718 as 3.7875 g/cm^3. The ADs of stainless steel and nickel-based superalloy powders discussed here are shown in Figure 4.2.

4.2.1.1.2 Carney Funnel Method

The AD of non-free-flowing metal powders that do not flow freely through the Hall flowmeter funnel is measured using the Carney funnel method. The test procedure of this method is same as the Hall flowmeter funnel method, but the only difference is that the diameter of the Carney funnel's opening is twice the diameter of the Hall flowmeter's opening which allows the non-free-flowing metal powders to flow easily through Carney funnel and the schematic sketch of the Carney funnel method is shown in the Figure 4.3. Along with that, this method also uses the length of wire to push or agitate the powder while filling the density cup but be careful not to let the wire enter the density cup. The formula to calculate the AD of non-free-flowing metal powders using this method is given by.

$$AD_C = \frac{M}{V} \tag{4.2}$$

where M = mass of powder in density cup in grams (g),
V = volume of the density cup in cubic centimeters (cm^3), and
AD_C = Apparent density of Carney funnel method in g/cm^3.

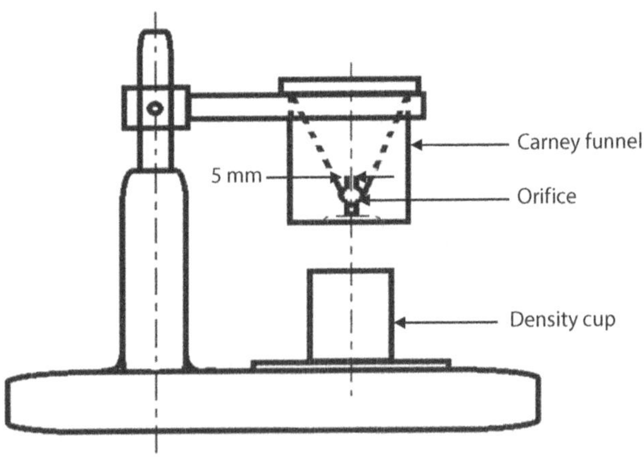

Figure 4.3 Carney funnel method apparatus.

Further detailed description to measure the AD by using the Carney funnel method for non-free-flowing metal powders is explained in ASTM B417 [5]. Mitterlehner *et al.* [6] used this method to determine the AD of Inconel 718 powders and calculated it as 3.77 g/cm^3.

4.2.1.1.3 Scott Volumeter Method

The AD of metal powders and compounds can be determined by using this method and it is also called "paint pigment volumeter method". In this method, measure the weight of the density cup first before placing it concentrically below the Scott volumeter. Load the powder into the top funnel and allow it to completely fill the density cup to overflow by forming a mound on top of the density cup. Brush the sample, through the top receiving funnel screen gently without startling the volumeter, if the powder is not free flowing. Excess powder can be removed by smoothly moving the spatula blade parallel to, and on top of the density cup. To avoid packing or pulling out the powder, the spatula must always be positioned level. Following the leveling process, lightly tap the density cup to settle the powder and avoid spilling while transferring the density cup. Weigh the density cup filled with powder to determine the mass. The schematic representation of Scott volumeter apparatus used to measure the apparent density of metal powders and compounds is shown in Figure 4.4. The formula to find the AD of metal powder and compounds using this method is given by,

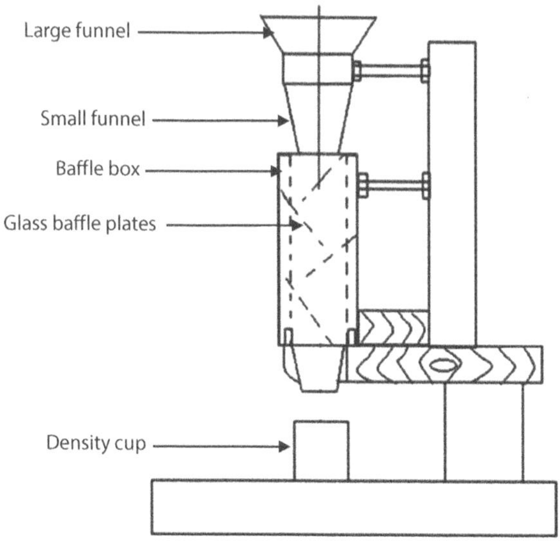

Figure 4.4 Scott volumeter method apparatus.

$$AD_S = \frac{M}{V} \tag{4.3}$$

where M = mass of powder in density cup in grams (g),
V = volume of the density cup in cubic centimeters (cm^3), and
AD_S = Apparent density of Scott volumeter method in g/cm^3.

Further detailed description to measure the AD by using the Scott volumeter method for metal powders and compounds is explained in ASTM B329 [7] and in DIN ISO 3923, also, powder AD measurement by using the Scott volumeter is described. Baesso *et al.* [8] measured the AD of quartz sand powder of different particle sizes by using the Scott volumeter method and calculated that the AD increases as particle size increases. This can be attributed to better flowability.

4.2.1.1.4 Arnold Meter Method

The AD of non-free-flowing metal powders, free-flowing metal powders, lubricated powder blends, and related powder compounds are all measured using the Arnold meter method. In this method, weigh the weighing paper and place it on a flat surface first. To get rid of any loose powder particles, thoroughly wipe the steel die block and the powder supply cylinder using a dry cotton cloth. Lay the steel die block in the middle of the weighing paper. Place the empty delivery cylinder upright on either side of the die cavity, on the steel die block. Fill the delivery cylinder with the specified amount of powder. Slowly and gently slide the delivery cylinder across the cavity while simultaneously applying downward pressure on it. Rotate the delivery cylinder while sliding to produce cascading action. Continue these

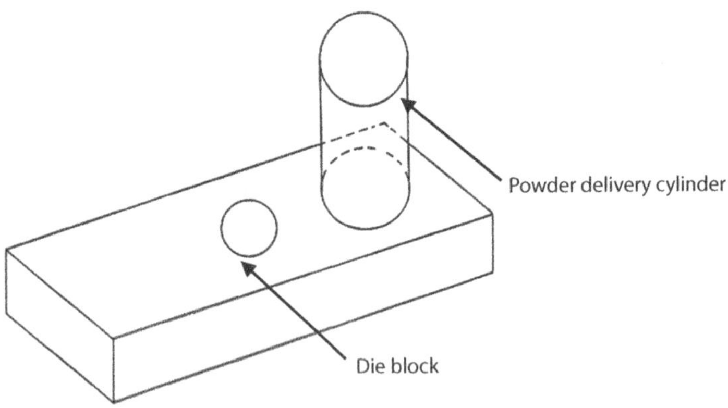

Figure 4.5 Arnold meter method apparatus.

movements until the cylinder passes over the cavity completely. Then, slide the cylinder straight back to the starting location by continuously maintaining downward pressure. Lift the die block slowly along with the partially emptied cylinder from the weighing paper. Weigh the pre-weighed paper with powder and measure the collected powder mass. The schematic representation of Arnold meter apparatus used to measure the apparent density of metal powders and compounds is shown in Figure 4.5. The formula to find the AD of powder using the Arnold method is given by,

$$AD_A = \frac{M}{V} \tag{4.4}$$

where M = mass of powder in grams (g),
V = volume of the die cavity in cubic centimeters (cm^3), and
AD_A = Apparent density of Arnold method in g/cm^3.

Further detailed description to measure the AD by using the Arnold meter method for metal powders and compounds is explained in ASTM B703 [9] and in MPIF standard 48. Also, powder AD measurement by using the Arnold apparatus is described. Altug-Peduk *et al.* [10] used this method to measure the AD of Ti-55 wt.% Ni and calculated as 3.506 g/cm^3. The AD of metallic powders can also be determined by using ISO standards namely ISO 3923-1 and ISO 3923-2.

4.2.1.2 *Tap Density (TD)*

As per the ASTM B243 standard, the tap density of the powder is defined as "the apparent density of the powder in a container that has been tapped under specified conditions". Powder packing can be controlled by many variables such as particle characteristics (size, shape, mass, resilience, interparticle friction, etc.), container dimensions, container wall friction, intensity of tapping, particle deposition velocity, and vibration after deposition. The tap density can be measured by using a standard test method B527. In this method, clean the graduated cylinder with a clean suitable brush or rinse it with a solvent like acetone first, if necessary. Pour the specified amount of test portion into the graduated cylinder. Start tapping the cylinder by using the tapping apparatus until there is no further decrease in the powder's volume. Read the volume directly if the surface is level. If the surface is not level, calculate the mean value between the highest and lowest readings of the tapped surface to estimate the tapped volume. The apparatus used to measure the Tap density of metal powders and compounds is shown in Figure 4.6. The formula to calculate the tap density is given by,

$$TD = \frac{M}{V} \tag{4.5}$$

Where M = mass of powder in grams (g), V = volume of the tapped powder in cubic centimeters (cm^3), and TD = tap density in g/cm^3.

Further detailed description to measure the tap density of metal powders and compounds is explained in ASTM B527 [11] and Metal Powder Industries Federation (MPIF). Young *et al.* [12] measured the tap densities of Cr-31Ni-3C and Cr-26Ni-3C by using this standard and calculated the tap densities as 4.4 g/cm^3 and 4.5 g/cm^3, respectively. Du *et al.* [13] measured the tap density of two different SiC powders (S1 & S2) by using this standard and measured the tap densities of S1 and S2 as 2.07 g/cm^3 and 2.19 g/cm^3, respectively. The tap densities of different powders used in the above-mentioned papers are given in Figure 4.7.

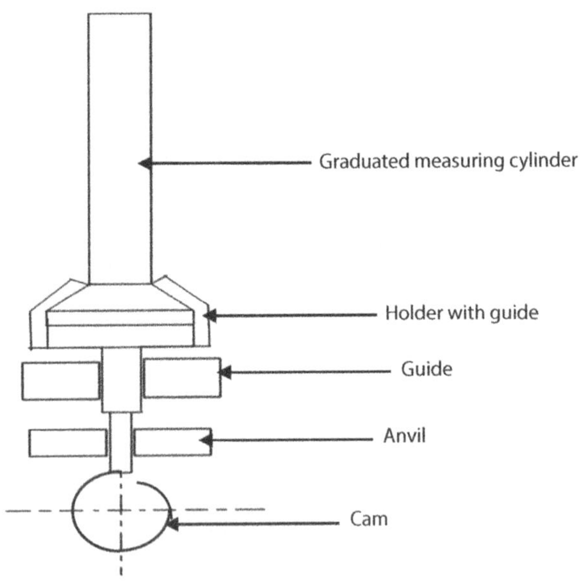

Figure 4.6 Tap density method apparatus.

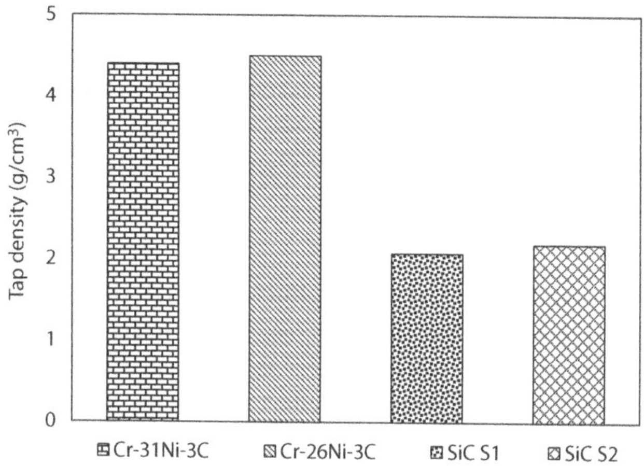

Figure 4.7 Tap densities of various powders.

4.2.1.3 Skeletal Density

As per ASTM B923 standard [14], skeletal density is defined as "the ratio of the mass of discrete pieces of solid material to the sum of the volumes of the solid material in the pieces and closed pores within the pieces" and it can be measured by using the Helium or Nitrogen pycnometer method. In this method, weigh the clean empty sample holder nearest to 0.0001 g first. Remove any adsorbed species in the open pores of powder by outgassing, so that the probe gas molecules can enter open pores easily. Place the filled sample holder in the pycnometer and close the sample chamber. Nitrogen, Helium, or any other high-purity gas for flushing and analysis. Sample cleaning may take place inside the pycnometer, also, so for best results reweigh the sample after analysis. This method uses the ideal gas law to calculate the skeletal volume of powder. The formula to calculate sample volume, if the used instrument has two pressure chambers, is given by,

$$V_s = V_c - V_e \cdot \left(\frac{P_2}{P_1 - P_2} \right) \tag{4.6}$$

where Vs = sample volume,

V_c = calibrated sample chamber volume,

V_e = calibrated expansion chamber volume,

P_1 = gas pressure when only the sample chamber is filled with gas, and

P_2 = gas pressure after expansion into the expansion chamber.

After finishing the analysis, weigh the sample holder with the sample nearest 0.0001 g. To get the sample mass, subtract the empty sample holder's mass from the mass of the sample holder with the sample. Further detailed description to determine the skeletal density of metal powders is explained in ASTM B923. Slotwinski *et al.* [15] measured the skeletal density of stainless steel (17-4 SS) and cobalt chrome (CoCr) alloy powders by using a Helium pycnometer and calculated skeletal densities as 7.878 g/cm^3 and 8.305 g/cm^3, respectively.

4.2.2 Flowability

Flowability is characterized by flowrate and as per the ASTM B243 standard, the flowrate is defined as "the time required for a powder sample of standard weight to flow through an orifice in a standard instrument according to a specified procedure". Flowrate can be affected by many factors like interparticle friction, friction between funnel walls and the particles, the density of the particles, and the relationship between funnel orifice size and orifice size. Methods available to find the flowrate of powder are the Hall flowmeter funnel, Carney funnel, and Arnold meter and Hall flowmeter funnel.

4.2.2.1 *Hall Flowmeter Funnel Method*

This method is used to determine the flowrate of free-flowing metal powders which flow through a specified funnel without any aid. There are two types of methods to find the flowrate by using this method. One is the static flow method. In this method, carefully pour the specified amount of sample powder into the funnel while simultaneously blocking the orifice with a dry finger. Start recording time the moment you remove your finger from the orifice and stop recording the moment the last of the powder exits the orifice. Record the total elapsed time which is used to find the flowrate. Another one is the dynamic flow method. The procedure is same, except the dry finger is not placed to block the orifice, instead, after pouring the specified amount of metal powder, start recording the time the moment the powder exits the orifice and stop recording the time the moment last of the powder exits the orifice. The apparatus used to measure the flowrate in Hall flowmeter funnel method is shown in Figure 4.8 and it is similar to the apparatus in Figure 4.1 but the only difference is in powder collecting beakers. The recorded time is used to calculate the Hall flowrate by using the given formula,

$$FH_H = \frac{t}{m} \tag{4.7}$$

where t = total elapsed time in seconds (s),
M = specified mass in grams (g), and
FH_H = Hall flowrate in s/g.

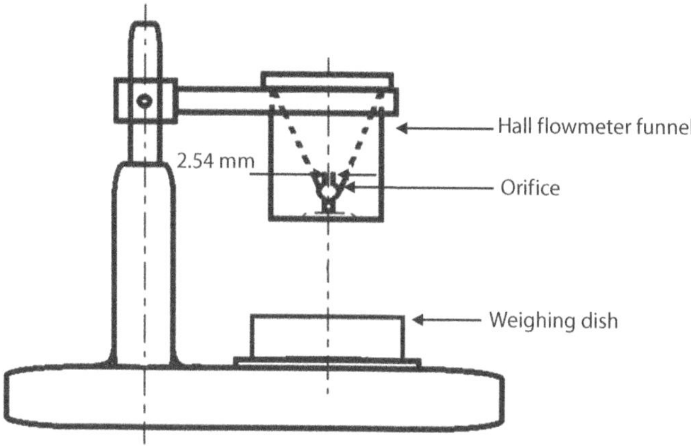

Figure 4.8 Hall flowmeter funnel method apparatus.

Further detailed description to measure the flowrate by using a Hall flowmeter funnel for metal powders is explained in ASTM B527 [16] and the flowrate of free-flowing metal powders can also be determined by using MPIF standard. Snow *et al.* [17] measured the flowtime of three grades of Al-10Si-0.5Mg using the Hall flowmeter funnel and reported that the flowtime of one grade is 34.9 s and the other two grades of powder did not flow through the funnel. Cordova *et al.* [18] measured the flowrate of Ti6Al4V powder using this method and reported the flowtime as 37 s. Lim *et al.* [19] used this method to measure the flowrate of 15-45 μm-sized IN

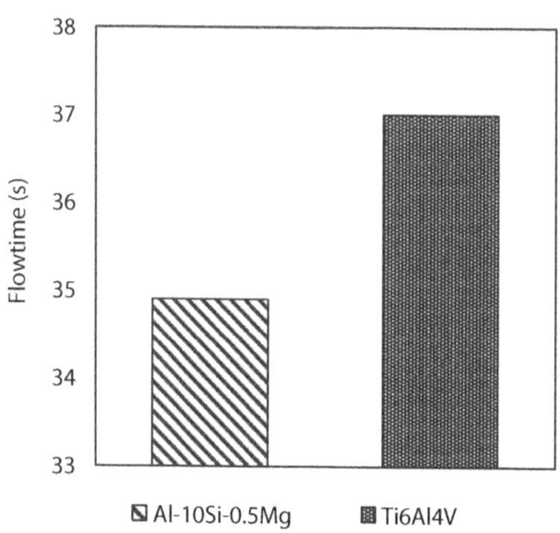

Figure 4.9 Flowrates of various powders using the Hall flowmeter funnel method.

625 powder and reported the flowrate as 2.847 g/s. Flowtimes of different powders used in the above-mentioned papers are given in Figure 4.9.

4.2.2.2 *Carney Funnel Method*

The two methods which are described in the Carney funnel method are similar to the two methods described in the Hall flowmeter funnel method. The only differences in this method are different metallic powders have different specific volumes and non-free-flowing metallic powders are used in this method. The apparatus used to measure the flowrate using carney funnel method is shown in Figure 4.10.

Further detailed description to measure the flowrate by using the Carney funnel for non-free-flowing metal powders is explained in ASTM

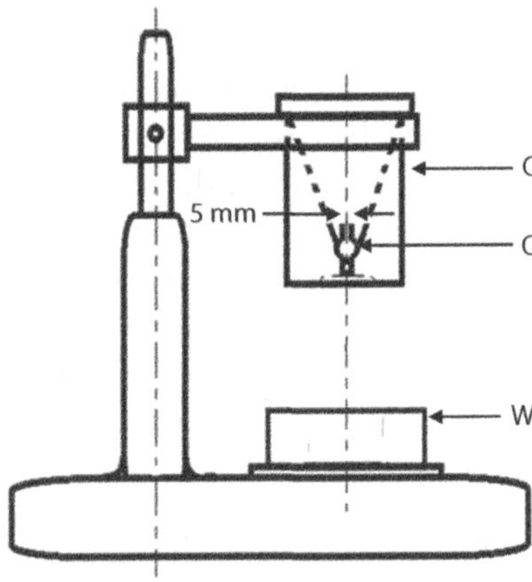

Figure 4.10 Carney funnel method apparatus.

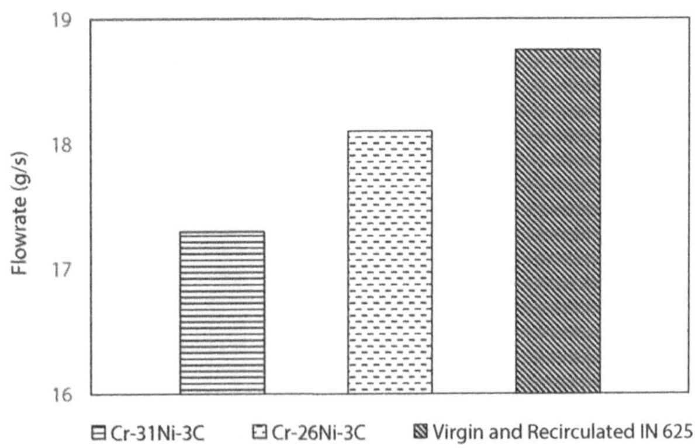

Figure 4.11 Flowrates of various powders using Carney funnel method.

B964 [20]. Young *et al.* [12] measured the flowrates of Cr-31Ni-3C and Cr-26Ni-3C by using the Carney flowmeter method and reported the values as 17.3 g/s and 18.1 g/s, respectively. Condruz *et al.* [21] measured the flowrate of both virgin and recirculated IN 625 and reported the flowrate as 18.75 g/s. The flowrates of different powders used in the above-mentioned papers are given in Figure 4.11.

4.2.2.3 *Arnold Meter and Hall Flowmeter Funnel Method*

This method is used to find the volumetric flowrate for both free-flowing and non-free-flowing metal powders. In this method, an Arnold meter is used to prepare a specified quantity of powder by following the test method ASTM B703. Measure the total elapsed time by using either of the test method described in ASTM B213. At least three measurements should be taken to minimize the error, and the volumetric flowrate (in sec/cc) is calculated by using the given formula.

$$Arnold/Hall\ flowrate = \frac{Avg.\ flow\ time \times Funnel\ correction\ factor}{specified\ volume} \tag{4.8}$$

Further detailed description to determine the flowrate for both free-flowing and non-free-flowing metal powders by using the Arnold meter and Hall flowmeter funnel is explained in ASTM B855 [22].

4.2.3 Particle Size and Particle Size Distribution

As per the ASTM B243 standard, particle size is defined as "the controlling lineal dimension of an individual particle as determined by analysis with sieves or other suitable means" and the particle size distribution is defined as "the percentage by weight, or by number, of each fraction into which a powder sample has been classified with respect to sieve number or microns". The standard test methods used to measure the particle size and distribution are sieve analysis and light scattering method. To supplement the suitable methods, non-standard methods such as image analysis, namely, optical microscopy and secondary electron microscopy, can also be used to determine the particle size.

4.2.3.1 *Sieve Analysis*

The sieve analysis method is used to measure the powder particle size from 45 μm to 850 μm. In this method, clean the sieves and check for any damages first. The group of sieves should be assembled with the coarsest sieve at the top and in sequential order based on the size of their openings. A sturdy collecting pan should be placed beneath the bottom sieve to finish the assembly. Put the specified test portion in the top sieve and cover it with a lid. Fasten the assembly in a sieve shaker machine and run it for at least 15 min. Note the actual time of sieving. Remove the screened powder fractions from the assembly of sieves by starting with the coarsest sieve first. Brush the bottom surface of the sieve and frame to collect any powder adhering to the next finer sieve. Collect the powders on glazed paper. The total mass of all the fractions should not be less than 99% of the specified test portion. Further detailed description to measure the metal powder's particle size by using sieve analysis is explained in ASTM B214 [23]. Similar test procedures are also described in MPIF standard 05 as well as International Organization for Standardization (ISO) 4497.0. Many researchers have also been using light microscope and scanning electron microscope (SEM) to measure the particle size of powders. In the SEM, the electron beam raster scans the surface of the specimen and signals of the emissions vary according to the specimen topography, which then produce a two-dimensional image by the imaging system. In the light microscope, since the probing radiation is light, it has less resolution compared to the electron microscope. Mordas *et al.* [24] measured the particle size of Co-Cr-Mo superalloy powders using SEM and reported that the particle size ranged from 10 μm to 80 μm.

4.2.3.2 *Light Scattering Method*

The light scattering method is used to find the particle size distribution in terms of the volume percentage of particles. This method can be used for both metals and compounds, in both aqueous and non-aqueous mediums. If the materials are hygroscopic, analysis can also be done by dispersing materials in a gaseous medium. This test method can be used to measure materials with particle sizes between 0.4 μm and 2000 μm. When a dispersed particle passes through the light beam, the light gets scattered, and the scattering angle is inversely proportional to the particle size. The scattered pattern is converted to size distribution by using Mie scattering or Fraunhofer diffraction, or a combination of both. A spherical model is assumed to analyze the scattering pattern. Hence, calculated particle sizes are interpreted as equivalent spherical diameters. Altug-Peduk *et al.* [10] measured the particle size distribution of Ti-55 wt.% Ni powder and obtained a median of size distribution of 92.48 μm. Further detailed description of finding the particle size distribution of metal powders and related compounds by using light scattering is explained in ASTM B822 [25].

4.2.4 Particle Morphology

As per ASTM E7 standard [26], morphology is defined as "the shape characteristics of a structure, the form, and orientation of specific phase or constituent". The particle morphology can be determined by using an SEM or light microscope. Slotwinski *et al.* [15] determined the morphology of 17-4SS and CoCr powders using an SEM and reported that the particles are roughly spherical.

4.2.5 Compressibility

As per ASTM E243 standard, compressibility is defined as "the capacity of a metal powder to be densified under a uniaxially applied pressure in a closed die". The densification behavior of bulk metal powders can be characterized in terms of compaction pressure and compaction density. With a specific compaction pressure, the equilibrium density of the entire compact represents an average over numerous differential volume elements. Thus, not only compaction pressure that is applied, but also on filling, the compact-die wall interactions and the compact dimensions have a role in determining the mean density. The uniaxial compaction method can be used to measure the compressibility of metal powders. This method is used for both metal powders and metal powder mixtures under the specified conditions. Compressibility test results might vary depending on the type of lubrication, method of lubrication, amount of lubrication, and rate of compaction pressure. Compressibility tests of metal powder can be conducted in three ways.

Method 1: "Compressibility is defined as the green density obtained at a selected compacting pressure".
Method 2: "Compressibility as a graph relating density as a function of at least four compacting pressures".
Method 3: "Compressibility is defined as the compacting pressure required to attain a target green density".

Compact the test specimens of specified shape and release the load soon after it reaches the maximum pressure because pressure dwell may increase the compressibility. Record the load used and calculate the compacting pressure. To remove flash, compacted samples may be sanded lightly with fine emery papers. Measure the mass of the compacted specimen nearest to 0.001 g and dimensions nearest to 0.005 mm. The green density can be calculated by using the following formulas.

For cylindrical test specimen, $D_g = \dfrac{(0.0777 \times M)}{D^2 \times T}$ (4.9)

where Dg = green density of cylindrical specimen in g/cm^3,
M = mass of cylindrical specimen in g,
D = diameter of cylindrical specimen in inches,
T = thickness of the cylindrical specimen in inches.

For rectangular test specimen, $D_G = \dfrac{(0.0610 \times M)}{L \times W \times T}$ (4.10)

Dg = green density of rectangular specimen in g/cm^3,
M = mass of rectangular specimen in g,
L = length of rectangular specimen in inches.,
W = width of rectangular specimen in inches,
T = thickness of the rectangular specimen in inches.

Compressibility can be measured in terms of green density by using above formula. Further detailed description to measure the compressibility of metal powders by using uniaxial compression is explained in ASTM B331 [27].

4.2.6 Crystal Structure

As per the ASTM standard E7, crystal is defined as "a solid composed of atoms, ions, or molecules arranged in a pattern which is periodic in three dimensions". Unit cell is defined as "a parallelepiped element of crystal structure, containing a certain number of atoms, the repetition of which through space will build up the complete crystal", and lattice parameter is defined as "the term used for the fractional coordinates x, y, z of lattice points when these are variable. It is also used to indicate the lengths of the axes a, b, c, and their included angles α, β, γ". Unit cells can be described by their lattice parameters. Thus, determining the lattice parameters can give the crystal structure. X-ray diffraction (XRD) may be used to determine crystal structure. When an x-ray is incident on a crystal, they diffract, and they may interfere constructively or destructively with each other. The constructive interference produces a diffraction point, and the set of all diffraction points is called a diffraction pattern. It contains information regarding the arrangement of atoms in the crystal. Braggs law defines the criterion for constructive interference and is given by,

$$n\lambda = 2\,d\,sin\,\theta \tag{4.11}$$

where λ = wavelength of x-rays,
D = interplanar distance,
θ = incident angle,
n = order of diffraction, integer.

Mordas *et al.* [24] used XRD to find the phase structure of cobalt-chromium-molybdenum powder and found that the $Cr_{0.32}Mo_{0.04}Co_{0.64}$ and $Mo_{0.1}Co_{0.9}$ phases present in the specimen powders.

4.2.7 Elemental Composition

The chemical composition of the specimen can be determined by using an energy-dispersive x-ray spectroscopy, x-ray photoelectron spectroscopy, etc.

4.2.7.1 Energy-Dispersive X-Ray Spectroscopy (EDS)

In energy-dispersive x-ray spectroscopy (EDS), when an electron beam interacts with the specimen, a characteristic x-ray material is generated, which provides a qualitative bulk chemistry. Quantitative bulk chemistry can be achieved with appropriate calibration. The interaction volume is dependent on the accelerating voltage, and it is approximately 1 μm for steel at 15 kV accelerating voltage. EDS is the most commonly and widely used technique for quantitative analysis. Mordas *et al.* [24] used SEM equipped with an energy-dispersive spectroscope to determine the chemical composition of Co-Cr-Mo superalloy powders and the compositions are reported in the article.

4.2.7.2 X-Ray Photoelectron Spectroscopy (XPS)

X-ray photoelectron spectroscopy (XPS) is an analytical tool that provides surface chemical composition and chemical bonding state from the outermost layers (10 nm) of a solid material. When x-rays interact with matter, they may emit electrons called photoelectrons. These photoelectrons come from a discrete energy level in an atom. The energy of the emitted photoelectron is measured by the detector, which is equal to the difference between the energy of incident x-rays and the binding energy of the particular electron in the atom, which determines the binding energy, a characteristic of the atom. Slotwinski *et al.* [15] used XPS to determine the chemical composition of the surface of 17-4SS and CoCr powders and the compositions are reported in the article.

4.3 Characterization of Liquid Form Feedstock

The following are some of the properties of liquid feedstock which must be studied before AM to get AM parts with consistent and predictable properties.

- Viscosity
- Surface tension
- Density
- Thermal stability
- Curing characteristics

4.3.1 Viscosity

Viscosity is a "measure of resistance of fluid to deformation when shear or an external force is applied". It is expressed in terms of Pascal-second (Pa.s). Liquid resin is often used as feedstock in all liquid-based AM systems and its compatibility with the AM process can be determined by its viscosity. In stereo lithography (SL), leveling problems during printing and shrinking can be minimized by using low viscous liquid resin. Similarly, in inkjet printing, actuating force to jet the liquid resin through the nozzle can be minimized by using low viscous liquid resin. Therefore, determining the optimum viscosity for resin in different AM processes is crucial. Viscosity can be measured by using the following methods.

- Capillary viscometers
- Orifice/cup viscometers

4.3.1.1 Capillary Viscometers

Viscosity of the Newtonian fluids like liquid resins is generally measured by using capillary viscometers. By noting the time taken for a known amount of liquid resin to flow through a caliper of known dimensions under the influence of gravitational force, viscosity of the liquid resin can be determined. Further detailed description to determine the kinematic viscosity of the liquid resin is explained in ASTM D446. This standard covers the measurement of kinematic viscosity by using an Ostwald viscometer, reverse flow viscometer, and suspended level viscometer.

4.3.1.2 Orifice/Cup Viscometers

This method can be used for both near Newtonian and Newtonian fluids. In this type of viscometers, the orifice is located at the bottom and gravitational force is used to draw the resin through it. The standard test procedure to determine the kinematic viscosity of liquid resins by using dip-type cups is explained in detail in ASTM D4212. In this method, immerse the viscometer in liquid resin to fill up the cup first, then slowly lift the cup from the resin. Note the time from the moment the cup leaves the resin surface to the moment when resin flow changes from continuous to drops. As per the standard, the readings should be taken at least twice and the difference between them should be less than 11%.

4.3.2 Surface Tension

As per the ASTM D459 standard, surface tension is defined as "the force existing in a liquid-vapor phase interface that tends to diminish the area of the interface. This force acts at each point on the interface in the plane tangent at that point". The surface tension is represented in terms of Newton per meter. In liquid-based AM systems, the wetting behavior of resin on the substrate and droplet size of resin in various processes can be affected by its surface tension. The surface tension of resins can be measured by using Wilhelmy plate method. In this method, the force (F) required to pull an aluminum plate or a clean glass of width (L) and weight (W) until it detaches can be measured by using a tensiometer, which is used to measure interface tensions. The following equation 4.12 is used to calculate surface tension with 0.1% accuracy when plate surfaces are completely wet.

$$\gamma = \frac{F - W}{2L} \tag{4.12}$$

where γ = surface tension in N/m,
F = force required to pull the plate in N,
W = weight of the plate in N,
L = width of the plate in m.

The measurements should be taken at least twice and if the variation is significant, more measurements should be taken. Further detailed description to determine the surface tension of resins is described in ASTM D1331.

4.3.3 Density

Density is mass per unit volume of the material at a specified temperature. The liquid resin's density can be measured by using ASTM D3505. This test method can be used only when the thermal expansion functions of the resins are known. In this test method, calibrate the pycnometer first, measure its weight, and draw the resin into it. The pycnometer is allowed to reach equilibrium before measuring the weight level of the sample pycnometer. This sample weight is used to calculate the density and relative density. Further detailed description to determine the density and relative density of the resin is explained in ASTM D3505.

4.3.4 Thermal Stability

One of the important characteristics of liquid resin is aging. The chemistry and rheological properties change as the resin ages. For rheological properties, viscosity increases with aging time. The viscosity can be determined by using the already discussed techniques. For chemistry change, inverse gas chromatography (IGC) and Fourier transforms infrared spectroscopy (FT-IR) can be used. In IGC, probe gas at a constant flowrate is injected into a stationary resin, the retention time is measured with gas chromatography detectors. The physiochemical properties can be determined from probe molecule size, concentration, and retention time. In FT-IR, when an infrared ray passes through a sample, the sample absorbs some radiation, and some radiation gets transmitted, resulting in a molecular absorption and transmission spectrum, which is a fingerprint molecular structure.

4.3.5 Curing Characteristics

Curing refers to toughening or hardening a polymer material by converting double-bonded monomers into stable cross-link bonds by using heat or chemical additives or even an electron beam. UV radiation is normally used as a heat source for curing. ASTM D4473 is used to determine curing characteristics. In this method, by applying free and forced vibration, the cure behavior of thermosetting resin over a temperature range can be determined. The properties of the cured materials can be related to the degree of cure by measuring the modulus. Further detailed description to determine the curing characteristics is explained in ASTM D4473. Differential scanning calorimetry (DSC) and FT-IR can also be used to determine the curing characteristics.

4.4 Characterization of Additively Manufactured Parts

It is reported that AM technique is employed in many fields like aerospace, automotive, construction, medical, etc., to fabricate the required parts or components. Recently, some critical components in different fields are fabricated by AM. Thus, it is important to characterize them. There are many properties that must be characterized like density, tensile properties, impact strength, hardness, fracture toughness, fatigue strength, wear, corrosion, residual stress, microstructure, crystal structure, composition, surface roughness, etc. The testing or characterization of the different properties of the additively manufactured parts are explained in detail in the following sections.

4.4.1 Density

The density of a solid metal is defined as the mass per unit volume of the solid metal. The density of a solid metal can be measured by using Archimedes' principle method. This test method uses Archimedes' principle to measure solid metal density. In this method, measure the mass of the test specimen and note it as A. Surface-connected porosity must be sealed by oil impregnation method or with a suitable material then measure this mass as B. Place the test specimen support which is immersed in water. Measure the combined mass of specimen support and test specimen and note it as C. Measure the specimen support immersed in water at a depth same as before and note it as E. While performing the test, water, atmosphere, and balance should be of the same temperature. The surface of the water should be free of dust particles. The density can be calculated from the following formula,

$$D = \frac{A \times \rho_w}{B - (C - E)} \tag{4.13}$$

where D = density of the solid metal,
A = mass of test specimen in the air in g,
B = mass of oil-impregnated test specimen in g,
C = mass of the oil-impregnated test specimen and specimen support in water in g,
E = mass of specimen support in water in g,
ρ_ω = density of water in g/cm^3.

Further detailed description to measure the density of sintered or compacted powder metallurgy parts by using Archimedes principle is described in ASTM B962 [28]. Gong *et al.* [29] uses the ASTM B962 method to measure the density of Ti-6Al-4V components made by SLM and EBM and reported the values.

4.4.2 Mechanical Properties

Mechanical properties are defined as "those properties of a material that are associated with the elastic and inelastic reaction when force is applied, or that involve the relationship between stress and strain". The most common mechanical properties of the additively manufactured parts, which are to be tested to measure the limit of the properties so that we can define the working conditions of the parts, are yield strength, ultimate tensile strength, ductility, strain, compressive strength, impact strength, wear, hardness, fracture toughness, fatigue strength, etc. Yield strength, ultimate tensile strength, ductility, and strain can be determined by using a tension test.

4.4.2.1 Tensile Properties

This test method provides information about the strength and ductility of the test specimen in any shape at room temperature under uniaxial tensile stresses. In this method, measure the dimensions of the test specimen in any form first, according to the procedure specified in the standard. Mark the gauge length to determine elongation after fracture and these marks should not affect the fracture location. Place the test specimen in their relevant grips in such a way that the longitudinal axis of the specimen coincides with the center line of the grips to eliminate the bending stresses since bending stresses are not included in stress calculations. Start the machine by defining the speed of the testing in terms of crosshead speed, rate of strain, rate of stressing, elapsed time for completing the part, or free-running crosshead speed. Record the elongations and forces, calculate strains and stresses, and draw the stress-strain diagram. Yield strength can be calculated by using the offset method, extension under load method, and other methods specified in the standard. The tensile strength can be calculated by dividing the maximum force in the test by the original cross-sectional area. Either elongation-after-fracture measurements or direct elongation-at-fracture measurements can be used to calculate elongation. The minimum area of the cross-section from fracture to the location can be used to calculate the reduction of area.

Further detailed description to find the tensile properties of metallic materials is described in ASTM E8/E8M [30]. Edwards *et al.* [31] used the ASTM E8 test method to measure the tensile properties of Ti-6Al-4V components prepared by electron beam additive manufacturing (EBAM) and reported that the ultimate tensile strength values as 833 and 851 MPa, yield strength values as 783 and 812 MPa, and percentage of elongation values as 2.7 and 3.6 in horizontal and vertical directions. Carroll *et al.* [32]

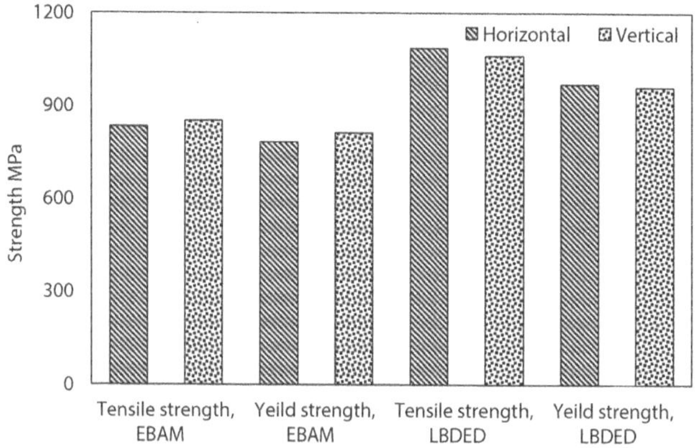

Figure 4.12 Tensile properties of Ti-6Al-4V fabricated by EBAM and LBDED methods.

measured the tensile properties of the laser-based direct energy deposited (LBDED) Ti-6Al-4V component by using the ASTM E8 standard and reported the average tensile strength as 1060 MPa in both transverse and longitudinal directions. Yield strengths and elongations are also reported in their studies. The tensile properties of Ti-6Al-4V fabricated by the different processes are given in Figure 4.12.

4.4.2.2 Compressive Properties

For some components or parts, design for compressive strength may also be necessary. In such cases, determining the compressive strength is also important. It can be determined by using the compression test method. This test method provides information about compressive yield strength, compressive strength, stress-strain curve, etc., under compression load at room temperature. In this method, measure the dimensions of the test specimen first, as specified in the standard. To remove any oil or grease at the ends of the specimen fixture and test specimen, clean them with acetone or suitable solvents. Lubrication may be applied to reduce the friction at the contact points. Mark the gauge length in the test specimen and install it in the fixture. Defined testing speed in terms of strain rate, force rate, or with crosshead speed. Start the machine at a constant rate and record the elongation and force values. These values can be used to draw the stress-strain curve and to calculate compression properties.

Further detailed description to find the compression properties for metallic materials is described in ASTM E9.

4.4.2.3 *Hardness*

As per the ASTM E6 standard, hardness is defined as "the resistance of a material to deformation, particularly permanent deformation, indentation, or scratching". There are many hardness test methods to determine the resistance to penetration. They are the Vickers hardness test, Brinell hardness test, Rockwell hardness test, Knoop hardness test, etc.

4.4.2.3.1 Vickers Hardness Test

This test method gives micro indentation hardness. In this method, a 136° face angle, highly polished, pointed, square-based pyramidal diamond indenter diameter is brought into contact with the surface of the test specimen, and applied force in perpendicular direction to the surface. The test force is maintained for a predetermined dwell duration before being released. The mean of the length of the two diagonals is used to calculate the Vickers hardness value by using the equation 4.14.

$$VHN = \frac{1854.4 \times F}{d^2} \tag{4.14}$$

where VHN = Vickers hardness number,
F = applied load in gf,
D = mean of indentation diagonal in μm.

Further detailed description to determine Vickers hardness for metallic materials is described in ASTM E92 [33]. Zhu *et al.* [34] followed the Vickers hardness test method to find the microhardness profile of the substrate zone to laser PBF additive manufacture zone of Ti alloy (TC11) and reported that

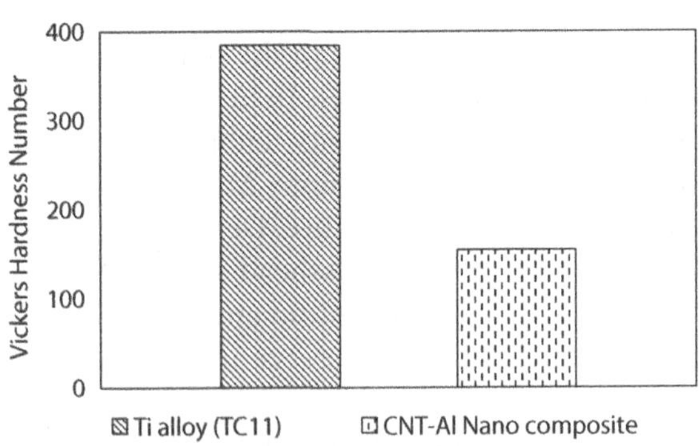

Figure 4.13 Hardness of two materials using the Vickers hardness method.

the hardness value of the additive manufacture zone is 50 HV higher than the substrate zone. Gu *et al.* [35] used the Vickers hardness test method to find the microhardness profile along the building direction of the cross-section of the laser additively manufactured CNT-Al nanocomposite part and reported a high microhardness of 154.12 HV, obtained for a sample manufactured with scanning speed of 2.0 m/s. Vickers hardness numbers of materials used in the above-mentioned papers are given in Figure 4.13.

4.4.2.3.2 Brinell Hardness Test

The test procedure is similar to the Vickers method, but the only difference is the Tungsten carbide ball indenter of specified diameter. The indentation is assumed to be spherical and at least two perpendicular directions are used to measure the indentation's diameter. The mean of the diameter measurements is used to calculate the Brinell hardness value "by using the equation 4.15.

$$BHN = \frac{2P}{\Pi D\left(D - \sqrt{D^2_- d^2}\right)} \tag{4.15}$$

where BHN = Brinell hardness number,
 P = applied load in kgf,
 D = diameter of the indenter ball in mm,
 D = average diameter of indentation in mm.

Further detailed description to measure Brinell hardness for metallic materials is described in ASTM E10 [36].

4.4.2.3.3 Rockwell Hardness Test

Rockwell hardness test follow three steps of applying and removing force. In the first step, the indenter of the specified material diameter is brought in contact with the surface of the test specimen, and a primary test force is applied in a direction perpendicular to the surface. The primary test force is maintained for a predetermined dwell duration and measures the base indentation depth. In the second step, with the help of additional test force, F_1, the force on the indenter is increased at a specific rate to obtain a total test force, F, and it is maintained for a predetermined dwell duration. In the third step, return to the primary test force by removing the additional test force. The final indentation depth is measured after holding the primary test force for a specified dwell time. Difference indentation depths, h, is used to calculate Rockwell hardness value by using the equations 4.16 and 4.17.

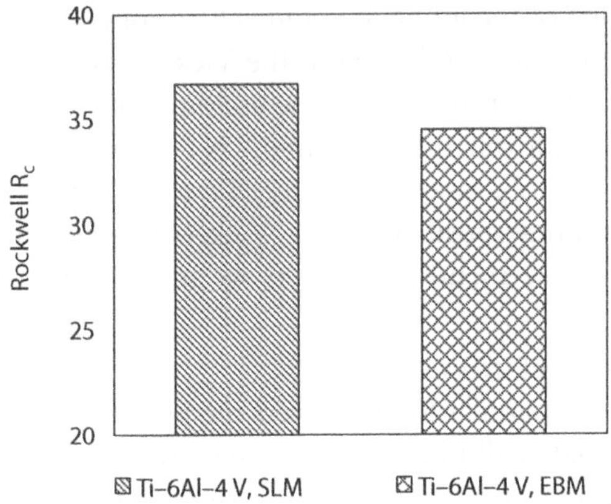

Figure 4.14 Hardness of two materials using the Rockwell hardness method.

$$\textit{Rockwell hardness number, ball indenter} = 130 - \frac{h}{0.002} \tag{4.16}$$

$$\textit{Rockwell hardness number, spheroconical indenter} = 100 - \frac{h}{0.002} \tag{4.17}$$

where h is in mm. Further detailed description to measure Rockwell hardness for metallic materials is described in ASTM E18 [37]. Gong *et al.* [29] measured the hardness of Ti-6Al-4V components prepared by both EBM and SLM using the ASTM E18 method, plotted a bar chart, and reported that the porosity affected hardness decrement. Rockwell hardness of Ti-6Al-4V manufactured by SLM and EBM are given in Figure 4.14.

4.4.2.3.4 Knoop Hardness Test

This test method also gives micro indentation hardness. The test method is similar to the Vickers test, but the only difference is the indenter. In this method, a highly polished, pointed, rhombic-based pyramidal diamond having included angles 172° 30' and 130° indenter is used. The length of the long diagonal is used to calculate the Knoop hardness value by using the equation 4.18.

$$KN = \frac{14229 \times F}{d^2} \tag{4.18}$$

where KN = Knoop hardness number,
F = applied load in gf,
D = length of long diagonal in μm.

Further detailed description to measure Knoop hardness for metallic materials is described in ASTM E92 [33].

4.4.2.4 *Impact Strength*

Absorbed energy is an indication of impact strength, and as per ASTM E6 standard, it is defined as the "work spent to fracture a specimen in a single pendulum swing, as measured by a compensated indicating device". Notch bar impact test can be used to find the absorbed energy. Notch bar impact testing can be done by two methods, namely, the Charpy test and the Izod test. In the Charpy impact test method, the specimen is made according to the specified dimensions, treat thermally if required, and positioned against the anvil in a simple beam support. Strike the specimen by releasing the pendulum without causing vibrations. Record the information from the machine to calculate the absorbed energy along with the broken specimen for analysis. The 8-mm striker should be used unless specified otherwise. If the unbroken specimen does not pass between the anvils, this indicates that the result exceeds the machine's capacity. In the Izod impact test, the test procedure is similar to the Charpy impact test procedure, but the only difference is, the specimen is positioned in cantilever beam support by clamping it in the holding fixture. Since the clamping fixture is a part of the machine, it cannot be readily cooled or heated. Consequently, the Izod impact test is carried out at room temperature only. Further detailed description to measure notched bar impact strength for metallic materials is described in ASTM E23 [38]. Hack *et al.* [39] measured the impact energy of laser powder bed 625 Ni alloy specimens in both longitudinal and transverse directions by using the Charpy impact test method and reported the values as (92 J, 115 J, and 149 J) for longitudinal direction and (131 J, 144 J, 150 J) for transverse direction. Shassere *et al.* [40] used the ASTM E23 method to find the impact strength, plotted a graph, and reported that the transition temperature changes from -36°C to -17°C with build height. The impact strengths of the two materials used in the above-mentioned papers are given in Figure 4.15.

4.4.2.5 *Fracture Toughness*

As per ASTM E1823 standard [41], fracture toughness is a "generic term for measures of resistance to extension of a crack". There are many methods to measure fracture toughness, one such method is the linear-elastic

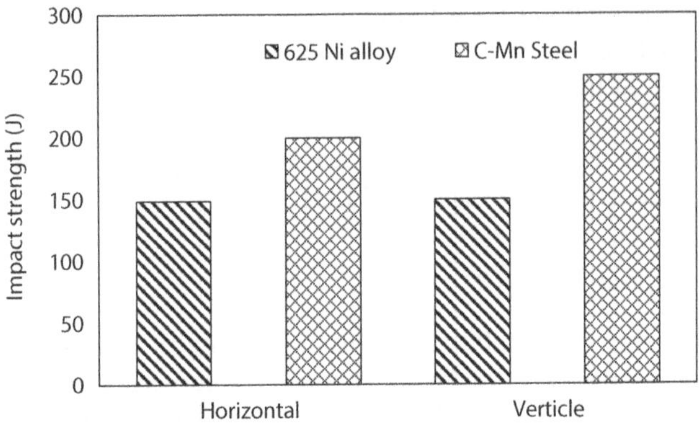

Figure 4.15 Impact strengths of two materials using the ASTM E23 method.

plane-strain method. This test method uses pre-cracked specimens of specified dimensions subjected to slowly or rapidly increasing crack displacement force under linear-elastic plane-strain conditions. In this method, measure the specimen dimensions like thickness first, width before the fracture, and crack length after the fracture as per the specified procedure. Apply load in such a way that the rate of increment in stress intensity factor is between 0.55 and 2.75 MPa $\sqrt{m}$ /s for slow force tests. The test must go on until the specimen can no longer withstand an increase in applied force. Note the maximum force which can be used to find fracture toughness.

Further detailed description to measure linear-elastic plane-strain fracture toughness of metallic materials is described in ASTM E399 [42]. A standard test procedure to measure plain strain (Chevron-notch) fracture toughness is given in ASTM E1304. ASTM E1820 is also used to measure fracture toughness. ASTM E992 describes a procedure to measure the fracture toughness of steels by using equivalent energy methodology. ASTM E1221 describes a test procedure to measure plane-strain crack arrest fracture toughness (K_{Ia}) of ferritic steels. ASTM E813 standard test method to measure JIc which is a measure of fracture toughness. Davis *et al.* [43] measured the fracture toughness of wire arc additively manufactured Ti alloy-alloy composite by using ASTM E399 and reported that the fracture toughness is 77.8 MPa $\sqrt{m}$ and 82.8 MPa $\sqrt{m}$ in the longitudinal direction and transverse direction, respectively.

4.4.2.6 Fatigue Life

As per ASTM E1823 standard, fatigue life is defined as "the number of cycles of a specified character that a given specimen sustains before failure of a specified nature occurs. Fatigue life or the logarithm of fatigue life, is a dependent variable". Force-controlled constant amplitude axial fatigue test is one of the most common fatigue tests to determine fatigue. This technique is restricted to the fatigue testing of axial notched and unnotched specimens under a periodic forcing function with constant amplitude in room-temperature air. In this method, the most important consideration is good alignment between the grips to prevent bending stress. To get more repeatable test results, the bending stresses should be minimum, especially for materials with low ductility. Since bending stresses influence fatigue behavior, especially in long-life regimes, they must be reported. Continue the tests until the specimen failure criteria are met or the specimen has undergone a set number of cycles. While reporting the results, also state the failure criteria.

Further detailed description to measure axial fatigue under force-controlled constant amplitude conditions for metallic materials is described in ASTM E466 [44]. ASTM E606 standard test method is used to measure strain-controlled fatigue. ASTM E647 method is used to measure growth rates of fatigue crack. Chang *et al.* [45] used ASTM 466 test method to determine the fatigue strength of SLM additively manufactured alloy in both lateral and longitudinal directions and reported that the fatigue strength in the lateral direction is 546 MPa whereas in the longitudinal direction it is 530 MPa.

4.4.2.7 Wear

Wear is one of the most common phenomena which happens in almost all mechanical systems. Thus, determining wear properties is essential

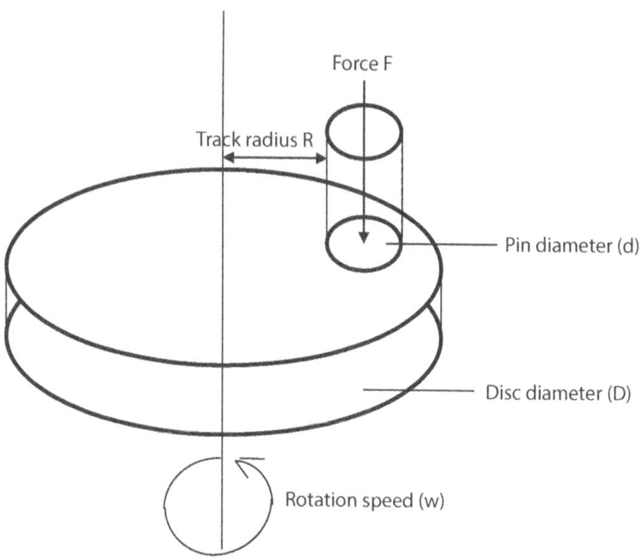

Figure 4.16 Pin-on-disc wear method apparatus.

in designing a component or part. Pin on disc is one of the most common methods to determine wear properties. In this method, clean the pin and the disc first to remove all the foreign matter. Measure the specimen's dimensions and weigh the specimen's mass. Insert the pin and the disc in their respective positions in which the pin should be perpendicular to the disc and the disc should be perpendicular to the axis of revolution as shown in the Figure 4.16. Add the mass to the lever to develop the required force to press the pin against the disc. Set the desired speed by disconnecting the pin with the disc. Set the required number of revolutions. Start the test, continue to run under constant load, and the test stops when set revolutions are achieved. Clean the specimens to remove any loose debris. Remeasure the specimen's dimensions and reweigh the specimen's mass. These values are used to measure the wear.

Further detailed description of doing a wear test by using pin-on-disc apparatus is described in ASTM G99 [46]. Attar *et al.* [47] used ASTM G99 test method to measure wear properties of SLM additively manufactured Ti and Ti-TiB composite parts and reported that the wear resistance of Ti-TiB composite is higher when compared to pure Ti.

4.4.3 Corrosion

Corrosion is defined as "the deterioration of a material, usually a metal, that results from a chemical or electrochemical reaction with its environment". For metals at or near their corrosion potential, polarization resistance can be used to determine how quickly general corrosion occurs. The general corrosion rate can be measured by using polarization resistance measurements. This test can provide potentiodynamic polarization resistance and corrosion potential of the test specimen. In this method, first, prepare the standard test solution and pour it into the test cell. Purge the test cell with oxygen-free gas like hydrogen, nitrogen, or argon. Start the purge 30 min before the specimen immersion and continue throughout the test. Polish the specimen, clean it with acetone, and rinse it with distilled water. Immerse the sample and note the corrosion potential after 5 and 55 min. Apply 30 mV potential more negatively to the recorded corrosion potential at 55 min. Start the anodic potential scan 1 min after applying -30mV potential, with a sweep rate of 0.6 V/h. Record the current and potential continuously until the potential reaches 30 mV more positive than the corrosion potential at 55 min, then terminate the sweep. Plot the potentiodynamic polarization curve and draw a tangent to the curve at $i = 0$ which gives the polarization resistance.

Further detailed description to measure the potentiodynamic polarization resistance is described in ASTM G59-97 [48]. ASTM G48 describes

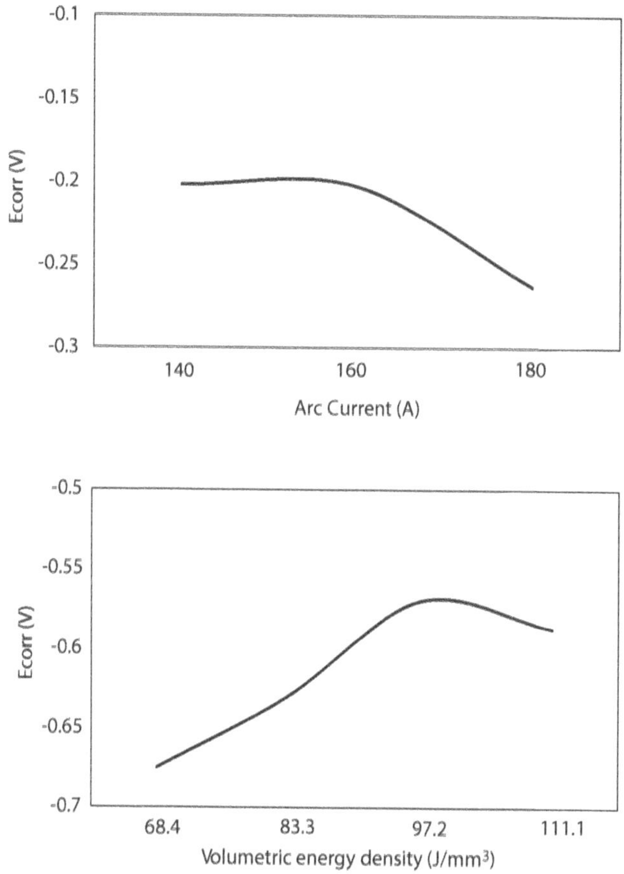

Figure 4.17 Corrosion resistance of (a) 316L SS and (b) AlCoCrFeNi.

the test method to determine the pitting and crevice corrosion resistance of stainless steels and other related alloys by using a ferric chloride solution. Ke *et al.* [49] studied the corrosion behavior of 316L SS additively manufactured part by using the ASTM G48 standard and reported that the corrosion resistance decreases as arc current increases in the process. Niu *et al.* [50] used the potentiodynamic polarization method to determine the corrosion resistance of selective laser-printed AlCoCrFeNi high entropy alloy and reported that the corrosion resistance increases with an increase in volumetric energy density. This may be because of fast cooling and more Ni-Al-ordered BCC phases. The corrosion resistance of the two materials used in the above-mentioned papers is given in Figure 4.17.

Natural salt spray tests can also be used to determine the corrosion resistance of materials. This test is especially used to find discontinuities such as cracks, pores, and other defects. This test does not specify specimen area and exposure time. A further detailed description of the apparatus, reagents, and test method is described in ISO 9227. ASTM A262 standard is used to identify the sensitivity to intergranular corrosion in austenitic stainless steel.

4.4.4 Residual Stress

As per the ASTM standard E6, residual stress is defined as the "stress in a body which is at rest and in equilibrium and at uniform temperature in the absence of external and mass forces". The presence of residual stresses in the component affects the properties. Thus, it is advisable to measure the residual stresses of the component.

ASTM E2860 describes the test method to calculate the residual stresses of bearing steels by using XRD. In this method, collect the diffraction profile at least at seven or more angles within the required angular range for a given plane. Correct the XRD data for background, Lorentz-polarization-absorption, and instrument-specified conditions. Determine the angles for all XRD peaks and calculate the d-spacings using Bragg's law. Plot the d-spacing values against their sin_{ψ}^2 or sin_{β}^2 values which are used to calculate residual stress.

ASTM E837 describes the test method to measure the residual stresses of isotropic linear-elastic material near the surface. The strains obtained from the strain gauge after drilling a hole at the center of the rosette are used to calculate the residual stresses. Gordon *et al.* [51] used XRD to measure residual stresses in wire arc additively manufactured 304L SS part as heat treated and as printed condition and reported that the residual stresses present on the part surface are compressive in nature, and these compressive stresses may help to increase the fatigue crack growth resistance of part at both conditions. Simonelli *et al.* [52] studied the effect of residual stress relieving of selective laser-melted Ti-6Al-4V part on mechanical properties by using XRD and reported a residual stress tensor, strength values are reduced slightly, and ductility increased slightly. This might be because of change in phase composition after stress relieving.

4.4.5 Crystal Structure

The lattice parameters describes the unit cell which in turn describes the crystal structure and XRD can be used to determine the crystal structure. The criterion for the constructive interference is given by the Braggs law and it is explained in detailed in section 4.2.6. XRD is commonly used to determine the crystal structure and the fallowing are some of the references where XRD was used to determine the crystal structure.

Agrawal *et al.* [53] determined the phase structure of Ti-6Al-4V additive friction stir deposition part by using XRD at different process parameters and reported that the part has major amounts of α phase with minor amounts of β phase at all process conditions, and there is no major change in phase quantities with respect to different process conditions. Yao *et al.* [54] used XRD to determine the crystal structure of powder, as cast and selective laser-melted AlCrFeNiV high entropy alloy and reported that the crystal structure of all the samples is the same and found to be Face Centered Cubic (FCC).

4.4.6 Microstructure

As per the ASTM E7 standard, the microstructure is defined as "the structure of a suitably prepared specimen as revealed by a microscope". Not only in AM but also in all other manufacturing techniques, properties depend on microstructure. Microstructure can be affected by many variables like manufacturing strategies, process parameters, postprocessing methods, and environment. Thus, controlling microstructure is very important to achieve the required properties. The microstructure has many features which are listed below.

4.4.6.1 Grain Size

As per the ASTM standard E7, grain is defined as "an individual crystallite in metals". Grain boundary can obstruct the dislocation motion and cause the strength to improve. Grain boundary gas has higher interfacial energy which makes them a potential site for any reactions. Thus, finding the grain size is important to understand the properties.

ASTM E112 standard describes three test procedures to determine the grain size by using a light microscope. They are grain size measurements by comparing to the standard chart method, grains per unit area method, and intercepting grains or grain boundaries with a known length of the straight-line method. The procedures determine the grain size of single-phase structures but can also be applied to determine the average grain size of multi-phase structures.

ASTM E2627 standard describes the test procedure to measure the average grain size using electron backscattered diffraction (EBSD). The diffracted electrons from the specimen strike a phosphor screen to produce a pattern. This diffraction pattern may be used to determine crystal structure and orientation as a function of the sample surface. ASTM E1382 describes the test procedure to determine the average grain size by analyzing the optical microscope image of grain structure using semiautomatic or automatic image analysis. ASTM E930 describes the test procedure to determine the largest grain size in the test section. Kreitcberg *et al.* [55] studied the effect of post-treatment conditions namely hot isostatic pressing, solution treatment, recrystallization annealing, and stress relief annealing on grain structure by using ASTM E112 and reported that the grains are equiaxed in the horizontal plane, elongated in the vertical plane, and the effect of post-treatment conditions on grain structure were also discussed in the article. McLouth *et al.* [56] used EBSD to measure the grain size of selective laser-melted IN718 at different conditions and reported the values.

4.4.6.2 *Crystal Orientation*

The texture is defined as the distribution of the grains' crystallographic orientations. The back-reflection Laue method for determining a metal crystal's orientation is described in ASTM E82/E82M. Between the x-ray source and the specimen, there is a pinhole aperture through which an x-ray incoming beam travel. It then contacts the specimen's crystal structure and is diffracted back to the detector. The x-ray beams that were diffracted by the atomic planes in the specimen crystals are observed as a pattern of white spots on the detector. This pattern is compared with the stereographic projection of standard crystals to determine the test specimen crystal orientation. Stoudt *et al.* [57] use EBSD to determine the texture of additively manufactured IN625 in the longitudinal and transverse directions and reported that the inverse pole figure (IPF) in the Z direction of the longitudinal sample predominantly consists of larger grains of (101) planes and less amount of smaller grains of (100) planes, IPF in Z of the transverse sample shows both dominant orientations of both (101) and (100) planes. Nguejio *et al.* [58] determine the grain texture of selective laser-melted and laser melting-deposited IN625 by using EBSD, and reported that IPF -Z map shows that the elongated grains of (100) planes present dominantly in the SLM sample, and in LMD sample, the grains are oriented in build direction in inverted v-shape with (100), (110), (111) planes IPF maps.

4.4.6.3 *Intermetallic Phases and Inclusions*

ASTM E562 describes the test procedure to determine the volume fraction of constituent phases using a light microscope or SEM with the help of a point grid. A test grid is superimposed over a micrograph and counts the test points which fall within the constituent phase of interest. Circular or square grid arrangements are typically used in this method. As per ASTM standard E7, inclusions are defined as the "foreign material held mechanically, usually referring to non-metallic particles, such as oxides, sulfides, silicates, etc." and they are mostly contained in the grain boundaries which affect the properties. Precipitates are secondary phases formed within the microstructure and help in strengthening the metal when they are at an optimum size. Thus, it is important to characterize both inclusion and precipitates. ASTM E1245 describes the method to determine the morphological characteristics of inclusions in steel or other metals by using a light microscope or SEM with an auto image analyzer. This standard measures the number of inclusions per unit area of each type of inclusion, the mean diameter of inclusions, and the area of each inclusion. Carroll *et al.* [32] used an optical microscope to determine the α (HCP) phase and β (BCC) phase titanium alloy and reported β phase grains and α phase grain boundaries in the micrograph. Chang *et al.* [45] used micro focused computer tomography to determine the internal defects of selective laser-melted Ti6Al4V alloy and reported that the sample has irregular voids and near-spherical-shaped impurities. The transmission electron microscope (TEM) is also a microstructure evaluation technique that uses an electron beam to pass through a thin specimen. The interaction of the electron beam with the specimen generates a pattern, which is used to generate an image. The TEM evaluates lattice defects and precipitates. Attar *et al.* [47] used a TEM to find the phases in the microstructure of selective laser-melted Ti-TiB composite and reported the bright field images of the composite showing TiB particles of needle-like morphology distributed in α-Ti matrix. McLouth *et al.* [56] used a TEM to determine the size of the γ precipitate at different conditions in selective laser-melted IN718 and reported the size as approximately 25 nm.

4.4.6.4 Voids, Microcracks, Delamination, and Porosity

Voids in the microstructure, cracks, and delamination greatly influence the properties. Thus, it is important to characterize them. There are two types of methods to evaluate them, one is the destructive method and the other is non-destructive method. Porosity has a great influence on properties, and it can be determined by using an optical microscope or a secondary electron microscope. In destructive evaluation, prepare the sample according to the standard ASTM E3 and analyze the microstructure for voids, microcracks, and delamination using either a light microscope or an SEM. Several factors should be considered to evaluate voids, microcracks, and delamination through non-destructive testing, like material and material property.

ASTM E543 listed different material properties measured from different non-destructive evaluation methods. Edwards *et al.* [31] used the optical microscope to determine the presence of crack initiation sites at the surface of electron beam additively manufactured Ti-6AI-4V alloy. Ren *et al.* [59] used an optical microscope to determine the effect of inter-pass temperature on porosity in wire additively manufactured Al-Mg alloy and reported that the porosity is discretely distributed with size <100 μm and pore count decreases as interpass temperature decreases. Gangireddy *et al.* [60] measured the porosity of additively manufactured WE43-Mg alloy at different processing parameters by using optical microscopy along with image analyzing software, reported the values, and studied the effect of porosity on the mechanical properties.

4.4.7 Elemental Composition

The chemical composition of the specimen can be determined by using an SEM equipped with energy-dispersive x-ray spectroscope, XPS, x-ray fluorescence spectroscopy, etc.

4.4.7.1 Energy-Dispersive X-Ray Spectroscopy

When an electron beam interacts with the specimen, a characteristic x-ray material is generated which provides a qualitative bulk chemistry. Quantitative bulk chemistry can be achieved with appropriate calibration. Zhang *et al.* [61] used the spot EDS method to determine the elemental composition of inclusion in fracture surface area and reported that the inclusion mainly consists of manganese and oxygen. Sridharan *et al.* [62] used EDS spot analysis to determine the oxygen content in the interfacial region of an Al-steel ultrasonic dissimilar welded joint and reported that there is no oxygen content at the interface.

4.4.7.2 X-Ray Photoelectron Spectroscopy

X-ray photoelectron spectroscopy is an analytical tool that provides surface chemical composition and chemical bonding state from the outermost layers (10 nm) of a solid material. When x-rays interact with matter, they may emit some electrons called photoelectrons. These photoelectrons come from a discrete energy level in an atom. The energy of the emitted photoelectron is measured by the detector which is equal to the difference in energies of incident x-rays and the binding energy of the particular electron in the atom, which determines the binding energy, a characteristic of the atom.

4.4.7.3 *X-Ray Fluorescence Spectroscopy*

X-ray fluorescence spectroscopy is an analytical tool that provides surface chemical composition from the topmost layers. When x-rays interact with matter, consequently, matter emits characteristic x-rays of each element present. The energy and intensity of x-rays are measured to determine the composition. Li *et al.* [63] used x-ray fluorescence spectroscopy to determine the chemical composition of the additively manufactured stainless steel and reported the composition at different acquisition times.

4.4.8 Surface Roughness

Surface roughness is defined as the closely spaced irregularities on the surface of the specimen. Both contact and non-contact methods are available to measure surface roughness. Non-contact optical profilometry is one of the many techniques used to measure surface roughness. Chen *et al.* [64] used a stylus contact method surface profilometer to study the effect of process parameters on the surface roughness of Ti-6Al-4V alloy fabricated by SLM, reported different surface roughness characteristic values and discussed the effect of process parameters on them. Nakatani *et al.* [65] used three-dimensional optical profilometers to study the effect of surface roughness on fatigue strength of electron beam-melted and direct metal laser-sintered Ti-6Al-4V alloy, reported different surface roughness characteristic values, and discussed the effects.

4.5 Conclusion

The different methods used for characterization of the feedstock and AM fabricated parts are discussed elaborately. It is essential to understand the characteristics of feedstock materials to build a successful finished component through AM processes. The researcher should have prior idea of these important properties, which will influence the final properties before starting the fabrication. The final properties are completely dependent on the quality of the feedstock materials. The service life of any components lies in how its properties match with the in-service requirements. It warrants the evaluation of properties of the additively manufactured parts to qualify for the specific application requirements. The detailed description of the various techniques that can be employed for characterization of final fabricated parts were presented in this chapter. The readers can look at the cited references for further details and procedures for measurement of any particular property. This chapter helps the readers to select the characterization technique based on their requirements and constraints. There are many characterizations that do not have standards, it warrants more standardization in AM in the near future.

References

1. Standard terminology of powder metallurgy, ASTM B243-22.
2. Standard test method for apparent density of free-flowing metal powders using the hall flowmeter funnel, ASTM B212-21.
3. Zhai, W., Zhou, W., Nai, S.M.L., Wei, J., Characterization of nanoparticle mixed 316 L powder for additive manufacturing. *J. Mater. Sci. Technol.*, 47, 162–168, 2020.
4. Nguyen, Q.B., Nai, M.L.S., Zhu, Z., Sun, C.N., Wei, J., Zhou, W., Characteristics of inconel powders for powder-bed additive manufacturing. *Engineering*, 3, 5, 695–700, 2017.
5. Standard test method for apparent density of non-free-flowing metal powders using the carney funnel, ASTM B412-22.
6. Mitterlehner, M. *et al.*, Comparative evaluation of characterization methods for powders used in additive manufacturing. *J. Mater. Eng. Perform.*, 30, 9, 7019–7034, 2021.
7. Standard test method for tap density of metal powders and compounds, ASTM B329-20.
8. Baesso, I., Karl, D., Spitzer, A., Gurlo, A., Günster, J., Zocca, A., Characterization of powder flow behavior for additive manufacturing. *Addit. Manuf.*, 47, August, 102250, 2021.
9. Standard test method for apparent density of powders using arnold meter, ASTM B703-21.
10. Altug-Peduk, G.S., Dilibal, S., Harrysson, O., Ozbek, S., West, H., Characterization of Ni–Ti Alloy powders for use in additive manufacturing. *Russ. J. Non-Ferrous Met.*, 59, 4, 433–439, 2018.
11. Standard test method for tap density of metal powders and compounds, ASTM B527-23.
12. Young, B., Heelan, J., Langan, S., Siopis, M., Walde, C., Birt, A., Novel characterization techniques for additive manufacturing powder feedstock. *Metals (Basel)*, 11, 5, 1–19, 2021.
13. Du, W., Singh, M., Singh, D., Binder jetting additive manufacturing of silicon carbide ceramics development of bimodal powder feedstocks by modeling and experimental methods. *Ceram. Int.*, 46, 12, 19701–19707, 2020.
14. Standard test method for metal powder skeletal density by helium or nitrogen, ASTM B923-22.
15. Slotwinski, J.A., Garboczi, E.J., Stutzman, P.E., Ferraris, C.F., Watsonm, S.S., Peltz, M.A., Characterization of metal powders used for additve manufacturing. *J. Res. Natl. Inst. Stand. Technol.*, 119, 460–493, 2014.
16. Standard test methods for flow rate of metal powders using the hall flowmeter, ASTM B213-20.

17. Snow, Z., Martukanitz, R., Joshi, S., On the development of powder spreadability metrics and feedstock requirements for powder bed fusion additive manufacturing. *Addit. Manuf.*, 28, November 2018, 78–86, 2019.
18. Cordova, L., Sithole, C., Macía Rodríguez, E., Gibson, I., Campos, M., Impact of powder reusability on batch repeatability of Ti6Al4V ELI for PBF-LB industrial production. *Powder Metall.*, 66, 2, 129–138, 2022.
19. Lim, J.H. and Khan, N.A., Effects of particle size distribution on the surface finish of selective laser melting parts. *Solid Free. Fabr. 2019 Proc. 30th Annu. Int. Solid Free. Fabr. Symp. - An Addit. Manuf. Conf. SFF 2019*, pp. 1990–2003, 2019.
20. Standard test methods for flow rate of metal powders using the carney funnel, ASTM B964-23.
21. Condruz, M.R., Matache, G., Paraschiv, A., Characterization of in 625 recycled metal powder used for selective laser melting. *Manuf. Rev.*, 7, 1–8, 2020.
22. Standard test method for volumetric flow rate of metal powders using the arnold meter and hall flowmeter funnel, ASTM B855-22.
23. Standard test method for sieve analysis of metal powders, ASTM B214-22.
24. Mordas, G. *et al.*, Characterisation of CoCrMo powder for additive manufacturing. *Int. J. Adv. Manuf. Technol.*, 111, 11–12, 3083–3093, 2020.
25. Standard test method for particle size distribution of metal powders and related compounds by light scattering, ASTM B822-20.
26. Standard terminology relating to metallography, ASTM E7-22.
27. Standard test method for compressibility of metal powders in uniaxial compaction, ASTM B331-20.
28. Standard test methods for density of compacted or sintered powder metallurgy (PM) products using archimedes' principle, ASTM B962-17.
29. Gong, H., Rafi, K., Gu, H., Janaki Ram, G.D., Starr, T., Stucker, B., Influence of defects on mechanical properties of Ti-6Al-4V components produced by selective laser melting and electron beam melting. *Mater. Des.*, 86, 545–554, 2015.
30. Standard test methods for tension testing of metallic materials, ASTM E8/E8M-22.
31. Edwards, P., O'Conner, A., Ramulu, M., Electron beam additive manufacturing of titanium components Properties and performance. *J. Manuf. Sci. Eng.*, 135, 6, 1–8, 2013.
32. Carroll, B.E., Palmer, T.A., Beese, A.M., Anisotropic tensile behavior of Ti-6Al-4V components fabricated with directed energy deposition additive manufacturing. *Acta Mater.*, 87, 309–320, 2015.
33. Standard test methods for vickers hardness and knoop hardness of metallic materials, ASTM E92-17.
34. Zhu, Y., Li, J., Tian, X., Wang, H., Liu, D., Microstructure and mechanical properties of hybrid fabricated Ti-6.5Al-3.5Mo-1.5Zr-0.3Si titanium alloy by laser additive manufacturing. *Mater. Sci. Eng. A*, 607, 427–434, 2014.
35. Gu, D., Rao, X., Dai, D., Ma, C., Xi, L., Lin, K., Laser additive manufacturing of carbon nanotubes (CNTs) reinforced aluminum matrix nanocomposites

Processing optimization, microstructure evolution and mechanical properties. *Addit. Manuf.*, 29, July, 100801, 2019.
36. Standard test method for brinell hardness of metallic materials, ASTM E10-18.
37. Standard test methods for rockwell hardness of metallic materials, ASTM E18-22.
38. Standard test methods for notched bar impact testing of metallic materials, ASTM E 23-18.
39. Hack, H., Link, R., Knudsen, E., Baker, B., Olig, S., Mechanical properties of additive manufactured nickel alloy 625. *Addit. Manuf.*, 14, 105–115, 2017.
40. Shassere, B., Nycz, A., Noakes, M.W., Masuo, C., Sridharan, N., Correlation of microstructure and mechanical properties of metal big area additive manufacturing. *Appl. Sci.*, 9, 787, 2019.
41. Standard terminology relating to fatigue and fracture testing, ASTM E1823-21.
42. Standard test method for plane-strain fracture toughness of metallic materials ASTM E399-22.
43. Davis, A.E. *et al.*, Mechanical performance and microstructural characterisation of titanium alloy-alloy composites built by wire-arc additive manufacture. *Mater. Sci. Eng. A*, 765, July, 138289, 2019.
44. Standard practice for conducting force controlled constant amplitude axial fatigue tests of metallic materials, ASTM E466-21.
45. Chang, K. *et al.*, Microstructural feature and mechanical property in different building directions of additive manufactured Ti6Al4V alloy. *Mater. Lett.*, 267, 127516, 2020.
46. Standard test method for wear testing with a pin-on-disk standard test method for wear testing with a pin-on-disk apparatus, ASTM G99-17.
47. Attar, H. *et al.*, Nanoindentation and wear properties of Ti and Ti-TiB composite materials produced by selective laser melting. *Mater. Sci. Eng. A*, 688, December 2016, 20–26, 2017.
48. Standard test method for conducting potentiodynamic polarization resistance measurements, ASTM G59-97.
49. Yang, K., Wang, Q., Qu, Y., Jiang, Y., Bao, Y., Microstructure and corrosion resistance of arc additive manufactured 316L stainless steel. *J. Wuhan Univ. Technol. Mater. Sci. Ed.*, 35, 5, 930–936, 2020.
50. Niu, P.D. *et al.*, Microstructures and properties of an equimolar AlCoCrFeNi high entropy alloy printed by selective laser melting. *Intermetallics*, 104, July 2018, 24–32, 2019.
51. Gordon, J.V., Haden, C.V., Nied, H.F., Vinci, R.P., Harlow, D.G., Fatigue crack growth anisotropy, texture and residual stress in austenitic steel made by wire and arc additive manufacturing. *Mater. Sci. Eng. A*, 724, December 2017, 431–438, 2018.
52. Simonelli, M., Tse, Y.Y., Tuck, C., Effect of the build orientation on the mechanical properties and fracture modes of SLM Ti-6Al-4V. *Mater. Sci. Eng. A*, 616, 1–11, 2014.

53. Agrawal, P. *et al.*, Processing-structure-property correlation in additive friction stir deposited Ti-6Al-4V alloy from recycled metal chips. *Addit. Manuf.*, 47, July, 102259, 2021.
54. Yao, H. *et al.*, High strength and ductility AlCrFeNiV high entropy alloy with hierarchically heterogeneous microstructure prepared by selective laser melting. *J. Alloys Compd.*, 813, 152196, 2020.
55. Kreitcberg, A., Brailovski, V., Turenne, S., Effect of heat treatment and hot isostatic pressing on the microstructure and mechanical properties of Inconel 625 alloy processed by laser powder bed fusion. *Mater. Sci. Eng. A*, 689, 1–10, 2017.
56. McLouth, T.D. *et al.*, Variations in ambient and elevated temperature mechanical behavior of IN718 manufactured by selective laser melting via process parameter control. *Mater. Sci. Eng. A*, 780, March, 139184, 2020.
57. Stoudt, M.R. *et al.*, Location-specific microstructure characterization within IN625 additive manufacturing benchmark test artifacts. *Integr. Mater. Manuf. Innov.*, 9, 1, 54–69, 2020.
58. Nguejio, J., Szmytka, F., Hallais, S., Tanguy, A., Nardone, S., Godino Martinez, M., Comparison of microstructure features and mechanical properties for additive manufactured and wrought nickel alloys 625. *Mater. Sci. Eng. A*, 764, July, 138214, 2019.
59. Ren, L. *et al.*, Effects of interpass cooling on material properties of wire arc additive manufactured Al-6.3Mg alloy. *3D Print. Addit. Manuf.*, 6, 6, 344–353, 2019.
60. Gangireddy, S., Gwalani, B., Liu, K., Faierson, E.J., Mishra, R.S., Microstructure and mechanical behavior of an additive manufactured (AM) WE43-Mg alloy. *Addit. Manuf.*, 26, September 2018, 53–64, 2019.
61. Zhang, X. *et al.*, Study on microstructure and tensile properties of high nitrogen Cr-Mn steel processed by CMT wire and arc additive manufacturing. *Mater. Des.*, 166, 107611, 2019.
62. Sridharan, N., Wolcott, P., Dapino, M., Babu, S.S., Microstructure and mechanical property characterisation of aluminium–steel joints fabricated using ultrasonic additive manufacturing. *Sci. Technol. Weld. Join.*, 22, 5, 373–380, 2017.
63. Li, D., Shen, X., Yang, H., Liu, Z.x., Zhao, L., Wang, H., Quantitative distribution characterization of gradient composition of additive-manufactured stainless steel using micro-beam X-ray fluorescence. *Spectrochim. Acta - Part B At. Spectrosc.*, 183, July, 106268, 2021.
64. Chen, Z., Wu, X., Tomus, D., Davies, C.H.J., Surface roughness of selective laser melted Ti-6Al-4V alloy components. *Addit. Manuf.*, 21, July 2017, 91–103, 2018.
65. Nakatani, M., Masuo, H., Tanaka, Y., Murakami, Y., Effect of surface roughness on fatigue strength of Ti-6Al-4V alloy manufactured by additive manufacturing. *Proc. Struct. Integr.*, 19, 294–301, 2019.

5

Additive Manufacturing of Polymeric Materials: Process and Properties

Devarajan Balaji

Department of Mechanical Engineering, KPR Institute of Engineering and Technology, Coimbatore, Tamil Nadu, India

Abstract
Polymers are the most important technological advancement of the last century. Polymers' low cost, easiness of production, resistance to water, and flexibility make them useful in a wide variety of settings, including the home and the workplace. Three-dimensional (3D) printing or additive manufacturing (AM), is a digital fabrication technique that uses a digital model to produce a physical object. This chapter explores the topic of rapid manufacturing with polymers in greater depth. Polymers that remember their previous shapes is one type of "smart material" that has been extensively researched [four-dimensional (4D) printing materials]. Additive manufacturing uses a wide range of polymeric materials, including bio-polymers and polymers infused with biological matter in addition to the more well-known thermoplastics, elastomers, and thermosets. The density, thermal stability, and mechanical properties of the product are analyzed to establish its quality. Polyether-ether ketone (PEEK), polyetherimide (ULTEM), acrylonitrile butadiene styrene (ABS), nylon, and polycarbonates (PC) are all commonly used polymers for applications requiring thermoplastics. An updated list of polymeric materials used in 3D printing's strength parameters was developed and tests showed that materials like polybutylene terephthalate have a high yield strength. Some 3D printing polymers had their strength parameters revised. For instance, new tests revealed that polybutylene terephthalate's yield strength is higher and polycaprolactone's ultimate tensile resilience is higher than previously thought.

Keywords: Polymeric materials, manufacturing process, properties, additive manufacturing

Email: balaji.ntu@gmail.com

R. Rajasekar, C. Moganapriya and P. Sathish Kumar (eds.) Additive Manufacturing with Novel Materials: Processes, Properties and Applications, (163–196) © 2024 Scrivener Publishing LLC

5.1 Introduction

One of the most influential technological developments since the industrial revolution, additive manufacturing (AM) has dramatically altered the structure of the industrial segment [1–3]. It is an enhanced method for composites processing and collection of methods that build up complex shaped parts one layer at a time. Common descriptions include rapid prototyping (RPT), free-form fabrication (FFF), minimal fuel utilization, and high aesthetic adaptability [4–6]. The AM method is highly flexible, scalable, and customizable, making it applicable to a wide range of manufacturing settings [7]. The structure, medical services, dentures, and aviation industries [8, 9] are just a few that have recognized the great value added by these traits, along with others like them. When compared to traditional manufacturing methods, AM's greatest strengths lie in the fact that it does not necessitate the use of molds or other special tools to create personalized parts and that it can produce these components in a shorter amount of time. In order to synthesize mock-ups, AM technology depends solely on computer-aided design (CAD) software, which allows for the creation of unique virtual models and 3D models [5]. These programs slice 3D models into horizontal and vertical planes that denote coatings for polymer laser printing. These design documents feature details regarding the properties of the substance and the printing parameters utilized by different printing methods as well [10]. When compared to the conventional, subtractive manufacturing methods currently in use, AM is fundamentally distinct. Instead of using foundry or molding techniques, this method employs a bottom-up or top-down approach, in which custom form is built layer by layer. Components such as materials, ceramic materials, polymers, and their mixtures, composites, and blends can all be used in the AM process. In order to make AM useful for diagnostic and reproductive purposes, a great deal of research must be done to overcome the obstacles connected with its two initial technologies, components and calibration [7]. However, because of its capacity to generate actual, intricately structured, novel, and durable components, AM technology has seen an exponential increase in interest at the industrial scale. An up-and-coming field of study in 3D printing (3DP) is the creation of complex geometries using state-of-the-art computer design software. Since fiber polymers have restrictions on printability and mechanical properties, the scientific community has been working to create elevated achievement 3D-printed polymer mixtures. These materials have many benefits like low price, high accuracy, reduced raw material waste and chemical utilization, and the ability to print intricate configurations according to individual specifications [11]. These refined geometries have a wide range of possible uses in industries like wrapping, healthcare, synthetic biology, transportation, aviation, and electronic parts design [12–14]. Particle, fiber, and nanopolymer composites will be used to improve the properties of each and every polymer [15–17] in a variety of 3DP processes.

Oil depletion issues in the production of non-decay conventional polymer have grown drastically, and the accumulation of these materials in the environment as a result of commercially available production techniques has unfortunately disrupted the ecosystem [18, 19]. Concerned about the environmental impact of using synthetic polymers and traditional fabrication methods like injection molding, compression molding, and thermoforming, scientists have been working to develop more sustainable materials and adaptable techniques as possible replacements. There has been some interest in this area since feasible and biopolymers can be made from sustainable energy resources and reused multiple times without breaking down. While 3DP, also referred to as AM, is the most cutting-edge method currently available and one of the best answers to the issues surrounding the accumulation, open dumping, and removal of polymer wastes, the latter three are still very much an issue.

Utilized artificial plastics can be blended with recycled particulate fillers like starch, [20] and Musa fibers [21] to fulfill the multiple necessary characteristics based on focused implementation, and this is an emerging field. Powdered conventional polymers like polyethylene terephthalate (PET) [22] and sustainable energy biobased materials like sawdust, rice hull, and grass fibers [23] have been used extensively in recent years for 3DP via selective laser sintering (SLS) and inkjet 3DP, respectively, to produce complex 3D geometries. As a result, powder-based 3D printer is one of the methods that could usher in revolutionary changes on an industrial scale as a result of its ability to rapidly tool and prototype in a variety of production domains. Nontoxic, biomaterial, quickly moldable, and structurally sturdy are all desirable qualities in a plastic wrapping or transplant substance [24–26]. Fixation devices, scaffolds, bone-like complicated systems, augmentations, wrappings, and auto components are just some of the many examples of application-specific polymer objects that 3D printers can produce with greater efficiency, improved precision, least mold design, and less time than conventional fabrication methods like injection molding and thermoforming [27, 28]. Recently, materials and ceramic materials are broadly used in industrial 3DP [29–31]. However, the handling screen, fragility, lesser mechanical characteristics, suitability with other traditional polymers, and poor reliability of sustainable polymers limit their use at the industrial level. Poly lactic acid and polycaprolactone are two biobased and/or biodegradable polymers that have been investigated for use in 3DP to create biomaterial structures for biomedical implementations like bone tissue biocompatibility [32, 33].

Additive manufacturing, as the term is commonly used, encompasses a number of distinct but related technologies. As the final product is created by subtracting the material, other traditional methods like compression and injection molding, and heating of a thermoplastic material are classified as subtractive manufacturing approaches [34, 35]. Alternatively, in AM, new material is added during the manufacturing process. Layer-by-layer material addition in 3DP is highly efficient due to the design's versatility [31]. Depending on the materials used and the underlying principle, different 3DP technologies have been developed. All technological advances, regardless of type, are cumulative. Many subsets of AM exist because of the wide range of possible customizations. 'It is worth noting that AM technologies are classified in different ways by the American Society for Testing and Materials (ASTM). The scientific and technological effect of AM has grown since 1980, when the initial marketing device relying on 3DP was introduced. Research on 3DP continued steadily for the next 10 years. Over the past four to five years, however, focus has shifted to biodegradable polymers. With a market size of $4.4 billion in 2013, 3DP has since grown to $12 billion in 2018. In addition, by 2021, analysts anticipate a significant rise to $21 billion. Researchers and manufacturers are becoming increasingly interested in 3DP for a wide range of AM-related applications, including those in the automobile, medical, aviation, power, and consumer-goods industries [36]. Three-dimensional printing has advanced to the point where it is not only a reliable technique, but also allows for the fabrication of bespoke products from a wide range of materials, including metal alloys, ceramic materials, and polymeric materials [13, 37, 38].

5.1.1 Overview of AM Process for Polymeric Materials

Fused deposition modeling (FDM), SLS, and stereolithography (SLA) are polymer-based 3DP techniques used to produce polymeric parts. This 'section's main objective is to provide a comparative analysis by comparing polymers' material, processing, and properties processing, and properties of polymers for three different 3DP techniques. Additionally, using magneto-active polymers as an example, the advanced material-process necessities for 4D printing via those print methods are covered. Overall, this section discusses various aspects of these printing techniques and acts as a guide for choosing a suitable print medium and 3DP technique for the intended applications based on polymeric materials. It also covers implementation strategies for 4D printing of polymer-based systems using the most up-to-date methodology as shown in Figure 5.1 [117].

The specific variants to print polymeric materials like SLA and FDM are discussed in this segment.

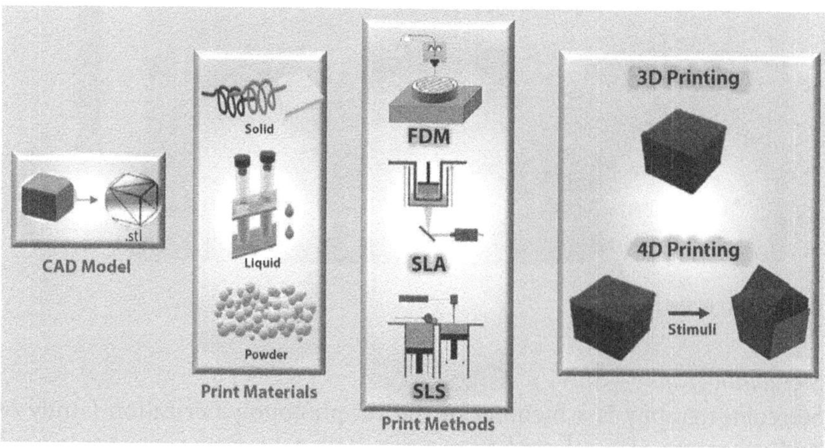

Figure 5.1 Printing of polymeric materials [117].

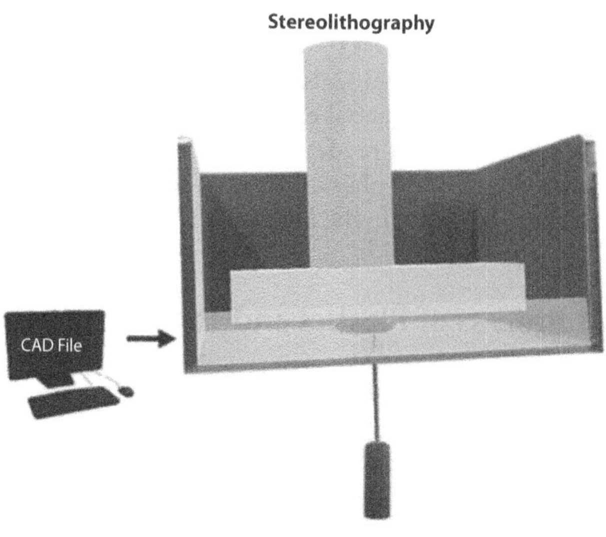

Figure 5.2 SLA [118].

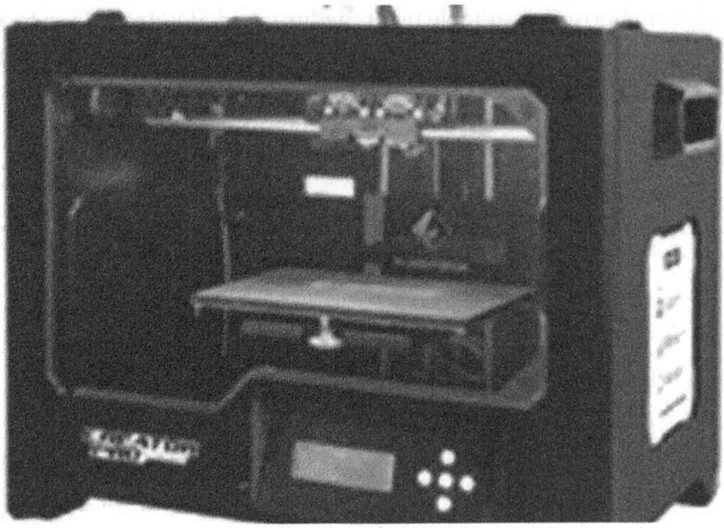

Figure 5.3 FDM [119].

Stereolithography—SLA

Stereolithography is a member of the vat photopolymerization family of AM processes, also referred to as resin 3DP. A laser or projector is used to cure liquid resin to hardened polymeric materials. The positioning of essential parts like the resin tank, build platform, and light source distinguishes one product from another physically as represented in Figure 5.2.

Fused deposition modeling (FDM)

Fused filament fabrication (FFF), also referred to as FDM is an AM technique used in the field of material extrusion. In FDM, melted material is selectively deposited along a predetermined path to build parts layer by layer. To create the finished physical objects, filaments of thermoplastic polymers are used, as shown in the Figure 5.3.

5.2 AM Process of Plastics and Their Properties

Acrylonitrile butadiene styrene, polypropylene (PP), polyethylene (PE), and PET can be used in AM which can be recycled as petroleum-based polymers, and allows for greater material sustainability and reusability at large manufacturing facilities [39, 40]. As well as cutting down on production time and costs, the widespread adoption of AM [3, 41] has also drastically decreased the amount of polymer waste created both during

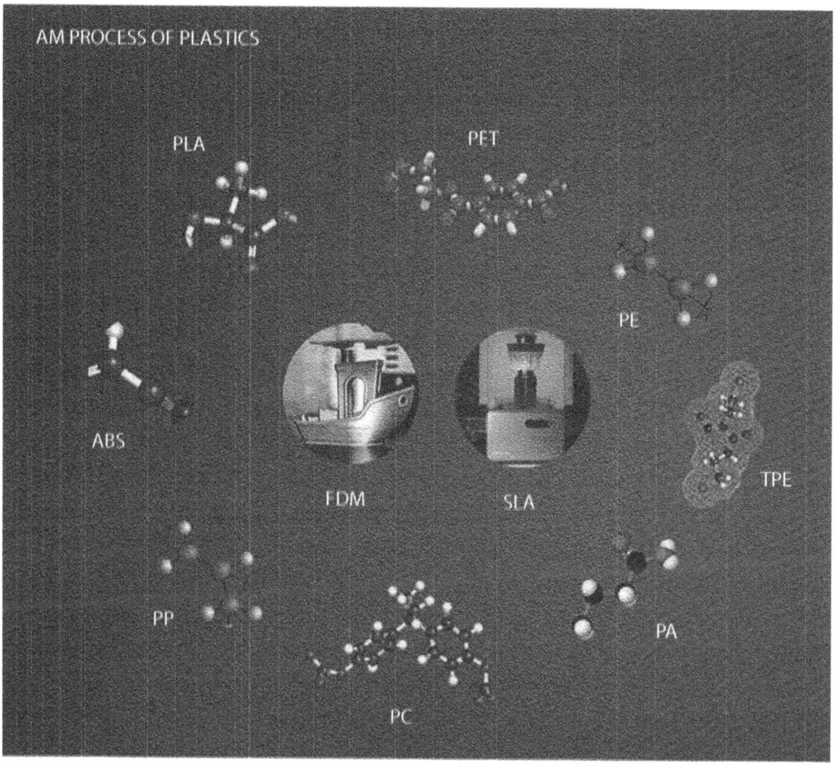

Figure 5.4 AM process of plastics.

and after processing. It also lends credence to the idea of PE composting, which is the main focus in the drive to cut down on waste sent to landfills. Fewer polymers can be recycled due to the dynamic and heat characteristics needed for 3D-printed products [42]. Alternative plastics to traditional petro-based plastics that can be used in AM without sacrificing performance are in high demand [43]. These include polylactic acid (PLA), polybutylene succinate, and polyhydroxyalkanoates. By reducing plastics waste and the need for landfills, the use of biodegradable polymers can help decrease the carbon footprint [44]. Using 3DP, this section explains how regular plastic composites can be used in new ways as shown in Figure 5.4.

5.2.1 Acrylonitrile Butadiene Styrene (ABS)

The amorphous thermoplastic known as ABS is made by reacting styrene and acrylonitrile in the existence of polybutadiene. An extremely tough and lightweight polymer, acrylonitrile butadiene styrene has superior influences resilience, solidity, smooth, and good electrical characteristics. Focused hydrochloric acid, sulphuric acid, ethyl alcohol, alkaline solutions, and vegetable oils do not harm it covalently [45]. But when combined with glacial acetic acid (CH_3COOH), carbon tetrachloride (CCl_4), and aromatic hydrocarbons, ABS can expand. Propylene chlorine, amides, and ketone bodies all can disintegrate with ABS with relative ease. Due to its high printability, it has become one of the most popular polymers for 3DP. ABS printability studies have looked at how handling with heat, bed heat, and scan speed. ABS specimens printed with an FFF-based 3D printer have had their mechanical properties analyzed by Morales *et al.* [46]. It was found that as the rate of cooling between adjacent layers decreases, so does the effective yield strength. ABS's mechanical properties can be enhanced through ultrasonication of 3D printable samples [47]. Tensile properties and elastic modulus of ABS specimens were improved by 11.3% and 16.7%, respectively, after undergoing ultrasonic treatment with tuneable reinforcement stress and duration. Tofangchi *et al.* [48] conducted similar studies to examine the impact of pulse sensations on ABS during printing. It was determined that ultrasonic vibrations increase stress relief of the polymerization from secondary interactions at the interaction among two parallel cells of ABS, resulting in a 10% increase in the interlayer adhesion strength. Further, after incorporating conductive zinc oxide (ZnO), the impact of filler content and printing pattern on mechanical and thermo-electrical properties of ABS were researched, as represented in Figure 5.5 [49]. When the infill density increased, the material's tensile, elastic modulus, vibrant bulk modulus, and conductance also increased. Nonetheless, as infill density increased from 50% to 100%, break elongation loss modulus, and dissipative factor all decreased. In addition to styrene-butadiene styrene and ultrahigh molecular weight polyethylene [50], montmorillonite clay [51], layered silicates [52], and other petroleum-based and biobased materials, ABS can be combined with a range of other resources. New findings from studies on 3DP with ABS are presented here. There is evidence that composites made from ABS have greater in-plane tensile strength than the virgin ABS specimen. Matrix cracking, fiber pulling, and debonding at the matrix-fiber interface were all observed, however. It was found that the hardness value of ABS had decreased by introducing the moisture-sliced carbon fibers and printing the specimens using FFF-based 3DP method [53]. Along with the addition of ABS, 3DP becomes a viable option for processing metal-organic frameworks (MOFs), the crystal materials most utilized in vapor adsorption, catalyst supports, filtration, and detecting. Kreider *et al.* [54] used ABS-MOF composites in similar ways to create hydrogen storage devices by means of 3DP. In order

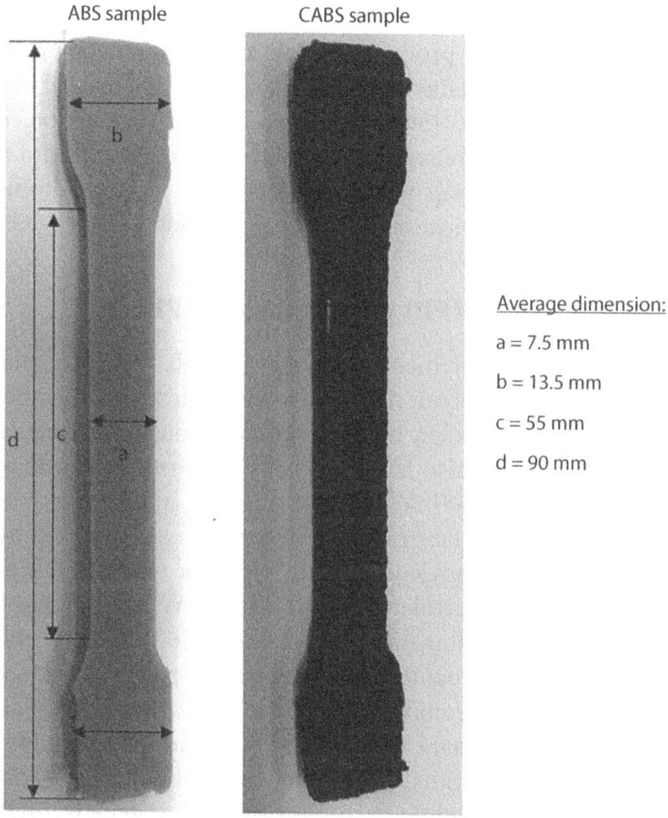

Figure 5.5 Conductive ABS printed samples [49].

to create the composite, an MOF composition of 10% or less was used in the ABS. It is observed that 3DP has even found a use in the culinary world, in the form of a most accurate viscous chocolate printer. As part of this field of study, 3D printers that use electrostatic inkjet technology were created. Changes to the printer's nozzle design using ABS resin microfibers which are harmless and resistant to current fields have resulted in more precise material printing.

Researchers and industrialists are focusing on conventional polymer recycling as a means to combat the environmental problems caused by the excessive production and disposal of polymer wastes. When it comes to making complex recycled objects, 3DP is among the most popular and straightforward methods. In their research, Woern *et al.* [55] used the identical polymer recycling strategy, printing polymers (PET, PP, PLA, and ABS), and the most common disposal plastics to compare the robotic characteristics of retrieved polymers to those of original PLA printed using conjoined element production 3DP at streamlined characteristics. Recycled thermoplastic resins, such as ABS and PLA, were used in the 3DP filaments used by Lee *et al.* [56], which help to lower production costs and carbon emissions. To increase the durability and conductivity of recycled PLA and ABS, the materials were combined with carbon nanotubes and polydopamine in the fabrication of high-value goods.

5.2.2 Polypropylene (PP) vs Polyethylene (PE)

Tough thermoplastic polymer produced through the molecular polymer via chain-growth polymerization, polypropylene is a member of the polyolefin family. Non-polar, half-crystallinity, and highly unaffected to chemicals, this polymer also has marginally improved hardness and heat resistance than PE [57, 58]. Wrapping, labeling, automotive, healthcare, research lab plasticware, fabrics, and many other uses [59] require it regularly. Powdered polypropylene is used in SLS, a type of 3DP that uses commercially available petroleum-based plastics. When it comes to treating wastewater in the water industry, Tan *et al.* [60] used PP fine particles to create net-like structures used in circular pattern layer segments. Because of its rigidity, flexibility, and chemical resistance, PP-based feed path stabilizers are necessary to sustain and defend the lining from damage and eroding. The optimal processing constraints for SLS-based 3DP of PP-based samples with a feed connection spacers type of infill design included laser intensity, scan structure, coating texture, temperature difference, and power density. Conclusions about the relationship between the Young's modulus of PP materials and the precision of 3D-printed objects' dimensions were also reached. Edge energy density was found to positively correlate with both yield strength and dynamic behavior. In 3DP ductile composites with a net type structure, the ductility and mechanical resilience enhanced sequentially with edge energy density. Fused filament fabrication was also used to create PP/FH and PP/PH specimens [61] using PP in conjunction with hydrocarbon resins fully and partially (FH, PH). Warping issues were clearly observed when 3DP with natural PP, but not when FH and PH were introduced at varying percentages (5, 10, 15, and 20 wt%) into the PP matrix, as shown by the specimen 3D printed with these modifications. A decrease in warping was observed as FH and PH content was varied; this was attributed to improved adhesion between the build platform and the first printed layer. No appreciable difference in Young's modulus was found between 0 and 45 degrees of raster angle for 3D-printed PP/PH blends. Using polarized optical microscopy (POM), we compared natural PP to PP/PH (80/20) and PP/FH (80/20) mixtures and found that the former had better nucleation sites and the latter had clearer interference in their morphology due to smaller spherulites. Due to its pesticide resistance, PP is also employed as response wares manufactured via 3DP for synthetic fabrication. For the purpose of creating a PP-based 3D-printed encased fuel rods that is capable of carrying out a multiphase organic synthesis, Kitson *et al.* [62] established a thorough procedure for the advancement of reaction ware device. Chemical experiments necessitate the use of custom-made reaction ware, made by means of 3DP technology. Multiple positions for initiating materials, reagents, and purification columns are available on the concocted response clothing gadget. In addition, an SLS-related 3DP technique was utilized to prepare PP composites with other fillers. There is a need for study into the viability of AM using blends/composites with improved analytical properties. Carbon fibers [63], bamboo fibers [38], cellulose [64], and glass spheres [66] were among the fillers added to PP in this direction, while the printability of PP-based composites was also evaluated. The structural resilience of the matrix was improved by reinforcing it with short CF in a PP matrix [63]. Short carbon fibers embedded in a PP matrix were used to increase the matrix's structural resilience [63]. The preferable orientation of CF in the printing direction resulted in an anisotropic behavior in the fabricated composites, which improved their physical behavior and thermal conductivity. Different complex designed objects were fabricated by Parandoush *et al.* [65] using PP as a mixture with crystal fibers as reinforcement. Outstanding interfacial bonding without void formation was observed in manufactured objects. It was discovered that SLS-related 3DP composites had maximum mechanical properties assessed to FFF-related 3DP PP/short fiber composites and were on par with compression molded samples in terms of mechanical properties. Spoerk *et al.* [64] fabricated 3D-printed objects with a PP matrix and borosilicate glass spheres of varying sizes as a reinforcement phase using FFF-related AM, and they studied the impact of additive region of space and barrel heat on dynamic, heat, sudden load, microstructure, anatomy, and surface roughness characteristics as represented in Figure 5.6. In this case, the glass spheres (12 m) were discovered to be better than all others due to their superior properties. When the temperature in the printing chamber was raised to 55°C, 3D-printed parts were annealed in the process, leading to greater precision in their dimensions. Blends of PP with polystyrene [67], styrene-(ethylene-butylene)-styrene (SEBS) copolymer [68], polycaprolactone [59], thermoset polychloroprene [69], and biopolymers, in specific PLA [70], were also utilized to improve 3D printability and related properties. The printed composite was given durability, physical, and temperature resilience through FFF-related 3DP with the help of a mixture of PP cured bamboo fibers. Terephthalate polypropylene, which acts as a compatibilizer to improve the adhesion strength among the polymer matrix and the treated bamboo fibers, was added to improve the compatibility between the two. Three-dimensional-printed PP/PLA/treated bamboo fiber saw increases in tensile strength by 13, flexural strength by 11.7%, and notched impact strength by 23.5%.

Porous polymer structures were developed using FFF, and one type of PP called expanded polypropylene (EPP) was used. Expanded polypropylene is a flexible closed-cell filament foam containing CO2 gas. Lightweight, thermal efficiency, shock resistance, chemical stability, and high durability are just some of the many desirable properties shared by EPP and PP. The dynamic pressure of PP, scan speed, laminate width, and nozzle temperature are all factors that can affect the mentioned properties. The growth in foaming ratio was due to a rise in pore number and pore volume without any corresponding rise in porous structure. Polypropylene's thermoplastic cousin, PE, shares many of the same properties as PP. Due to its machinability, impact resilience, good water obstruction, and chemical resistance [71], thermoplastic petroleum-based polymer PE is widely used in the production of thermoplastics, tissues, and boxes. Polyethylene comes in a wide variety of chemical structures, densities, and branching patterns, including low density (LDPE), linear low density (LLDPE), high density (HDPE), high molecular weight (HMWPE), and ultra-high molecular weight (UHMWPE) [73]. Ultra-high molecular weight PE is just one type of PE that is used as a resource in 3DP to make things with complicated geometries, such as –anti-friction goods used mechanical and healthcare engineering [72]. Intricately shaped products are created using a variety of 3DP techniques, including FFF and SLS. High density PE powder containing copper was used as a feedstock to create flexible 3D circuitry with semiconducting tracks [74]. It is possible that the cooling of HDPE during SLS-based 3DP, which requires a greater input signal and a slower scanning speed, is to blame for the material's granular structure and, by extension, its poor conductivity. In order to turn the plastic wastes into worthy goods, reusing petroleum-based plastics is presently a top priority. When compared to conventional manufacturing methods, 3DP offers many advantages, including reduced waste and a low cost of processing. There is widespread agreement that recycled HDPE is superior to ABS as a material for use in manufacturing a variety of products. When compared to ABS pellets or powder, HDPE is a viable alternative due to its superior thermal properties, moisture refusal, and chemical stability [75].

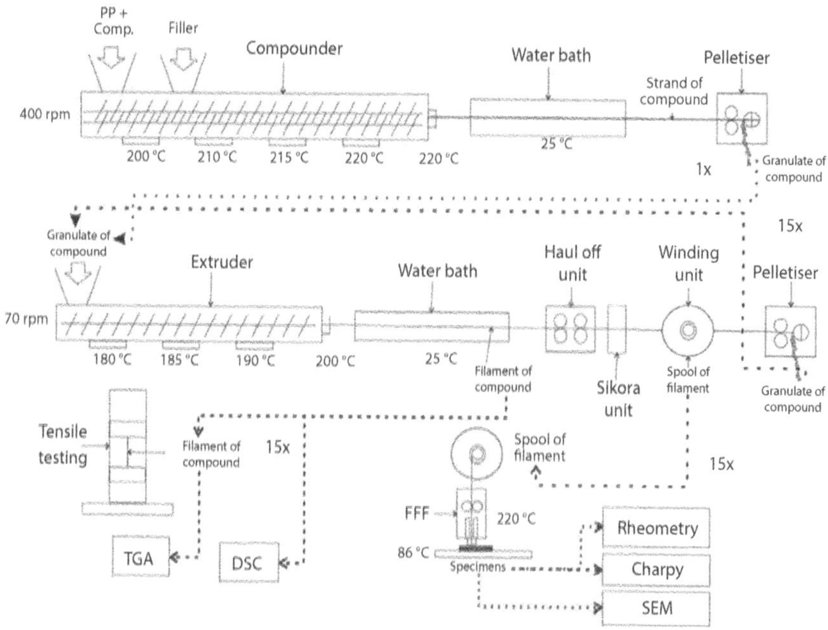

Figure 5.6 Schematic reparation of preparation of filament and testing [64].

5.2.3 Polyethylene Terephthalate (PET)

For those unfamiliar, PET is a polyester that is both colorless and semicrystalline, meaning it can be used in a wide range of temperatures. Textile fibers, liquid and food containers, thermoplastic goods, construction PC in circuit boards, and solar cell substances are just a few of the many uses for PET [76]. Due to its non-biodegradable nature, PET is typically recycled after its end use and is produced via condensation polymerization (followed by esterification and transesterification). With its superior yield stress, low coefficient of friction, less wear and greater stiffness [77, 78], PET is frequently used as a feedstock in AM methods for the fabrication of intricately designed objects. PET's bulk tensile strength is between 47 and 90 MPa, but in 3DP it is only between 27 and 45 MPa. However, the reused PET demonstrates a mean ductile stability of 40 MPa, indicating that the strength of the material is unaffected by the recycling process when using FFF-based 3DP. Additive manufacturing's potential to recycle plastic scrap into useful new products while simultaneously lowering operational costs and boosting material sustainability is compelling. Recycled PET was also found to have an elongation percentage that was on par with that of injection-molded materials. Differences in crystallinity have a major impact on the properties of 3D-printed components, including their ability to act as a barrier, transmit light clearly, and bend. Adding fillers, compatibilizers, and impact modifiers can help solve issues like distorting and disconnection in liquid state fiber that arise during 3DP [79]. These issues arise because of the material's crystallization and the 'filament's unequal size. On the same principle, Idrees *et al.* [80] used 3DP technology to create a new, minimum priced, 3D recycled PET specimen with enhanced thermal and mechanical properties. Aquafina water bottle recycling centers provided the source for the PET polymer, while packaging landfills provided the biochar. The biochar (100 m) was used as filler, and the recycled post-consumer PET bottles served as the matrix. Biochar was made by subjecting the packaging waste to a high temperature (1100°C) and induction pressure (150 bar). A CLTE value of 53.2 ppm °C 1 was measured for virgin recycled PET. Recycled PET/biochar composites, on the other hand, showed CLTE values between 37 and 50.8 ppm °C 1, indicating an increase in dimensional stability due to the biochar's addition. Reducing the CLTE value during 3DP means less deformation as a result of cooling. Fracture images obtained from a scanning electron microscope (SEM) revealed that virgin recycled PET was quite brittle. PET forms a fibrous framework and intercalated through biochar pore spaces, resulting in excellent interactions in composites. Powdered PET was also used in SLS-based 3DP to create complex shapes like spur gears [81]. After reviewing the available research, we have concluded that polyamide 12 -PA12 is the most suitable plastic resource for SLS-based 3DP. Recently, however, PE has been included because of its enhanced performance. Highly crystalline PET was found to have good print quality with SLS-based 3DP in the absence of a need for optimized particle shape and size. When compared to PA12, the printed parts made from highly crystalline PET maintained a higher quality surface finish and gloss. Withstanding extended storage in the build chamber (4 h), PET powder kept important characteristics such as molecular mass and crystalline nature [82].

5.2.4 Polycarbonates (PC)

Thermoplastic polymers with carbonate groups in their molecular forms are known as polycarbonates (PCs). Bisphenol A and phosgene react to create PCs, which are hard, robust, heat resistant and optically crystal plastics. Devices, concrete structures, data storage gadgets, intricate shaped parts for automobiles, healthcare, and mobile bodies are just some of the many engineering applications that make use of PCs' thermoformable and moldable properties. Plasticizer-challenged PC can be plasticized extensively without cracking, unlike most other thermoplastics. For these and other uses, the 3DP technique is fabricated on personal computers. Shemelya *et al.* [83] created x-ray radiation sheltering for space-based devices by 3DP tungsten-doped PC composites. After only 0.3% tungsten was added, the x-ray radiation-sheltering properties of 3DP tungsten-doped PC specimens increased by 10%. Incorporating more tungsten into PC enhanced its compressibility and complex viscosity without changing its tensile, ductility, impact, or glass transition heat. As was previously stated, PC has many uses, including those in the medical industry. For medical bone applications, Alaboodi *et al.* [84] used colorjet 3DP technology to create biocompatible PC-based structures. Compression tests were run on scaffolds with varying porosity (0%, 1%, 5%, 11%, 19%, and 30%) to determine ductile stability, Young's modulus, and shear modulus. Increases in porosity were found to decrease ductile stability, Young's modulus, and shear modulus. Using FFF-based 3DP and PC and PLA as raw material, Blaya *et al.* [85] developed a recyclable and eco-friendly orthopedic product (a custom-designed arm splint). It was also possible to create polymer blends with enhanced properties by combining PCs with other polymers. To determine the processing window, Blanco *et al.* [86] looked at how polyetherimide (PEI) affected the heat performance of 3D-printed PC samples. The results of the heat analysis and the microstructure revealed that PC and PEI are partially miscible. Still, incorporating PC into the mixture reduced viscosity and, in turn, expanded the accessible computing time. Peng *et al.* [87] combined PC and Surlyn in three different weight percentages (25%, 45%, 55%) to create filaments for 3DP, then 3D printed impact and tensile samples in 0°-, 15°-, 30°-, and 45°-degree orientations. The impact performance of PC/Surlyn blends was found to be enhanced by efficient energy dissipation, thanks to the formation of a simple core-shell morphology. In the stress-strain diagram for uniaxial stretching of pure PC samples, failure occurred right after the yield strength was reached. In contrast to PC/Surlyn blends, 3DP Surlyn exhibited lower ductile stress and elastic constants. The 3D-printed Surlyn had the lowest modulus related to original PC and PC/Surlyn blends because of the exceptional deflection of printable sample at comparatively minimal applied loads. Still, increases of 25% and 45% Surlyn caused a considerable reduction in the modulus of –3D-printed PC and a remarkable enhancement of toughness characteristics with core-shell shaped fibers. Three-dimentional-printed PC/Surlyn (55%) mix was found to have the highest toughness related to natural PC and other combinations due to its core-shell structure and enhanced convergence among PC and Surlyn. Liu *et al.* [88] developed CAD models and made use of porous PC material to 3D print an entirely different device with varying perforation angles (0°, 15°, 30°, and 45°) to increase its sound-absorbing capacity. This set of printed samples features perforations at a variety of angles. Researchers found that by tweaking the perforation angle, porosity, and air gap during printing, a significant improvement in the material's ability to absorb sound at low frequencies is possible. Mechanical specimens were 3D printed using PC and ABS by Rohde *et al.* [89] to test and improve shear properties. The composites were printed in flat, on edge, and upright alignments with varying raster sizes. When comparing ABS and PC, it was found that the ABS-related composites exhibited much greater aeolotropy depending on the build orientation. Whereas ABS-based raster orientation specimens showed little to no anisotropy, PC-based specimens showed significant anisotropy. In addition, 3DP was used to test the impact of recycling on physical and heat characteristics, and the usefulness of recycled PCs in a variety of implementations. By optimizing nozzle temperature and molten polymer velocity, Reich *et al.* [42] successfully 3D printed with recycled PC in particle/granule form. Recycled PC particle-based 3D-printed parts had a ductile stability of 64.9 Mega Pascal, which was found to be comparable to commercially printed parts. In addition, it was determined that 3D-printed components made from recycled PC particles can maintain greater physical resilience and increased thermal resistance without breaking the bank.

5.2.5 Thermoplastic Elastomers (TPE)

Thermoplastic elastomers are a type of copolymer in which one of the components is typically plastic and the other is rubber. To put it another way, TPE possesses both thermoset and elastic characteristics [90]. Alternatively, thermosets include most materials used to make elastomers. Thermoplastic elastomers outperform conventional rubbers and plastics in terms of durability, service life, and physical properties [91]. Using 3DP, Farstad *et al.* [92] created prototype wheels made of different grades of TPEs to test the friction of their surfaces and found that TPE composite materials could be a viable option for functional situations requiring high friction. The ability of 3D-printed components made from TPEs to bend without breaking or permanently deforming was also mentioned. A comparison of friction properties between standard rigid polymers like ABS and PLA and softer materials revealed that the latter have inferior properties. In this research, the lowest-density material, NinjaFlex, was found to have the highest friction coefficient among all forms of printable rollers. To save time and effort over the conventional micromolding process, Wang *et al.* [93] used FFF-based 3DP methods to manufacture pulsatile microchannels approaches (horizontal and vertical diaphragm with the aid of TPEs). In comparison to traditional polydimethylsiloxane (PDMS)-based systems, it was found that the 3D-printed microfluidic system made from TPEs can accurately identify minimum levels of insulin composition. The hydrophilic surface of 3D-printed TPEs is yet another benefit over PDMS that bolsters their biocompatibility. Applications require big heat able to operate variations and self-healing characteristics to stimulate the interlayer adhesion led Ellson *et al.* [94] to develop hard thiourethane thermoplastics as feed resource for FFF-based 3DP. Thiol-isocyanate polymerization was used to create the tough thiourethane thermoplastic. The synthesized material was extremely durable, with a low melting point and suitable thermal properties for the 3DP process. Applying stress while printing had no effect on polythiourethane's yield strength. Under stress, the commercial material's yield strength dropped by 45%. Furthermore, biomass was integrated into TPEs to enhance their sustainability and lend credence to the circular economy strategy. Using hot-melt extrusion and 3DP, Bi *et al.* [95] fabricated specimens with enhanced mechanical, thermal, and shape memory characteristics by combining TPU with wood flour (WF) and varying loadings of ethylene propylene diene monomer-grafted-maleic anhydride (EPDM-g-MAH) as compatibilizer. The tensile elongation and thermal stability of TPU/WF composites were improved by the addition of 4% EPDM-g-MAH to the formulation. The mechanism by which TPU/WF composites undergo a shape-memory effect is when the TPU/WF composite is heated to 60°C, this causes the material to deform and then spring back into shape. At elevated temperatures and in the presence of external forces, TPU's soft segment polymer chains were extended and oriented, allowing the material to take on any desired form. Cooling the deformed material below its temperature of glass transition of the soft segment, however, preserved the short-term form, and the atoms remained in a transparent position. The material's temporary shape was also preserved after application of external forces was ceased. Because of conformation hysteresis, the printed material gradually recovered after being heated above its glass transition temperature again. To get the 3D-printed material back to its original shape after being deformed, the temperature had to be raised above the melting point, at which point the soft segment began melting. The shape-memory effect was tested in a humanoid framework.

5.2.6 Polyamide (PA)

A polymer, or polyamide (PA), is a large mega atom in which amide bonds connect cyclic pieces. Step growth polymerization is used to create polyamides, a material that finds widespread use in a wide range of industries and products because of its resilience and reliability in environments ranging from automobiles to textiles to kitchenware to carpeting and sportswear. Nylons (PA6 and PA66) and aramids are two common types of polyamides used in 3DP technology. Extensive research was conducted on polyamides and their performances using a wide range of fillers, including bioceramics [97], graphene, and its oxide [98], copper [99], carbon black [22], crystal fibers [100]. There was also a notable increase in cell viability in the prepared composites. If graphene nanoplatelets are added to PA12 before 3DP using FFF technology, the material's thermal conductivity can increase by 51.4% and its elastic modulus can increase by 7% compared to compression-molded parts [98]. As a result of the graphene nanoplatelets being aligned in the printing direction, the final ductile resilience increased. With SLS-based 3DP, polyamides are used up faster than with FFF-based 3DP. In order to study rupture mechanism, Brugo *et al.* [98] utilized SLS-based 3DP to create polyamide-based intricate shapes. They found that printed specimens with a crack in every layer had superior mechanical properties. In a different study [100], SLS-based 3DP was used to prepare specimens made from PA12, a material with a distinct synthetic configuration than PA6. Isothermal and non-isothermal crystallization kinetics were used to explain issues similar to deforming of PA12-based specimens at the time of printing. The printed components had warped due to the nonuniform crystallization that happens when the liquid state PA12 emerges from the nozzle jet. This is because the microcrystalline polymers lose volume as they cool from their molten state. In addition to this issue, the physical nature of SLS-based 3D-printed PA12 samples was analyzed, as well as the visco-elasto-plastic reaction witnessed through use of tensile and compression, and relaxation testing with a variety of equipment [101]. The results showed that, with the exception of the deformation at break and stress relief behavior, SLS-based 3D-printed PA12 specimens were reusable below regulated printing factors. The PA12 was confirmed to be a bimodular substance through compression testing. The trend in 3DP innovation is toward creating lightweight, multipurpose materials. In this direction, a composite of polymeric material wrapped with graphene oxide nanostructure was developed using SLS 3DP technique for a metal-free electric motor [99]. Compared to pure PA, the moduli of compression and tensile stress were slightly diminished. In other studies [102], researchers used commercial PA6/carbon fibers, PA6/glass fibers, and PA6/Kevlar fibers to create samples for testing in both high- and low-pressure environments. Photographs of the vacuum chamber used to create low pressure during printing, as well as micrographs of cross sections of filaments made of PA6/

carbon fibers, PA6/glass fibers, and PA6/Kevlar fibers. Low-pressure 3DP of specimens was found to increase interlayer shear strength by as much as 33%, while decreasing porosity by as much as 12.7% in PA6/Kevlar fibers, 1.7% in PA6/glass fibers, and 5.7% in PA6/carbon fiber-based composites. Carbon black is a popular filler used to increase the heat and electromechanical permeability of polymeric composites. SLS-based 3DP was used by Espera Jr. *et al.* [22] to create dynamically strong intricate shapes made of PA12/carbon black particle specimens. The carbon black content had a linear relationship with the composites' bulk resistivity. When carbon black was added to PA12 at concentrations of 1.5% and 3%, it helped prevent cracks from propagating in response to stress. By increasing the 3% carbon black to PA12, there is a drop in its mechanical properties because the black pigment prevents the polymer and the carbon black from making physical contact. In contrast, its thermal stability was enhanced in comparison to PA12.

5.2.7 Polylactic Acid (PLA)

Polylactic acid is a type of thermoplastic aliphatic polyester that can be made through a number of different polymerization techniques [106]. Traditional fermentation methods can extract lactic acid from wheat, potato, cocoyam, pulp from sugar beets and tapioca [103]. Changing the orientation of the polymer chains affects the material properties of PLA. Depending on the structure of its chains, PLA can be categorized as either amorphous, semicrystalline, or highly crystalline. As an industrially compostable and recyclable material, PLA has found widespread use as a filler resource for 3DP. To increase impact strength and heat resistance, it experimentally optimized FFF-based 3DP factors for PLA and discovered a connection between chemical structure, properties, and processing. In this study, we analyzed the effects of annealing samples, melting temperatures, bed temperatures, and raster angles on PLA's crystalline phase and found that the heat-treated samples exhibited better properties. Higher bed temperatures (105°C) were found to result in noticeable increases in yield stress, flexural strength, impact resistance (80%), and heat distortion temperature. The impact of different fillers on the quantitative properties of PLA 3DP was explored. Wu *et al.* [104] combined the biodegradable properties of PLA and rice husk (RH) through melt extrusion and FFF-based 3DP to create biocomposites with the desired characteristics. Treating rice husk with a coupling agent and then fusing PLA with acrylic acid (AA) were the methods of experimentation (TRH). Lastly, melt extrusion and 3DP were applied to two composites (i.e., PLA/RH and AA-g-PLA/TRH). In order to improve the compatibility between the matrix and the reinforcement phase, we swapped out the PLA and RH for AA-g-PLA and TRH, respectively. Increased branching and a 3D network structure were found to be responsible for the enhanced mechanical behavior of AA-g-PLA/TRH. Similarly, Lebedev *et al.* [105] compared the mechanical characteristics of hot-pressed specimens to those of 3D-printed specimens using PLA with various fillers, including single-walled nanotubes, WF, bronze, and copper. The results indicated that the 3D-printed samples' %modulus of elasticity and yield strength were lower than the hot-pressed samples by 15%-60%. When it comes to mechanical properties, however, the deposition method makes no difference. Few types of complex molecular materials, such as carbon nanotubes (CNT), were used as filler or reinforcing agent because they can perform more than one function. This resulted in significant increases in conductivity, mechanical strength, and modulus since multi-walled carbon nanotubes (MWCNT) are extremely rich in terms of carbon content.

Researchers also used FFF-based 3DP to create capacitive composite intricately shaped goods made of PLA and various proportions of MWCNTs-based nanocomposites. By adding just 5% MWCNTs to 3D-printed PLA, the material's permeability, ductile strength, and deformation all improved significantly, making it suitable for use in the production of everyday consumables made with AM. The ability to print quality and other properties of PLA with natural fibers as additive or conditioning were investigated in a manner analogous to that of nanotubes. Wood fibers were used by Dong *et al.* [106] as filler in a PLA matrix to print specimens that were then tested for mechanical properties. Several printing parameters, including layer height, infill density, and shell count, were optimized via a design of experiments. The mechanical properties of the specimen were found to be maximised when the material type and the number of shells used in the specimen's design were optimized. Researchers have looked into the possibility of using PLA-based materials and composites as fire retardants to lessen the risk of fire in a variety of contexts [107]. Researchers used a cone calorimeter to compare the ignition times of 3D-printed PLA and injection-molded PLA, and found that the 3D-printed samples had a significantly lower ignition time than the injection-molded samples due to a more even distribution of the fire retardant and two types of phyllosilicates. Higher porosity in 3DP specimens related to injection-molded specimens helped to reduce amount of heat lost during the manufacturing process. Nanoclays have demonstrated significant synergy with fire retardants, making them a leading candidate for use as a fire retardant in PLA. Even binary blends of PLA, both with and without additive or reinforcement, showed promise in performance and could be utilized in 3DP to expand the range of implementations for this material. By combining

PLA and PBS in varying proportions, Ou-Yang *et al.* [108] were able to 3D print complex components using FFF. Increasing the proportion of PLA in PBS from 20% to 40% increased the melt viscosity, elastic modulus, and modulus with only a small increase in distortion. Dimensional stability and sheen were also present in the blend ratio. It is important to note that the blends ratio presented is optimal for 3DP because of its reduced crystallite size, medium viscosity, and high stiffness. Pellets were fabricated using optimized preparation for FFF-based 3DP to make comparisons for mechanical properties [109], and compression-molded samples were analyzed for thermal, mechanical, and morphological properties.

Cork is a renewable and biodegradable material that was mixed in several compositions with PLA to construct pellets. A decrease in tensile strength and a rise in density were ascribed to the addition of cork to PLA-based composites, it was suggested. Cork's closed-cell morphology gives it a very low density, which is why PLA/cork composites are lighter. However, there was an increase in the material's specific modulus, specific tensile strength, and impact strength. The addition of tributyl citrate (TBC), a biodegradable plasticizer, can alleviate PLA's brittleness. When TBC was added to PLA/cork-based composites, ductility increased while ductile yield different strength and modulus decreased. The physical findings, ductile yield strength and modulus of the 3D-printed PLA/cork-based composite specimens, are lower than those of the compression-molded samples. The amorphous phase in the composites was increased by the preferred printing conditions and the closed cell morphology of cork, so the %elongation was slightly better for 3D-printed samples than for compression molded samples. As opposed to short fibers, the quantitative characteristics of 3D-printed samples with continuous fibers reinforced polymer matrix showed marked improvement. Tensile, compressive, and flexural properties of 3D-printed CFRHCs (continuous fiber reinforced heterogeneous composites) were examined in relation to fiber arrangement design. Fibers made up 15% of the total volume and were separated into four groups (G1, G2, G3, and G4) [110]. G1 indicated that the fiber content of the lower 3D-printed layer was higher than that of the upper layer. The fiber distribution in G3 was the inverse of G1, with more fibers concentrated in the top 3D-printed layer and fewer fibers concentrated in the bottom layer. Both the upper and lower layers of G2 had low fiber content, while the middle layer had a high fiber content. G4, on the other hand, demonstrated how fibers were evenly dispersed across all the different layers. When loaded with G1 fiber content, elastic modulus strength and modulus were enhanced. The simulation model was also used to investigate the reasons why CFRHCs sometimes fail. Researchers examined the morphology of printed CFRHCs and found improved dispersion. Researchers also used sugarcane bagasse (SCB) cellulose fibers extracted via chemical treatment methods as filler in PLA matrices to fabricate filaments, which were then used in FFF-based 3D-printed specimens for analysis of printing orientation's effect on mechanical behavior. According to the results of 3D-printed PLA/SCB-based composites, the best tensile strength was achieved by including 6 wt% SCB in the PLA matrix. However, increasing the SCB content of the PLA matrix had a negative effect on the material's mechanical properties, especially its flexural modulus, tensile strength, and flexural strength. It was also determined that the printing performance suffered as a result of the G code change. Therefore, the "parallel" sample had greater tensile strength than the "vertical" sample. SCB boosted composites' thermal stability, which aided PLA melt processing.

The mechanical characteristics of FFF-based 3D-printed test samples were investigated in another study [111], which examined the impact of using both long and short poplar fibers derived from biomass as reinforcement in PLA matrix. The dispersion of PLA/poplar fibers increased their mechanical strength, which ranged from 34 to 54 MPa depending on fiber size. One of the most promising sustainable fillers, biocarbon, is created by pyrolyzing biomass at various temperatures in an inert gas environment. Biomass is transformed into biocarbon at elevated temperatures. Under an inert atmosphere, the maximum temperature for converting biomass into biocarbon can be as high as 700°C [113]. Biocarbon solids, gases, and oil are all produced during the pyrolysis process. The pyrolysis temperature affects the exterior framework of the biocarbon that is created. A review of the available literature revealed a lack of extensive investigation into biocarbon. Biocarbon's elevated yield strength and high carbon composition can be useful in automobiles and optoelectronics, and it is also used as a filler in a wide range of petroleum products and biodegradable plastics polymeric matrix for 3DP. Researchers Ertane *et al.* [112] took a step in this direction by incorporating biogenic carbon (5, 15, and 30 vol%) as an additive into PLA and then 3DP the resulting filaments to study their handling and wear behavior. After pyrolyzing wheat stems at 800°C in a vacuum oven, it was discovered that the resulting biocarbon had a higher content, which caused nozzle blocking issues and slowed down printing. In contrast, it was found that biocarbon particles were uniformly distributed in the PLA matrix, which reduced nozzle wear. The adherence nature of biocarbon granules from the heating element surface was the root cause of the increase in nozzle wear as biocarbon content rose. Ultimately, the biodegradable nature of the PLA/biocarbon composite allowed it to be used as a building material that enhanced soil quality. Also, Gao *et al.* [114] found that the tensile

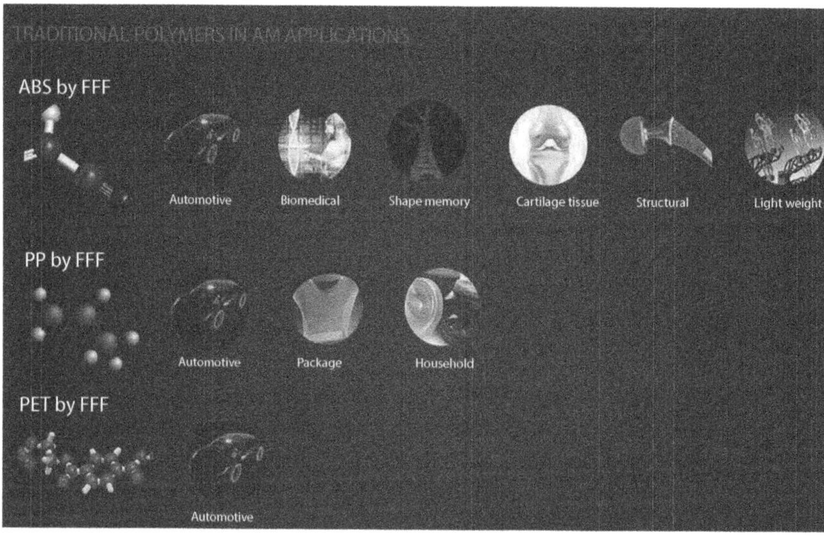

Figure 5.7 Polymers in versatile AM implications.

strength of PLA with different fillers (talc and carbon fibers) was significantly lower than that of injection-molded parts. Micrographs of printed parallel lines and corners from 3D-printed cubes made from neat PLA, PLA/talc, and PLA/carbon fibers portray the same. The filler in the PLA matrix altered the material's mechanical anisotropy. The filler's dispersion helped molecules move more freely through the bond's porous interface. Kalinke *et al.* [115] developed a non-enzymatic method for determining glucose by fabricating an electrode out of activated biochar (AB) that was supported by nickel oxyhydroxide and then coupling that electrode to an electroanalytical device. Biochar was created by pyrolyzing castor oil cake with nitric acid at °C; this process also increased the cake's contact area, operational collectives, and pore volume. When it came to the unexpected assessment of Ni (II) ions, AB performed admirably [116–119]. Figure 5.7 reveals the various implications of AM of conventional polymers.

5.3 Conclusion

Polymeric materials are predominantly used as an AM material. Acrylonitrile butadiene styrene, polypropylene vs polyethylene, polyethylene terephthalate, polycarbonate, thermoplastic elastomers, polyamide, and polylactic acid were discussed. The most suitable AM variant to print polymeric materials are fused deposition model (FDM) and Free form fabrication (FFF). The chapter revealed the method of processing the polymeric materials along with the characteristics. Widely available printers can print PLA and ABS, even the cheapest variants. Other materials including PET, PC, TPE, and PA require higher end printers. Assessing the strength of these materials, PLA is weakest and PC is stronger. Accordingly, the printer parameters should be altered for the requirements.

References

1. Wang, X., Zhao, L., Fuh, J.Y.H., Lee, H.P., Effect of porosity on mechanical properties of 3D printed polymers: Experiments and micromechanical modeling based on X-ray computed tomography analysis. *Polymers*, 11, 7, 1154, 2019.
2. Huang, S.H., Liu, P., Mokasdar, A., Hou, L., Additive manufacturing and its societal impact: A literature review. *Int. J. Adv. Manuf. Technol.*, 67, 5, 1191–1203, 2013.
3. Ford, S. and Despeisse, M., Additive manufacturing and sustainability: An exploratory study of the advantages and challenges. *J. Cleaner Prod.*, 137, 1573–1587, 2016.
4. Daminabo, S.C., Goel, S., Grammatikos, S.A., Nezhad, H.Y., Thakur, V.K., Fused deposition modeling-based additive manufacturing (3D printing): Techniques for polymer material systems. *Mater. Today Chem.*, 16, 100248, 2020.
5. Hari, B.S., Mahesh Kumar, K.V., Krishnamurthy, K., Sathish Kumar, P., Gobinath, V.K., Sachinbala, R., Rajasekar., R., Influence of graphene oxide on the morphological and mechanical behaviour of compatibilized low density polyethylene nanocomposites. *Mater. Today: Proc.*, 39, 1487–1493, 2021.
6. Melenka, G.W., Cheung, B.K., Schofield, J.S., Dawson, M.R., Carey, J.P., Evaluation and prediction of the tensile properties of continuous fiber-reinforced 3D printed structures. *Compos. Struct.*, 153, 866–875, 2016.
7. Tofail, S.A., Koumoulos, E.P., Bandyopadhyay, A., Bose, S., O'Donoghue, L., Charitidis, C., Additive manufacturing: Scientific and technological challenges, market uptake and opportunities. *Mater. Today*, 21, 1, 22–37, 2018.
8. Lille, M., Nurmela, A., Nordlund, E., Metsä-Kortelainen, S., Sozer, N., Applicability of protein and fiber-rich food materials in extrusion-based 3D printing. *J. Food Eng.*, 220, 20–27, 2018.
9. Liu, J., Sun, L., Xu, W., Wang, Q., Yu, S., Sun, J., Current advances and future perspectives of 3D printing natural-derived biopolymers. *Carbohydr. Polym.*, 207, 297–316, 2019.
10. Somireddy, M., Czekanski, A., Singh, C.V., Development of constitutive material model of 3D printed structure via FDM. *Mater. Today Commun.*, 15, 143–152, 2018.

11. Engel, K.E., Kilmartin, P.A., Diegel, O., Recent advances in the 3D printing of ionic electroactive polymers and core ionomeric materials. *Polym. Chem.*, 13, 4, 456–473, 2022.
12. Ramesh M, Rajeshkumar L, Balaji D, Bhuvaneswari V., Polymer composites for enzyme sensors, in: *Polymeric Nanocomposite Materials for Sensor Applications*, 343–366, Woodhead Publishing, USA, 2023.
13. Wolff, R., Ehrmann, K., Knaack, P., Seidler, K., Gorsche, C., Koch, T., Stampfl, J., Liska, R., Photo-chemically induced polycondensation of a pure phenolic resin for additive manufacturing. *Polym. Chem.*, 13, 6, 768–777, 2022.
14. Xu, X., Zhang, J., Jiang, P., Liu, D., Jia, X., Wang, X., Zhou, F., Direct ink writing of aluminum-phosphate-bonded Al2O3 ceramic with ultra-low dimensional shrinkage. *Ceram. Int.*, 48, 1, 864–871, 2022.
15. Raveendran, N., Ivanovski, S., Vaquette, C., The effect of culture conditions on the bone regeneration potential of osteoblast-laden 3D bioprinted constructs. *Acta Biomater.*, 156, 190–201, 2022.
16. Caldona, E.B., Dizon, J.R.C., Espera Jr., A.H., Advincula, R.C., On the economic, environmental, and sustainability aspects of 3D printing toward a cyclic economy, in: *Energy Transition: Climate Action and Circularity*, 507–525, American Chemical Society, Washington D.C., 2022.
17. Zhou, Y. and Schweizer, K.S., Theory for the elementary time scale of stress relaxation in polymer nanocomposites. *ACS Macro Lett.*, 11, 2, 199–204, 2022.
18. Surendren, A., Mohanty, A.K., Liu, Q., Misra, M., A review of biodegradable thermoplastic starches, their blends and composites: Recent developments and opportunities for single-use plastic packaging alternatives. *Green Chem.*, 24, 8606–8636, 2022.
19. Datta, D., Shetty, A., Das, B., Comparative analysis of the quantity and quality of both synthetic and biodegradable films on kitchen waste. *Mater. Today: Proc.*, 57, 1798–1804, 2022.
20. Noè, C., Cosola, A., Tonda-Turo, C., Sesana, R., Delprete, C., Chiappone, A., Hakkarainen, M., Sangermano, M., DLP-printable fully biobased soybean oil composites. *Polymer*, 247, 124779, 2022.
21. Cafiso, D., Septevani, A.A., Noè, C., Schiller, T., Pirri, C.F., Roppolo, I., Chiappone, A., 3D printing of fully cellulose-based hydrogels by digital light processing. *Sustain. Mater. Technol.*, 32, e00444, 2022.
22. Espera Jr., A.H., Dizon, J.R.C., Valino, A.D., Chen, Q., Silva, I.G., Nguyen, D.V., Rong, L., Advincula, R.C., On the 3D printability of silicone-based adhesives via viscous paste extrusion. *MRS Commun.*, 13, 1, 1–9, 2023.
23. Francis, J.K., Pullan, T.T., Joseph, M., Thomas, M.J., Joseph, K., Investigations on mechanical properties of wood composite for sustainable manufacturing. *Mater. Today: Proc.*, 72, 3111–3115, 2022.
24. Parkkinen, I., Development and optimisation of tools for preclinical studies on Parkinson's disease, Dissertationes Scholae Doctoralis Ad Sanitatem Investigandam Universitatis Helsinkiensis, Finland, 2022

25. Khan, E., Soomro, A.H., Saleem, M., Siyal, Z.H., Ali, S., Regenerative dentistry in Pakistan: Awareness, challenges, and advancements among general dental practitioners and specialists. *Eur. J. Gen. Dent.*, 11, 02, 108–113, 2022.
26. Kumar, K.V.M., Krishnamurthy, K., Rajasekar, R., Kumar, P.S., Pal, K., Nayak, G.C., Influence of graphene oxide on the static and dynamic mechanical behaviour of compatibilized polypropylene nanocomposites. *Mater. Test.*, 61, 10, 986–990, 20192019.
27. Ho, C.M.B., Hu, K., Mishra, A., Noh, J., Kim, J., Lee, S., Yoon, M., Yoon, Y.J., Printing of woodpile scaffold using fresnel lens for tissue engineering. *Int. J. Precis. Eng. Manuf.-Green Technol.*, 9, 2, 507–522, 2022.
28. Wu, C.Y., Chen, J.R., Su, C.K., 4D-printed pH sensing claw. *Anal. Chim. Acta*, 1204, 339733, 2022.
29. Arulmurugan B, Sasikumar GK, Rajeshkumar L. Nanostructured metals: Optical, electrical, and mechanical properties, in: *Mechanics of Nanomaterials and Polymer Nanocomposites*, 69–85, Springer Nature, Singapore, 2023.
30. Xiong, S., Liu, J., Cao, J., Li, Z., Idrees, M., Lin, X., Long, Z., Liu, Z., Wang, P., Liu, C., Chen, Z., 3D printing of crack-free dense polymer-derived ceramic monoliths and lattice skeletons with improved thickness and mechanical performance. *Addit. Manuf.*, 57, 102964, 2022.
31. Mulongo, N.Y., A review of value drivers of Industry 4.0 in the manufacturing sector using prisma approach. *Int. J. Inf. Syst. Comput. Sci. (IJISCS)*, 6, 1, 43–56, 2022.
32. Qin, L., Liu, S., Qin, S., Liao, L., He, M., Yu, J., MOF derived porous Ni/Co@C nanocomposite as electromagnetic wave absorber with optimized impedance matching. *Compos. Commun.*, 33, 101196, 2022.
33. Ghilan, A., Nita, L.E., Pamfil, D., Simionescu, N., Tudorachi, N., Rusu, D., Rusu, A.G., Bercea, M., Rosca, I., Ciolacu, D.E., Chiriac, A.P., One-step preparation of carboxymethyl cellulose—Phytic acid hydrogels with potential for biomedical applications. *Gels*, 8, 10, 647, 2022.
34. Dong, J., Hu, C., Holmes, J., Qin, Q.H., Xiao, Y., Structural optimisation of cross-chiral metamaterial structures via genetic algorithm. *Compos. Struct.*, 282, 115035, 2022.
35. Yoon, M. and Lim, C.S., Comparative experiments on amine vs. acid anhydride curing agents for epoxy resin required for automotive parts. *J. Polym. Res.*, 30, 1, 9, 2023.
36. Azlin, M.N.M., Ilyas, R.A., Zuhri, M.Y.M., Sapuan, S.M., Harussani, M.M., Sharma, S., Nordin, A.H., Nurazzi, N.M., Afiqah, A.N., 3D printing and shaping polymers, composites, and nanocomposites: A review. *Polymers*, 14, 1, 180, 2022.
37. Du, J., Zhang, Y., Wang, J., Xu, M., Qin, M., Zhang, X., Huang, D., Highly resilient porous polyurethane composite scaffolds filled with whitlockite for bone tissue engineering. *J. Biomater. Sci., Polym. Ed.*, 34, 7, 1–15, 2022.

38. Salifu, S., Ogunbiyi, O., Olubambi, P.A., Potentials and challenges of additive manufacturing techniques in the fabrication of polymer composites. *Int. J. Adv. Manuf. Technol.*, 122, 2, 577–600, 2022.
39. Vidakis, N., Petousis, M., Tzounis, L., Maniadi, A., Velidakis, E., Mountakis, N., Papageorgiou, D., Liebscher, M., Mechtcherine, V., Sustainable additive manufacturing: Mechanical response of polypropylene over multiple recycling processes. *Sustainability*, 13, 1, 159, 2020.
40. Little, H.A., Tanikella, N.G., J. Reich, M., Fiedler, M.J., Snabes, S.L., Pearce, J.M., Towards distributed recycling with additive manufacturing of PET flake feedstocks. *Materials*, 13, 19, 4273, 2020.
41. Mikula, K., Skrzypczak, D., Izydorczyk, G., Warchoł, J., Moustakas, K., Chojnacka, K., Witek-Krowiak, A., 3D printing filament as a second life of waste plastics—A review. *Environ. Sci. Pollut. Res.*, 28, 10, 12321–12333, 2021.
42. Reich, M.J., Woern, A.L., Tanikella, N.G., Pearce, J.M., Mechanical properties and applications of recycled polycarbonate particle material extrusion-based additive manufacturing. *Materials*, 12, 10, 1642, 2019.
43. Mohan, D., Teong, Z.K., Bakir, A.N., Sajab, M.S., Kaco, H., Extending cellulose-based polymers application in additive manufacturing technology: A review of recent approaches. *Polymers*, 12, 9, 1876, 2020.
44. Folino, A., Karageorgiou, A., Calabrò, P.S., Komilis, D., Biodegradation of wasted bioplastics in natural and industrial environments: A review. *Sustainability*, 12, 15, 6030, 2020.
45. Wang, S., Hu, Y., Song, L., Wang, Z., Chen, Z., Fan, W., Preparation and thermal properties of ABS/montmorillonite nanocomposite. *Polym. Degrad. Stab.*, 77, 3, 423–426, 2002.
46. Morales, N.G., Fleck, T.J., Rhoads, J.F., The effect of interlayer cooling on the mechanical properties of components printed via fused deposition. *Addit. Manuf.*, 24, 243–248, 2018.
47. Li, G., Zhao, J., Jiang, J., Jiang, H., Wu, W., Tang, M., Ultrasonic strengthening improves tensile mechanical performance of fused deposition modeling 3D printing. *Int. J. Adv. Manuf. Technol.*, 96, 5, 2747–2755, 2018.
48. Tofangchi, A., Han, P., Izquierdo, J., Iyengar, A., Hsu, K., Effect of ultrasonic vibration on interlayer adhesion in fused filament fabrication 3D printed ABS. *Polymers*, 11, 2, 315, 2019.
49. Aw, Y.Y., Yeoh, C.K., Idris, M.A., Teh, P.L., Hamzah, K.A., Sazali, S.A., Effect of printing parameters on tensile, dynamic mechanical, and thermoelectric properties of FDM 3D printed CABS/ZnO composites. *Materials*, 11, 4, 466, 2018.
50. Rocha, C.R., Perez, A.R.T., Roberson, D.A., Shemelya, C.M., MacDonald, E., Wicker, R.B., Novel ABS-based binary and ternary polymer blends for material extrusion 3D printing. *J. Mater. Res.*, 29, 17, 1859–1866, 2014.

51. Weng, Z., Wang, J., Senthil, T., Wu, L., Mechanical and thermal properties of ABS/montmorillonite nanocomposites for fused deposition modeling 3D printing. *Mater. Des.*, 102, 276–283, 2016.
52. Ramesh, M., Balaji, D., Rajeshkumar, L., Sivalingam, S., Aerogels of sodium alginate and their behavior toward wastewater treatment. In: *Sodium Alginate-Based Nanomaterials for Wastewater Treatment*, 273–287, Elsevier, UK, 2023.
53. Young, D., Wetmore, N., Czabaj, M., Interlayer fracture toughness of additively manufactured unreinforced and carbon-fiber-reinforced acrylonitrile butadiene styrene. *Addit. Manuf.*, 22, 508–515, 2018.
54. Kreider, M.C., Sefa, M., Fedchak, J.A., Scherschligt, J., Bible, M., Natarajan, B., Klimov, N.N., Miller, A.E., Ahmed, Z., Hartings, M.R., Toward 3D printed hydrogen storage materials made with ABS-MOF composites. *Polym. Adv. Technol.*, 29, 2, 867–873, 2018.
55. Woern, A.L., Byard, D.J., Oakley, R.B., Fiedler, M.J., Snabes, S.L., Pearce, J.M., Fused particle fabrication 3-D printing: Recycled materials' optimization and mechanical properties. *Materials*, 11, 8, 1413, 2018.
56. Lee, D., Lee, Y., Lee, K., Ko, Y., Kim, N., Development and evaluation of a distributed recycling system for making filaments reused in three-dimensional printers. *J. Manuf. Sci. Eng.*, 141, 2, p.021007, 2019.
57. Natta, G. and Corradini, P., Structure and properties of isotactic polypropylene, in: *Stereoregular Polymers and Stereospecific Polymerizations*, pp. 743–746, Pergamon, UK, 1967.
58. Maddah, H.A., Polypropylene as a promising plastic: A review. *Am. J. Polym. Sci.*, 6, 1, 1–11, 2016.
59. Graziano, A., Garcia, C., Jaffer, S., Tjong, J., Sain, M., Novel functional graphene and its thermodynamic interfacial localization in biphasic polyolefin systems for advanced lightweight applications. *Compos. Sci. Technol.*, 188, 107958, 2020.
60. Tan, W.S., Chua, C.K., Chong, T.H., Fane, A.G., Jia, A., 3D printing by selective laser sintering of polypropylene feed channel spacers for spiral wound membrane modules for the water industry. *Virtual Phys. Prototyping*, 11, 3, 151–158, 2016.
61. Das, A., Marnot, A.E., Fallon, J.J., Martin, S.M., Joseph, E.G., Bortner, M.J., Material extrusion-based additive manufacturing with blends of polypropylene and hydrocarbon resins. *ACS Appl. Polym. Mater.*, 2, 2, 911–921, 2019.
62. Kitson, P.J., Glatzel, S., Chen, W., Lin, C.G., Song, Y.F., Cronin, L., 3D printing of versatile reactionware for chemical synthesis. *Nat. Protoc.*, 11, 5, 920–936, 2016.
63. Spoerk, M., Savandaiah, C., Arbeiter, F., Traxler, G., Cardon, L., Holzer, C., Sapkota, J., Anisotropic properties of oriented short carbon fibre filled polypropylene parts fabricated by extrusion-based additive manufacturing. *Compos. Part A: Appl. Sci. Manuf.*, 113, 95–104, 2018.

64. Spoerk, M., Arbeiter, F., Raguž, I., Holzer, C., Gonzalez-Gutierrez, J., Mechanical recyclability of polypropylene composites produced by material extrusion-based additive manufacturing. *Polymers*, 11, 8, 1318, 2019.
65. Parandoush, P., Tucker, L., Zhou, C., Lin, D., Laser assisted additive manufacturing of continuous fiber reinforced thermoplastic composites. *Mater. Des.*, 131, 186–195, 2017.
66. Zander, N.E., Gillan, M., Burckhard, Z., Gardea, F., Recycled polypropylene blends as novel 3D printing materials. *Addit. Manuf.*, 25, 122–130, 2019.
67. Banerjee, S.S., Burbine, S., Kodihalli Shivaprakash, N., Mead, J., 3D-printable PP/SEBS thermoplastic elastomeric blends: Preparation and properties. *Polymers*, 11, 2, 347, 2019.
68. Agkathidis, A., Berdos, Y., Brown, A., Active membranes: 3D printing of elastic fibre patterns on pre-stretched textiles. *Int. J. Archit. Comput.*, 17, 1, 74–87, 2019.
69. Sauerbier, P., Anderson, J., Gardner, D.J., Surface preparation and treatment for large-scale 3D-printed composite tooling coating adhesion. *Coatings*, 8, 12, 457, 2018.
70. Yoo, C.J., Shin, B.S., Kang, B.S., Gwak, C.Y., Park, C., Ma, Y.W., Hong, S.M., A study on a new 3D porous polymer printing based on EPP beads containing CO2 gas. *Proc. Eng.*, 184, 10–15, 2017.
71. Panin, S.V., Kornienko, L.A., Alexenko, V.O., Buslovich, D.G., Dontsov, Y.V., Extrudable polymer-polymer composites based on ultra-high molecular weight polyethylene. *AIP Conf. Proc.*, 1915, 1, 020005, 2017, December. AIP Publishing LLC.
72. Khanam, P.N. and AlMaadeed, M.A.A., Processing and characterization of polyethylene-based composites. *Adv. Manuf.: Polym. Compos. Sci.*, 1, 2, 63–79, 2015.
73. Hou, S., Qi, S., Hutt, D.A., Tyrer, J.R., Mu, M., Zhou, Z., Three dimensional printed electronic devices realised by selective laser melting of copper/high-density-polyethylene powder mixtures. *J. Mater. Process. Technol.*, 254, 310–324, 2018.
74. Chong, S., Pan, G.T., Khalid, M., Yang, T.C.K., Hung, S.T., Huang, C.M., Physical characterization and pre-assessment of recycled high-density polyethylene as 3D printing material. *J. Polym. Environ.*, 25, 2, 136–145, 2017.
75. Wojtyła, S., Klama, P., Baran, T., Is 3D printing safe? Analysis of the thermal treatment of thermoplastics: ABS, PLA, PET, and nylon. *J. Occup. Environ. Hyg.*, 14, 6, D80–D85, 2017.
76. Ji, L.N., Study on preparation process and properties of polyethylene terephthalate (PET), in: *Applied Mechanics and Materials*, vol. 312, pp. 406–410, Trans Tech Publications Ltd, Switzerland, 2013.
77. Woern, A.L., Byard, D.J., Oakley, R.B., Fiedler, M.J., Snabes, S.L., Pearce, J.M., Fused particle fabrication 3-D printing: Recycled materials' optimization and mechanical properties. *Materials*, 11, 8, 1413, 2018.

78. Zander, N.E., Gillan, M., Lambeth, R.H., Recycled polyethylene terephthalate as a new FFF feedstock material. *Addit. Manuf.*, 21, 174–182, 2018.
79. Gu, J., Wensing, M., Uhde, E., Salthammer, T., Characterization of particulate and gaseous pollutants emitted during operation of a desktop 3D printer. *Environ. Int.*, 123, 476–485, 2019.
80. Idrees, M., Jeelani, S., Rangari, V., Three-dimensional-printed sustainable biochar-recycled PET composites. *ACS Sustainable Chem. Eng.*, 6, 11, 13940–13948, 2018.
81. Bashir, Z., Gu, H., Yang, L., Evaluation of poly (ethylene terephthalate) powder as a material for selective laser sintering, and characterization of printed part. *Polym. Eng. Sci.*, 58, 10, 1888–1900, 2018.
82. Parvin, M. and Williams, J.G., The effect of temperature on the fracture of polycarbonate. *J. Mater. Sci.*, 10, 11, 1883–1888, 1975.
83. Shemelya, C.M., Rivera, A., Perez, A.T., Rocha, C., Liang, M.I.N., Yu, X., Kief, C., Alexander, D., Stegeman, J., Xin, H.A.O., Wicker, R.B., Mechanical, electromagnetic, and X-ray shielding characterization of a 3D printable tungsten–polycarbonate polymer matrix composite for space-based applications. *J. Electron. Mater.*, 44, 8, 2598–2607, 2015.
84. Alaboodi, A.S. and Sivasankaran, S., Experimental design and investigation on the mechanical behavior of novel 3D printed biocompatibility polycarbonate scaffolds for medical applications. *J. Manuf. Processes*, 35, 479–491, 2018.
85. Blaya, F., Pedro, P.S., Silva, J.L., D'Amato, R., Heras, E.S., Juanes, J.A., Design of an orthopedic product by using additive manufacturing technology: The arm splint. *J. Med. Syst.*, 42, 3, 1–15, 2018.
86. Blanco, I., Cicala, G., Ognibene, G., Rapisarda, M., Recca, A., Thermal properties of polyetherimide/polycarbonate blends for advanced applications. *Polym. Degrad. Stab.*, 154, 234–238, 2018.
87. Peng, F., Zhao, Z., Xia, X., Cakmak, M., Vogt, B.D., Enhanced impact resistance of three-dimensional-printed parts with structured filaments. *ACS Appl. Mater. Interfaces*, 10, 18, 16087–16094, 2018.
88. Liu, Z., Zhan, J., Fard, M., Davy, J.L., Acoustic properties of a porous polycarbonate material produced by additive manufacturing. *Mater. Lett.*, 181, 296–299, 2016.
89. Rohde, S., Cantrell, J., Jerez, A., Kroese, C., Damiani, D., Gurnani, R., DiSandro, L., Anton, J., Young, A., Steinbach, D., Ifju, P., Experimental characterization of the shear properties of 3D–printed ABS and polycarbonate parts. *Exp. Mech.*, 58, 6, 871–884, 2018.
90. Gao, Y., Li, Y., Hu, X., Wu, W., Wang, Z., Wang, R., Zhang, L., Preparation and properties of novel thermoplastic vulcanizate based on bio-based polyester/polylactic acid, and its application in 3D printing. *Polymers*, 9, 12, 694, 2017.
91. Woern, A.L. and Pearce, J.M., Distributed manufacturing of flexible products: Technical feasibility and economic viability. *Technologies*, 5, 4, 71, 2017.

92. Farstad, J.M.G., Netland, Ø., Welo, T., Surface friction of rapidly prototyped wheels from 3D-printed thermoplastic elastomers: An experimental study. *Proc. CIRP*, 60, 247–252, 2017.
93. Wang, J., McMullen, C., Yao, P., Jiao, N., Kim, M., Kim, J.W., Liu, L., Tung, S., 3D-printed peristaltic microfluidic systems fabricated from thermoplastic elastomer. *Microfluid. Nanofluid.*, 21, 6, 1–13, 2017.
94. Ellson, G., Carrier, X., Walton, J., Mahmood, S.F., Yang, K., Salazar, J., Voit, W.E., Tough thiourethane thermoplastics for fused filament fabrication. *J. Appl. Polym. Sci.*, 135, 6, 45574, 2018.
95. Bi, H., Xu, M., Ye, G., Guo, R., Cai, L., Ren, Z., Mechanical, thermal, and shape memory properties of three-dimensional printing biomass composites. *Polymers*, 10, 11, 1234, 2018.
96. Rahim, T.T., Abdullah, A.M., Akil, H.M., Mohamad, D., Rajion, Z.A., The improvement of mechanical and thermal properties of polyamide 12 3D printed parts by fused deposition modelling. *Express Polym. Lett.*, 11, 12, 963–982, 2017.
97. de Leon, A.C., Rodier, B.J., Bajamundi, C., Espera Jr., A., Wei, P., Kwon, J.G., Williams, J., Ilijasic, F., Advincula, R.C., Pentzer, E., Plastic metal-free electric motor by 3D printing of graphene-polyamide powder. *ACS Appl. Energy Mater.*, 1, 4, 1726–1733, 2018.
98. Brugo, T.O.M.M.A.S.O., Palazzetti, R., Ciric-Kostic, S., Yan, X.T., Minak, G., Zucchelli, A., Fracture mechanics of laser sintered cracked polyamide for a new method to induce cracks by additive manufacturing. *Polym. Test.*, 50, 301–308, 2016.
99. Neugebauer, F., Ploshikhin, V., Ambrosy, J., Witt, G., Isothermal and non-isothermal crystallization kinetics of polyamide 12 used in laser sintering. *J. Therm. Anal. Calorim.*, 124, 2, 925–933, 2016.
100. Lammens, N., Kersemans, M., De Baere, I., Van Paepegem, W., On the visco-elasto-plastic response of additively manufactured polyamide-12 (PA-12) through selective laser sintering. *Polym. Test.*, 57, 149–155, 2017.
101. O'Connor, H.J. and Dowling, D.P., Low-pressure additive manufacturing of continuous fiber-reinforced polymer composites. *Polym. Compos.*, 40, 11, 4329–4339, 2019.
102. Garlotta, D., A literature review of poly (lactic acid). *J. Polym. Environ.*, 9, 2, 63–84, 2001.
103. Komesu, A., de Oliveira, J.A.R., da Silva Martins, L.H., Maciel, M.R.W., Maciel Filho, R., Lactic acid production to purification: A review. *BioResources*, 12, 2, 4364–4383, 2017.
104. Wu, C.S. and Tsou, C.H., Fabrication, characterization, and application of biocomposites from poly (lactic acid) with renewable rice husk as reinforcement. *J. Polym. Res.*, 26, 2, 1–9, 2019.
105. Lebedev, S.M., Gefle, O.S., Amitov, E.T., Zhuravlev, D.V., Berchuk, D.Y., Mikutskiy, E.A., Mechanical properties of PLA-based composites for fused

deposition modeling technology. *Int. J. Adv. Manuf. Technol.*, 97, 1, 511–518, 2018.
106. Dong, Y., Milentis, J., Pramanik, A., Additive manufacturing of mechanical testing samples based on virgin poly (lactic acid)(PLA) and PLA/wood fibre composites. *Adv. Manuf.*, 6, 1, 71–82, 2018.
107. Regazzi, A., Pucci, M.F., Dumazert, L., Gallard, B., Buonomo, S., Ravel, R., Lopez-Cuesta, J.M., Controlling the distribution of fire retardants in poly (lactic acid) by fused filament fabrication in order to improve its fire behaviour. *Polym. Degrad. Stab.*, 163, 143–150, 2019.
108. Ou-Yang, Q., Guo, B., Xu, J., Preparation and characterization of poly (butylene succinate)/polylactide blends for fused deposition modeling 3D printing. *ACS Omega*, 3, 10, 14309–14317, 2018.
109. Daver, F., Lee, K.P.M., Brandt, M., Shanks, R., Cork–PLA composite filaments for fused deposition modelling. *Compos. Sci. Technol.*, 168, 230–237, 2018.
110. Hou, Z., Tian, X., Zhang, J., Zhe, L., Zheng, Z., Li, D., Malakhov, A.V., Polilov, A.N., Design and 3D printing of continuous fiber reinforced heterogeneous composites. *Compos. Struct.*, 237, 111945, 2020.
111. Zhao, X., Tekinalp, H., Meng, X., Ker, D., Benson, B., Pu, Y., Ragauskas, A.J., Wang, Y., Li, K., Webb, E., Gardner, D.J., Poplar as biofiber reinforcement in composites for large-scale 3D printing. *ACS Appl. Bio Mater.*, 2, 10, 4557–4570, 2019.
112. Ertane, E.G., Dorner-Reisel, A., Baran, O., Welzel, T., Matner, V., Svoboda, S., Processing and wear behaviour of 3D printed PLA reinforced with biogenic carbon. *Adv. Tribol.*, 2018, 11 pages, 2018. https://doi.org/10.1155/2018/1763182
113. He, X., Liu, Z., Niu, W., Yang, L., Zhou, T., Qin, D., Niu, Z., Yuan, Q., Effects of pyrolysis temperature on the physicochemical properties of gas and biochar obtained from pyrolysis of crop residues. *Energy*, 143, 746–756, 2018.
114. Gao, X., Zhang, D., Qi, S., Wen, X., Su, Y., Mechanical properties of 3D parts fabricated by fused deposition modeling: Effect of various fillers in polylactide. *J. Appl. Polym. Sci.*, 136, 31, 47824, 2019.
115. Kalinke, C., Wosgrau, V., Oliveira, P.R., Oliveira, G.A., Martins, G., Mangrich, A.S., Bergamini, M.F., Marcolino-Junior, L.H., Green method for glucose determination using microfluidic device with a non-enzymatic sensor based on nickel oxyhydroxide supported at activated biochar. *Talanta*, 200, 518–525, 2019.
116. Mohammed, M.I., Wilson, D., Gomez-Kervin, E., Tang, B., Wang, J., Investigation of closed-loop manufacturing with acrylonitrile butadiene styrene over multiple generations using additive manufacturing. *ACS Sustainable Chem. Eng.*, 7, 16, 13955–13969, 2019.
117. Kafle, A., Luis, E., Silwal, R., Pan, H.M., Shrestha, P.L., Bastola, A.K., 3D/4D Printing of polymers: Fused deposition modelling (FDM), selective laser sintering (SLS), and stereolithography (SLA). *Polymers*, *13*, 18, 3101, 2021.

118. Lakkala, P., Munnangi, S.R., Bandari, S., Repka, M., Additive manufacturing technologies with emphasis on stereolithography 3D printing in pharmaceutical and medical applications: A review. *Int. J. Pharm.: X*, 5, 100159, 2023.
119. Ahmad, M.N., Ishak, M.R., Taha, M.M., Mustapha, F., Leman, Z., Mechanical, thermal and physical characteristics of oil palm (Elaeis Guineensis) fiber reinforced thermoplastic composites for FDM–Type 3D printer. *Polym. Test.*, *120*, 107972, 2023.

6

Additive Manufacturing of Metal-Matrix and Polymer-Matrix Composites

Sandhyarani Biswas* **and Anurag Jasti**

Department of Mechanical Engineering, NIT Rourkela, Rourkela, Odisha, India

Abstract

Additive manufacturing (AM) is a process that builds solid parts by joining materials in a layer-by-layer fashion to make the part. It has an ability to build complex structures without using complicated manufacturing techniques and expensive tooling that are impossible or difficult to make by traditional methods. The low cost and flexibility in making complex products have increased the use of AM in the automotive industry, defense, aerospace and biomedical applications, arts and designs, architecture, etc. However, inconsistent product quality, limited materials availability, various manufacturing defects, and speed of manufacturing are the challenges faced in AM. This chapter gives an overview of different AM processes mainly used for fabricating metal matrix composites (MMCs) and the polymer matrix composites (PMCs). Different steps involved in 3D printing, form of materials used, and their working principle are discussed. The influence of various factors on the quality of the 3D-printed components is emphasized. Further, the recent applications of 3D-printed MMC and PMC products in various areas are explored. Finally, the research opportunities and challenges in AM are provided in the chapter.

***Keywords*:** Polymeric materials, manufacturing process, properties, additive manufacturing

**Corresponding author*: srb@nitrkl.ac.in

R. Rajasekar, C. Moganapriya and P. Sathish Kumar (eds.) Additive Manufacturing with Novel Materials: Processes, Properties and Applications, (197–236) © 2024 Scrivener Publishing LLC

List of Abbreviations

AM	Additive Manufacturing
BJ	Binder Jetting
DED	Directed Energy Deposition
DIW	Direct Ink Writing
DLP	Digital Light Processing
DMLS	Direct Metal Laser Sintering
EBM	Electron Beam Melting
FDM	Fused Deposition Modelling
FFF	Fused Filament Fabrication
LOM	Laminated Object Manufacturing
ME	Material Extrusion
MJ	Material Jetting
MMC	Metal Matrix Composite
PBF	Powder Bed Fusion
PMC	Polymer Matrix Composite
SL	Sheet Lamination
SLA	Stereolithography
SLM	Selective Laser Melting
SLS	Selective Laser Sintering
STL	Standard Triangle Language
VP	Vat Photopolymerization

6.1 Introduction

Additive manufacturing (AM), also called 3D printing, is a process that builds layers to make a physical part from a digital model. In contrast to subtractive manufacturing, 3D objects are manufactured by successively depositing material in layers. It offers many advantages including material flexibility, design freedom, reduction in material waste and an increased range of complex structures etc. Unlike traditional processes, there is no requirement of any tooling and dies in AM to fabricate parts, which also reduces material waste and labor costs. Apart from the above advantages, this method mainly comes under green manufacturing process as the scrap and noise pollution is significantly reduced in their usage. This technology has many potential applications in aerospace, automotive, biomedical, consumer products etc. However, its wider application is hindered by several barriers and the most important of which are the limited material choices, relatively high initial investment, not suitable for mass production, part quality issues, and lack of expertise. The present study mainly focuses on the 3D printing techniques used for metal matrix composites (MMCs), polymer matrix composites (PMCs) and their applications. The functional principles of AM methods, different material form used, steps associated in 3D printing are discussed. The research done on the impact of reinforcements, print parameters on the different properties of 3D-printed objects are highlighted. Finally, the recent applications of MMCs and PMCs made by 3D printing, the research opportunities and challenges for AM of composites are highlighted in the chapter.

6.1.1 General Steps Involved in 3D Printing

Additive manufacturing includes many steps starting from the design making to manufacturing of the final part, which differs as per the manufacturing methods used. Figure 6.1 shows the different steps involved in making parts using AM. The AM techniques mainly used the computer, the 3D modeling software, printing machine and materials. After computer-aided design (CAD), the 3D printing machine reads out the data from the CAD file and finally, the 3D part is produced. The first step is to develop a digital model using CAD software. The second step is to convert the digital model into standard triangle language (.STL) file. The third step is slicing, which is mainly converting the 3D model into printing instructions. This step is very important as it affects the printed part quality. It is mainly used to

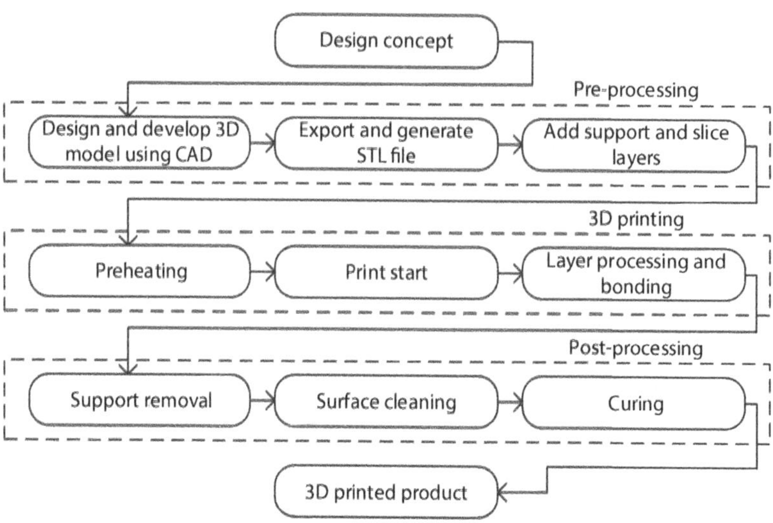

Figure 6.1 Steps involved in 3D printing.

create specific 3D printer code as per the .STL file information. The fourth step is to print the part.

6.1.2 Types of AM Processes

Today, many AM techniques are used to fabricate products of different materials. These methods can be classified according to different criteria. According to the ASTM-F42 committee, it can be classified as (1) powder bed fusion (PBF), (2) vat photopolymerization (VP), (3) material extrusion (ME), (4) binder jetting (BJ), (5) material jetting (MJ), (6) directed energy deposition (DED) and (7) sheet lamination (SL) [1]. The ME process includes techniques like fused filament fabrication (FFF), fused deposition modeling (FDM) and direct ink writing (DIW). Examples of VP are digital light processing (DLP) and stereolithography (SLA). Based on the material and the heat source used, there are various types of PBF such as electron beam melting (EBM), selective laser melting (SLM), direct metal laser sintering (DMLS) and selective laser sintering (SLS). One example of SL is laminated object manufacturing (LOM). Table 6.1 presents the classification of AM processes, their working principle, and materials used. Figure 6.2 shows the schematic diagram of different AM techniques.

6.1.3 Materials for Additive Manufacturing

Raw material plays a vital role for any AM technique. During the initial phase of AM, materials like polymers, paper laminates, and waxes are commonly used as raw materials. However, with technological advancement and research progress, other materials like ceramics, metals, and composites are currently being used. Nowadays, an impressive range of material exists for manufacturing through AM techniques. The type and form of material to be used generally depends on the AM process type. Materials in different forms such as powder, filaments, paste, wire, inks, and sheets are mainly used for AM. Figure 6.3 shows the broad classification of materials used for AM.

Table 6.2 shows the most frequently used materials for different AM techniques. Among the different materials, polymers are the most commonly used for fast prototyping. Different thermoplastic filaments, thermosetting powders and different photopolymer resins are the most commonly used polymers for AM. Metals are mainly used in the form of powders or wires for AM. A limited range of metals suitable for AM resulted in the demand for a wider range of alloys and composites. Polymer matrix composites and

Table 6.1 Classification of AM processes.

Methods	Working principle	Main principle	Material type	Material				
				Polymer	Metal	Ceramic	Paper	Composites
FDM (Figure 6.2a)	Material is mainly extruded through a heated nozzle to build the object.	Filament extrusion	Solid	✓	×	✓	×	✓
DIW (Figure 6.2b)	Ink is extruded through a nozzle to create a model.	Inkjet printing	Liquid	✓	✓	✓	×	✓
SLS (Figure 6.2c)	A laser is mainly used to fuse together layers of powder to produce a solid model.	Laser sintering	Powder	✓	✓	✓	×	✓

(Continued)

Table 6.1 Classification of AM processes. (*Continued*)

Methods	Working principle	Main principle	Material type	Material				
				Polymer	Metal	Ceramic	Paper	Composites
BJ (Figure 6.2d)	A binder is used to bond together layers of powder to create a solid model.	Binder jetting bound	Powder	✓	✓	✓	×	✓
VP (Figure 6.2e)	A vat of liquid polymer is solidified by ultraviolet (UV) light to create a model.	Photopolymerization (light cured)	Liquid	✓	×	×	×	✓
MJ (Figure 6.2f)	Droplets of material are deposited to build a model.	Inkjet printing	Liquid	✓	×	×	×	✓

(*Continued*)

Table 6.1 Classification of AM processes. (*Continued*)

Methods	Working principle	Main principle	Material type	Material				
				Polymer	Metal	Ceramic	Paper	Composites
LOM (Figure 6.2g)	Layers of material are bonded together to create a solid model.	Lamination (layer by layer)	Solid	✓	×	✓	✓	✓
DED (Figure 6.2h)	Energy is used to melt and deposit material to create a model.	Energy beam deposition	Powder	×	✓	✓	×	✓
DMLS (Figure 6.2i)	A laser is mainly used to fuse together layers of metal powder to create a solid model.	Laser sintering	Powder	×	✓	×	×	✓

(*Continued*)

Table 6.1 Classification of AM processes. (*Continued*)

Methods	Working principle	Main principle	Material type	Material				
				Polymer	Metal	Ceramic	Paper	Composites
SLM (Figure 6.2j)	A laser is mainly used to melt layers of metal powder to create a solid model.	Laser melting	Powder	×	✓	✓	×	✓
EBM (Figure 6.2k)	An electron beam is mainly used to melt layers of metal powder to create a solid model.	Electron beam melting	Powder	×	✓	×	×	✓

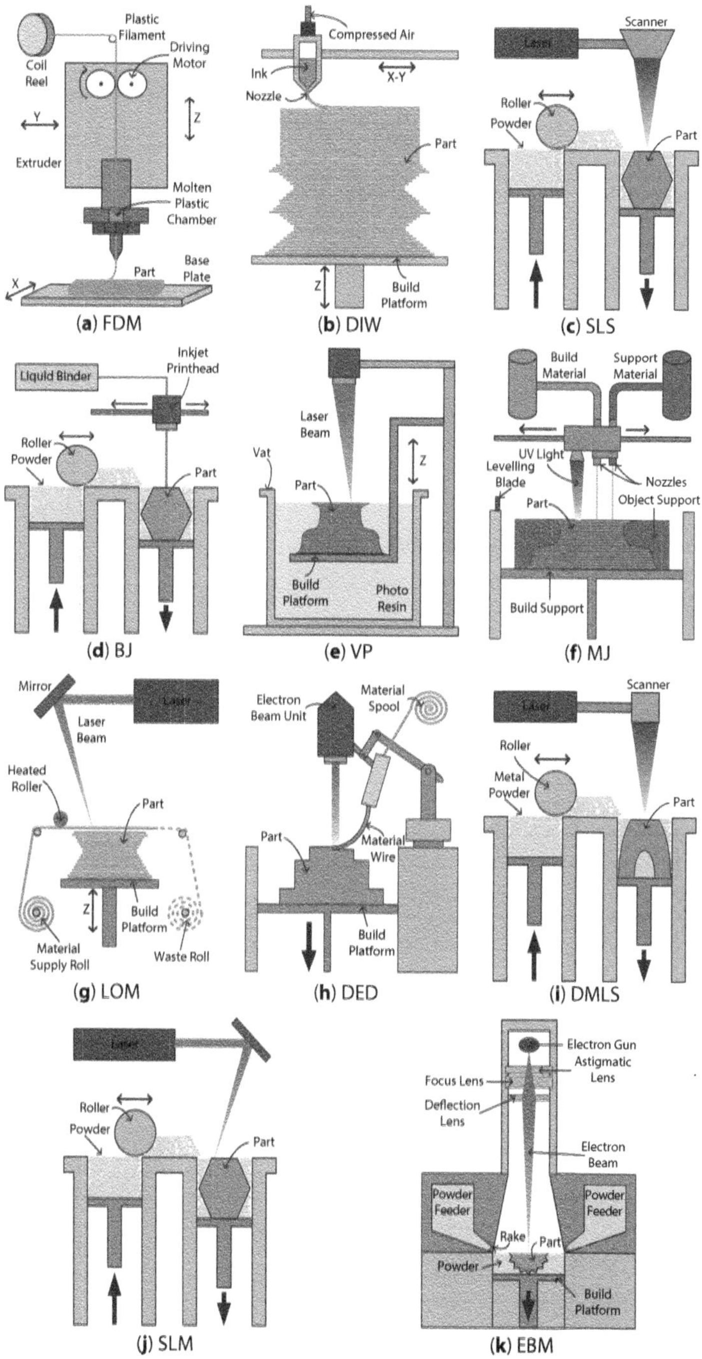

Figure 6.2 AM techniques.

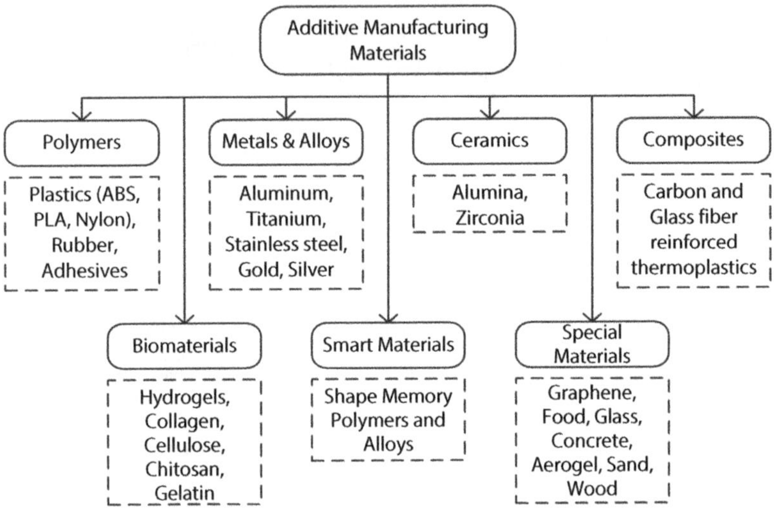

Figure 6.3 Classification of AM materials.

Table 6.2 Materials used in different AM techniques [2].

Process	Materials
VP	Photopolymers, bioresins
PBF	Titanium alloys ceramics, polymers, metal, glass and starch powder
ME	Thermoplastics, hydrogel, ceramic powder, bioinks PC, ABS, medical-grade PC, wax, chocolate syrup, metal powder, dough
MJ	Bioinks, photopolymers, organic and inorganic electronic inks
BJ	Polymeric materials, metal powder, sand, ceramics, metal alloys, metal ceramic composites
DED	Nylon, metal
SL	Paper, metals, ceramics, fabrics, plastics, composites

metal matrix composites are the special class of materials and have better properties as compared to pure metals or polymers.

6.2 AM of Composite Materials

Composite materials are widely used in many industries for making lightweight components and structures as they have advantages over monolithic counterparts. In composites, the matrix material acts as a continuous phase to hold the reinforcement together, to transfer the load and provide a protective environment, whereas reinforcement gives strength and stiffness to the composites. The matrix material can be metallic, polymeric or can even be ceramic. The reinforcements usually exist in fibrous forms or particulate form. Composites made by AM processes have many advantages such as adjustable functional reinforcements, material formulation, arrangements of phases, and the complex geometries. Each AM process has different advantages as per the base matrix (polymer, metal, ceramics) and the reinforcement type.

6.2.1 AM of Metal Matrix Composites

An important class of composite material is MMC. It mainly contains a continuous metal matrix and a reinforcement dispersed in the matrix. In comparison to the pure metals and alloys, MMC gives superior materials properties like high specific strength and stiffness and high wear resistance, which makes them useful in a wide range of applications. Metal matrix composites mainly consist of metal matrix (e.g., copper, aluminum, titanium, etc.) combined with another phase which is mainly in the form of fibers, filaments, or fillers. Enhanced properties can be obtained when the ductile metal matrix is reinforced with the hard phase. There are various 3D printing methods available for producing MMCs. Table 6.3 presents some of the important research work done on the MMCs fabricated using different AM processes.

6.2.2 AM of Polymer Matrix Composites

When the matrix is a polymer, then the composite is called a polymer matrix composite (PMC). The development of printable PMCs with better performance is required as 3D-printed neat polymer parts show inferior mechanical performance. Most of the polymer 3D printing process can also be used for fabricating PMCs. Various kinds of reinforcements such as natural/synthetic fibers, ceramic, metallic fillers, and organic additives, are currently being used for PMCs. There is a wide range of polymers such as elastomers, thermoplastic, hydrogels, functional polymers, and biomaterials that are used for 3D printing of customized products. Various AM techniques have been used to fabricate PMCs. Each process has its own advantages and limitations. The selection of such technique mainly depends

Table 6.3 Research on MMCs fabricated using different AM processes.

Process	Matrix material	Reinforcement type	Study	Reference
DED	Ti-6Al-4V alloy	SiC (powder)	Mechanical properties	[3]
	Al-Mg-Sc-Zr alloy	TiC (powder)	Mechanical properties	[4]
	Aluminium 2024	TiB_2 (powder)	Physical characterization, friction and wear behavior	[5]
	Ti-6Al-4V alloy	SiC (powder)	Physical and mechanical characterization	[6]
	Titanium	nano-Al_2O_3 (powder)	Effect of filler on microhardness	[7]
DMLS	Al-4.5Cu-3Mg	SiC (particulates)	Surface crack and wear behavior	[8]
	Titanium	Carbon nanotubes	Microstructure, wear, microhardness and tensile behavior	[9]
	Ti-6Al-4V alloy	SiC (particulates)	Physical, mechanical and wear behavior	[10]
	Stainless steel 316	TiN (powder)	Microstructure, density, porosity, microhardness and wear behavior	[11]
SLM	Iron	SiC (powder)	Microstructure and tensile behavior	[12]
	Inconel 718	TiC and B_4C (powder)	Microhardness and wear behavior	[13]
	S136 mold steel	GO (powder)	Microstructure, microhardness and tensile behavior	[14]

(*Continued*)

Table 6.3 Research on MMCs fabricated using different AM processes. (*Continued*)

Process	Matrix material	Reinforcement type	Study	Reference
EBM	NiBSi	WC (powder)	Hardness, flexural, impact and compressive strengths	[15]
	Nb-TiAl	TiC (nanoparticles)	Microhardness, fracture toughness, tensile and compressive strengths	[16]
	NiBSi	WC (powder)	Microstructure, mechanical and corrosion properties	[17]
PBF	Aluminium	ZrC and TiC (powder)	Microstructure, microhardness wear and tensile behavior	[18]
	AlSi12	Fe_2O_3 (powder)	Microstructure and tensile behavior	[19]
	X3NiCoMoTi 18-9-5 steel	Tungsten (particles)	Tensile behavior	[20]
	AlSi10Mg	SiC (particles)	Density, microhardness and tribological behavior	[21]
BJ	Ti	TiC (powder)	Microstructure and hardness behavior	[22]
	Tungsten carbide	Co (powder)	Density, hardness and fracture toughness	[23]
	Aluminium	TiC (powder)	Hardness and density	[24]
LOM	TiC	Ni (powder)	Microstructure and mechanical behavior	[25]
	TiC	SiC	Microstructure, hardness and tensile behavior	[26]

on the part to be manufactured, its quality, and material to be used. Additive manufacturing needs proper polymer formulation and reinforcement selection as they have significant impact on build speed, quality, accuracy, surface finish, and other properties of printed products. In polymer, the most commonly used reinforcements are carbon fiber, glass fiber, carbon nanotubes, and many other metallic and non-metallic fillers. The incorporation of these reinforcement phases enhances the different properties of the printed composite parts fabricated. Lot of research work has been done on the 3D-printed PMCs by many researchers. Table 6.4 shows some of the important research on the PMCs fabricated by different AM techniques.

Table 6.4 Research on the PMCs fabricated using different AM techniques.

Process	Matrix material	Reinforcement	Study	Reference
BJ	Hydroxyapatite	Carboxymethyl chitosan, polyvinylpyrrolidone, dextrin	Optimize the distribution of particle size to enhance the printed part quality and increase the strength	[27]
	Biphasic calcium phosphate	Polycaprolactone	Enhance mechanical properties and biocompatibility	[28]
	MXene particles	Poly(vinyl alcohol)	For strain sensing and energy storage application	[29]
DLP	AlSi10Mg powder	Photocurable acrylate	Influence of particle surface modification on the process	[30]
	Sacrificial salt particulates	Photocurable resin	Fabrication of porous polymeric materials using photocurable resin	[31]
	Graphene powder	UV-curable PLA-PUA resin	Influence of graphene nanofillers on wettability and resin viscosity with printing parameters and mechanical behavior	[32]

(*Continued*)

Table 6.4 Research on the PMCs fabricated using different AM techniques. (*Continued*)

Process	Matrix material	Reinforcement	Study	Reference
	Carbon nanotube	Acrylate-based UV-curable resin	Mechanical properties of carbon nanotube-based composites	[33]
FDM	PLA	Pine wood fibers (short fiber), inorganic clays	Mechanical and thermal properties	[34]
	PLA, Onyx	Short carbon fiber	Influence of polymers and fiber to strong magnetic field	[35]
	Nylon	Carbon fiber, glass fiber	Mechanical properties of composites	[36]
	ABS	Short carbon fiber	Effect of various factors on tensile properties	[37]
	ABS	Glass fiber	Properties of short glass fiber/ABS composites	[38]
	PLA	Continuous carbon fiber	Mechanical and thermal properties	[39]
	ABS	Carbon fiber powder	Effect of various factors on mechanical behavior of composites	[40]
FFF	PPS	Continuous carbon and Kevlar fibers	Strength properties	[41]
	PLA	Carbon fiber	Thermal conductivity of composites	[42]

(*Continued*)

Table 6.4 Research on the PMCs fabricated using different AM techniques. (*Continued*)

Process	Matrix material	Reinforcement	Study	Reference
LOM	Glass fiber	Epoxy	Interfacial characteristics of composites	[43]
	Carbon fiber	PA6	Mechanical properties of composites	[44]
	Hollow glass microspheres	Polyamide 12	Thermal, physical and mechanical tests	[45]
	Carbon fiber	PA6	Mechanical properties of printed parts	[46]
MJ	VeroMagentaV particles	Agilus30	Mechanical and fractographic analysis of 3D-printed composites	[47]
	Vero white	Tango Plus	Modeling of multimaterial component and analysis of the stress-strain behavior	[48]
VP	Graphene oxide (GO)	PEGDA	3D bioprinted graphene oxide (GO)-doped gelatine-based scaffold fabrication and biomedical use	[49]
	TiO_2 nanoparticles	Acrylic-urethane resin	Improvement of mechanical and thermal behavior	[50]

(*Continued*)

Table 6.4 Research on the PMCs fabricated using different AM techniques. (*Continued*)

Process	Matrix material	Reinforcement	Study	Reference
	Silver nanoparticles	Acrylic photocurable resin	Physical and mechanical behavior	[51]
	Magnetic particles	Urethane acrylate resin	Effect of the different factors on dimensions of parts and resin curing behavior	[52]
	Silica nanoparticles	Cationic/free-radical resin	Influence of filler addition on mechanical properties	[53]
SLS	Polyether ether ketone (PEEK)	Milled carbon fiber	High temperature rheological behaviors	[54]
	Nanographene platelets	PA11	Electrical and mechanical properties of PA11	[55]
	Hydroxyapatite (HA)	Polycaprolactone	Influence of HA reinforcement on the wettability, surface roughness, mechanical behavior and surface morphology	[56]
	HA	Polyvinyl alcohol	Development of scaffolds	[57]

6.3 Effect of Parameters on Printed MMCs

The process parameters of AM determines the quality, consistency, and performance of 3D-printed MMCs. Temperature, pressure, printing speed, printing pattern, feed rate, layer thickness, etc., significantly affect the different properties, accuracy, and surface finish of the components made by AM. Other factors influencing the quality of printed parts include fiber/filler types, concentration, and form of materials used. The effectiveness of

AM methods depends upon the careful choice of parameters that will give the desired properties.

6.3.1 Influence of Parameters on DMLS-Printed Parts

The factors that influence the quality of the part in DMLS are laser power, pulse width, layer thickness, hatch distance, scan speed, powder size, build orientation, temperature, powder bed compaction, etc. [58–60]. Laser power and pulse width mainly influence the sintering rate and the final density of the printed object. The scanning speed mainly affects the surface finish and time to build the object. The layer thickness determines the thickness of each layer of the part and affects the accuracy and resolution of the final part. Powder size is another important factor that determines the size of the metal powders used in the process and affects the resolution of the part. Build orientation is the angle at which the part is built and it significantly affects the strength properties. The build platform temperature and the surrounding environment also have influence on the component's quality. Finally, the powder bed compaction determines density and porosity of the final part.

Gupta *et al.* [61] evaluated the influence of scan speed and the laser power on the density, porosity, hardness, and coefficient of friction of the hybrid B_4C-cBN-reinforced Ti-based MMCs. It was observed that the porosity of the composites decreases with the increase in laser power, while the other three increased (Figure 6.4). However, scan speed showed

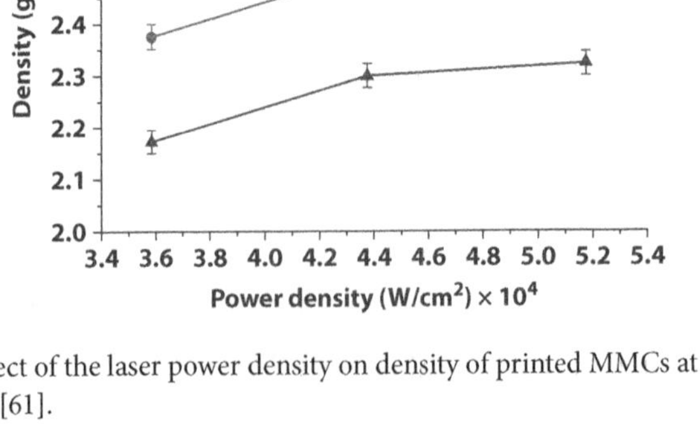

Figure 6.4 Effect of the laser power density on density of printed MMCs at scan speed of 3500 mm/min [61].

an opposite effect. Kundu *et al.* [62] and Hussain *et al.* [63] analyzed the impact of different factors on the TiN-reinforced Ti6Al4V alloy and TiN particulates-reinforced SS316-based MMCs, respectively. They also found similar results as Gupta *et al.* [61]. Ghosh *et al.* [64] analyzed the impact of laser power, pulse width and the layer thickness on SiC particulates-reinforced Al–4.5Cu–3Mg MMCs.

6.3.2 Influence of Parameters on SLM-Printed Parts

In SLM process, various process parameters significantly influence the resulting part characteristics [65–69]. One of these parameters is the laser power, and adjusting the laser power affects the final quality of the part such as its density and the mechanical behavior. Generally, the inadequate laser power causes partial melting of the powder grains and excessive laser power creates keyholes due to the evaporation of the material. The scan speed can be adjusted to control the rate of energy delivery. Increasing the scan speed and hatch distance can lead to a faster build rate, but it can also increase the risk of defects and porosity. The layer thickness can be adjusted to control the part quality. The size of the metal powders also has effect on the mechanical behavior and surface finish. The build temperature can be adjusted to prevent warping and cracking during the building process. The arrangement of the powder is also influence the mechanical behavior of SLM printed objects.

Li *et al.* [70] analyzed the impact of the laser power, scanning speed and hatch distance on density of SiC- and TiN-reinforced Al12Si composites. They reported that the composite density increases with an increase in the laser power, while it decreases with an increase in both the scan speed and hatch distance. Shi *et al.* [19] analyzed the effect of the laser power on the hardness of Fe_2O_3/AlSi12 composites. They concluded that hardness and density decreased as laser power increases. Jue *et al.* [71] analyzed the effect of laser scan speed on the mechanical behavior of Al-Al_2O_3 composites. The significant influence of laser scan speed was observed from their study. Ghodsi *et al.* [72] evaluated the influence of the scanning speed and laser power on the hardness of zirconia-reinforced Inconel 625 composites. They reported that the hardness and density increase with the increase in the laser power, but decrease with the increase in the scan speed. Xi *et al.* [18] analyzed the influence of laser power on the mechanical and wear behavior of ZrC/TiC composites. They reported that with an increase in the laser power, tensile strength and hardness increase but wear rate decreases. Mandal *et al.* [13] analyzed the impact of scanning speed on microhardness of B_4C- and TiC-reinforced IN718 MMCs, and they reported that with increase in the scanning speed, the hardness of the part decreased (Figure 6.5).

6.3.3 Influence of Parameters on EBM-Printed Parts

The EBM process involves several factors to improve the part characteristics [73]. The scanning speed impacts the melting rate and final part resolution. The layer thickness mainly impacts the final resolution and build time. The powder feed rate affects the build rate and resolution. The electron beam current controls the delivered heat and melting rate. The vacuum level is necessary for avoiding oxidation, contamination, and improving the beam-to-powder interaction. Temperature control is necessary to avoid thermal stresses, distortion, and increase the part quality. Kurzynowski *et al.* [74] analyzed the impact of scan speed, electron beam power and current on the quality of Ti-5Al-5Mo-5V-1Cr-1Fe alloy. They saw that with the increase in the scan speed and beam current, the density decreases (Figure 6.6). Ma *et al.* [75] investigated the impact of scanning speed on the surface morphology of the TiB_2/Al-Cu-Mg-Fe-Ni composites. They concluded that with an increase in scan speed, the roughness of the printed part increases, whereas slower scan speeds produce smooth surface composites.

6.3.4 Influence of Parameters on BJ-Printed Parts

Binder jetting-printed part quality and properties are influenced by factors such as powder material size and type, binder composition and viscosity,

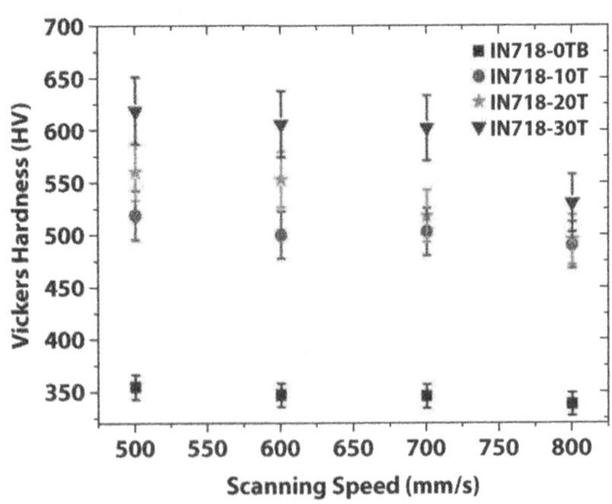

Figure 6.5 Effect of scanning speed on microhardness [13].

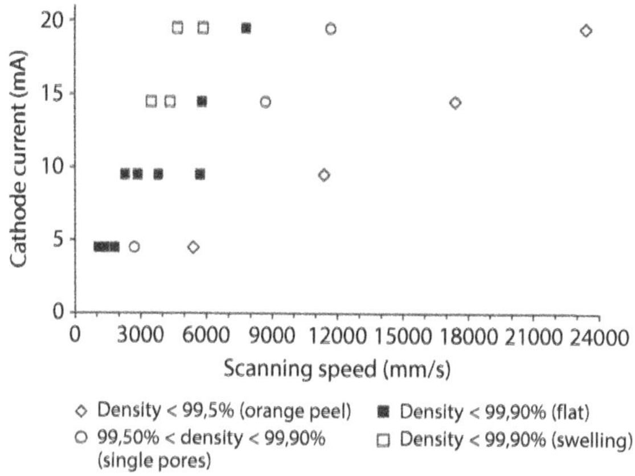

Figure 6.6 Effect of scan speed on density [74].

layer thickness, printing resolution, print head speed, build platform, and print head temperature. Smaller powder size leads to better resolution and surface finish. Higher binder viscosity also leads to improved surface finish. Resolution and surface finish depend on layer thickness, printing resolution, and print head speed, and higher resolution and faster speed results in improved surface finish but longer production time. Build platform and print head temperature affects production time, resolution, and surface finish, with higher temperature shortening the production time but potentially lowering the resolution and a rough surface finish. Enneti *et al.* [23] analyzed the impact of sintering temperature on density of WC/Co composites and concluded that density increases with increase in sintering temperature (Figure 6.7).

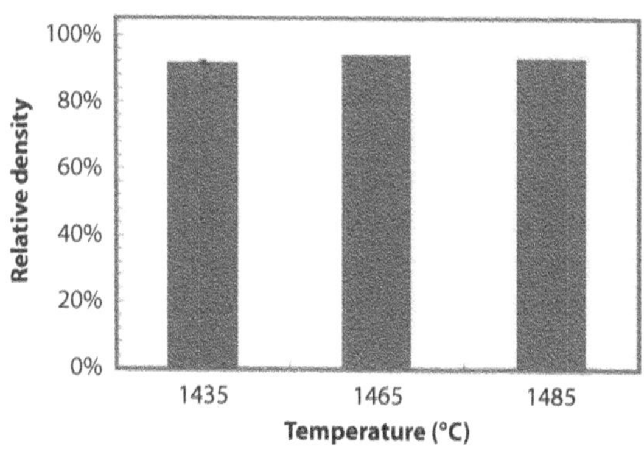

Figure 6.7 Effect of temperature on density [23].

6.3.5 Influence of Parameters on DED-Printed Parts

Directed energy deposition is a process used for adding material to a substrate. This process involves several parameters that affect the quality of the part being built. The selection of an energy source primarily relies on the characteristics of the material. The traverse speed, feed rate, laser energy density, and laser power mainly affect the build rate and the final properties of the part. The distance between the substrate and nozzle also influence the printed part characteristics. Farayibi *et al.* [76] analyzed the impact of traverse speed and laser power on the hardness of TiB_2-reinforced TiAl V composites. They saw that the hardness decreases as laser power increases. Traxel and Bandyopadhyay [77] analyzed the impact of scan speed, laser power, and laser energy density on the hardness and density of CPTi-B_4C-BN composites. They concluded that hardness decreases, and density increases with the increase in energy input. No clear trend was observed after varying the laser power and scan speed. Ang *et al.* [78] analyzed the effect of scan speed, laser power, hopper speed, and energy density on the hardness and density of TiB_2-reinforced 316L stainless steel composites.

6.4 Effect of Parameters on Printed PMCs

The processing parameters of AM play a vital role to get the desired performance from the printed objects. The mechanical characteristics of the 3D-printed parts are extremely dependent on the proper selection of these parameters. The influence of various print parameters on the quality of 3D-printed PMC parts are investigated by many researchers.

6.4.1 Influence of Parameters on FDM-Printed Parts

There are many factors influencing the FDM-printed part characteristics such as infill density, nozzle head temperature, print bed temperature, nozzle diameter, layer thickness, print speed, and the raster angle. Studies on the influence of various factors on the printed part characteristics have been made by many researchers [35–37, 79–82]. Muhamedagic *et al.* [79] examined the impact of layer thickness, raster angle, printing speed, and wall thickness on the mechanical behavior of short carbon fiber/PA composites produced using FDM. They concluded that the tensile strength of the material is influenced by the layer thickness and raster angle. Peng *et al.* [82] studied the influence of factors such as printing directions and build-plate temperature on the mechanical behavior of PA composites reinforced with carbon fiber. Tensile properties of composites in the three different printing directions were analyzed. It was observed that composites printed along the direction of tensile loading exhibits better tensile properties compared to the other two directions (Figure 6.8). Further, it was seen that the tensile strength of printed objects increases with the increase in build plate temperature. Tekinalp *et al.* [37] studied the impact of fiber loading on the mechanical behavior of carbon/fiber ABS composites. Short carbon fiber/ABS composites at various fiber loading were fabricated by FDM and analyzed its advantages and limitations as compared to the conventional compression molding process.

6.4.2 Influence of Parameters on VP-Printed Parts

To obtain the quality printed part using VP processes, the parameters to be controlled are layer thickness, scanning speed, printing direction, printing resolution, UV intensity, exposure time, hatch cure depth, hatch spacing, reinforcement type, polymer type, reinforcing particle size/shape, and content. Study of the impact of various factors on the different properties of SLA-printed and DLP-printed parts have been done by many researchers [30, 49, 83–85]. Maalihan *et al.* [84] investigated the thermal and mechanical behavior of SLA-printed chitin nano whiskers/methacrylate (MA) resin composites. They reported that reinforcement of chitin nano whiskers not only enhances the thermal and mechanical properties, but also provides better resolution and geometrical accuracy of composite parts. The

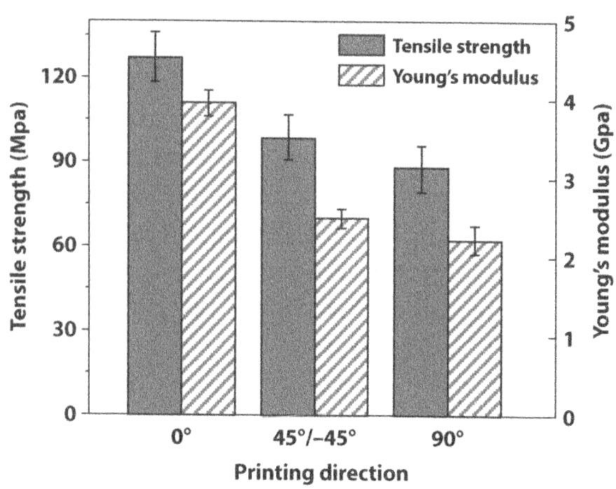

Figure 6.8 Tensile properties of CF/PA6 composites with the three different deposition patterns (0°, 45°/-45° and 90°) [82].

incorporation of up to 1.0 wt.% filler, increases the strain at break, tensile strength, modulus, and the maximum thermal degradation temperatures of the composites. Gowda *et al.* [83] analyzed the impact of layer thickness, orientation, and hatch spacing on the mechanical behavior of SLA-printed parts. The Taguchi experimental design was used to select the optimum parameter setting. Among these parameters, layer thickness and print orientation are found to be the main contributing factors for tensile strength. Similarly, orientation and hatch spacing are found to be the key contributing factors for flexural strength, and orientation is found to be the more significant factor for impact strength. Hanon *et al.* [85] analyzed the influence of graphene and other factors (print orientation and layer thickness) on the tribological behavior of DLP-printed graphene/resin composites. The optimal graphene concentration for effective tribological performance was determined to be 0.5% by weight. Based on their study, it was reported that the graphene-reinforced PMCs can be used as functional materials as it can enhance the wear behavior of DLP-printed composites significantly.

6.4.3 Influence of Parameters on SLS-Printed Parts

The parameters influencing SLS-printed parts are print speed, hatch space, powder bed temperature, laser power, energy density, and the layer thickness. Other factors influencing the printing quality includes polymer used, filler type, size, and content. Numerous works were done by researchers on the effect of various parameters on the SLS-printed composite parts [56, 86, 87]. Kısasöz *et al.* [86] analyzed the influence of filler content on the mechanical behavior of SLS-printed hollow glass microspheres-filled polyamide 12. Idriss *et al.* [87] analyzed the impact of SLS print parameters like preheating temperature, laser speed, and laser power on the mechanical behavior of SLS-printed sisal fiber/polyethersulfone composites. It was described that when the laser power and the preheating temperature increase, the mechanical performance of the printed objects significantly increases. However, the impact of scanning speed showed a decreasing trend for the strength properties. Eosoly *et al.* [56] analyzed the impact of filler content on the mechanical behavior of polycaprolactone-hydroxyapatite scaffolds fabricated using SLS.

6.4.4 Influence of Parameters on DIW-Printed Parts

The DIW-printed parts are influenced by many factors such as pressure, print speed, nozzle size, and direct writing medium. A few researchers have studied the impact of various parameters on the printed part quality

[88–90]. Tandel and Gozen [88] fabricated composites using the graphene flakes, spherical eutectic gallium indium fillers and poly(ethylene oxide) using DIW. The effects of the filler on the process mechanisms, ink rheology, and electrical properties of the composites were studied. It was reported that the electrical conductivity value decreases as strain increases due to the displacement of the fillers along the direction of printing. This research can be utilized to regulate the properties of conductive PMCs and the DIW procedure. Haney *et al.* [89] analyzed the impact of particle concentration and printing speed on the orientation of graphene nanoplatelets and electrical properties of DIW-printed graphite reinforced epoxy composites. A sharp increase in conductivity was clearly observed when the print speed was increased from 5 to 40 mm/s (Figure 6.9). Compton and Lewis [90] studied the mechanical behavior of DIW-printed epoxy-based composites incorporation with milled carbon fiber and silicon whisker. The fiber-filled epoxy established excellent mechanical properties because of the alignment of particles along the print direction. Tensile fracture surfaces of transversely and longitudinally printed specimens were also analyzed.

6.4.5 Influence of Parameters on BJ-Printed Parts

Many factors influence the BJ-printed parts, such as characteristics of the powder (e.g., material, distribution, and size), the print parameters including print direction, speed of printing, cure time and layer thickness,

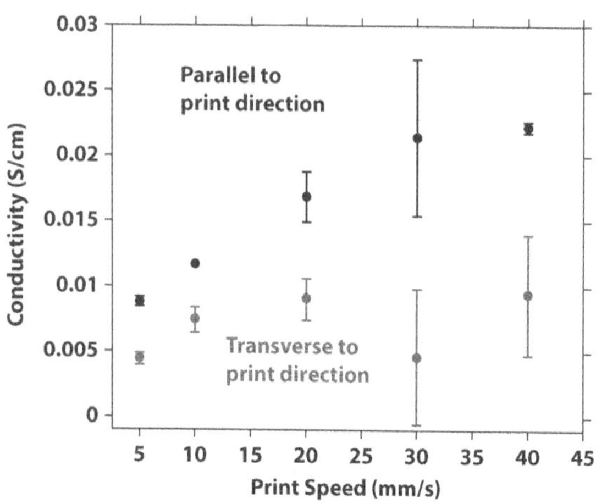

Figure 6.9 Effect of printing speed on the electrical conductivity [89].

sintering, which includes holding time and temperature, and the postprocessing. Many researchers have done a study on the influence of various factors on the quality of BJ-printed PMCs [28, 29, 91, 92]. Kim *et al.* [91] examined the mechanical behavior of BJ-printed ceramic/polymer composites. Two contrasting designs were chosen to compare a conventional lightweight metal design with the newly proposed structure. Dini *et al.* [27] analyzed the impact of powder composition, particle size distribution, and polymer adhesive on the mechanical behavior of hydroxyapatite/carboxymethyl chitosan/polyvinyl pyrrolidone/dextrin composites. The compressive strength as well as Young's modulus of printed objects improved after being treated in a chitosan solution. Ahn *et al.* [28] proposed a biodegradable PMC scaffolds with high ceramic content made by BJ conjugated with capillary rise infiltration and studied their mechanical properties and biocompatibility. The proposed PMC scaffold shown the improved mechanical properties.

6.4.6 Influence of Parameters on MJ-Printed Parts

Factors such as print direction, layer thickness, reinforcement type and content, UV intensity, temperature, and exposure time have significant impact on the mechanical behavior and geometrical accuracy of the 3D-printed components [47, 48, 93]. Tee *et al.* [47] analyzed the mechanical behavior of MJ-printed composites. The mechanical behavior of PMCs with various compositions and the arrangement of reinforced particles were studied. The results of compression tests revealed that the stiffness of printed composites is significantly affected by both the geometry of reinforced particles and the orientation in which they are printed. It was observed from the tensile tests that composite samples oriented at 45° have a lower elongation at break and ultimate tensile strength, in comparison to the composites oriented in the parallel direction [Figure 6.10 (a) and (b)]. It was also reported that the hard particles strengthen the part, however, soft particles mainly act as a crack initiation site.

Sugavaneswaran and Arumaikkannu [48] fabricated the randomly oriented multimaterial considering the plastic reinforcement distribution in the matrix elastomer. Stress-strain behavior of 3D-printed parts is carried out in UTM. It was reported that randomly oriented multimaterial with the reinforcement of plastic provides a significant improvement in stiffness compared to the pure elastomer component. Another study conducted an experimental and analytical examination of the elastic modulus of 3D-printed structures reinforced with materials [93]. A study on the influence of reinforcement addition and its orientation on the mechanical

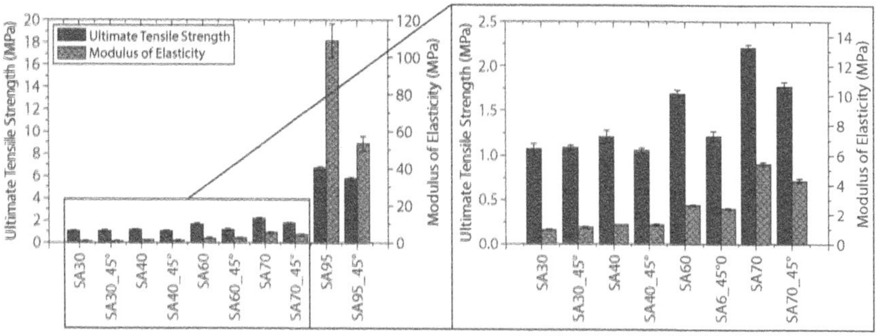

Figure 6.10 (a) Stress-strain diagram of tensile specimens with parallel and 45° print directions (b) modulus of elasticity and ultimate tensile strength [47].

behavior and fracture surface morphology was carried out. It was concluded that the addition of reinforcement increases the modulus of elasticity significantly.

6.4.7 Influence of Parameters on LOM-Printed Parts

Parameters influencing the quality of printed parts include part orientation, layer thickness, time between two consecutive exposures, exposure time, and the reinforcement and polymer used. Although substantial research work has been done on polymer composites fabricated by other AM techniques, limited work has been done on the LOM-printed composites. Chang *et al.* [46] suggested a novel AM technique based on ultrasonic-assisted LOM to fabricate CF/PA6 composites and analyzed the influence of different parameters on the mechanical behavior like tensile, bending, and inter-laminar shear strength. It was reported that composites made the with proposed method exhibit high tensile strength and stiffness. Klosterman *et al.* [43] analyzed the interfacial characteristics of glass fiber-reinforced epoxy composites fabricated by LOM. It was reported that three issues need to be addressed such as printing parameters, material preforms, and postprocessing methods to achieve the good interlaminar adhesion.

6.5 Applications of 3D-Printed Polymer Matrix and Metal Matrix Composites

Over the years, improvement in AM processes and materials coupled with the development in design and modeling has resulted in effective commercial applications. Figure 6.11 shows the applications of AM products. AM techniques are now widely used in numerous applications such as biomedical, aerospace and aviation, automotive, construction, and manufacturing industries. Table 6.5 shows the applications of 3D-printed MMC and PMC products.

6.6 Future Scope and Challenges

Although AM processes are becoming established and there are new opportunities to use it in a wide range of applications, some limitations still need to be addressed. The limitation of printing materials, which can be adapted for the AM of a wide range of applications of high-performance 3D-printed composite parts, still remains a key challenge. Fabrication of composites using AM can provide many advantages like customized and complex parts, flexible processing to produce quality parts, tailorable, and precise control of different properties. However, the distribution of

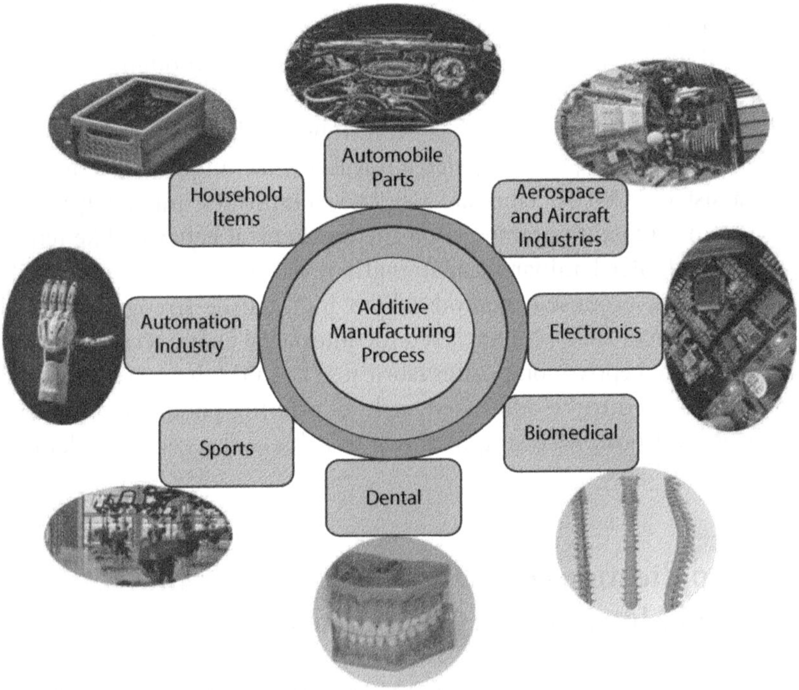

Figure 6.11 Applications of AM in different industries.

Table 6.5 Applications of 3D-printed MMC and PMC products.

Applications	Parts	References
Biomedical	Soft scaffold, blood vessels, human osteosarcoma, cartilage, bone, soft tissues, drug delivery devices and artificial organs, peripheral nerve repair, regeneration of bones, hard tissue engineering (bone, cartilage), soft tissue engineering (neural tissue, peripheral nerve restoration, blood vessels/platelets), lightweight prosthetic leg, dental splints, orthopedic insoles, dental restoration, endoscopic surgery, dental braces, hearing aids, artificial cosmetic organs such as ear, breast, nose, hips and denture	[40, 94, 95–102, 103]
Aerospace	Helicopter blades mold, unmanned aerial vehicle, full-wing structure, airduct, fuselage of Airbus A350 aircraft, honeycomb structure, fuel turbopump, valves, fuel injector, different parts of engine, aircraft components like brackets, inspection ports, harness clips and hatch cover	[40, 100, 104–106]
Automobile	Automotive components, automobile inlet manifold, air conditioning ducts, battery covers, and functional mounting brackets, piston parts, brakes, clutches, car body	[100, 102, 107]
Electronics	Actuated robotic limbs, smart structures with sensors, dielectrics, battery and supercapacitor sensor, electrical circuit and robot, robotic arm, stimuli-responsive electrical devices, lithium-ion batteries, supercapacitors, electrostatic discharge, force, strain and liquid sensors, actuator, broadband resonant transducer, circuit boards, antennas, capacitors, resistors, inductors, flexible displays, solar cells, sandwich strain sensors	[95, 96, 98, 99, 102–105, 107–109]

(Continued)

Table 6.5 Applications of 3D-printed MMC and PMC products. (*Continued*)

Applications	Parts	References
Industrial and manufacturing	Impeller, casting patterns, molds and cores, tools for injection molding, small series parts, tools prototypes, living cells, tools for vehicle maintenance, molds and prototypes, tool parts, molds for injection molding, bearing, blades of wind, turbine, turbine nozzle segments, nozzle, jigs, fixtures, molds, and dies, windmill blades, mold, large trim tool, fashion and textile printing	[99,101–104, 107, 109–112]
Structural	Footbridge, bicycle frames, textile, furniture, ropes, roofing, door panels, building boards, roofing sheets, insulation boards	[104, 111]
Others	Membrane for water treatment, adsorption, bioreactor operation, surface texturing, magnetic particle imaging phan, bullet proof vests, socket prosthesis	[96, 111, 113, 114]

reinforcements, uniformity in microstructure, matrix-reinforcement bonding, feature resolution, wear and blocking of printing heads, curing techniques, etc., are possible issues that need to be addressed for the fabrication of PMC and MMC parts using AM. The factors influencing the quality of the printed fiber-reinforced polymer composite parts include feedstock materials, printing speed, chemical composition, fiber content, aspect ratio, and melting temperature. The arrangement of long fibers and the distribution of short fibers are mainly controlled by the material formulation and the process pattern, which also has significant influence on the design approaches to improve the performance of 3D-printed parts. 3D printing of both PMCs and MMCs offer significant potential for fabricating parts with excellent functionality and mechanical properties. There is always a need for further research in order to explore the more suitable composites and applications for 3D printing materials. One emerging material for AM is nanomaterials. There is substantial potential for the fabrication of nanocomposites when nanofillers are reinforced into host matrix materials using AM. Though significant research has been done on nanofiller-reinforced 3D-printed composites, the agglomeration of nanoparticles due to its high content is still the key challenge for the fabrication of this novel material. The print parameters have a vital role in determining the quality and performance of printed parts. These parameters have a significant influence on the properties of the printed objects. Therefore, careful selection of these parameters based on the materials used and end use application is a key challenge currently. It is also essential to develop optimization and modeling tools in order to predict the properties of the printed components. Though extensive research has been done on the AM of composite materials for making prototypes, the commercial application of this class of material is very narrow due to its limited materials choice, high porosity, and poor interfacial bonding to modify the properties of composites as per the desired requirements. Fabrication of defect-free 3D-printed composite part is the biggest challenge. Thus, it is important to detect the origin of the defects in 3D-printed objects and eliminating them. Finally, postprocessing of the 3D-printed parts is also a key challenge for AM. Despite being an innovative manufacturing process for customised products, AM requires more research and development to compete with the traditional manufacturing processes in mass production because of its lower speed and higher cost. Worldwide research would result in a fast transition from the traditional manufacturing methods to AM in the near future.

6.7 Summary and Conclusion

This chapter provides an overview of the recent literature on AM of composites. Specifically, the fabrication of MMCs and PMCs by the incorporation of different reinforcements were discussed in terms of their working principle, materials used, and their advantages. The influence of different factors of AM techniques on the quality of the parts were highlighted. Further, the recent applications of 3D-printed MMC and PMC products were explored. Finally, the research opportunities and challenges of AM were highlighted in the chapter. The properties and quality of the printed objects were found to be significantly impacted by the printing parameters. The final product's performance is also dependent on factors such as the reinforcement type used, its concentration, fiber length, and the nature of the fiber-matrix interface. Thus, it is very important to optimize these factors in order to produce the composite products with the desired quality. AM technologies for PMCs have specific requirements for material property like rheology properties and melting point. It is therefore important to design and choose the polymers for 3D printing as per the polymer characteristics. The capability of 3D printing in converting AM from a basic prototyping process to a strong composite fabrication method is remarkably evident in the production of MMCs and PMCs.

References

1. Froes, F. and Boyer, R., *Additive Manufacturing for the Aerospace Industry*, Elsevier Inc, Amsterdam, Netherlands, 2019.
2. Al Rashid, A., Khan, S.A., G. Al-Ghamdi, S., Koç, M., Additive manufacturing: Technology, applications, markets, and opportunities for the built environment. *Autom. Constr.*, **118**, 103268, 2020.
3. Shalnova, S.A., Volosevich, D.V., Sannikov, M.I., Magidov, I.S., Mikhaylovskiy, K.V., Turichin, G.A., Klimova-Korsmik, O.G., Direct energy deposition of SiC reinforced Ti–6Al–4V metal matrix composites: Structure and mechanical properties. *Ceram. Int.*, **48**, 23, 35076–35084, 2022.
4. Xu, R., Li, R., Yuan, T., Zhu, H., Li, P., Microstructure and mechanical properties of TiC-reinforced Al–Mg–Sc–Zr composites additively manufactured by laser direct energy deposition. *Acta Metall. Sin. (Engl. Lett.*, **35**, 3, 411–424, 2022.
5. Wang, Q.Z., Lin, X., Wen, X.L., Kang, N., Huang, W.D., Microstructure and wear behavior of nano-TiB2p/2024Al matrix composites fabricated by laser direct energy deposition with powder feeding. *J. Tribol.*, **143**, 5, 051101, 2021.
6. de Rojas Candela, C.S., Riquelme, A., Bonache, V., Rodrigo, P., Rams, J., Ti6Al4V/SiC metal matrix composites additively manufactured by direct laser deposition. *Met. Mater. Int.*, **28**, 12, 3120–3144, 2022.
7. Lanfant, B., Bär, F., Mohanta, A., Leparoux, M., Fabrication of metal matrix composite by laser metal deposition—A new process approach by direct dry injection of nanopowders. *Mater. (Basel)*, **12**, 21, 3584, 2019.
8. Ghosh, S.K. and Saha, P., Crack and wear behavior of SiC particulate reinforced aluminium based metal matrix composite fabricated by direct metal laser sintering process. *Mater. Des.*, **32**, 1, 139–145, 2011.
9. Chang, K. and Gu, D., Direct metal laser sintering synthesis of carbon nanotube reinforced Ti matrix composites: Densification, distribution characteristics and properties. *J. Mater. Res.*, **31**, 2, 281–291, 2016.
10. Misra, S., Hussain, M., Gupta, A., Kumar, V., Kumar, S., Das, A.K., Fabrication and characteristic evaluation of direct metal laser sintered SiC particulate reinforced Ti6Al4V metal matrix composites. *J. Laser Appl.*, **31**, 1, 12005, 2019.
11. Mandal, V., Hussain, M., Kumar, V., Kumar Das, A., Singh, N.K., Development of reinforced TiN-SS316 metal matrix composite (MMC) using direct metal laser sintering (DMLS) and its characterization. *Mater. Today Proc.*, **4**, 9, 9982–9986, 2017.

12. Song, B., Dong, S., Coddet, P., Zhou, G., Ouyang, S., Liao, H., Coddet, C., Microstructure and tensile behavior of hybrid nano-micro SiC reinforced iron matrix composites produced by selective laser melting. *J. Alloys Compd.*, **579**, 415–421, 2013.
13. Mandal, V., Tripathi, P., Kumar, A., Singh, S.S., Ramkumar, J., A study on selective laser melting (SLM) of TiC and B4C reinforced IN718 metal matrix composites (MMCs). *J. Alloys Compd.*, **901**, 163527, 2022.
14. Wen, S., Chen, K., Li, W., Zhou, Y., Wei, Q., Shi, Y., Selective laser melting of reduced graphene oxide/S136 metal matrix composites with tailored microstructures and mechanical properties. *Mater. Des.*, **175**, 107811, 2019.
15. Peng, H., Liu, C., Guo, H., Yuan, Y., Gong, S., Xu, H., Fabrication of WCp/NiBSi metal matrix composite by electron beam melting. *Mater. Sci. Eng. A*, **666**, 320–323, 2016.
16. Gao, B., Peng, H., Liang, Y., Lin, J., Chen, B., Electron beam melted TiC/high Nb–TiAl nanocomposite: Microstructure and mechanical property. *Mater. Sci. Eng. A*, **811**, 141059, 2021.
17. Liu, C., Peng, H., Zhao, Y., Yuan, Y., Guo, H.B., Xu, H.B., Microstructure, mechanical and corrosion properties of electron-beam-melted and plasma-transferred arc-welded WCP/NiBSi metal matrix composites. *Rare Met.*, **38**, 9, 814–823, 2019.
18. Xi, L., Feng, L., Gu, D., Wang, R., Sarac, B., Prashanth, K.G., Eckert, J., ZrC+TiC synergically reinforced metal matrix composites with micro/nanoscale reinforcements prepared by laser powder bed fusion. *J. Mater. Res. Technol.*, **19**, 4645–4657, 2022.
19. Shi, Q., Mertens, R., Dadbakhsh, S., Li, G., Yang, S., *In-situ* formation of particle reinforced Aluminium matrix composites by laser powder bed fusion of Fe2O3/AlSi12 powder mixture using laser melting/remelting strategy. *J. Mater. Process. Technol.*, **299**, 117357, 2022.
20. Jäcklein, M., Pfaff, A., Hoschke, K., Developing tungsten-filled metal matrix composite materials using laser powder bed fusion. *Appl. Sci.*, **10**, 24, 8869, 2020.
21. Xue, G., Ke, L., Liao, H., Chen, C., Zhu, H., Effect of SiC particle size on densification behavior and mechanical properties of SiCp/AlSi10Mg composites fabricated by laser powder bed fusion. *J. Alloys Compd.*, **845**, 156260, 2020.
22. Levy, A., Miriyev, A., Elliott, A., Babu, S.S., Frage, N., Additive manufacturing of complex-shaped graded TiC/steel composites. *Mater. Des.*, **118**, 198–203, 2017.
23. Enneti, R.K., Prough, K.C., Wolfe, T.A., Klein, A., Studley, N., Trasorras, J.L., Sintering of WC-12%Co processed by binder jet 3D printing (BJ3DP) technology. *Int. J. Refract. Met. Hard Mater.*, **71**, 28–35, 2018.
24. Porter, Q., Pei, Z., Ma, C., Binder jetting and infiltration of metal matrix nanocomposites. *J. Manuf. Sci. Eng.*, **144**, 7, 074502, 2022.

25. Zhang, Y., Han, J., Zhang, X., He, X., Li, Z., Du, S., Rapid prototyping and combustion synthesis of TiC/Ni functionally gradient materials. *Mater. Sci. Eng. A*, **299**, 1–2, 218–224, 2001.
26. Krinitcyn, M., Fu, Z., Harris, J., Kostikov, K., Pribytkov, G.A., Greil, P., Travitzky, N., Laminated object manufacturing of *in-situ* synthesized MAX-phase composites. *Ceram. Int.*, **43**, 12, 9241–9245, 2017.
27. Dini, F., Ghaffari, S.A., Javadpour, J., Rezaie, H.R., Binder jetting of hydroxyapatite/carboxymethyl chitosan/polyvinylpyrrolidone/dextrin composite: The role of polymeric adhesive and particle size distribution on printability of powders. *J. Mater. Eng. Perform.*, **31**, 7, 5801–5811, 2022.
28. Ahn, J.H., Kim, J., Han, G., Kim, D.E., Cheon, K.H., Lee, H., Kim, H.E., Kim, Y.J., Jang, T.S., Jung, H.D., 3D-printed biodegradable composite scaffolds with significantly enhanced mechanical properties via the combination of binder jetting and capillary rise infiltration process. *Addit. Manuf.*, **41**, 101988, 2021.
29. Li, T., Chen, T., Shen, X., Shi, H.H., Jabari, E., Naguib, H.E., A binder jet 3D printed MXene composite for strain sensing and energy storage application. *Nanoscale Adv.*, **4**, 3, 916–925, 2022.
30. Zhang, Y., Li, S., Zhao, Y., Duan, W., Liu, B., Wang, T., Wang, G., Digital light processing 3D printing of AlSi10Mg powder modified by surface coating. *Addit. Manuf.*, **39**, 101897, 2021.
31. Mu, X., Bertron, T., Dunn, C., Qiao, H., Wu, J., Zhao, Z., Saldana, C., Qi, H.J., Porous polymeric materials by 3D printing of photocurable resin. *Mater. Horiz.*, **4**, 3, 442–449, 2017.
32. Feng, Z., Li, Y., Xin, C., Tang, D., Xiong, W., Zhang, H., Fabrication of graphene-reinforced nanocomposites with improved fracture toughness in net shape for complex 3D structures via digital light processing. *C*, **5**, 2, 25, 2019.
33. Cortés, A., Sánchez-Romate, X.F., Jiménez-Suárez, A., Campo, M., Ureña, A., Prolongo, S.G., Mechanical and strain-sensing capabilities of carbon nanotube reinforced composites by digital light processing 3D printing technology. *Polym. (Basel)*, **12**, 4, 975, 2020.
34. Slejko, E.A., Sesto Gorella, N., Makaya, A., Gallina, P., Scuor, N., Seriani, S., Vacuum 3D printing of highly filled polymeric matrix composites. *Acta Astronaut.*, **204**, August 2022, 25–33, 2023.
35. Pearson, C., Hawi, S., Lira, C., Goel, S., Yazdani Nezhad, H., Magnetic field assisted 3D printing of short carbon fibre-reinforced polymer composites. *Mater. Today Proc.*, **64**, 1403–1411, 2022.
36. Justo, J., Távara, L., García-Guzmán, L., París, F., Characterization of 3D printed long fibre reinforced composites. *Compos. Struct.*, **185**, July 2017, 537–548, 2018.
37. Tekinalp, H.L., Kunc, V., Velez-Garcia, G.M., Duty, C.E., Love, L.J., Naskar, A.K., Blue, C.A., Ozcan, S., Highly oriented carbon fiber-polymer composites via additive manufacturing. *Compos. Sci. Technol.*, **105**, 144–150, 2014.

38. Zhong, W., Li, F., Zhang, Z., Song, L., Li, Z., Short fiber reinforced composites for fused deposition modeling. *Mater. Sci. Eng. A*, **301**, 2, 125–130, 2001.
39. Li, N., Li, Y., Liu, S., Rapid prototyping of continuous carbon fiber reinforced polylactic acid composites by 3D printing. *J. Mater. Process. Technol.*, **238**, 218–225, 2016.
40. Ning, F., Cong, W., Qiu, J., Wei, J., Wang, S., Additive manufacturing of carbon fiber reinforced thermoplastic composites using fused deposition modeling. *Compos. Part B Eng.*, **80**, 369–378, 2015.
41. Parker, M., Ezeokeke, N., Matsuzaki, R., Arola, D., Strength and its variability in 3D printing of polymer composites with continuous fibers. *Mater. Des.*, **225**, 111505, 2023.
42. Elkholy, A. and Kempers, R., An accurate steady-state approach for characterizing the thermal conductivity of Additively manufactured polymer composites. *Case Stud. Therm. Eng.*, **31**, January, 101829, 2022.
43. Klosterman, D., Chartoff, R., Graves, G., Osborne, N., Priore, B., Interfacial characteristics of composites fabricated by laminated object manufacturing. *Compos. Part A Appl. Sci. Manuf.*, **29**, 9–10, 1165–1174, 1998.
44. Parandoush, P., Zhou, C., Lin, D., 3D printing of ultrahigh strength continuous carbon fiber composites. *Adv. Eng. Mater.*, **21**, 2, 1800622, 2019.
45. Özbay, B. and Serhatlı, İ.E., Processing and characterization of hollow glass-filled polyamide 12 composites by selective laser sintering method. *Mater. Technol.*, **37**, 4, 213–223, 2022.
46. Chang, B., Parandoush, P., Li, X., Ruan, S., Shen, C., Behnagh, R.A., Liu, Y., Lin, D., Ultrafast printing of continuous fiber-reinforced thermoplastic composites with ultrahigh mechanical performance by ultrasonic-assisted laminated object manufacturing. *Polym. Compos.*, **41**, 11, 4706–4715, 2020.
47. Tee, Y.L., Tran, P., Leary, M., Pille, P., Brandt, M., 3D Printing of polymer composites with material jetting: Mechanical and fractographic analysis. *Addit. Manuf.*, **36**, 101558, 2020.
48. Sugavaneswaran, M. and Arumaikkannu, G., Modelling for randomly oriented multi material additive manufacturing component and its fabrication. *Mater. Des.*, **54**, 779–785, 2014.
49. Zhou, X., Nowicki, M., Cui, H., Zhu, W., Fang, X., Miao, S., Lee, S.J., Keidar, M., Zhang, L.G., 3D bioprinted graphene oxide-incorporated matrix for promoting chondrogenic differentiation of human bone marrow mesenchymal stem cells. *Carbon N. Y.*, **116**, 615–624, 2017.
50. Mubarak, S., Dhamodharan, D., Divakaran, N., Kale, M.B., Senthil, T., Wu, L., Wang, J., Enhanced mechanical and thermal properties of stereolithography 3D printed structures by the effects of incorporated controllably annealed anatase TiO2 nanoparticles. *Nanomaterials*, **10**, 1, 79, 2020.
51. Taormina, G., Sciancalepore, C., Bondioli, F., Messori, M., Special resins for stereolithography: *In situ* generation of silver nanoparticles. *Polym. (Basel)*, **10**, 2, 212, 2018.

52. Nagarajan, B., Mertiny, P., Qureshi, A.J., Magnetically loaded polymer composites using stereolithography—Material processing and characterization. *Mater. Today Commun.*, **25**, 101520, 2020.
53. Invernizzi, M., Suriano, R., Muscatello, A., Turri, S., Levi, M., Near-visible stereolithography of a low shrinkage cationic/free-radical photopolymer blend and its nanocomposite. *J. Appl. Polym. Sci.*, **137**, 5, 48333, 2020.
54. Yan, M., Tian, X., Peng, G., Li, D., Zhang, X., High temperature rheological behavior and sintering kinetics of CF/PEEK composites during selective laser sintering. *Compos. Sci. Technol.*, **165**, June, 140–147, 2018.
55. Gaikwad, S., Tate, J., Theodoropoulou, N., Koo, J., Electrical and mechanical properties of PA11 blended with nanographene platelets using industrial twin-screw extruder for selective laser sintering. *J. Compos. Mater.*, **47**, 23, 2973–2986, 2013.
56. Eosoly, S., Vrana, N.E., Lohfeld, S., Hindie, M., Looney, L., Interaction of cell culture with composition effects on the mechanical properties of polycaprolactone-hydroxyapatite scaffolds fabricated via selective laser sintering (SLS). *Mater. Sci. Eng. C*, **32**, 8, 2250–2257, 2012.
57. Chua, C.K., Leong, K.F., Tan, K.H., Wiria, F.E., Cheah, C.M., Development of tissue scaffolds using selective laser sintering of polyvinyl alcohol/hydroxyapatite biocomposite for craniofacial and joint defects. *J. Mater. Sci. Mater. Med.*, **15**, 10, 1113–1121, 2004.
58. Gajera, H.M., Dave, K.G., Darji, V.P., Abhishek, K., Optimization of process parameters of direct metal laser sintering process using fuzzy-based desirability function approach. *J. Braz. Soc. Mech. Sci. Eng.*, **41**, 3, 1–23, 2019.
59. Ghosh, S.K., Bandyopadhyay, K., Saha, P., Development of an *in-situ* multi-component reinforced Al-based metal matrix composite by direct metal laser sintering technique-optimization of process parameters. *Mater. Charact.*, **93**, 68–78, 2014.
60. Gu, D., Shen, Y., Xiao, J., Influence of processing parameters on particulate dispersion in direct laser sintered WC-Cop/Cu MMCs. *Int. J. Refract. Met. Hard Mater.*, **26**, 5, 411–422, 2008.
61. Gupta, A., Hussain, M., Misra, S., Das, A.K., Mandal, A., Processing and characterization of laser sintered hybrid B4C/cBN reinforced Ti-based metal matrix composite. *Opt. Lasers Eng.*, **105**, January, 159–172, 2018.
62. Kundu, S., Hussain, M., Kumar, V., Kumar, S., Das, A.K., Direct metal laser sintering of TiN reinforced Ti6Al4V alloy based metal matrix composite: Fabrication and characterization. *Int. J. Adv. Manuf. Technol.*, **97**, 5–8, 2635–2646, 2018.
63. Hussain, M., Mandal, V., Kumar, V., Das, A.K., Ghosh, S.K., Development of TiN particulates reinforced SS316 based metal matrix composite by direct metal laser sintering technique and its characterization. *Opt. Laser Technol.*, **97**, 46–59, 2017.
64. Ghosh, S.K., Saha, P., Kishore, S., Influence of size and volume fraction of SiC particulates on properties of *ex situ* reinforced Al-4.5Cu-3Mg metal matrix

composite prepared by direct metal laser sintering process. *Mater. Sci. Eng. A*, **527**, 18–19, 4694–4701, 2010.
65. Liao, H., Zhu, H., Zhu, J., Chang, S., Zeng, X., Effect of process parameters on selective laser melting al2o3-al cermet material. *Solid Free. Fabr. 2019 Proc. 30th Annu. Int. Solid Free. Fabr. Symp. - An Addit. Manuf. Conf. SFF 2019*, pp. 960–966, 2019.
66. Song, B., Wang, Z., Yan, Q., Zhang, Y., Zhang, J., Cai, C., Wei, Q., Shi, Y., Integral method of preparation and fabrication of metal matrix composite: Selective laser melting of *in-situ* nano/submicro-sized carbides reinforced iron matrix composites. *Mater. Sci. Eng. A*, **707**, July, 478–487, 2017.
67. Du, Z., Chen, H.C., Tan, M.J., Bi, G., Chua, C.K., Effect of nAl2O3 on the part density and microstructure during the laser-based powder bed fusion of AlSi10Mg composite. *Rapid Prototyp. J.*, **26**, 4, 727–735, 2020.
68. Zhang, Y., Yu, Y., Wang, L., Li, Y., Lin, F., Yan, W., Dispersion of reinforcing micro-particles in the powder bed fusion additive manufacturing of metal matrix composites. *Acta Mater.*, **235**, 118086, 2022.
69. Shuai, L., Qinghong, J., Sai, G., Bi, Z., Cong, Z., Effect of spattering on formation mechanisms of metal matrix composites in laser powder bed fusion. *J. Mater. Process. Technol.*, **304**, December 2021, 117533, 2022.
70. Li, X., Astfalack, L., Drew, P., Schliwa, Y., Sercombe, T.B., Selective laser melting of aluminium metal matrix composites. *Proc. Int. Conf. Prog. Addit. Manuf.*, vol. **Part F1290**, pp. 433–438, 2016.
71. Jue, J., Gu, D., Chang, K., Dai, D., Microstructure evolution and mechanical properties of Al-Al2O3 composites fabricated by selective laser melting. *Powder Technol.*, **310**, 80–91, 2017.
72. Ghodsi, M.Z., Khademzadeh, S., Marzbanrad, E., Razmpoosh, M.H., De Marchi, N., Toyserkani, E., Development of Yttria-stabilized zirconia reinforced Inconel 625 metal matrix composite by laser powder bed fusion. *Mater. Sci. Eng. A*, **827**, March, 142037, 2021.
73. Yang, G., Yang, P., Yang, K., Liu, N., Jia, L., Wang, J., Tang, H., Effect of processing parameters on the density, microstructure and strength of pure tungsten fabricated by selective electron beam melting. *Int. J. Refract. Met. Hard Mater.*, **84**, July, 105040, 2019.
74. Kurzynowski, T., Madeja, M., Dziedzic, R., Kobiela, K., The effect of EBM process parameters on porosity and microstructure of Ti-5Al-5Mo-5V-1Cr-1Fe alloy. *Scanning*, **2019**, 2903920, 2019.
75. Ma, S., Li, Y., Kan, W., Zhang, J., Wang, M., Wang, L., Wu, Y., Wang, H., Chen, Z., Enhancement of grain refinement and heat resistance in TiB2-reinforced Al-Cu-Mg-Fe-Ni matrix composite additive manufactured by electron beam melting. *J. Alloys Compd.*, **924**, 166395, 2022.
76. Farayibi, P.K., Abioye, T.E., Kennedy, A., Clare, A.T., Development of metal matrix composites by direct energy deposition of 'satellited' powders. *J. Manuf. Process.*, **45**, July, 429–437, 2019.

77. Traxel, K.D. and Bandyopadhyay, A., Designing high-temperature oxidation-resistant titanium matrix composites via directed energy deposition-based additive manufacturing. *Mater. Des.*, **212**, 110205, 2021.
78. Ang, Y.T., Sing, S.L., Lim, J.C.W., Process study for directed energy deposition of 316L stainless steel with TiB2 metal matrix composites. *Mater. Sci. Addit. Manuf.*, **1**, 2, 13, 2022.
79. Muhamedagic, K., Berus, L., Potočnik, D., Cekic, A., Begic-Hajdarevic, D., Cohodar Husic, M., Ficko, M., Effect of process parameters on tensile strength of FDM printed carbon fiber reinforced polyamide parts. *Appl. Sci.*, **12**, 12, 6028, 2022.
80. Belei, C., Joeressen, J., Amancio-Filho, S.T., Fused-filament fabrication of short carbon fiber-reinforced polyamide: Parameter optimization for improved performance under uniaxial tensile loading. *Polym. (Basel)*, **14**, 7, 1292, 2022.
81. Liao, G., Li, Z., Cheng, Y., Xu, D., Zhu, D., Jiang, S., Guo, J., Chen, X., Xu, G., Zhu, Y., Properties of oriented carbon fiber/polyamide 12 composite parts fabricated by fused deposition modeling. *Mater. Des.*, **139**, 283–292, 2018.
82. Peng, X., Zhang, M., Guo, Z., Sang, L., Hou, W., Investigation of processing parameters on tensile performance for FDM-printed carbon fiber reinforced polyamide 6 composites. *Compos. Commun.*, **22**, 100478, 2020.
83. Gowda, R.B.S., Udayagiri, C.S., Narendra, D.D., Studies on the process parameters of rapid prototyping technique (stereolithography) for the betterment of part quality. *Int. J. Manuf. Eng.*, **2014**, 1–11, 2014.
84. Maalihan, R.D., Pajarito, B.B., Advincula, R.C., 3D-printing methacrylate/chitin nanowhiskers composites via stereolithography: Mechanical and thermal properties. *Mater. Today Proc.*, **33**, 1819–1824, 2020.
85. Hanon, M.M., Ghaly, A., Zsidai, L., Klébert, S., Tribological characteristics of digital light processing (DLP) 3D printed graphene/resin composite: Influence of graphene presence and process settings. *Mater. Des.*, **218**, 110718, 2022.
86. Özbay Kısasöz, B., Serhatlı, I.E., Bulduk, M.E., Selective laser sintering manufacturing and characterization of lightweight PA 12 polymer composites with different hollow microsphere additives. *J. Mater. Eng. Perform.*, **31**, 5, 4049–4059, 2022.
87. Idriss, A.I., Li, J., Guo, Y., Wang, Y., Li, X., Zhang, Z., Elfaki, E.A., Sintering quality and parameters optimization of sisal fiber/PES composite fabricated by selective laser sintering (SLS). *J. Thermoplast. Compos. Mater.*, **35**, 10, 1632–1646, 2022.
88. Tandel, R. and Gozen, B.A., Direct-ink-writing of liquid metal-graphene-based polymer composites: Composition-processing-property relationships. *J. Mater. Process. Technol.*, **302**, 117470, 2022.
89. Haney, R., Tran, P., Trigg, E.B., Koerner, H., Dickens, T., Ramakrishnan, S., Printability and performance of 3D conductive graphite structures. *Addit. Manuf.*, **37**, August 2020, 101618, 2021.

90. Compton, B.G. and Lewis, J.A., 3D-printing of lightweight cellular composites. *Adv. Mater.*, **26**, 34, 5930–5935, 2014.
91. Kim, D.H., Lee, J., Bae, J., Park, S., Choi, J., Lee, J.H., Kim, E., Mechanical analysis of ceramic/polymer composite with mesh-type lightweight design using binder-jet 3D printing. *Mater. (Basel)*, **11**, 10, 1941, 2018.
92. Utela, B., Storti, D., Anderson, R., Ganter, M., A review of process development steps for new material systems in three dimensional printing (3DP). *J. Manuf. Process.*, **10**, 2, 96–104, 2008.
93. Sugavaneswaran, M. and Arumaikkannu, G., Analytical and experimental investigation on elastic modulus of reinforced additive manufactured structure. *Mater. Des.*, **66**, PA, 29–36, 2015.
94. Singh, S., Ramakrishna, S., Singh, R., Material issues in additive manufacturing: A review. *J. Manuf. Process.*, **25**, 185–200, 2017.
95. Park, S., Shou, W., Makatura, L., Matusik, W., Fu, K., 3D printing of polymer composites: Materials, processes, and applications. *Matter*, **5**, 1, 43–76, 2022.
96. Singh, N. and Singh, G., Advances in polymers for bio-additive manufacturing: A state of art review. *J. Manuf. Process.*, **72**, October, 439–457, 2021.
97. Sharma, V. and Pandey, P. M., *Additive and Subtractive Manufacturing Processes: Principles and Applications*, CRC Press, Boca Raton, 2022.
98. Li, Y., Feng, Z., Huang, L., Essa, K., Bilotti, E., Zhang, H., Peijs, T., Hao, L., Additive manufacturing high performance graphene-based composites: A review. *Compos. Part A Appl. Sci. Manuf.*, **124**, October 2018, 105483, 2019.
99. Wu, H., Fahy, W.P., Kim, S., Kim, H., Zhao, N., Pilato, L., Kafi, A., Bateman, S., Koo, J.H., Recent developments in polymers/polymer nanocomposites for additive manufacturing. *Prog. Mater. Sci.*, **111**, 100638, January, 2020.
100. Yuan, S., Shen, F., Chua, C.K., Zhou, K., Polymeric composites for powder-based additive manufacturing: Materials and applications. *Prog. Polym. Sci.*, **91**, 141–168, 2019.
101. Sun, J., Ye, D., Zou, J., Chen, X., Wang, Y., Yuan, J., Liang, H., Qu, H., Binner, J., Bai, J., A review on additive manufacturing of ceramic matrix composites. *J. Mater. Sci. Technol.*, **138**, 1–16, 2023.
102. Dave, H. K. and Davim, J. P., *Fused Deposition Modeling Based 3D Printing*, pp. 23–42, Springer, Cham, Switzerland, 2021.
103. Izdebska-Podsiadły, J., *Polymers for 3d printing: Methods, properties, and characteristics*. Elsevier, Oxford, United Kingdom, 2022.
104. Li, J., Durandet, Y., Huang, X., Sun, G., Ruan, D., Additively manufactured fiber-reinforced composites: A review of mechanical behavior and opportunities. *J. Mater. Sci. Technol.*, **119**, 219–244, 2022.
105. Ngaruiya, J.K. and Ngigi, M.M., *Check point threat extraction secured this document get original*, 2020.
106. Balasubramanian, K.R. and Senthilkumar, V., *Additive manufacturing applications for metals and composites*. IGI Global, Pennsylvania, USA, 2020.
107. Park, S. and Fu, K., Polymer-based filament feedstock for additive manufacturing. *Compos. Sci. Technol.*, **213**, June, 108876, 2021.

108. Gao, W., Zhang, Y., Ramanujan, D., Ramani, K., Chen, Y., Williams, C.B., Wang, C.C.L., Shin, Y.C., Zhang, S., Zavattieri, P.D., The status, challenges, and future of additive manufacturing in engineering. *CAD Comput. Aided Des.*, **69**, 65–89, 2015.
109. Bourell, D., Kruth, J.P., Leu, M., Levy, G., Rosen, D., Beese, A.M., Clare, A., Materials for additive manufacturing. *CIRP Ann. - Manuf. Technol.*, **66**, 2, 659–681, 2017.
110. Rafiee, M., Farahani, R.D., Therriault, D., Multi-material 3D and 4D printing: A survey. *Adv. Sci.*, **7**, 12, 1–26, 2020.
111. Mishra, T., Mandal, P., Rout, A.K., Sahoo, D., A state-of-the-art review on potential applications of natural fiber-reinforced polymer composite filled with inorganic nanoparticle. *Compos. Part C Open Access*, **9**, May, 100298, 2022.
112. El Moumen, A., Tarfaoui, M., Lafdi, K., Additive manufacturing of polymer composites: Processing and modeling approaches. *Compos. Part B Eng.*, **171**, April, 166–182, 2019.
113. Mohd Yusoff, N.H., Irene Teo, L.R., Phang, S.J., Wong, V.L., Cheah, K.H., Lim, S.S., Recent advances in polymer-based 3D printing for wastewater treatment application: An overview. *Chem. Eng. J.*, **429**, August 2021, 132311, 2022.
114. Jafferson, J.M. and Chatterjee, D., A review on polymeric materials in additive manufacturing. *Mater. Today Proc.*, **46**, 1349–1365, 2021.

7

Postprocessing of Additively Manufactured Polymeric and Metallic Parts—A Review

Vishal Mishra, Navya Ann Cherian, Sushant Negi* and Simanchal Kar

Department of Mechanical Engineering, NIT Silchar, Assam, India

Abstract

Additive manufacturing (AM) which is popularly known as 3D printing is an advanced manufacturing process which produces complex geometries in reduced manufacturing time. Moreover, it is easy to use because of the flexibility in the operation and compatibility with polymeric as well as metallic parts. Although this advanced technique is gaining more popularity, it also faces criticisms like poor surface finish, formation of unwanted microstructural phases, delamination, warping, and wear tracks. Due to the presence of the mentioned defects, there is a reduction in the part's physical and mechanical properties. Therefore, it becomes a necessity to opt for postprocessing techniques to resolve these issues. This is an important step in the 3D printing process which is required to be performed on the printed part to enhance its properties and provide an aesthetic look. This chapter includes the basics and a literature review of the various postprocessing techniques like coating, surface polishing, vapor smoothing, chemical, laser, ultrasound, and heat treatment, which are used for property enhancement of the printed parts. Lastly, this review chapter will help the reader to know the future perspectives of these techniques and will open an additional domain for future research.

Keywords: Additive manufacturing, 3D printing, postprocessing techniques, mechanical properties, polymers, metals

**Corresponding author*: sushant@mech.nits.ac.in; sushant.negi@hotmail.com

R. Rajasekar, C. Moganapriya and P. Sathish Kumar (eds.) Additive Manufacturing with Novel Materials: Processes, Properties and Applications, (237–266) © 2024 Scrivener Publishing LLC

7.1 Introduction

Additive manufacturing (AM) is a rapid prototyping technique commonly known as three-dimensional (3D) printing which generates 3D parts in a layer-by-layer mode based on a computer-aided-design (CAD) model [1–3]. This process in 2010 was classified in seven categories by the American Society of Testing and Materials (ASTM) group "ASTM F42-additive manufacturing" which is presented in Table 7.1. Additive manufacturing possesses several advantages which include short manufacturing cycle and reasonable fabrication cost in minimal batches [4]. This technology has eliminated the prerequisite of sophisticated tooling or molds and produces parts by using mere raw materials and equipment. This results in significant savings in processing as well as assembly time. Moreover, AM processes introduce near-net part generation with minimal machining allowance and first-rate material utilization [5].

Additive manufacturing enables the fabrication of lightweight products with a revolutionary method of industrial production. These processes are used in a variety of technological domains and frequently enhance current designs and features [7–9]. Aside from that, 3D printing is progressively used in high performance and high throughput industrial applications, such as rapid prototyping, rapid tooling, rapid manufacturing [10, 11], advanced electronics [12], health-related applications [13], water filtration [14, 15], and gas sensing [16]. This necessitates and justifies the technology needed to supply high-quality parts with the necessary characteristics to be used in practical and demanding applications. These include establishing process control, minimizing human error, and guaranteeing the performance of used materials [17–19]. Larger pieces, intricate geometries, and the consolidation of many part components can all be made via 3D printing. It can also be utilized to create parts with integrated form, fit, and function. Finally, 3D-printed objects and prototypes are evaluated for their aesthetic value in addition to the applications and features.

Although AM processes have the advantages mentioned, it also faces several criticisms like poor surface finish, debonding, and development of residual stresses, which are of significant concern [20]. The surface quality of the end product varies with varying AM processes, which means that AM alone cannot fabricate parts which can meet both mechanical and surface roughness requirements [21]. In metallic 3D printing, the interaction mechanisms between the powder bed and the melt pool, the powder bed and laser beam, and the melting processes are difficult because of the complex metallurgical and thermophysical phenomena. For example, in case of the SLM process, it is necessary to assess the rapid solidification process

Table 7.1 Additive manufacturing processes as per ASTM F42 [6].

AM processes	Description	Energy source	Related technology	Materials	
				Metals	Polymer
Vat polymerization	Selectively curing liquid photopolymer deposit by light activated polymerization	Laser	Stereolithography (SLA), digital light processing (DLP)		☑
Material jetting	Selectively depositing build material droplets	UV light	Polyjet technology, nanoparticle jetting, multijet modeling		☑
Binder jetting	Selectively depositing a liquid bonding agent	Binder	Powder bed and ink jet head (PBIH), plaster-based 3D printing (PP)	☑	☑
Material extrusion	Depositing semi solidus material through nozzle on print bed in layer-by-layer fashion	Melt extrusion	Fused deposition modeling (FDM)/ fused filament fabrication (FFF)		☑
Sheet lamination	Bonding the sheets of material	Adhesive & Laser	Laminated object manufacturing (LOM), ultrasonic consolidation		☑
Powder bed fusion	Fusing of powder bed region using thermal energy	Laser	Electron beam melting (EBM), selective laser sintering (SLS), selective heat sintering (SHS), and direct metal laser sintering (DMLS)	☑	☑
Direct energy deposition	Focusing thermal energy to fuse the material	Laser	Laser metal deposition (LMD)	☑	

under an extremely high temperature gradient as internal defects such as balling effect, porosity, cracks, development of internal stress, and thermal stresses frequently occur. Moreover, these defects greatly impact the micro-structural as well as physical structural mechanics of the final product. To avoid such harsh impacts, postprocessing techniques plays as vital role and are frequently needed. This technique is intended to enhance the mechanical and surface characteristics of the 3D-printed parts [22]. Similarly for the 3D-printed polymeric parts, finishing/postprocessing helps to improve mechanical, chemical and aesthetic characteristics, thus, making the final product competitive with injection-molded parts and achieve added durability [12]. It improves mechanical, geometric, and other high-value features while producing aesthetic effects, enabling the delivery of a product with the greatest value and impact feasible. The chapter is divided into two main sections. The first section (Section 7.2) discusses the different defects that occur in 3D-printed polymeric and metallic parts during production, what is the cause of these defects, and what are the appropriate measures that must be taken to avoid these defects. In the second section (Section 7.3) different basic and literature reviews on different postprocessing techniques are clearly discussed.

7.2 Defects Associated with Additively Manufactured Parts

7.2.1 Defects in Additively Manufactured Polymeric Parts

There have been numerous studies done that clearly point the significant 3D printing defects like unwanted surface roughness, porosity, and poor adhesion. These 3D printing defects affect the physical and mechanical characteristics of the final product. There have been several techniques developed to identify these defects and SEM image analysis was found to be a suitable technique to easily assist the researcher to identify the defects and its causes [23]. Table 7.2 summarizes the different defects associated with the 3D-printed polymeric parts. It also describes the troubleshooting steps to resolve the defects, but there are some defects or irregularities that can be solved using postprocessing techniques that are mentioned in the upcoming section.

Table 7.2 Brief summary of different 3D printing polymeric defects.

SLS process [24]				
Sr. no.	**3D printing defects**	**Defective printed parts**	**Cause**	**Measures to be taken**
1	Delamination		Insufficient laser power; contaminant present over optical surface	Cleaning of optical cassette; thermal treatment
2	Warping		Uneven cooling; part is removed from the build chamber prior to cooling the part completely	Provide sufficient time for printed parts to remain within the build camber so that it can cool to below 45 °C.
3	Poor surface finish on vertical region of 3D-printed part	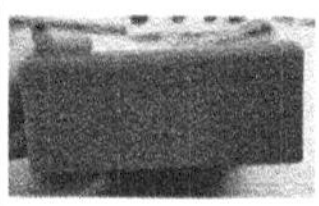	Uneven recycled to fresh powder ratio which leads to improper sintering	Maintain proper refresh rate of polymeric powder; dispose degraded powder
4	Pitting	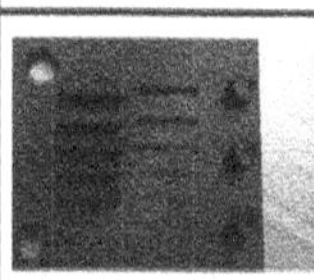	Too high print bed temperature; insufficient supply of laser onto smaller cross sections resulting in unsintered powder	Proper focusing of laser onto the smaller region; proper inspection of IR sensor to avoid laurolactam.
5	Melting		Heating of powder beyond its melting point due to malfunction of heaters or IR sensors; problem with the wall heaters mounted in the build chamber	Proper inspection of IR sensor

(Continued)

Table 7.2 Brief summary of different 3D printing polymeric defects. (*Continued*)

Sr. no.	3D printing defects	Defective printed parts	Cause	Measures to be taken
FDM process [23, 25, 26]				
1	Warping		Lack of material adhesion over the print bed; thermal contraction of the upper layer of final part	Use print bed adhesives; maintain hot environment by providing hot air to reduce thermal contraction
2	Delamination/ cracking		Poor adhesion between two adjacent layers; use of poor quality filament	Increase printing temperature; maintain proper fan speed
3	Nozzle clogging		Build-up of excessive carbon inside the print nozzle; deposition of dirt and dust in nozzle tip that are present in filament	Use high quality filaments; use recommended printing temperature
4	Layer shift		Malfunction of electronics like motor drives; mechanical failure; lack of power supply to driving equipment	Install heat sink and fans to cool drivers; maintain proper voltage supply; regular check of mechanical devices
5	Stringing		Inappropriate selection of printing processing parameters like nozzle temperature, retraction distance and speed	Proper selection of 3D printing process parameters

(*Continued*)

Table 7.2 Brief summary of different 3D printing polymeric defects. (*Continued*)

Sr. no.	3D printing defects	Defective printed parts	Cause	Measures to be taken
6	Curling/over heating		Insufficient time given for part to cool	Select recommended printing process parameters; provide sufficient time for part to cool
SLA process [27]				
1	Raft silhouetting	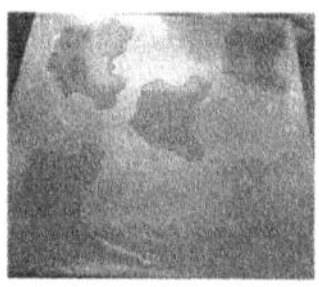	Impairment in resin tank or presence of debris; contamination of optical surfaces	Proper inspection of resin tank and removal of failed print/cured resin from tank bottom; keep optical glasses, films, etc., tidy
2	Delamination		Support problem; use of older resin tank; loose fitting of build platform; contamination of optical surfaces; frequent pauses during printing	Proper check for frequent pauses in printing; proper inspection of resin tank
3	Cupping blowout		Problem with model orientation during printing; absence of drainage holes in hollow parts	Add drain holes to the cup in CAD model before 3D printing
4	Undeveloped features	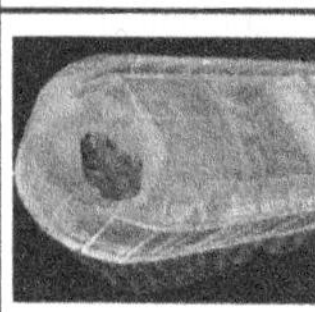	Impairment in resin tank; presence of debris in resin tank bottom; problem with model orientation; dirty optical surfaces	Proper inspection of resin tank and removal of failed print/cured resin from tank bottom; keep optical glasses, films, etc., tidy

(*Continued*)

Table 7.2 Brief summary of different 3D printing polymeric defects. (*Continued*)

Sr. no.	3D printing defects	Defective printed parts	Cause	Measures to be taken
5	Pinholes/cuts	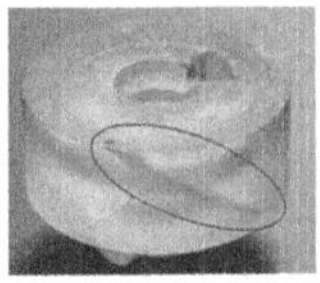	Presence of scratches on the resin tank bottom surface; presence of scratches on the optical window/ surfaces	frequent checking of resin tank; keep optical glasses, films, etc., cleaned
6	Poor surface quality i.e., ragging		Use of expired resin material; damage in resin tank; dirty optical surfaces; Non-uniform resin flow due to overly dense support or issue with model orientation	Inspection of manufacturing time, purity content, etc., of the resin material check resin tank

Table 7.3 Brief summary of different defects associated with 3D-printed metallic parts.

Sr. no.	Defects	Defective 3D printed parts	Cause	Measures to be taken	Ref.
1	Porosity	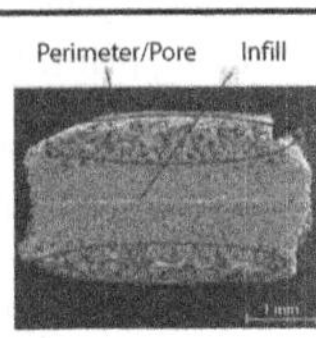	Factors like processing parameters, impurities that are present in base material, temperature gradient of combustion chamber, low laser energy absorption capability of material and wettability problem in solid materials.	Use hot isostatic pressing (HIP) as post processing technique; Use thermo-capillary force during part fabrication	[30–32]

(*Continued*)

Table 7.3 Brief summary of different defects associated with 3D-printed metallic parts. (*Continued*)

Sr. no.	Defects	Defective 3D printed parts	Cause	Measures to be taken	Ref.
2	Crack		Extensive material solidification. When there is lack of liquid supply in the inter dendritic spaces during solidification.	Preheat the print bed; thermal post processing; Select proper input process parameters to provide sufficient melting temperature; Give sufficient time for build part to cool	[33]
3	Poor Surface roughness	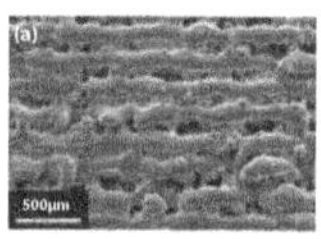	Improper selection of process parameters, balling phenomena resulting in formation of coarse melt pool, etc.	Use post processing techniques like laser polishing, heat treatment techniques, and conventional machining like micro milling and grinding.	[34-36]
4	Layer delamination	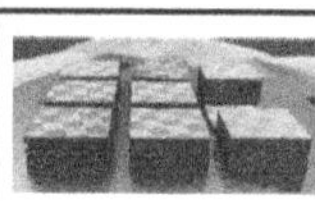	Lack of fusion between the sedimentary layers; Incomplete melting of metal powders.	Clean the hot end; Lower print speed. Raise hot end temperature	[37]
5	Spatter	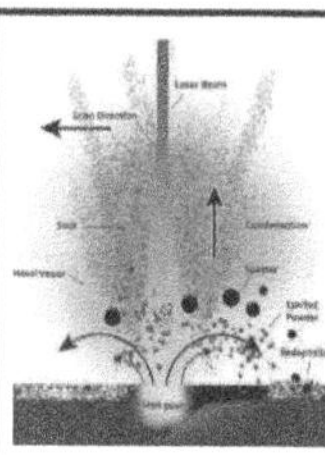	Lateral flow of inert gases; Fluctuation of molten pool; Backflushing pressure.	Use high speed inert gas flow	[38]

(*Continued*)

Table 7.3 Brief summary of different defects associated with 3D-printed metallic parts. (*Continued*)

Sr. no.	Defects	Defective 3D printed parts	Cause	Measures to be taken	Ref.
6	Anisotropy of mechanical charac-teristics	-	Improper selection of part build orientation; Complicated thermal history that the part has underwent during its fabrication.	-	-
7	Geometric deformation	-	Heat accumulation; Development of residual stress; Improper selection of input process parameters.	Using proper processing parameters	[38]

7.2.2 Defects in Additively Manufactured Metallic Parts

The most used alloys as a feedstock for AM are aluminium, titanium, stainless steel, and nickel. The primary factor that causes the defects in Al alloy are high reflectivity and thermal conductivity [28]. Some of the popular AM defects that are associated with metallic 3D-printed parts are porosity, cracks, anisotropy, residual stresses, thermal stresses, laser splattering, poor surface finish, and distortion in shape [29]. Table 7.3 summarizes the different types of defects associated with metallic 3D-printed parts.

7.3 Postprocessing Techniques for AM Processes

7.3.1 Postprocessing Techniques for 3D-Printed Polymers

7.3.1.1 Heat Treatment

It is a common postprocessing technique used to improve the surface as well as mechanical characteristics of 3D-printed parts. The schematic of heat treatment as postprocessing process is shown in Figure 7.1 [39]. There are two most common types of heat treatment process for plastic namely annealing and normalizing. Annealing helps the polymers in improvement of interlaminar toughness, ductility and wear resistance [40], whereas, normalizing provides polymers a higher resistance to internal stresses and structural

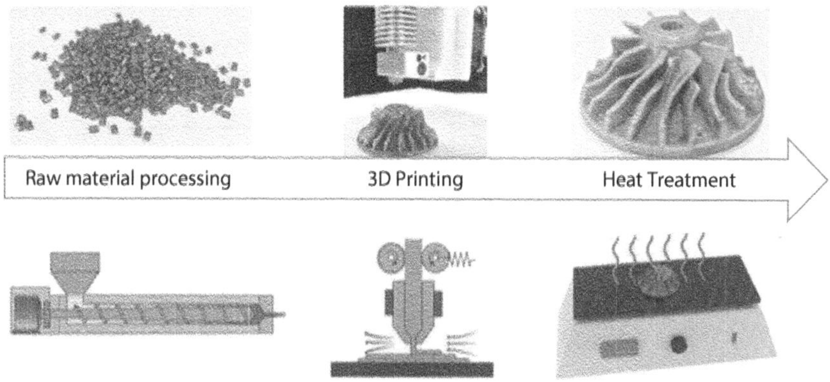

Figure 7.1 Heat treatment as postprocessing technique for 3D-printed polymeric parts [39].

uniformity. In one of the studies, FDM printed ABS part was annealed and it was observed that there was significant reduction in staircase effect and surface roughness. It was also noticed that heating enhanced the part density and filled up the unwanted voids and provided a smooth surface finish. Moreover, there was reduction in viscosity of polymer as the part was heated at glass transition temperature, which has minimized the molecular surface tension and made the material reflow over the part surface. This reflow of materials results in filling up of gaps present over the 3D-printed part surface [41]. Hong *et al.* [42] performed annealing as a post heat treatment operation to improve the characteristics of 3D-printed PLA parts. The part was heated at 140°C for 600 sec and it was observed that heating improved the bonding between the intermediate layers and made the raster stronger as well. However, this processing improved the mechanical strength of material but sacrificed the ductility. Similarly, in another study, 3D-printed short CF-reinforced PLA and PETG composite were annealed at 65°C and 85°C, respectively, for an hour. It was found that the tensile strength of CF/PLA and CF/PETG parts were enhanced by 16.8% and 12.4%, respectively.

7.3.1.2 Ultrasound Treatment

Ultrasound treatment is the fastest growing method to enhance the surface quality of 3D-printed polymeric components. It is a nonchemical and nonthermal post-treatment method which does not affect the performance of the end product. This method uses ultrasonic vibration which is supplied during part fabrication which reduces the staircase effect and layer thickness, resulting in the improvement of the final product surface quality [43].

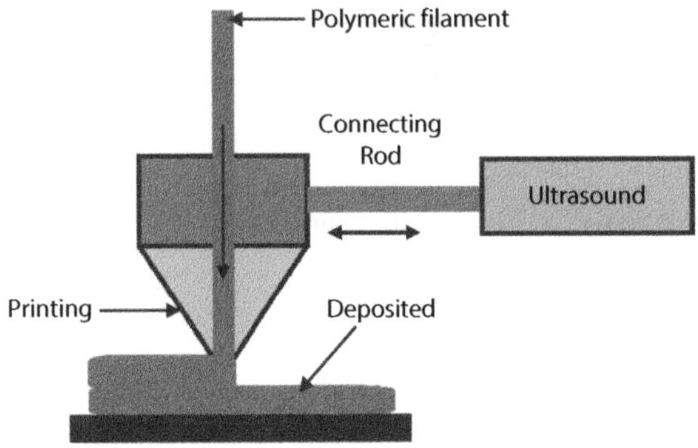

Figure 7.2 Schematic of ultrasound treatment process [47].

The schematic of this process is shown in Figure 7.2. Maidin *et al.* [44] explored the effect of using multiple piezoelectric transducer to improve the surface quality of the printed parts. The test was conducted at different ultrasonic frequencies of 27, 40, and 50 kHz, and the surface quality of the 3D-printed part was measured using optical microscope aided with pro VIS V2.90 software. It was found that at 1 piezo with frequency of 50 kHz produced maximum surface finish with a generated surface roughness reading of 13.66 μ. The improvement in surface quality was due to the occurrence of lesser layer thickness defect and the generation of finer layer thickness. Similarly, in another study, the effect of 34.4 kHz-ultrasonic frequency during 3D printing of ABS parts was investigated. It was observed that this treatment enhanced the interlayer adhesion by 10% when compared to untreated 3D-printed ABS [45]. Qiao *et al.* [46] conducted a study to investigate the effect of supplying ultrasonic vibration during impregnation of continuous carbon fiber and PLA for 3D printing. The results concluded that ultrasonic vibration reduced the viscosity and surface tension of PLA which in turn resulted in better coating of PLA around the carbon fiber.

7.3.1.3 Chemical Treatment

Chemical treatment is a post-treatment method that is commonly used to enhance the surface property of 3D-printed polymeric parts. Moreover, this technique is also used for customizing the fiber surface to improve the fiber and matrix bonding for 3D-printed composites. The common chemical that is used for this technique is acetone [48]. In this technique, 3D-printed parts were submerged in a bathtub which contained acetone to reduce the parts' surface roughness. It is an economical and safe process as compared to other post treatment techniques. Hambali *et al.* [49] investigated the effect of acetone treatment on the surface quality of a 3D-printed ABS part. The ABS part was submerged in bathtub containing acetone and after some time, it was observed that the surface roughness of parts was reduced by 97.2% and surface became shinier. However, it was also noted that this treatment sacrificed the part tensile strength and its strength was reduced by 42.58%. In another study, comparative investigation on 3D-printed ABS parts' surface quality was conducted by treating the part surface with acetone and dichloroethane, respectively. It was found that the parts treated with dichloroethane possessed a smoother surface with a recorded reduction in smoothness by 60 to 90%. But the problem of reduction in tensile strength continues in this case as well. The tensile strength of the chemically treated part was reduced by 48% as compared to untreated ABS part. The chemical dissolves the voids formed over the part surface and filled up the layer gaps by dissolving the polymers, thus resulted in a smooth surface finish [50]. Similarly, in one of the study, methylene chloride was used to treat carbon fiber (CF)-reinforced PLA filament. Due to this chemical treatment, PLA was dissolved and the CF obtained was modified by using emulsified surface-active antifoaming agent. Afterwards, it was found that chemically treated CF formed good bonding with PLA matrix during part fabrication using FDM printing [51]. Galantucci and Percoco [52], in a study, treated FDM part with a mixture of 90% dimethyl ketone and 10% water solution to enhance the part surface characteristics. It was noted that surface roughness was dramatically reduced by 91% with the material possessing good ductility.

Another technique which can be categorized in chemical treatment process (CMP) is chemical mechanical polishing, which is commonly used for SLA build parts. It is a hybrid process that uses a combination of mechanical material removal and chemical surface reactions process to enhance the surface property of the 3D-printed samples. An abrasive particle located on the real contact area is used between the polishing pad and the surface to be polished. Son *et al.* [53] compared two polishing steps which are used to improve the glossiness of SLA-printed ABS like resin. In one-step polishing, direct CMP was applied onto the part surface, whereas in two-step polishing, the part surface was first polished with sandpaper (#2000) and then CMP was performed. The surface assessment of the samples is shown in Figure 7.3. It was observed that two-step polishing enhanced the surface finish more and with a faster increase in part glossiness.

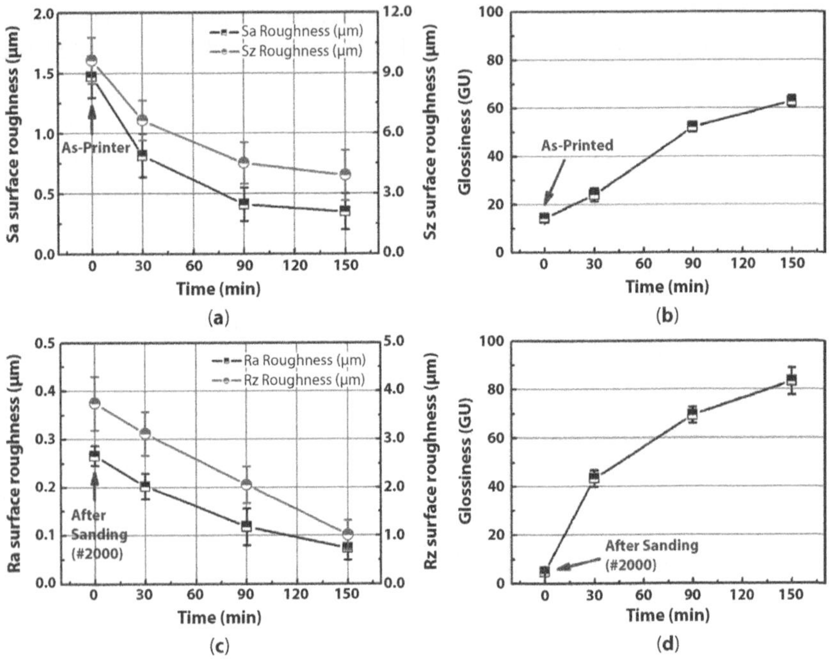

Figure 7.3 (a) S_a and S_b; (b) Glossiness as a function of CMP one step polishing; (c) R_a and R_z; (d) Glossiness as a function of CMP two step polishing [53].

7.3.1.4 Laser Treatment

Laser treatment is another more commonly used postprocessing method to improve the surface quality of 3D-printed polymeric parts. In this method, the 3D-printed part surface is exposed to a laser beam to remove undesirable irregularities. The focused laser beam breaks down the polymeric chain and melts the surface to fill the voids [54]. As per recent literature, it was found that 3D-printed PLA parts can easily be post treated by this method as compared to 3D-printed ABS parts [55–57]. Chen *et al.* [58] used 5W-power laser beam of 200-μ diameter to polish the surface of 3D-printed ABS/Cu composite part. It was noticed that laser polishing improved the surface finish and mechanical property of the end product. The surface finish and the tensile strength of the part were enhanced by 91% and 34.1%, respectively. Laser polishing improved the bond strength between the filler and the matrix and eliminated the voids that were present over the part surface.

7.3.2 Postprocessing Techniques for 3D-Printed Metallic Parts

This section illustrates the different postprocessing techniques used for 3D-printed metallic parts.

7.3.2.1 Laser Shock Peening (LSP)

Laser shock peening is a lateral expansion process which causes plastic compression of materials, perpendicular to the target surface. It creates a compressive residual stress of required magnitude, underneath the preserved surface of metallic components. With the introduction of compressive residual stress by LSP, there can be a significant improvement in the mechanical characteristics of the 3D-printed parts, like resistance to crack instigation and extended fatigue life, and fatigue strength [59]. This emerging technique acts as an alternative for surface treatment and has overcome the limitations of conventional shot peening technique [60]. Figure 7.4 shows the schematic of LSP technique. In this technique, a laser beam is focused onto the metallic surface for 30 ns and at the heated zone, the temperature reaches 10000°C, resulting in the formation of plasma. Until the laser-material interaction period is reached, the plasma created absorbs the laser energy. The pressure produced by the plasma is then transferred to the material through shock waves. This technique commonly uses a "confined mode" to produce a high amplitude of shock pressure. The metallic surface is covered with an opaque substance, such as aluminium foil or black paint, and protected from direct laser radiation by a

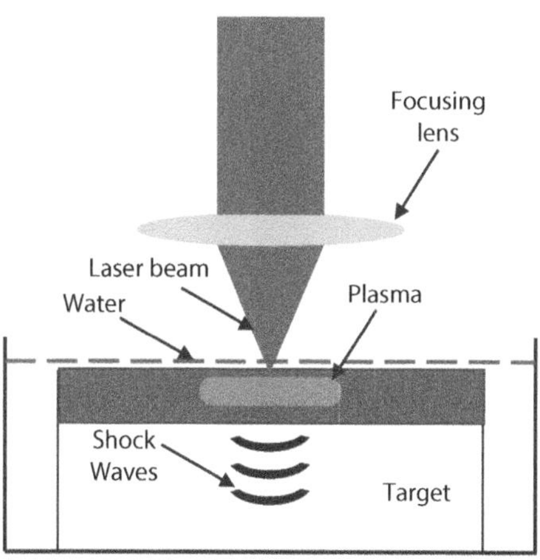

Figure 7.4 Schematic of laser shock peening technique.

transparent substance. Recent studies have shown that using the restricted mode can produce metallic surface plasma pressures of up to 10 GPa. A stronger pressure pulse may enhance the results of LSP to a greater depth when the amount of the compressive residual stress is high [61].

Laser shock peening is typically used to increase any component's fatigue life, and it has recently been employed to increase the fatigue life of aviation components [62]. Additionally, LSP has been utilized to improve the characteristics of maraging steels [61, 63, 64]. In order to create more realistic aerodynamic models, it has also been utilized to bend and stretch aircraft fenders [62]. Local plastic deformations are brought about by short but powerful laser pulses that create a plasma within the restricted geometry and lead to pressure pulses. The pressure that is created can be increased to help with effective operation [62]. Liang Lan [61] studied the influence of LSP on the microstructural and mechanical performance of Ti-6Al-4V (Ti64) titanium alloy parts fabricated by SLM process. It was noticed that there was significant refinement in grains of α phase [a type of hexagonal closed packed (hcp) phase in Ti alloy] along with the generation of maximum compressive residual stress of 180 MPa and 250-μm depth. The electron backscattering diffraction (EBSD) maps in Figure 7.5 shows the cross section of SLM-manufactured Ti64 samples with and without LSP postprocessing. From Figure 7.5, it was observed that prior to LSP processing, the SLM-built part exhibits hcp phase with no body centred

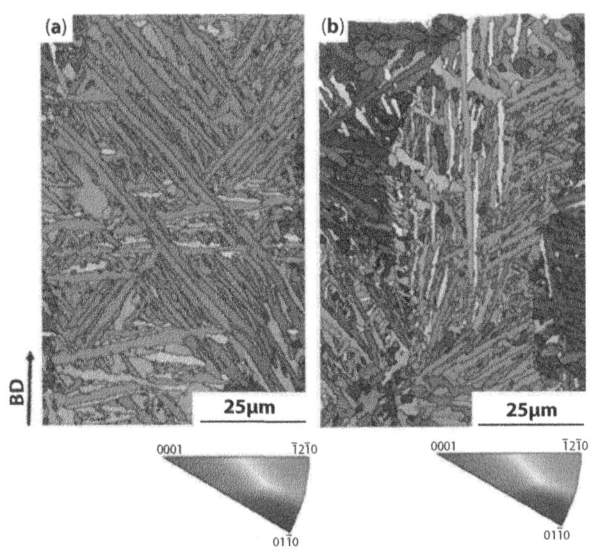

Figure 7.5 Electron backscattering diffraction (EBSD) maps of SLM-fabricated Ti64 specimens. (a) Before LSP; (b) After LSP [61].

cubic (bcc) β phase which was due to rapid cooling of each layer, resulting in the transformation of diffusion less martensitic bcc β phase to hcp α phase. In contrast, after LSP treatment, refinement in hcp phase can be seen with a decrease in long and straight laths marks. Similarly, in another study, 3D laser metal deposited (LMD) TC17 (Ti-5Al-2Sn-2Zr-4Mo-4Cr) alloy was subjected to LSP as a postprocessing approach to investigate the microstructural, residual stress, and fatigue performance. It was found that after LSP processing, material can endure maximum compressive strength of 240 MPa to the depth of 200 μm and fatigue cycle of LMD + LSP samples were enhanced by 212% [65]. Navid *et al.* [66] studied fatigue performance on additively manufactured Zr-based bulk metallic glass (BMG) composite after subjected to LSP postprocessing. Similarly, Lu *et al.* [64] studied the effect of LSP on mechanical characteristics of laser additively manufactured 316L steel. It was observed that after LSP treatment, surface hardness and yield strength of specimens increased by 7% and 16%, respectively.

7.3.2.2 Laser Polishing (LP)

During the manufacturing of mechanical components, it is crucial to adhere to specifications like the surface morphology and the dimensional or geometric tolerances. Specifically, from a scientific point of view, it is obligatory to understand how to create an appropriate surface according to their specific application. Therefore, for the metallic additively manufactured parts, laser polishing technique is observed as one of the postprocessing techniques that can improve surface roughness. In this technique, energy from a laser is irradiated onto the material surface which then attains a quick melting temperature and forms a melt pool. After the formation of the melt pool, due to gravitation action and surface tension, the molten or liquified material gets reorganized to the same level. Once the task is completed, the laser beam is stopped, resulting in a rapid reduction of the heat affected zone (HAZ), solidification of the melt pool, and enhancement of the part surface quality [34–36]. The schematic of the LP process is shown in Figure 7.6.

Cwikla *et al.* [67] studied the influence of LP on the surface quality of 3D-printed Inconel 718 parts. The studied was conducted under the effect of air as well as argon gas. Figure 7.7 illustrates the physical appearance of LP-processed samples.

It was noticed that samples prepared under argon atmosphere show a glossy effect whereas samples prepared under air atmosphere appear to be dark which was due to the formation of an oxide layer over the surface. Moreover, it was also found that LP processing reduced the surface

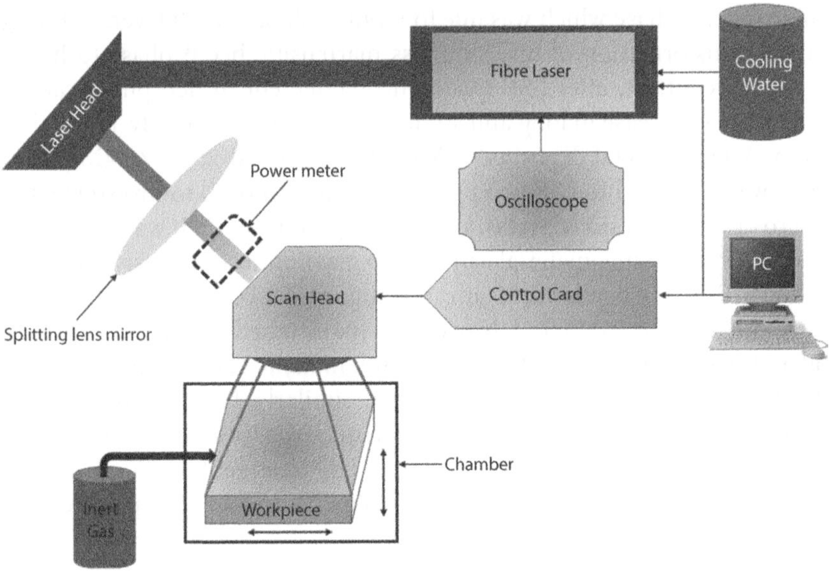

Figure 7.6 Schematic of laser polishing technique.

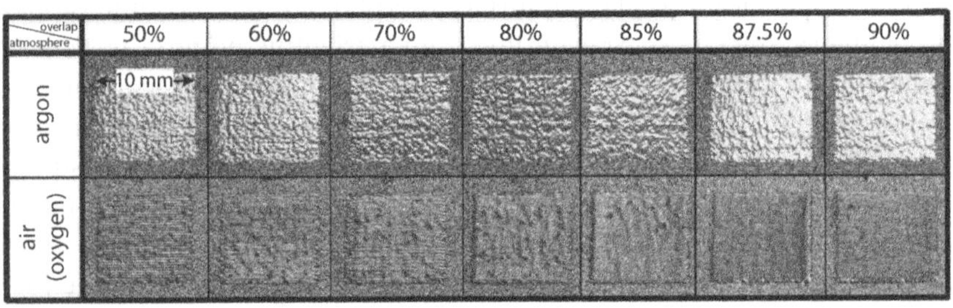

Figure 7.7 Physical illustration of LP samples [67].

roughness by 82.8% and 87.9% under air and argon atmosphere, respectively. In another study, Ti6Al4V parts manufactured by SLM process was laser polished to improve the surface quality. It was found that the surface of the SLM samples after LP processing was enhanced due to the laser beam neatly eliminating the metallic globules and repairing the cracks and pores from the part surface. From the study, it was also observed that during LP processing, there are two surface process mechanisms, namely surface shallow melting (SSM) and surface over melting (SOM). In SSM, melting of the top surface (crests) is allowed to reach lower roughness value by pouring molten metal into the valley. Whereas in SOM, excessive surface is melted, resulting in the removal of the original surface, leading to the formation of capillary waves developed in the direction of laser displacement due to gradient in surface tension behind the progressing laser beam [68, 69]. Obeidi *et al.* [70] studied the LP of 316L stainless steel specimens fabricated by SLM process and found that there was maximum reduction in surface roughness by 74.038%. The reduction in surface roughness can be easily seen from Figure 7.8.

7.3.2.3 *Conventional Machining Processes (CMP) Like Milling, Rolling, Chemical, and Abrasive Machining*

Conventional machining processes are the traditional manufacturing processes which are employed to enhance the dimensional accuracy and surface quality of the manufactured parts. Due to its popularity and global acceptability, these processes were employed as postprocessing techniques for AM parts. In a recent study, additively manufactured ASTM A131steel was postprocessed using computer numerical control (CNC) milling machine to improve part surface characteristics. It was noticed that CNC milling smoothened the AM part surface from 22.7 to 0.6 μm without affecting the hardness of the final part [71]. Lopes *et al.* [72] studied the effect of milling parameters on microstructural and mechanical performance of HSLA (thin-walled high strength alloy) steel parts fabricated by wire and arc additive manufacturing (WAAM) technique. It was observed that milling does not have a significant effect on the mechanical characteristics of built parts. However, it improved the overall surface quality of the part by increasing the cutting speed and decreasing feed per tooth. During metal AM, the major challenge is residual stress and material distortion which weakens the end part performance. Therefore, to eliminate these

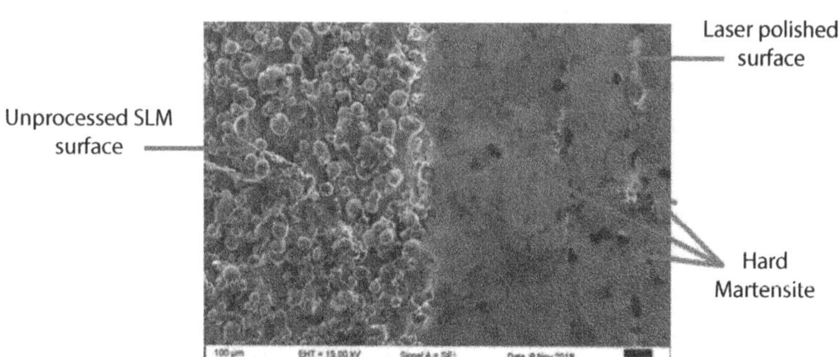

Figure 7.8 SEM images of 316L SLM specimen surface showing unprocessed surface on the left and LP surface on the right [70].

effects, in one of the studies, the impact of vertical interpass and post deposition side rolling on aluminium walls, which was fabricated via WAAM technique, was investigated. It was noted that using the rolling technique, there was considerable reduction in the residual stress as well as in part defects. Moreover, due to work hardening effect, the hardness of the built part was enhanced with increment in applied load [73].

Scherillo [74] applied chemical surface finish as a postprocessing technique for 3D-printed AlSi10Mg component. The aim of chemical machining was to remove the gross irregularities and increase the chemical brightening to further smoothen the built part surface. The 3D-printed specimens were treated with a mixture of HNO_3 and HF at 85°C for 75 min. The specimens were further rinsed in a bath containing a mixture of ultrasonic water and acetic acid to remove the unwanted insoluble products. With chemical machining, there was considerable decrease in surface roughness of the end product [75].

Like the above-mentioned processes, there are various abrasive machining processes that have been used for the postprocessing of AM parts. Zhang *et al.* [76] polished additively manufactured AISI 316L steel parts using magnetic abrasive finishing setup at multiple angles. It was observed that the surface roughness of the part was reduced by 76% and additional defects were also eliminated. Yamaguchi *et al.* [77], for the same abrasively polished specimen, observed that the surface roughness reduced from 100 to 0.1 μm and that compressive residual stress was formed after the specimen was subjected sanding, polishing, and ball burnishing. Similarly, in another study, ultrasonic abrasive polishing (UAP) was used to enhance the surface quality of Inconel 625 specimen fabricated by SLM technique. In order to simulate the impact action of abrasive particles on the AM parts, smoothed particle hydrodynamics methodology was adopted. Different abrasive concentrations were used to polish the SLM part for 20 min under 500W power, which can be seen in Figures 7.9 and 7.10. Figure 7.9 shows that increasing abrasive concentrations beyond 20% worsened the finish effect. It was found that UAP removed the partially melted particles and the surface roughness was reduced from 5.02 to 2.93 μm [78].

7.3.2.4 *Heat Treatment Techniques*

Heat treatment is a vital technique used to alleviate the residual stresses, reduce the cracks, and create a homogeneous microstructure of the built part. There are several heat treatment techniques like annealing, laser melting, solution heat treatment (SHT), and hot isotactic pressing (HIP), which are commonly used as postprocess for additively manufactured parts. Bai

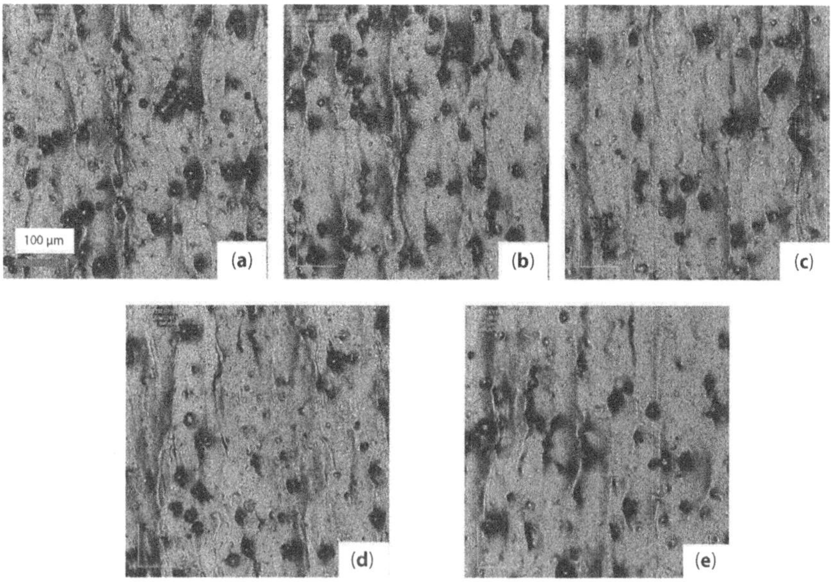

Figure 7.9 Laser microscopic image of SLM surfaces polished under 500 W for 20 min with abrasive concentrations of (a) 5%; (b) 10%; (c) 20%; (d) 30%; (e) 40% [78].

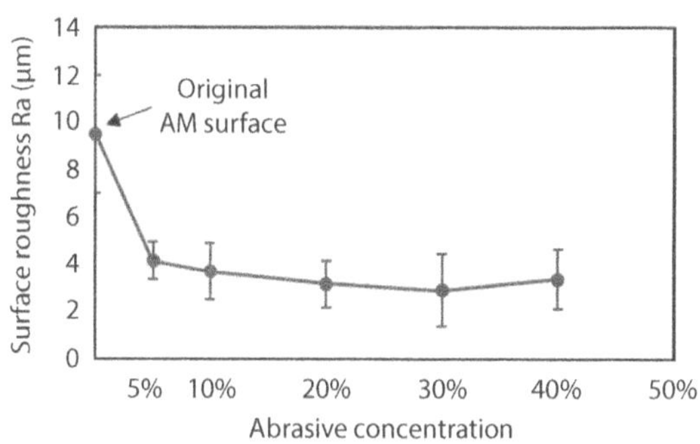

Figure 7.10 Plot between average surface roughness and % of abrasive concentration for UAP [78].

et al. [79] studied the effect different heat treatment experiments like solution treatment (ST), direct aging treatment (DAT) and solution + aging treatment (SAT) on the microstructure and mechanical properties of SLM-fabricated maraging steel parts. It was found that ST and SAT eliminated the cellular and lathe structures from the material. Moreover, by increasing DAT temperature and holding time, the austenitic content in the material also increased. It was also noticed that at DAT of 520°C for 6 h, SLM samples

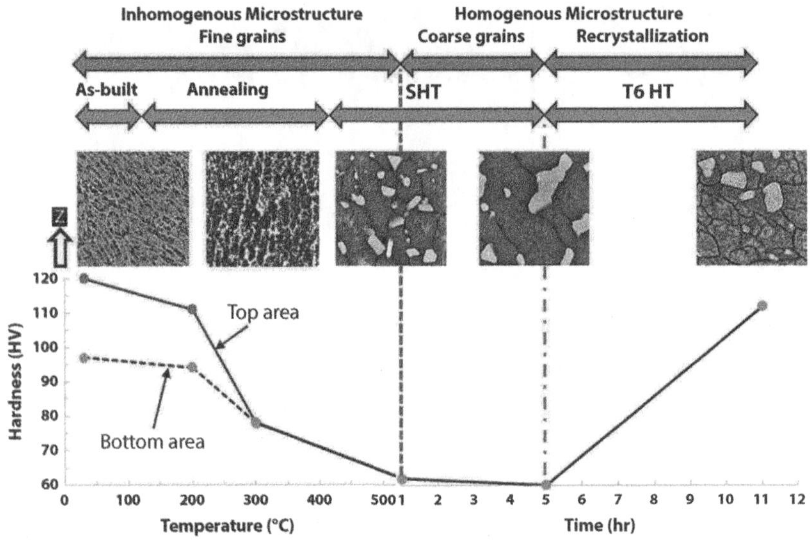

Figure 7.11 Microhardness map for thermal postprocessing of AlSi10Mg samples cross-section fabricated by SLM [81].

exhibit maximum mechanical characteristics (microhardness of 653.93 HV and ultimate tensile strength of 2126.30 MPa). Similarly, the effect of ST and aging duration (steps of precipitation hardening) on microstructure and hardness of SLM-based AlSi10Mg parts was investigated. It was found that after precipitation hardening, a coarse microstructure formed which becomes a solid solution of Si particles in α-Al (aluminium in its highest oxidation state) matrix, and the hardness of materials decreased [80]. In another study, SLM parts fabricated from AlSi10Mg recycled powder was subjected to three different thermal postprocessing techniques, namely, annealing, SHT, and T6 heat treatment (T6 HT). Figure 7.11 summarizes the effect of heat treatment on developed microstructure and microhardness of the built SLM samples. It noted that microhardness is reduced after annealing and SHT while homogeneity in microstructure was achieved along the built direction. However, after T6 HT, microhardness obtained was same as that of the built part, with retained homogeneity in microstructure for the XY plane and Z direction [81].

7.4 Conclusion

Additive manufacturing offers a ton of huge benefits when contrasted with conventional developmental assembling techniques which includes the rapid fabrication of prototypes, reduction in industrial expenses, minimal waste production, and flexibility in production flow. However, this technique faced several criticisms like poor surface finish, formation of unwanted microstructural phases, delamination, warping, and wear tracks, which then affected the part's performance characteristics and its aesthetic beauty. Therefore, researchers came up with the idea of postprocessing, where the 3D-printed part was postprocessed by using numerous post-treatment techniques like heat treatment, laser treatment, chemical treatment, and ultrasound treatment. Due to this postprocessing, there was considerable enhancement in the physical and mechanical performances of 3D-printed products. Moreover, it was observed that most of the postprocessing techniques for polymeric materials targeted the aesthetic beauty of the materials which resulted in enhancing the beauty of the parts but sacrificed mechanical strength. Similarly, for 3D-printed metallic parts, surface roughness and residual stress were major concerns. It was concluded that postprocessing techniques have very effectively solved the problems and enhanced the performance of the parts. This chapter demonstrated that postprocessing holds promise for improving the impact of AM by enhancing unique properties and adding high value to practical applications.

References

1. Mishra, V., Negi, S., Kar, S., Additive manufacturing: An emerging tool to fabricate bioinspired structures, in: *Recent Advancements in Mechanical Engineering. Lecture Notes in Mechanical Engineering*, Springer, Singapore, pp. 297–312, 2022.
2. Mishra, V., Negi, S., Kar, S., Sharma, A.K., Rajbahadur, Y.N.K., Kumar, A., Recent advances in fused deposition modeling fused deposition modeling based additive manufacturing of thermoplastic composite structures: A review. *J. Thermoplast. Compos. Mater.*, 36, 3094–3132, 2022, https://doi.org/10.1177/08927057221102857.
3. Mishra, V., Negi, S., Kar, S., FDM based additive manufacturing of recycled thermoplastics and associated composites. *J. Mater Cycles Waste Manage.*, 25, 758–784, 2023, https://doi.org/10.1007/s10163-022-01588-2.
4. Charalampous, P., Kostavelis, I., Tzovaras, D., Non-destructive quality control methods in additive manufacturing: A survey. *Rapid Prototyp J.*, 26, 777–790, 2020, https://doi.org/10.1108/RPJ-08-2019-0224.
5. Ko, H., Moon, S.K., Hwang, J., Design for additive manufacturing in customized products. *Int. J. Precis. Eng. Manuf.*, 16, 2369–2375, 2015, https://doi.org/10.1007/s12541-015-0305-9.

6. Negi, S., Dhiman, S., R.K.S., Basics, applications and future of additive manufacturing technologies: A review. *J. Manuf. Technol. Res.*, 5, 75–96, 2016.
7. Dizon, J.R.C., Espera, A.H., Chen, Q., Advincula, R.C., Mechanical characterization of 3D-printed polymers. *Addit. Manuf.*, 20, 44–67, 2018, https://doi.org/10.1016/j.addma.2017.12.002.
8. De Leon, A.C., Chen, Q., Palaganas, N.B., Palaganas, J.O., Manapat, J., Advincula, R.C., High performance polymer nanocomposites for additive manufacturing applications. *React Funct. Polym.*, 103, 141–155, 2016, https://doi.org/10.1016/j.reactfunctpolym.2016.04.010.
9. Manapat, J.Z., Mangadlao, J.D., Tiu, B.D.B., Tritchler, G.C., Advincula, R.C., High-strength stereolithographic 3D printed nanocomposites: Graphene oxide metastability. *ACS Appl. Mater. Interfaces*, 9, 10085–10093, 2017, https://doi.org/10.1021/acsami.6b16174.
10. Dizon, J.R.C., Valino, A.D., Souza, L.R., Espera, A.H., Chen, Q., Advincula, R.C., Three-dimensional-printed molds and materials for injection molding and rapid tooling applications. *MRS Commun.*, 9, 1267–1283, 2019, https://doi.org/10.1557/mrc.2019.147.
11. Dizon, J.R.C., Gache, C.C.L., Cascolan, H.M.S., Cancino, L.T., Advincula, R.C., Post-processing of 3D-printed polymers. *Technologies*, 9, 61, 2021, https://doi.org/10.3390/technologies9030061.
12. Espera, A.H., Dizon, J.R.C., Chen, Q., Advincula, R.C., 3D-printing and advanced manufacturing for electronics. *Prog. Addit. Manuf.*, 4, 245–267, 2019, https://doi.org/10.1007/s40964-019-00077-7.
13. Advincula, R.C., Dizon, J.R.C., Chen, Q., Niu, I., Chung, J., Kilpatrick, L., Newman, R., Additive manufacturing for COVID-19: Devices, materials, prospects, and challenges. *MRS Commun.*, 10, 413–427, 2020, https://doi.org/10.1557/mrc.2020.57.
14. Tijing, L.D., Dizon, J.R.C., Ibrahim, I., Nisay, A.R.N., Shon, H.K., Advincula, R.C., 3D printing for membrane separation, desalination and water treatment. *Appl. Mater. Today*, 18, 100486, 2020, https://doi.org/10.1016/j.apmt.2019.100486.
15. Tijing, L., Dizon, J.R., Cruz Jr., G., 3D-printed absorbers for solar-driven interfacial water evaporation: A mini-review. *Adv. Sustain Sci. Eng. Technol.*, 3, 0210103, 2021, https://doi.org/10.26877/asset.v3i1.8367.
16. Wei, P., Leng, H., Chen, Q., Advincula, R.C., Pentzer, E.B., Reprocessable 3D-printed conductive elastomeric composite foams for strain and gas sensing. *ACS Appl. Polym. Mater.*, 1, 885–892, 2019, https://doi.org/10.1021/acsapm.9b00118.
17. Chen, Q., Mangadlao, J.D., Wallat, J., De Leon, A., Pokorski, J.K., Advincula, R.C., 3D printing biocompatible polyurethane/poly(lactic acid)/graphene oxide nanocomposites: Anisotropic properties. *ACS Appl. Mater. Interfaces*, 9, 4015–4023, 2017, https://doi.org/10.1021/acsami.6b11793.
18. Palaganas, N.B., Mangadlao, J.D., De Leon, A.C.C., Palaganas, J.O., Pangilinan, K.D., Lee, Y.J., Advincula, R.C., 3D printing of photocurable

cellulose nanocrystal composite for fabrication of complex architectures via stereolithography. *ACS Appl. Mater. Interfaces*, 9, 34314–34324, 2017, https://doi.org/10.1021/acsami.7b09223.
19. Espera, A.H., Valino, A.D., Palaganas, J.O., Souza, L., Chen, Q., Advincula, R.C., 3D Printing of a robust polyamide-12-carbon black composite via selective laser sintering: Thermal and electrical conductivity. *Macromol. Mater. Eng.*, 304, 1–8, 2019, https://doi.org/10.1002/mame.201800718.
20. Zhang, J., Xiang Toh, A.Y., Wang, H., Lu, W.F., Fuh, J.Y.H., Vibration-assisted conformal polishing of additively manufactured structured surface. *Proc. Inst. Mech. Eng. Part C J. Mech. Eng. Sci.*, 233, 4154–4164, 2019, https://doi.org/10.1177/0954406218811359.
21. Tan, H., Fang, Y., Zhong, C., Yuan, Z., Fan, W., Li, Z., Chen, J., Lin, X., Investigation of heating behavior of laser beam on powder stream in directed energy deposition. *Surf. Coat. Technol.*, 397, 126061, 2020, https://doi.org/10.1016/j.surfcoat.2020.126061.
22. Chavez, L.A., Ibave, P., Wilburn, B., Alexander, D., Stewart, C., Wicker, R., Lin, Y., The influence of printing parameters, post-processing, and testing conditions on the properties of binder jetting additive manufactured functional ceramics. *Ceramics*, 3, 65–77, 2020, https://doi.org/10.3390/ceramics3010008.
23. Günaydın, K. and Türkmen, H.S., Common FDM 3D printing defects, in: *International Congress on 3D Printing (Additive Manufacturing) Technologies and Digital Industry*, 2018.
24. Diagnosing a print failure (SLS), in: *Formlabs Fuse 1 Gener*, 2020, https://support.formlabs.com/s/article/Diagnosing-a-print-failure-SLS?language=en_US. Accessed 23 Jan 2023.
25. Trapero, D., FDM 3D Printing: Common problems and how to solve them, in: *Bitfab via @diegotrap*, 2020, https://wikifactory.com/+bitfab/stories/fdm-3D-printing-common-problems-and-how-to-solve-them. Accessed 18 Nov 2021.
26. Luhin, [Help] Nozzle keeps clogging after some time. What could cause this?, in: *Reddit*, 2016, https://www.reddit.com/r/3Dprinting/comments/41e4ld/help_nozzle_keeps_clogging_after_some_time_what/. Accessed 8 Apr 2022.
27. Diagnosing a print failure (SLA), in: *Formlabs Form 3B*, 2020, https://support.formlabs.com/s/article/Diagnosing-a-print-failure?language=en_US. Accessed 23 Jan 2023.
28. Zhang, J., Song, B., Wei, Q., Bourell, D., Shi, Y., A review of selective laser melting of aluminum alloys: Processing, microstructure, property and developing trends. *J. Mater. Sci. Technol.*, 35, 270–284, 2019, https://doi.org/10.1016/j.jmst.2018.09.004.
29. Aboulkhair, N.T., Simonelli, M., Parry, L., Ashcroft, I., Tuck, C., Hague, R., 3D printing of aluminium alloys: Additive manufacturing of aluminium alloys using selective laser melting. *Prog. Mater. Sci.*, 106, 100578, 2019, https://doi.org/10.1016/j.pmatsci.2019.100578.

30. Wei, X., Behm, I., Winkler, T., Scharf, S., Li, X., Bähr, R., Experimental study on metal parts under variable 3D printing and sintering orientations using bronze/PLA hybrid filament coupled with fused filament fabrication. *Mater. (Basel)*, 15, 1–13, 2022, https://doi.org/10.3390/ma15155333.
31. Frazier, W.E., Metal additive manufacturing: A review. *J. Mater. Eng. Perform.*, 23, 1917–1928, 2014, https://doi.org/10.1007/s11665-014-0958-z.
32. Neal, B.O., Eliminating porosity in metal 3D printing through thermocapillary force, in: *3Dprint.com*, https://3dprint.com/250453/eliminating-porosity-in-metal-3d-printing-through-thermocapillary-force/. Accessed 24 Jan 2023.
33. Kogo, B., Xu, C., Wang, B., Chizari, M., Kashyzadeh, K.R., Ghorbani, S., An experimental analysis to determine the load-bearing capacity of 3D printed metals. *Mater. (Basel)*, 15, 1–13, 2022, https://doi.org/10.3390/ma15124333.
34. Wang, Q., Morrow, J.D., Ma, C., Duffie, N.A., Pfefferkorn, F.E., Surface prediction model for thermocapillary regime pulsed laser micro polishing of metals. *J. Manuf. Processes*, 20, 340–348, 2015, https://doi.org/10.1016/j.jmapro.2015.05.005.
35. Ukar, E., Lamikiz, A., Liébana, F., Martínez, S., Tabernero, I., An industrial approach of laser polishing with different laser sources: Industrielle methode zum laserpolieren mit verschiedenen laserstrahlquellen. *Materwiss. Werksttech.*, 46, 661–667, 2015, https://doi.org/10.1002/mawe.201500324.
36. Pfefferkorn, F.E., Duffie, N.A., Morrow, J.D., Wang, Q., Effect of beam diameter on pulsed laser polishing of S7 tool steel. *CIRP Ann. - Manuf. Technol.*, 63, 237–240, 2014, https://doi.org/10.1016/j.cirp.2014.03.055.
37. Connell, J.O., 3D print layer separation: 8 tips to avoid delamination, in: *All3DP*, 2022, https://all3dp.com/2/3d-print-delamination-tips-tricks-to-avoid-layer-separation/. Accessed 24 Feb 2023.
38. Typical defects of 3D printed metal parts, https://www.eplus3d.com/typical-defects-of-3d-printed-metal-parts.html. Accessed 24 Jan 2023.
39. Geng, P., Zhao, J., Wu, W., Wang, Y., Wang, B., Wang, S., Li, G., Effect of thermal processing and heat treatment condition on 3D printing PPS properties. *Polym. (Basel)*, 10, 1–12, 2018, https://doi.org/10.3390/polym10080875.
40. Hart, K.R., Dunn, R.M., Sietins, J.M., Hofmeister Mock, C.M., Mackay, M.E., Wetzel, E.D., Increased fracture toughness of additively manufactured amorphous thermoplastics via thermal annealing. *Polym. (Guildf)*, 144, 192–204, 2018, https://doi.org/10.1016/j.polymer.2018.04.024.
41. Singh, S. and Singh, R., Effect of annealing on surface roughness of additively manufactured plastic parts: A case study. *Proc Natl Conf Prod Eng (COPE-2016)*, vol. 9, Guru Nanak Dev Eng Coll Ludhiana, India, pp. 7–8, 2016.
42. Hong, J.H., Yu, T., Chen, Z., Park, S.J., Kim, Y.H., Improvement of flexural strength and compressive strength by heat treatment of PLA filament for 3D-printing. *Mod. Phys. Lett. B*, 33, 3–7, 2019, https://doi.org/10.1142/S0217984919400256.

43. Mohamed, A.S., Maidin, S., Mohamed, S.B., Muhamad, M.K., Wong, J.H.U., Romlee, W.F.A., Improvement of surface finish by multiple piezoelectric transducers in fused deposition modeling. *Int. J. Adv. Sci. Eng. Inf. Technol.*, 6, 764–769, 2016, https://doi.org/10.18517/ijaseit.6.5.957.
44. Maidin, S., Mohamed, A.S., Mohamed, S.B., Wong, J.H.U., Sivarao, Effect of multiple piezoelectric transducer on fused deposition modeling to improve parts surface finish. *J. Adv. Manuf. Technol.*, 12, 101–116, 2018.
45. Tofangchi, A., Han, P., Izquierdo, J., Iyengar, A., Hsu, K., Effect of ultrasonic vibration on interlayer adhesion in fused filament fabrication 3D printed ABS. *Polym. (Basel)*, 11, 15–17, 2019, https://doi.org/10.3390/polym11020315.
46. Qiao, J., Li, Y., Li, L., Ultrasound-assisted 3D printing of continuous fiber-reinforced thermoplastic (FRTP) composites. *Addit. Manuf.*, 30, 100926, 2019, https://doi.org/10.1016/j.addma.2019.100926.
47. Wickramasinghe, S., Do, T., Tran, P., FDM-based 3D printing of polymer and associated composite: A review on mechanical properties, defects and treatments. *Polym. (Basel)*, 12, 1–42, 2020, https://doi.org/10.3390/polym12071529.
48. Jayanth, N., Senthil, P., Prakash, C., Effect of chemical treatment on tensile strength and surface roughness of 3D-printed ABS using the FDM process. *Virtual Phys. Prototyp.*, 13, 155–163, 2018, https://doi.org/10.1080/17452759.2018.1449565.
49. Hambali, R.H., Cheong, K.M., Azizan, N., Analysis of the influence of chemical treatment to the strength and surface roughness of FDM. *IOP Conf. Ser. Mater. Sci. Eng.*, 210, 1–9, 2017, https://doi.org/10.1088/1757-899X/210/1/012063.
50. Jayanth, N., Senthil, P., Prakash, C., Effect of chemical treatment on tensile strength and surface roughness of 3D-printed ABS using the FDM process. *Virtual Phys. Prototyp.*, 13, 155–163, 2018, https://doi.org/10.1080/17452759.2018.1449565.
51. Li, N., Li, Y., Liu, S., Rapid prototyping of continuous carbon fiber reinforced polylactic acid composites by 3D printing. *J. Mater. Process. Technol.*, 238, 218–225, 2016, https://doi.org/10.1016/j.jmatprotec.2016.07.025.
52. Galantucci, L.M., Lavecchia, F., Percoco, G., Quantitative analysis of a chemical treatment to reduce roughness of parts fabricated using fused deposition modeling. *CIRP Ann. – Manuf. Technol.*, 59, 247–250, 2010, https://doi.org/10.1016/j.cirp.2010.03.074.
53. Son, J. and Lee, H., Preliminary study on polishing SLA 3D-printed ABS-like resins for surface roughness and glossiness reduction. *Micromachines*, 11, 1–14, 2020, https://doi.org/10.3390/MI11090843.
54. Chen, L. and Zhang, X., Modification the surface quality and mechanical properties by laser polishing of Al/PLA part manufactured by fused deposition modeling. *Appl. Surf. Sci.*, 492, 765–775, 2019, https://doi.org/10.1016/j.apsusc.2019.06.252.
55. Kumbhar, N.N. and Mulay, A.V., Finishing of fused deposition modelling (FDM) printed parts by CO2 laser. *6th Int 27th All India Manuf. Technol.*

Des. Res. Conf. (AIMTDR-2016), December 16-18, 2016, College Eng Pune, Maharashtra, INDIA, Finish pp. 63–67, 2016.

56. Chai, Y., Li, R.W., Perriman, D.M., Chen, S., Qin, Q.H., Smith, P.N., Laser polishing of thermoplastics fabricated using fused deposition modeling. *Int. J. Adv. Manuf. Technol.*, 96, 4295–4302, 2018, https://doi.org/10.1007/s00170-018-1901-5.
57. Moradi, M., Moghadam, M.K., Shamsborhan, M., Bodaghi, M., Falavandi, H., Post-processing of FDM 3D-printed polylactic acid parts by laser beam cutting. *Polym. (Basel)*, 12, 1–18, 2020, https://doi.org/10.3390/polym12030550.
58. Chen, L., Zhang, X., Wang, Y., Osswald, T.A., Laser polishing of Cu/PLA composite parts fabricated by fused deposition modeling: Analysis of surface finish and mechanical properties. *Polym. Compos.*, 41, 1356–1368, 2020, https://doi.org/10.1002/pc.25459.
59. Ding, K. and Ye, L., 3 - Simulation methodology, in: *Woodhead Publishing Series in Metals and Surface Engineering*, K. Ding, L.B.T. Ye, L.S.P. (Eds.), pp. 47–72, Woodhead Publishing, United Kingdom, 2006.
60. Ding, K. and Ye, L., 2 - Physical and mechanical mechanisms of laser shock peening, in: *Woodhead Publishing Series in Metals and Surface Engineering*, K. Ding, L.B.T. Ye, L.S.P. (Eds.), pp. 7–46, Woodhead Publishing, United Kingdom, 2006.
61. Lan, L., Xin, R., Jin, X., Gao, S., He, B., Rong, Y., Min, N., Effects of laser shock peening on microstructure and properties of Ti-6Al-4V titanium alloy fabricated via selective laser melting. *Mater. (Basel)*, 13, 1–11, 2020, https://doi.org/10.3390/MA13153261.
62. Hackel, L., Rankin, J.R., Rubenchik, A., King, W.E., Matthews, M., Laser peening: A tool for additive manufacturing post-processing. *Addit. Manuf.*, 24, 67–75, 2018, https://doi.org/10.1016/j.addma.2018.09.013.
63. Lan, L., Jin, X., Gao, S., He, B., Rong, Y., Microstructural evolution and stress state related to mechanical properties of electron beam melted Ti-6Al-4V alloy modified by laser shock peening. *J. Mater. Sci. Technol.*, 50, 153–161, 2020, https://doi.org/10.1016/j.jmst.2019.11.039.
64. Lu, Y., Sun, G.F., Wang, Z.D., Su, B.Y., Zhang, Y.K., Ni, Z.H., The effects of laser peening on laser additive manufactured 316L steel. *Int. J. Adv. Manuf. Technol.*, 107, 2239–2249, 2020, https://doi.org/10.1007/s00170-020-05167-3.
65. An, Z., He, W., Zhou, X., Zhou, L., Nie, X., On the microstructure, residual stress and fatigue performance of laser metal deposited TC17 alloy subjected to laser shock peening. *Mater. (Basel)*, 15, 1–12, 2022, https://doi.org/10.3390/ma15186501.
66. Sohrabi, N., Hamidi-Nasab, M., Rouxel, B., Jhabvala, J., Parrilli, A., Vedani, M., Logé, R.E., Fatigue performance of an additively manufactured Zr-based bulk metallic glass and the effect of post-processing. *Metals (Basel)*, 11, 1–19, 2021, https://doi.org/10.3390/met11071064.

67. Ćwikła, M., Dziedzic, R., Reiner, J., Influence of overlap on surface quality in the laser polishing of 3D printed inconel 718 under the effect of air and argon. *Mater. (Basel)*, 14, 1–20, 2021, https://doi.org/10.3390/ma14061479.
68. Gisario, A., Barletta, M., Veniali, F., Laser polishing: A review of a constantly growing technology in the surface finishing of components made by additive manufacturing. *Int. J. Adv. Manuf. Technol.*, Springer London, 120, 1433–1472, 2022.
69. Liang, C., Yazhou Hu, N.L., Zou, X., Wang, H., Zhang, X., Hu, Y.F., J., Laser polishing of Ti6Al4V fabricated by selective laser melting. *Metals (Basel)*, 10, 191, 2020.
70. Obeidi, M.A., McCarthy, E., O'Connell, B., Ahad, I.U., Brabazon, D., Laser polishing of additive manufactured 316L stainless steel synthesized by selective laser melting. *Mater. (Basel)*, 12, 1–15, 2019, https://doi.org/10.3390/ma12060991.
71. Bai, Y., Chaudhari, A., Wang, H., Investigation on the microstructure and machinability of ASTM A131 steel manufactured by directed energy deposition. *J. Mater. Process. Technol.*, 276, 116410, 2020, https://doi.org/10.1016/j.jmatprotec.2019.116410.
72. Lopes, J.G., Machado, C.M., Duarte, V.R., Rodrigues, T.A., Santos, T.G., Oliveira, J.P., Effect of milling parameters on HSLA steel parts produced by wire and arc additive manufacturing (WAAM). *J. Manuf. Processes*, 59, 739–749, 2020, https://doi.org/10.1016/j.jmapro.2020.10.007.
73. Hönnige, J.R., Colegrove, P.A., Ganguly, S., Eimer, E., Kabra, S., Williams, S., Control of residual stress and distortion in aluminium wire + arc additive manufacture with rolling. *Addit. Manuf.*, 22, 775–783, 2018, https://doi.org/10.1016/j.addma.2018.06.015.
74. Scherillo, F., Chemical surface finishing of AlSi10Mg components made by additive manufacturing. *Manuf. Lett.*, 19, 5–9, 2019, https://doi.org/10.1016/j.mfglet.2018.12.002.
75. Mahmood, M.A., Chioibasu, D., Rehman, A.U., Mihai, S., Popescu, A.C., Post-processing techniques to enhance the quality of metallic parts produced by additive manufacturing. *Metals (Basel)*, 12, 1–32, 2022, https://doi.org/10.3390/met12010077.
76. Zhang, J., Chaudhari, A., Wang, H., Surface quality and material removal in magnetic abrasive finishing of selective laser melted 316L stainless steel. *J. Manuf. Processes*, 45, 710–719, 2019, https://doi.org/10.1016/j.jmapro.2019.07.044.
77. Yamaguchi, H., Fergani, O., Wu, P., Manufacturing technology modification using magnetic field-assisted finishing of the surface roughness and residual stress of additively manufactured components. *CIRP Annals*, 66, 305–308, 2017, https://doi.org/10.1016/j.cirp.2017.04.084.
78. Wang, J., Zhu, J., Liew, P.J., Material removal in ultrasonic abrasive polishing of additive manufactured components. *Appl. Sci.*, 9, 5359, 1–13, 2019.

79. Bai, Y., Wang, D., Yang, Y., Wang, H., Effect of heat treatment on the microstructure and mechanical properties of maraging steel by selective laser melting. *Mater. Sci. Eng. A*, 760, 105–117, 2019, https://doi.org/10.1016/j.msea.2019.05.115.
80. Aboulkhair, N.T., Tuck, C., Ashcroft, I., Maskery, I., Everitt, N.M., On the precipitation hardening of selective laser melted AlSi10Mg. *Metall. Mater. Trans. A Phys. Metall. Mater. Sci.*, 46, 3337–3341, 2015, https://doi.org/10.1007/s11661-015-2980-7.
81. Maamoun, A.H., Elbestawi, M., Dosbaeva, G.K., Veldhuis, S.C., Thermal post-processing of AlSi10Mg parts produced by selective laser melting using recycled powder. *Addit. Manuf.*, 21, 234–247, 2018, https://doi.org/10.1016/j.addma.2018.03.014.

8

Additive Manufacturing of Nanoscale and Microscale Materials

D. Vasanth Kumar[1], N. Srinivasan[1*], S. J. Davis Hans[1], S. Gokul[2], B. Arulmurugan[3] and B. Sathishkumar[4]

[1]Department of Mechanical Engineering, Jansons Institute of Technology, Coimbatore, Tamil Nadu, India
[2]Department of Mechanical Engineering, Sri Eshwar College of Engineering, Coimbatore, Tamil Nadu, India
[3]Department of Mechanical Engineering, KPR Institute of Engineering and Technology, Coimbatore, Tamil Nadu, India
[4]Department of Mechanical Engineering, Sri Krishna Polytechnic College, Coimbatore, Tamil Nadu, India

Abstract

With the advent of additive manufacturing (AM) technologies, the fabrication of products with any dimensions, shapes, and complexities has become easier. It is the most widely followed technique which almost eliminates preprocessing and most of the postprocessing which was adopted in conventional manufacturing techniques. The dimensional scale of the materials used as precursors in AM methods plays a vital role in deciding the final quality of the product. Though microscale caters to the purpose in most cases, the advent of nanoscale and nanotechnologies has further revolutionized the AM of parts. Reducing to nanoscale from the microscale dimension was found to enhance the strength of the 3D-printed part but at the cost of structural intrinsic metastability, which results in thermally activated coarsening of the 3D-printed grains structures. Nowadays, the advancements in AM technologies give the provision of printing bulk nanostructured materials through rapid solidification and cooling, but the grain coarsening effect was seen to be much more common in those materials even at some normal thermal exposure, more like the typical metastable materials. Materials also play a significant role in dodging this effect. Various materials like metals, alloys, polymer-metal hybrids, and fiber-reinforced composites are used in AM technologies to manufacture various parts.

*Corresponding author: srini.prs@gmail.com

R. Rajasekar, C. Moganapriya and P. Sathish Kumar (eds.) Additive Manufacturing with Novel Materials: Processes, Properties and Applications, (267–294) © 2024 Scrivener Publishing LLC

This chapter comprehensively deals with the backlogs of micro AM, the advantages of nanoscale AM, and the materials used commonly. This chapter also deals with the properties and applications of parts built using nanoscale AM.

Keywords: Additive manufacturing, 3D metal printing, properties, nanocomposites, microscale additive manufacturing, photopolymers

8.1 Introduction

In recent years, additive manufacturing, also known as AM, has emerged as a production process that generates the least amount of waste. AM has reached the level of extreme manufacturing processes because of its ability to make any item within the allotted period and with nearly no downtime at all [1]. However, nanotechnology's enormous surface area and reactivity on a range of one to a few nanometer ranges, which are not visible at the tiny level, are enabling novel solutions to many problems. This is because nanotechnology operates on a scale that is not visible to the naked eye. Since the advent of AM, it has become possible to manufacture on a variety of scales, ranging from the macro to the tiny [2, 3]. The advent of nanomanufacturing and the use of nanomaterials paved the path for the development of functionalized nanomaterials, which in turn allowed AM to begin pushing the boundaries of what was previously considered possible [4]. To this day, nanomaterials have undergone functionalization through a variety of processes, such as synthesis, and have been used in the field of biomedicine. AM-functionalized nanoparticles or nanomaterials have increased the field's diversity and efficiency [5].

There are several approaches and accomplishments that AM has helped achieve through nanomaterials and their use in vital fields. For instance, nanomaterials that have been functionalized in multiple ways, also including nanographene and nanocarbon fiber, have garnered a lot of attention recently [6, 7]. Sensors use nanographene and carbon fibers, and AM is also used in medicine delivery systems. Nanomaterials in polymer matrices and three-dimensional (3D) printing will expand tissue engineering. This will be accomplished through the creation of individualized 3D structures that have engineered properties and multiple functions as shown in Figure 8.1 [8].

In addition, AM and nanoparticles were used in hard and soft natural polymers to create acellular and cellular 3D cell constructs using a variety of 3D printing and bioprinting technologies. Conventional methods developed acellular and cellular 3D cell constructs in hard and soft biodegradable polymers. Fused deposition modeling (FDM), stereolithography (SLA),

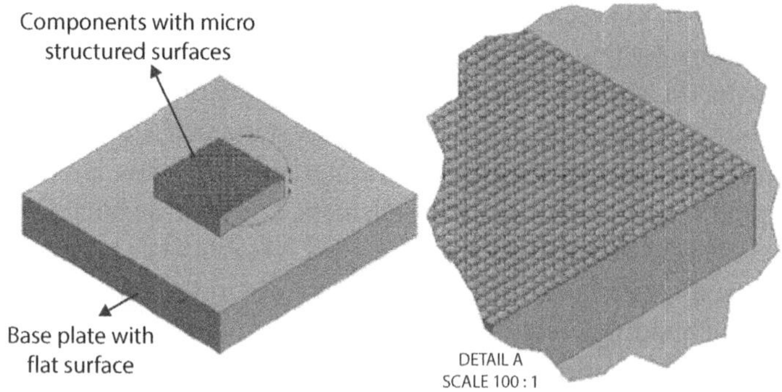

Figure 8.1 Micro-structured surfaces by additive manufacturing [8].

laser sintering, droplet, laser, and extrusion-assisted tissue regeneration are employed. Surface functionalization, scaffold degradability, and nanocomposite cytotoxicity are the key strategies that change things [9]. Additive manufacturing allows deep undercuts and complex shapes, and manufacturing technology makes these qualities possible. Conventional manufacturing requires a cutting tool, jigs, fixtures, coolant, molds, patterns, and external support devices to make components. Additive manufacturing has many benefits, including a lightweight design, automation, higher flexibility, no material waste, lower prices, and the elimination of trained artisans due to the automated process, and it also makes little noise and is environmentally friendly. It is regarded as a fundamentally forward-thinking and adaptable element of the third wave of the industrial revolution [10, 11].

In today's world, AM has become increasingly popular and is being utilized in a variety of countries for the manufacture of –open-sourced designs for a variety of industries, including the aerospace industry, the aviation industry, the healthcare business, and the automotive industry. Additive manufacturing technologies have some drawbacks, such as costly equipment costs, size constraints, increased unemployment, and slow development rates. However, current R&D has reduced the cost of 3D printing equipment, which has increased their use in schools, labs, libraries, households, and other institutions. Additionally, AM technology is capable of producing 3D prints of small numbers of bespoke gadgets or items at significantly lower costs than traditional methods [12].

The orientation in AM is currently shifting toward the creation of customized usable items and it has been predicted that commercial product manufacturing will account for 50% of AM by 2025. Because computed tomography reproductions can function as a distinct form of medical implant, pharmaceutical specialists are interested in these technologies. In product development, the authors have used selective laser sintering (SLS), stereolithography (SLA), electron beam melting (EBM), fusion filament fabrication (FFF), selective laser melting (SLM), and ink jet printing (IJP). Every one of these procedures is based on the same fundamental idea, which states that the ultimate desired item can be produced by depositing the material in successive layers [13]. AM uses ceramics, wax, alloys, concrete, polymers, acrylonitrile butadiene (ABS), adhesive-coated sheets, and polylactic acid (PLA) for manufacturing products. Ceramics and concrete are both put to heavy use in the production of 3DP scaffolds and buildings, respectively.

However, the potential of AM on a broad scale is constrained since the components that are produced by AM have poor mechanical qualities and an isotropic feature. In order to manage anisotropic behavior and fault sensitivity, a better pattern of AM is absolutely necessary [14]. Changes in the atmosphere of the printing process can have a major impact on the quality of completed products. The construction of micro- to macroscale parts in a variety of sizes can be accomplished using rapid prototyping. The precision of the components that are printed is primarily determined by the precision of the process that was utilized as well as the printing scale [15].

8.2 Microscale Additive Manufacturing

Layer bond and resolution difficulties necessitate postprocessing, making microscale additive printing risky. Applying this cutting-edge technique in many different industries is risky because AM materials are scarce. Thus, optimal material and process combinations for AM of various components for various applications must be chosen. Another development area is AM

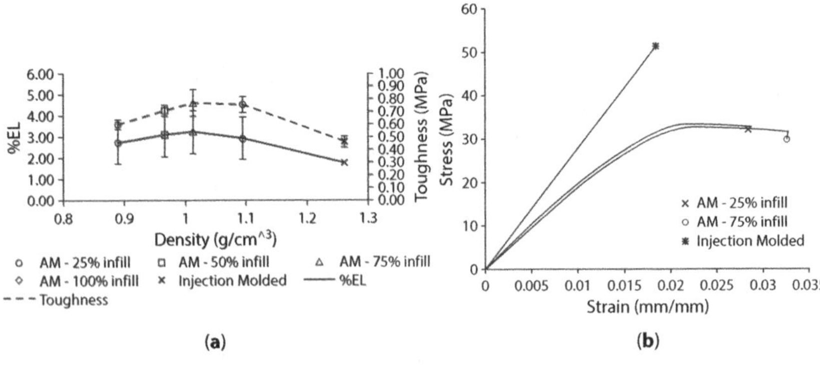

Figure 8.2 (a) % Elongation and toughness data (b) Stress-strain plots, for tensile specimens [16].

component mechanical quality. It also depends on infill geometry, possibly to increase the energy dissipation with increased elongation percentage and toughness as shown in Figure 8.2 [16]. If the right material and technique are used, any construction or piece will be more successful. Thus, this chapter reviews AM procedures, materials, and applications in numerous sectors.

The research establishment has been focusing on tissue-engineered, electronics, microelectronics, reconstructive surgery, health care, and materials, but as resolution improves, photovoltaic panels, optics, and nanodevices are emerging. Photolithography, which makes use of electromagnetic radiation with a wavelength that gets shorter and shorter with each passing decade, has been the only viable choice for full-scale industrial nanomanufacturing for decades [17]. However, the technology required to consistently attain nanoscale dimensions is prohibitively expensive. As a direct outcome of this, less extensive research and development centers, as well as small- and medium-sized enterprises (SMEs), are unable to capitalize on the developments made by more extensive businesses [18]. Photolithography and electron beam lithography (EBL) require multiple thin polymer film exposures for device manufacturing. These planar technologies function nearly solely in two dimensions, thus manufacturing processes that expand into 3D space appear like a clear solution to save space, cut costs, and add functionality. Resist-based production wastes a lot on lift-off and engraving [19, 20] since only a small percentage of materials are converted into usable pieces. Additive nanomanufacturing or ANM, adds functionality to resist-based planar production methods used to make integrated circuits (ICs). Most critically, it may change our nanoscale manufacturing mindset and increase design freedom [19].

When contrasting different methods of production on the micro- and nanoscales, it is common practice to concentrate on the level of resolution offered by each technology. This is because resolution is directly related to the level of performance offered by the end product. Commercial applications require resolution, accuracy/alignment, writing speed, and materials. Nanoscale manufacturing differs from macroscale production due to physical restrictions. Thus, obstacles vary greatly [21, 22]. To fabricate something at the nanoscale requires only the manipulation of a very small number of molecules or atoms at a time, and any variation in the manufacturing process, even of only one atom or molecule, can result in the failure of the device. Structures below 50 nm have a lower melting point, which affects mechanical stability. Due to the fact that a single atom can affect material properties, specifying alloy mixture, contamination levels, and crystallinity is difficult [23]. At the nanoscale, surface forces like van der Waals force dominate over gravity, which can cause undesired adhesion. In liquid, nonlinear velocity and advection forces make particle capture challenging. Fans, pumps, and traffic can prevent structural alignment even in gas or vacuum. Additive nanomanufacturing must use biomaterials, ceramics, and metals to create large-scale nanostructures with high precision, resolution, and cost. Finally, the production methods direct printing and particle placement are compared to forecast ANM's development into the following decade [24].

Direct writing on nano organic, insulator, and metal devices could help ANM applications. This chapter discusses the latest advances in three of the most important nanostructure direct writing technologies such as dip-pen nanolithography (DPN), electrohydrodynamic (EHD) inkjet printing, and direct laser writing (DLW). These methods can achieve sub-30 nm resolution, reasonable writing speeds, and 3D nanomanufacturing. DPN and EHD jet printing materials are already plentiful, which includes metals and biomaterials [25]. SPM-based flexible nanofabrication using DPN creates two-dimensional (2D) nanoscale features. SPM's direct-write soft-matter compatibility and atomic force microscopy (AFM)'s high resolution and registry are combined in a mask-free, biocompatible system. In the mid-1980s, scanning probe microscopes (SPMs) like the scanning tunneling microscope (STM) and AFM were introduced as nano imaging and spectroscopic tools. Scanning-probe lithography (SPL) soon followed, demonstrating that scanning probe microscopy could be used for microscopic printing or manufacturing [16, 19].

8.2.1 Photopolymers

Stereolithography, projection stereolithography (PSL), two-photonpolymerization (TPP), and micro-computed axial lithography (CAL) utilize photopolymers, polymer subunits and prepolymers, in vat photo

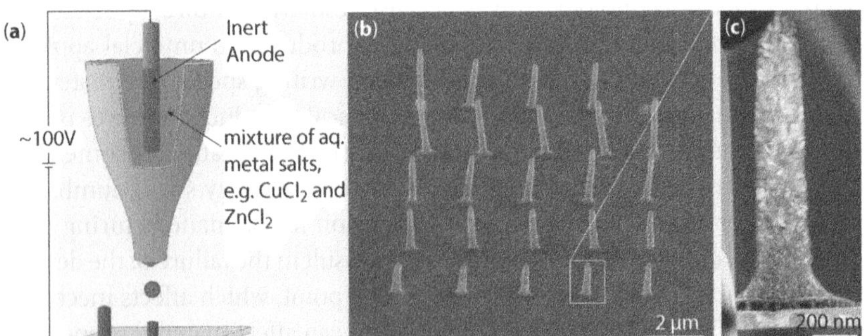

Figure 8.3 (a) Schematic of electrohydrodynamic redox (EHD-RP) 3D printing (b) Scanning electron microscopy (SEM) micrograph of array Cu structures (c) Transmission electron microscopy (TEM) image of Cu pillar [22].

polymerization [26, 27]. Excited molecules break down into active groups or free radicals when the photo initiator absorbs UV light at 250–300 nm. Polymerization, crosslinking, and other procedures cure prepolymers or energetic monomers fast. Epoxy acrylate (EA), hydroxypropyl methacrylate, and trimethoxysilane are fast-curing, strong, and flowable photopolymers. Photosensitive resin research encompasses organic modification and accuracy improvement which can also observed in Figure 8.3 [28].

Hydrogel Composites: Hydrogels are flexible, self-healing 3D networking structures made of hydrophilic polymers. Thus, hydrogel swelling coefficient manipulation allows functional materials to respond intelligently [29, 30]. Hydrogels are natural or manmade polymers. Stimulus response settings include pH-, temperature-, and photosensitive hydrogels. Polyacrylamides (PAM) and hydrolyzed PAM are hydrophilic polymers (HPAM). Direct ink writing (DIW) and FDM micro/nano additive manufacturing (MNAM) techniques for hydrophilic polymers increase material properties through deposition and extrusion. Robotics, actuators, and tissue engineering use hydrophilic polymers extensively [31].

Thermoplastics Polymer: Material extrusion (ME) and material jetting (MJ) print largely using low-melting thermoplastic polymers such as polycarbonate (PC), PLA, polycaprolactone (PCL), wax, and ABS [32]. A single molecular filament material cannot suit all needs, hence, the main research focus is on a composite and mixing system [33]. Thermosetting filaments are utilized as basis materials because of their good ductility, making them easier to manufacture [34]. Polymer tubes with superior manufacturing and triboelectric capabilities have replaced thread-like filaments and formed microstructures for triboelectric nanogenerators' friction layers (TENG) [35].

Metallic Material Composites: Metals are good in mechanical and chemical properties and works well with most steel alloys, titanium alloys, nickel- or cobalt-based composite material powder, aluminum alloy, and stainless steel (SS). Metal-associated composites with hydrophilic and thermoplastic polymers improve material performance [36, 37].

Metallic Materials: Most of the time, metals are divided into two groups: ferrous and nonferrous. Chromium, Mn, and Fe alloys are ferrous. Nonferrous alloys are those without iron. Titanium mixture powders resist temperature and corrosion best among nonferrous metals. Micro/nano functional devices (MNFDs) need lower density level and higher specific strength. Micro-SLS and micro-SLM reduce titanium processing time and are commonly utilized to construct complicated 3D structures [38, 39]. Finding the appropriate process variables for each metal requires many tests due to their various properties. In the present material development

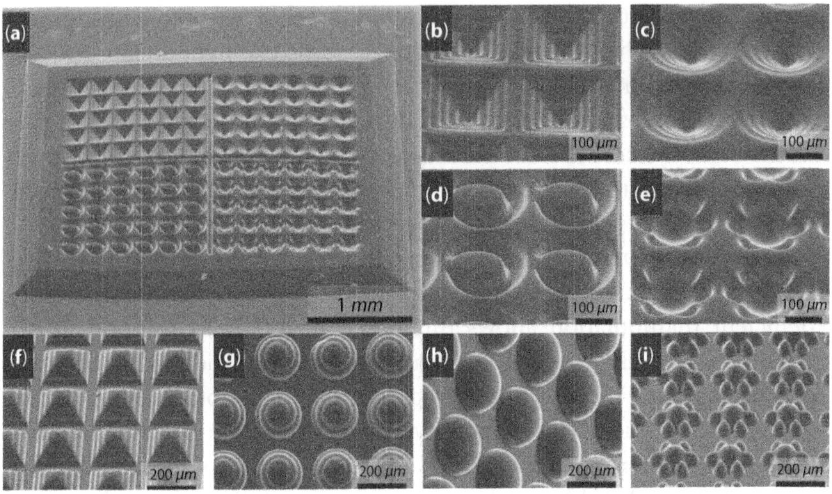

Figure 8.4 SEM images of the fabricated 100%-PEGDA microwells and microarchitectures. (a) The fabricated chip with various microstructured wells: (b) stepwise, (c) spiral, (d) embryo-like and (e) flower-like. (f-i) are the inverse of the wells, demonstrating the versatility of the biofabrication system [40].

process, polishing, electrolytic sharpening, and heat treatment increase quality of surface as shown in Figure 8.4, elimination of step effect and distribution of stress, and speed up material development [40, 41].

Metallic Hydrogel Composites: Hydrogel-metal composites often contain liquid metals (LMs). Alginate-tannic acid (TA) and polyvinyl alcohol (PVA)-agarose hydrogel-metal alloys are common. Hydrogels use gallium (Ga) for its fluidity and low toxicity [42]. Injecting 75% Ga and 25% In LMs into PVA-agarose hydrogel formed a hydrogel-metal composite. Mixing LM fluidity with biopolymers and polyphenol creates biocompatible hydrogel-metal composites (Alg and TA). Due to its flow, DIW is the major MNAM approach for hydrogel-metal composites [43, 44].

Thermoplastic Composites With Metal: In numerous engineering contexts, metal-filled thermoplastic composites have proven useful. It has high tensile strength and modulus and is resistant to corrosion, weight, and flame. Metal filler particles in a polymer matrix can be either randomly distributed or kept apart [45, 46]. It is the random contact between locations in the vicinity that ensures the continuation. Compression forming creates isolated structures while extrusion, injection molding, and internal mixers create random patterns. For this purpose, FDM is the MNAM process of choice and created heat-shrinkable composites with superior mechanical characteristics by including a metallic filler [47–49]. Some examples of FDM metal-filled composite materials include the more common Fe/PA composite, ABS-Cu, ABS-Fe, ABS coupled with 10% metal powders, and ionic polymer metal composite (IPMCs) [50]. Nanocomposites of polymers and metals have been developed with a focus on improved liquid behavior, thermal stability, and conductivity. The quality, efficiency, and performance of MNAM products are enhanced by metal-filled thermoplastic composites [51, 52].

8.2.2 Non-Organic Nonmetallic Composites

A material system includes inorganic components, nonmetal components, metallic components, and organic polymer components. These components often have rigidity, compressibility, resistance to high temperatures, and resistance to both corrosion and abrasion [53, 54]. Materials such as ceramics and quartz glass are examples of inorganic nonmetallic materials. Several different types of inorganic oxides, carbides, and nitrides are also used in composites [55, 56]. Particle- and fiber-reinforced composites can be manufactured by combining inorganic composite materials with deformable polymers and sensitive polymers. Ceramic, quartz glass, magnetic MNFDs, and conductivity MNFDs all benefit from the innovative functional material system provided by MNAM thanks to inorganic nonmetals materials and their composites [57].

Ceramics: Ceramics are used in aeronautical, industrial, and biomedical applications due to their compressive durability and high-temperature endurance. A new manufacturing strategy developed by MNAM, ceramic fabrication, becomes less cost-effective and more time-consuming [58]. Tricalcium phospahate (TCP's) composition is very close to that of a human bone, hence it is frequently used in MNAM methods to create bone tissue scaffolds and bioceramic scaffolds. Chemical synthesis is the primary method for producing preceramic polymers, which find widespread use in SLA, PSL, and continuous liquid interface production (CLIP) [59, 60]. Preceramic polymers excel as heat insulators because of their great compressive strength and tolerance to high temperatures [61]. Polycarbosilane (PCS) carbides have high resistance to oxidation, mechanical characteristics, chemical stability, and shock. The primary application for these supplies is in the production of ceramics and ceramic coatings. Alumina whiteware is a widely used oxide ceramic due to its great hardness and wear resistance [62–64].

Quartz Glass: A major area of investigation for MNAM's potential uses is in the production of quartz glass. Quartz glass has been made using MNAM, according to various research. Silica and SiO_2 are the primary materials, and they have greatly increased the variety of useful substances available [65]. MNAM is a straightforward approach to generating very accurate glass microstructures, especially when compared to more traditional production methods. Examples include the work on PSL to create molten silica glass with a 60-m resolution and a 2-nm surface roughness [66]. The printed quartz glass had the same chemical composition and physical qualities as regular quartz glass. More than 90% of visible light is allowed through, and the relative density is greater than 99%. However, MNAM still has a long way to go before it can successfully manufacture quartz glass, with obstacles including a suitable printable material system, pinpoint accuracy, and affordable prices. The hardest part is glass shrinkage during heat treatment, which is affected by printing resin solid content. Silica/SiO_2, a unique functional material, expands MNAM's usefulness [67, 68].

Particle-Reinforced Composites: Micro and nanoparticle reinforcement of a single material yields a composite with a wide range of applications. Magnetic materials, conductive materials, and piezoelectric materials are the three types of particle-reinforced composites. Functional materials in MNFDs often include magnetic materials [69]. However, there are still obstacles to overcome in the production of magnetic materials with a high degree of complexity in their 3D structures. The stability of a large concentration of particles determines the magnetic properties of functional materials. 30 wt% Fe_3O_4 magnetic nanoparticles printed sub-75 m fine features [70]. In printing magnetic particles using photopolymerization-based MNAM, this is the highest concentration to date. However, there are constraints on resin stability and printing quality. The inability to control spatial and temporal dimensions is another restriction of magnetic MNFDs in external magnetic fields [71].

As a result, magnetic composites with multiple uses is an important functional material for the future. Electronic gadgets rely on conductive materials as their foundation. Functional materials with conductive qualities have been obtained by incorporating magnetic nanoparticles (Fe_3O_4), carbon nanotubes (CNTs), nano silver powder, Ag-Au metals, and conductive polymers into polymers [72]. By combining CNTs or graphene with thermoplastics, conductive carbon filament composites could be made. One benefit of carbon composite filaments is that they can be used to create conductive electrodes, which is not the case with carbon nanotube-epoxy composite electrodes. Conductive materials must be evaluated for particle ratio, electrode balancing, and printing performance. Piezoelectric materials transform pressure, vibration, and elongation into electrical charges. Electric fields cause mechanical strain in piezoelectric materials [73]. Thus, piezoelectric materials can be used to construct electronic items, sensor, actuator, and energy devices, however, given their severe brittleness, this makes MMAM printing difficult. Piezoelectric polymers can print piezoelectric materials efficiently. Polyvinylidene fluoride (PVDF) outperforms ceramics for MNAM due to its flexibility and biocompatibility. Polyvinylidene fluoride or PVDF-$BatiO_3$ nanocomposites make up most 3D-printed piezoelectric devices. To improve piezoelectric properties, more PVDF functionalization, piezoelectric polymers, and MNAM research are needed [74].

Fiber-Reinforced Composites: There are several FDM materials available, including PC, ABS, PLA, and PA. Isolated thermoplastic polymers, however, are quite fragile. Carbon, glass, and aramid fibers are commonly used in the construction of fiber-reinforced composites. Fiber volume composition in carbon fiber-reinforced polymer composites was the most robust [75–77]. A few examples of fiber-reinforced composites are carbon fiber-reinforced plastic (CFRP), high-performance polyamide 12 (HAP-PA12), and dimethyl methylphosphonate (DMMP). Fiber-reinforced nylon composite can reach 600 MPa tensile strength by increasing fiber volume fraction [78]. Carbon fiber is the strongest, followed by glass and aramid. Short or continuous fibers depend on length which interconnected pore network polymer (IPN) matrix to make composite materials using carbon and glass fiber reinforcement, plasticizers, and compatibilizers. Continuous fiber reinforcement improves performance [79].

8.3 Properties of Materials

8.3.1 Mechanical Properties

Nanoparticles exhibit distinct mechanical properties in comparison to microparticles and bulk materials. As a result, they offer more efficient options for the surface treatment of many devices in the realm of mechanical strength, as well as for the improvement of the quality of nanomanufacturing and nanofabrication, amongst other things [80, 81]. To be more explicit, on the one hand, the mechanical properties of nanoparticles can influence the tribological characteristics of fluids with nanoparticles as well as enhance composite coatings [82]. This is because nanoparticles are extremely small. When the contact pressure is high enough, nanoparticles will either be sheared or indented first onto the surface of the contacting surface depending on the relative hardness of the nanoparticles and the surface they are rubbing against. This information might shed light on the way the particles act when they touch [83, 84]. In contrast hand, nanoparticles are typically utilized as abrasive particles in the nanopolish of liquid or semi surface by chemicals mechanical polishing (CMP). Chemicals mechanical polishing is the most efficient planarization process currently used in the production of ICs. In order to improve the surface quality and increase the amount of material removed, it is vital to have good control of the physical properties of the particles, as well as their interaction with the polished surface and other factors [85]. A comprehensive grasp of the fundamentals of the nanoparticles' mechanical characteristics, such as toughness and Young's modulus, interfacial bonding and frictional, mobility law, in addition to their size-dependent impacts, is typically essential for the successful use in these disciplines. Various testing methods, such as microhardness with AFM, *in situ* compression by a stress probe holder based on observations with transmission electron microscopy (TEM), and so on, have been developed in order to gather more of this information [72, 86]. Nevertheless, the results that were achieved are still insufficient, and some of them are debatable. For instance, a conclusive answer to the question of whether the Young's modulus of nanoparticles measured using AFM is influenced by the particle diameter and the indentation depth has not yet been found.

8.3.2 Optical Properties

Nanoscale dimensions and surface plasmon resonance make nonmaterial optical features extremely fascinating for investigation. These properties are affected by size, shape, surface modification, doping, interaction with the other materials, and so on. Size-dependent optical characteristics are caused by the optical energy band gap change [87]. This affects nanomaterial surface plasmon resonance. As particle size decreases, the optical band gap expands, especially in semiconductor nanomaterials. Thus, variable-sized colloidal gold and silver nanoparticles, notably 1–10 nm, produce diverse colors. Coherent activation of nanoparticle electrons in the conductivity zone and light-induced in-phase resonant oscillations create electronic structure. Surface plasmon resonance is only detectable if the material size of the particles is less than the light beam wavelength. Surface plasmon resonance occurs only in nanomaterials. The surface plasmon resonance interfere with stimulated ions and medium dielectric constant [88, 89].

8.3.3 Dielectric Properties

Due to their electronics applications, polymers and functional composites have garnered attention in recent years. This includes piezo-resistive/electric devices, and conductors. Generally speaking, the relative permittivity or dielectric constant of polymers is quite small [90]. Polymer composites are formed when a ceramic material with a high dielectric constant (such as

$BaTiO_3$) is infilled into the polymer at the micro- or nanoparticulate scale to increase the polymer's dielectric performance [91, 92]. The production of electronic components benefits greatly from these polymer composites' high dielectric achievement. As they could be used in conjunction with modern radio frequency (RF) devices that function at microwave frequencies, they have recently gained significant attention. Because of their high dielectric and low dielectric loss, ceramics have seen extensive usage for this purpose. A lot of time and effort has been invested in developing new ceramic materials [93, 94]. The processing at high temperatures required to obtain structural integrity and the desired dielectric characteristics also comes with a number of difficulties in the production of ceramic components, including high tooling costs and challenging implementation [95].

8.4 Nanocomposites Applications

8.4.1 Materials Compatible With Medical Implants

Biocompatibility must be carefully chosen and created in metals, ceramics, polymers, and their composites. Based on density, ductility, fractures and fatigue strength, etc., these implants must be chosen carefully [96]. Due to metals' strength and ductility, most implants have been metal. Stainless, CoCr metals, and Ti alloys are widely used because they resist corrosion. Cellular absorption by organic tissues often requires surface functionalization. For example, coating implants with hydroxyapatite (Hap) to ensure bone compatibility. Gold, silver, and platinum, which are castable and ductile, are intriguing dental materials [97, 98]. Tissue-engineered and regenerative medicine use biodegradable, biocompatible polymers such chitosan, gelatin, heparin, and collagen. However, the immune system swiftly degrades these materials. Issues at all levels, PCL, polyglycolide, and related copolymers are utilized as non-permanent scaffolds because they are quasi, biocompatible, biodegradable, and have high mechanical quality synthetic polymers [99, 100]. Poly ethylene glycol (PEG), a biocompatible polymer with hydrophilicity and solubility in many solvents, makes processing easier. Polyvinyl alcohol and poly acrylic acid (PAA) are also used, however, their non-degradability limit their use in implants [101].

Polyurethanes have been studied for their high mechanical strength. Hydrolyzable polymeric segments like PCL increase biodegradability [102]. Polypyrrole, polyaniline, and polythiophene, as well as its derivatives, are biocompatible, making them suitable for electrically stimulated cells like neuron, bone, muscle, and heart cells. However, these materials are weak mechanically, hydrophobic, and nondegradable, therefore novel composites with multifunctional properties are needed [103]. Bioceramics in implantable devices and tissue engineering are crucial. Titanium dioxide (TiO_2) has been used as a bone substitute regeneration due to its biocompatibility, osteoconductivity, antibacterial characteristics, and bone and vascular system improvement. This attention has been directed toward TiO_2 [104]. Calcium phosphates are another excellent option since they may imitate the structure of bone in its 3D form. Depending on the specific application, these bulk materials each have their own set of advantages and disadvantages [105]. The use of nanomaterials in conjunction with these materials can improve or tune the mechanical performances or functionalities of either of the materials [106, 107].

For biomedical applications, the types of nanoparticles that can be used depend on the specific situations of the device being used. These nanomaterials can be of a variety of different types. Nanocomposites can be essentially defined as matrices, which are most frequently polymers, that have particular structural features and are doped with fillers that have dimensions of less than 100 nm [108]. The structural features of the matrix can be altered using fillers, or additional functionality can be added by using fillers in the form of electrical conductivity, biocompatibility, and so on. It is particularly fascinating to make use of nanoscale fillers because this can raise the extent of the change that occurs in the characteristics of composites [109]. However, the difficulty lies in properly dispersing them throughout the matrices in order to ensure that the attributes are maximized while simultaneously preventing them from aggregating [110].

8.4.2 Nanocomposites With Metallic-Based Nanofillers

Metallic nanofillers like silver or gold nanoparticles may increase tissue electrical conductivity and antibacterial properties in biomedical applications. In musculoskeletal tissue engineering, gold nanoparticles reinforced extracellular matrices (ECM) and reduced inflammation, increasing cellularity [111]. Silver nanoparticles are known to be antimicrobial and can replace antibiotics. Silver nanoparticles infused the hydrogel. This hydrogel showed promise as an intriguing potential dressing for wounds because of its ability to inhibit microbial activity [112]. In a similar manner, the backbones of bio-based polyurethane were functionalized with magnetic nanoparticles (Fe_2O_3), which led to the creation of a material that possesses both antibacterial and magnetism capabilities. The graphene oxide was used as an intermediary to strengthen PCL that had been treated with silver [113]. Polycaprolactone composites containing graphene oxide and silver demonstrated considerable improvements in modulus, electrical conductivity, Ag ion tolerance, and human cell non-toxicity compared to PCL treated with Ag nanoparticles only [114].

8.4.3 Nanocomposites With Ceramic Nanofillers

One typical use of ceramic nanofillers in the polymer sector is seen in the reinforcement of chitosan using Hap nanopowders. Since unmodified chitosan has so few mechanical properties, it cannot be used for anything that requires bearing weight [115–117]. Hydroxyapatite, with its crystal structure that is close to the structure of genuine bone, was allowed to produce scaffolds with regulated pore topologies and excellent strength. Scaffolds for synthetic bone tissue have been developed utilizing nano-Hap-functionalized cellulose, which mimics the properties of chitosan [118]. Many scientists have looked into how additives like zinc affect Hap to find ways to increase the metal's cytocompatibility and resistance to corrosion. Titanium dioxide nanoparticles are also widely used as reinforcements for toughening plastics [119].

There is biocompatibility between PCL and TiO_2 nanoparticles for the purpose of bioactivating titanium implants via coating, hence increasing cell attachments [120]. Because of this, titanium can now be used in living systems. Diopside improves bioactivity, workability, and stability of bovine-hydroxyapatite covering titanium implants. Textile-reinforced rubber is great for surgical gloves, water temperature bags, as well as other medical devices because it is bacteria-resistant and mechanically strong. Nanocrystals and cellulosic fiber enhance rubber composites with antibacterial properties. ZnO disperses cellulosic fiber in the matrix and has antimicrobial effects [121, 122].

8.4.4 Nanocomposites With Carbon-Based Nanofillers

Polyurethanes, PCLs, collagen, and chitosan have been doped with CNTs in recent years. Nanotubes on biocompatible polymers enhance bone, nerve, and other cells, promising osteogenesis. In addition to CNT mechanical reinforcement, SWNTs increased osteoblast adherence to Hap-doped chitosan scaffolds [123]. Before using CNTs in the body, cytotoxicity must be studied. Another carbon-based substance, graphene, is being studied for medical uses. (i) Pure graphite sheets, (ii) graphene oxide (GO), and (iii) reduced graphene oxide (rGO). Graphene oxide was added to polypyrrole to generate a biocompatible implant covering.

8.4.5 Nanocomposites With Cellulose-Based Nanofillers

Cellulose nanomaterials are attractive because of their availability, mechanical qualities, capacity to self-assemble in distributed systems, biocompatibility, and lower toxicity. They are ideal for biological research and practice. However, hydrophilic particles are difficult to scatter in polymer-based matrices [124]. Cellulose nanomaterial (CNs), cellulose nano fibrils (CNFs), and bacterial cellulose (BC) are all fillers. There is evidence to suggest that mechanical strength can be improved by adding CNs to cement-based products as an additive. CNs are good candidates for usage in biomedical applications such as the repair and mending of tissues (such as skin and bones) and the implantation of medical implants due to their biocompatibility and biodegradability. Bacterial cellulose-based materials have found widespread application in the field of vascular implantology [125].

Bacterial cellulose is a nanofiber mesh of pure cellulose spun by bacteria. It can be designed at nano-, micro-, and macroscales and exhibits exceptional strength. Bacterial cellulose's biocompatibility, optimum tri- and microfibrillar shape, and barrier to limit bacterial infection make them ideal for medicinal nanocomposites and are also affordable [126, 127]. They are used in bone and cartilage regeneration (BC/Hap nanocomposites), dental grafts, artificial corneas (BC/PVA composites), wound dressing (BC/hyaluronan composite films), and more. Artificial vascular grafts use polyurethane-nanocellulose nanocomposites. Nanocomposites can endure 400 kPa hydraulic pressure and have 800–1200% elongation at break [128]. Because of its hydrophilic nature, biocompatible PVA has long been used to replace soft tissue. The replacement of cardiac tissue is one such example [129, 130].

8.5 Conclusion

Additive manufacturing methods, such as fused filament fabrication, fused deposition modeling, direct ink writing and stereolithography offer many advantages when compared to conventional fabrication methods like casting because 3D printing enable fabricating complex material specimen geometries without the use of a mold. This chapter discussed the advent of AM technologies towards the implementation of nanoscale materials over microscale materials. Nanoscale materials were proven to render better properties in the final 3D-printed part owing to their structure and accuracy. Additive manufacturing processes were shown to render promising results in the namufacture of high performance 3D-printed components.

The systems and devices embedded with such components were also seen to perform better owing to their unique features. However, with the available features and technologies, it may be difficult to evaluate the capability of the current technology. However, this chapter described the potential of AM technologies to accommodate nanoscale materials with substantial description of properties of the 3D-printed part, which can revolutionize various fields such as aerospace, marine, optical fabrication, tissue engineering, and biomedical applications. Even at micro- and nanoscale levels, the most promising results appear to only have occurred in lab settings rather than production environments. There are many unanswered questions and issues before this technology can be widely adopted. These issues include, but are not limited to, things such as index distribution, geometry, and volume shrinkage of the 3D-printed part. When these issues are addressed through research, AM technologies will irreversibly replace all the conventional methods of manufacturing in many applications.

References

1. Park, J.M., Asghari-Rad, P., Zargaran, A., Bae, J.W., Moon, J., Kwon, H., Choe, J., Yang, S., Yu, J.-H., Kim, H.S., Nano-scale heterogeneity-driven metastability engineering in ferrous medium-entropy alloy induced by additive manufacturing. *Acta Mater.*, **221**, 117426, 2021.
2. Muldoon, K., Song, Y., Ahmad, Z., Chen, X., Chang, M.-W., High precision 3D printing for micro to nano scale biomedical and electronic devices. *Micromachines*, **13**, 4, 642, 2022.
3. Weng, F., Yu, H., Chew, Y., Bi, G., Du, X., Tian, H., Chen, C., Microstructure and mechanical behavior of the laser synthesized composites modified by micro/nano scale rare earth oxides. *J. Alloys Compd.*, **895**, 162641, 2022.
4. Chen, C., Xie, Y., Yin, S., Li, W., Luo, X., Xie, X., Zhao, R., Deng, C., Wang, J., Liao, H., Ductile and high strength Cu fabricated by solid-state cold spray additive manufacturing. *J. Mater. Sci. Technol.*, **134**, 234–243, 2023.
5. Li, X., Zhang, G., Li, W., Yu, Z., Yang, K., Lan, H., The electric-field-driven fusion jetting 3D printing for fabricating high resolution polylactic acid/multi-walled carbon nanotube composite micro-scale structures. *Micromachines (Basel)*, 11, 12, 1132, 2020.
6. Ramesh, M., Janani, R., Deepa, C., Rajeshkumar, L., Nanotechnology-enabled biosensors: A review of fundamentals, design principles, materials, and applications. *Biosensors*, **13**, 1, 40, 2022.
7. Ramesh, M., Rajeshkumar, L., Balaji, D., Bhuvaneswari, V., Sustainable and renewable nano-biocomposites for sensors and actuators: A review on preparation and performance. *Curr. Anal. Chem.*, **19**, 1, 38–69, 2023.

8. Davoudinejad, A., Cai, Y., Pedersen, D.B., Luo, X., Tosello, G., Fabrication of micro-structured surfaces by additive manufacturing, with simulation of dynamic contact angle. *Mater. Des.*, **176**, 107839, 2019.
9. Wang, Z., Tang, S.Y., Scudino, S., Ivanov, Y.P., Qu, R.T., Wang, D., Yang, C., Zhang, W.W., Greer, A.L., Eckert, J., Additive manufacturing of a martensitic Co–Cr–Mo alloy: Towards circumventing the strength–ductility trade-off. *Addit. Manuf.*, **37**, 101725, 2021.
10. Anandraj, M.K., Rathanasamy, R., Rathinasamy, P., Palaniappan., S.K., Effect of cloisite 15A on the mechanical properties of an abaca-based composite. *Mater. Test.*, **64**, 1, 125–131, 2022.
11. Mahendran, R., Amarnath, V., Rajkumar, P., Nirmal raj, L., Karthikeyan, S., Rajeskumar, L., Productivity improvement in a manufacturing industry by using man–machine chart analysis, in: *Technology Innovation in Mechanical Engineering*, 1ed., pp. 603–613, Springer Singapore, Singapore, 2022.
12. Sudha, S.P. and Govardhan, K., Nano technology in engineering. *Front. Nanotechnol.*, 32.
13. Wang, Y., Gao, M., Wang, D., Sun, L., Webster, T.J., Nanoscale 3D bioprinting for osseous tissue manufacturing. *Int. J. Nanomedicine*, 15, 215–226, 2020.
14. Armstrong, R.W. and Elban, W.L., Crystal strengths at micro-and nano-scale dimensions. *Crystals*, **10**, 2, 88, 2020.
15. Souza, J.C.M., Sordi, M.B., Kanazawa, M., Ravindran, S., Henriques, B., Silva, F.S., Aparicio, C., Cooper, L.F., Nano-scale modification of titanium implant surfaces to enhance osseointegration. *Acta Biomater.*, **94**, 112–131, 2019.
16. Kamau-Devers, K., Yanez, V.R., Miller, S.A., Using internal micro-scale architectures from additive manufacturing to increase material efficiency. *J. Cleaner Prod.*, **291**, 125799, 2021.
17. Danielak, A., Ribo, M.M., Ustunyagiz, E., Almdal, K., Islam, A., Pedersen, D.B., Characterization techniques for evaluation of micro scale vat photopolymerization-based additive manufacturing. *ASPE Euspens 19th Int. Conf. Exhib. Nantes*, pp. 1–4, 2019.
18. Chi, J., Self-organizing post process for two color powder bed 3D printing of preceramic polymer, 2022.
19. Yin, J., Liu, Y., Zhao, P., Editorial for the special issue on laser additive manufacturing: Design, processes, materials and applications. *Micromachines*, **13**, 12, 2057, 2022.
20. Balaji, D., Ranga, J., Bhuvaneswari, V., Arulmurugan, B., Rajeshkumar, L., Manimohan, M.P., Devi, G.R., Ramya, G., Masi, C., Additive manufacturing for aerospace from inception to certification. *J. Nanomater.*, **2022**, 1–18, 2022.
21. Reed, N. and Kim, D., Development of embeddable additive manufacturing microsensors for structural health monitoring. *Smart Mater. Adapt. Struct. Intell. Syst.*, vol. **86274**, p. V001T05A006, 2022.
22. Nydegger, M., Porenta, N., Menétrey, M., Hammadi, S., Reiser, A., Spolenak, R., Electrohydrodynamic redox 3D printing: Confined electroplating of

alloys for additive manufacturing at the submicron scale. *Electrochem. Soc. Meet. Abstr.*, vol. 241(22), p. 1119, 2022.

23. Guven, E., Karpat, Y., Cakmakci, M., Improving the dimensional accuracy of micro parts 3D printed with projection-based continuous vat photopolymerization using a model-based grayscale optimization method. *Addit. Manuf.*, **57**, 102954, 2022.
24. Zare, E.N., Makvandi, P., Ersoy, S., Micro and nano sensors from additive manufacturing. *J. Nanomater.*, **2022**, 1–2, 2022.
25. Zhang, X., Zhang, K., Zhang, L., Wang, W., Li, Y., He, R., Additive manufacturing of cellular ceramic structures: From structure to structure-function integration. *Mater. Des.*, 1, 110470, 2022.
26. Schwartz, J.J., Additive manufacturing: Frameworks for chemical understanding and advancement in vat photopolymerization. *MRS Bull.*, **47**, 6, 628–641, 2022.
27. Kumar, H.K.M., Subramaniam, S., Rathanasamy, R., Pal, S.K., Palaniappan, S.K., Influence of dual filler reinforcement on the curing and tribo-mechanical behaviour of natural rubber nanocomposite for tire tread application. *Arch. Metall. Mater.*, **66**, 3, 893–899, 2021.
28. Raines, R., Day, J.B., Salary, R., Experimental characterization of the mechanical properties of medical-grade dental implants, fabricated using vat-photopolymerization additive manufacturing process. *Int. Manuf. Sci. Eng. Conf.*, vol. **85802**, p. V001T07A011, 2022.
29. Bahati, D., Bricha, M., El Mabrouk, K., Vat photopolymerization additive manufacturing technology for bone tissue engineering applications. *Adv. Eng. Mater.*, 25, 1, 2200859, 2022.
30. Siwal, S.S., Mishra, K., Saini, A.K., Alsanie, W.F., Kovalcik, A., Thakur, V.K., Additive manufacturing of bio-based hydrogel composites: Recent advances. *J. Polym. Environ.*, 30, 11, 1–16, 2022.
31. Zare, P., Pezeshki-Modaress, M., Davachi, S.M., Chahsetareh, H., Simorgh, S., Asgari, N., Haramshahi, M.A., Alizadeh, R., Bagher, Z., Farhadi, M., An additive manufacturing-based 3D printed poly ε-caprolactone/alginate sulfate/extracellular matrix construct for nasal cartilage regeneration. *J. Biomed. Mater. Res. Part A*, **110**, 6, 1199–1209, 2022.
32. Menzel, V.C. and Tudela, I., Additive manufacturing of polyaniline-based materials: An opportunity for new designs and applications in energy and biotechnology. *Curr. Opin. Chem. Eng.*, **35**, 100742, 2022.
33. Rajeshkumar, L., Biodegradable polymer blends and composites from renewable resources, in: *Biodegradable Polymers, Blends and Composites*, S. Mavinkere Rangappa, J. Parameswaranpillai, S. Siengchin, Ramesh Blends and Composites, M.B.T.-B.P. (Eds.), pp. 527–549, Elsevier, United Kingdom, 2022.
34. Told, R., Ujfalusi, Z., Pentek, A., Kerenyi, M., Banfai, K., Vizi, A., Szabo, P., Melegh, S., Bovari-Biri, J., Pongracz, J.E., A state-of-the-art guide to the

sterilization of thermoplastic polymers and resin materials used in the additive manufacturing of medical devices. *Mater. Des.*, **223**, 111119, 2022.
35. Klüver, Ė., Baltzer, M., Langer, A., Meyer, M., Additive Manufacturing with Thermoplastic Collagen. *Polym. (Basel)*, **14**, 5, 974, 2022.
36. Parsons, E.M. and Shaik, S.Z., Additive manufacturing of aluminum metal matrix composites: Mechanical alloying of composite powders and single track consolidation with laser powder bed fusion. *Addit. Manuf.*, **50**, 102450, 2022.
37. Saravanakumar, A., Rajeshkumar, L., Sanjay kumar, N.S., Sivakumar, R., Sriramthilak, S., Vignesh, R., Multi response optimization of aluminium alloy machining using full factorial design of experiments, p. 020002, 2020.
38. Wu, D., Song, C., Di, T., Niu, F., Ma, G., Intermetallic regulation mechanism of inconel 718/Ti6Al4V composite by novel follow-up ultrasonic assisted laser additive manufacturing. *Compos. Part B Eng.*, **235**, 109736, 2022.
39. Ross, N.S., Ananth, M.B.J., Jafferson, J.M., Rajeshkumar, L., Kumar, M.S., Performance assessment of vegetable oil-based MQL in milling of additively manufactured AlSi10Mg for sustainable production. *Biomass Convers. Biorefin.*, 1, 1–17, 2022.
40. Zhang, A.P., Qu, X., Soman, P., Hribar, K.C., Lee, J.W., Chen, S., He, S., Rapid fabrication of complex 3D extracellular microenvironments by dynamic optical projection stereolithography. *Adv. Mater.*, 24, 31, 4266–70, 2012 Aug 16, doi: 10.1002/adma.201202024.
41. Mahmood, M.A., Chioibasu, D., Ur Rehman, A., Mihai, S., Popescu, A.C., Post-processing techniques to enhance the quality of metallic parts produced by additive manufacturing. *Metals (Basel)*, **12**, 1, 77, 2022.
42. Ma, Z., Zhai, Q., Wang, K., Chen, G., Yin, X., Zhang, Q., Meng, L., Wang, S., Wang, L., Fabrication of Fe-based metallic glass reinforced FeCoNiCrMn high-entropy alloy through additive manufacturing: Mechanical property enhancement and corrosion resistance improvement. *J. Mater. Res. Technol.*, **16**, 899–911, 2022.
43. Ramesh, M., Rajeshkumar, L., D.B., Aerogels for insulation applications, in: *Aerogels II: Preparataion, Properties and Applications*, pp. 57–76, 2021.
44. Balaji, D., Sivalingam, S., Bhuvaneswari, V., Amarnath, V., Adithya, J., Balavignesh, V., Ganesh Surya, R., Aerogels as alternatives for thermal insulation in buildings – A comparative teeny review. *Mater. Today Proc.*, **62**, 5371–5377, 2022.
45. Saccone, M.A., Gallivan, R.A., Narita, K., Yee, D.W., Greer, J.R., Additive manufacturing of micro-architected metals via hydrogel infusion. *Nature*, 625, 1–2, 2022.
46. Sathishkumar, T.P., Rajeshkumar, L., Rajeshkumar, G., Sanjay, M.R., Siengchin, S., Thakrishnan, N., Improving the mechanical properties of jute fiber woven mat reinforced epoxy composites with addition of zinc oxide filler. *E3S Web Conf.*, vol. **355**, p. 02006, 2022.

47. Pinho, A.C. and Piedade, A.P., Stimuli-responsive smart materials for additive manufacturing, in: *Nanotechnology-Based Addit. Manuf. Prod. Des. Prop. Appl.*, vol. **1**, pp. 249–276, 2023.
48. Sathishkumar, T.P., Satheeshkumar, S., Rajeshkumar, L., Effect of zinc oxide filler on compressive and impact properties of jute fiber fabric-reinforced epoxy composites, in: *Cellulose Fibre Reinforced Composites*, A. Ramnath, M.R. Sanjay, S. Siengchin, V.F. (Eds.), pp. 219–229, Elsevier, United Kingdom, 2023.
49. Devarajan, B., LakshmiNarasimhan, R., Venkateswaran, B., Mavinkere Rangappa, S., Siengchin, S., Additive manufacturing of jute fiber reinforced polymer composites: A concise review of material forms and methods. *Polym. Compos.*, **43**, 10, 6735–6748, 2022.
50. Li, H., Chen, C., Yi, R., Li, Y., Wu, J., Ultrasonic welding of fiber-reinforced thermoplastic composites: A review. *Int. J. Adv. Manuf. Technol.*, **120**, 1–2, 29–57, 2022.
51. Kumar, R., Kumar, R., Ranjan, N., Chohan, J.S., On development of alternating layer acrylonitrile butadiene styrene-al composite structures using additive manufacturing. *J. Mater. Eng. Perform.*, **31**, 11, 9349–9361, 2022.
52. Liu, Z., Li, Y., Liu, Z., Yang, Y., Li, Y., Luo, Z., Ultrasonic welding of metal to fiber-reinforced thermoplastic composites: A review. *J. Manuf. Processes*, **85**, 702–712, 2023.
53. Huang, Z., Shao, G., Li, L., Micro/nano functional devices fabricated by additive manufacturing. *Prog. Mater. Sci.*, 131, 101020, 2022.
54. Díaz-García, Á., Law, J.Y., Felix, M., Guerrero, A., Franco, V., Functional, thermal and rheological properties of polymer-based magnetic composite filaments for additive manufacturing. *Mater. Des.*, **219**, 110806, 2022.
55. Hou, Y., Gao, M., An, R., Tey, W.S., Li, B., Chen, J., Zhao, L., Zhou, K., Surface modification of oriented glass fibers for improving the mechanical properties and flame retardancy of polyamide 12 composites printed by powder bed fusion. *Addit. Manuf.*, 62, 10, 103195, 2022.
56. Priyadharshini, M., Balaji, D., Bhuvaneswari, V., Rajeshkumar, L., Sanjay, M.R., Siengchin, S., Fiber reinforced composite manufacturing with the aid of artificial intelligence – A state-of-the-art review. *Arch. Comput. Methods Eng.*, **29**, 7, 5511–5524, 2022.
57. Jianghong, L.I., Liang, F., Biru, C., Jiang, X., Liang, J., Kaibang, X., Fuhua, L.I., Xin, C., Weicheng, X.U., Research progress of nonmetallic doped graphite carbon nitride composites in photocatalytic activation of peroxymonosulfate system for organic pollutants removal. *Environ. Chem.*, **41**, 10, 3457–3468, 2022.
58. Schuschnigg, S., Hufnagl, M., Holzer, C., Eder, F., Kroke, E., Schallmeiner, S., Regula, C., Hamedi, Y., Gorsche, C., Morganti, E., Novel additive manufacturing materials for waste heat deduction. *Adv. Eng. Mater.*, 25, 7, 2200671, 2022.

59. Sun, J., Ye, D., Zou, J., Chen, X., Wang, Y., Yuan, J., Liang, H., Qu, H., Binner, J., Bai, J., A review on additive manufacturing of ceramic matrix composites. *J. Mater. Sci. Technol.*, 138, 1, 1–16, 2022.
60. Bhuvaneswari, V., Rajeshkumar, L., Saravanakumar, R., Balaji, D., Synthesis and characterization of bioceramics reinforced aluminium matrix composites. *Arch. Metall. Mater.*, **67**, 4, 1217–1226, 2022.
61. Sathishkumar, T.P., de-Prado-Gil, J., Martínez-García, R., Rajeshkumar, L., Rajeshkumar, G., Mavinkere Rangappa, S., Siengchin, S., Alosaimi, A.M., Hussein, M.A., Redeemable environmental damage by recycling of industrial discarded and virgin glass fiber mats in hybrid composites—An exploratory investigation. *Polym. Compos.*, 44, 1, 318–329, 2022.
62. Grigoriev, S.N., Khmyrov, R.S., Gridnev, M.A., Tarasova, T.V., Gusarov, A.V., Optimizing the process parameters for additive manufacturing of glass components by selective laser melting: Soda-lime glass versus quartz glass. *J. Manuf. Sci. Eng.*, **144**, 6, 064503–064508, 2022.
63. Bhuvaneswari, V., Rajeshkumar, L., Nimel Sworna Ross, K., Influence of bioceramic reinforcement on tribological behaviour of aluminium alloy metal matrix composites: Experimental study and analysis. *J. Mater. Res. Technol.*, **15**, 2802–2819, 2021.
64. R kumar, L. and A., K.S., Corrosion and wear behaviour of nano Al2O3 reinforced copper metal matrix composites synthesized by high energy ball milling. *Part. Sci. Technol.*, **38**, 2, 228–235, 2020.
65. Lang, M., Ruan, X.-L., He, C., Chen, Z.-Q., Xu, T., Zhang, H.-B., Cheng, Y.-T., Efficient fabrication of quartz glass using laser coaxial powder-fed additive manufacturing approach. *3D Print. Addit. Manuf.*, 2022.
66. Lang, M., Ruan, X., Chen, G., He, C., Hu, R., Cheng, Y., Numerical analysis of the temperature field during single-pass two-layers laser additive manufacturing of fused silica glass. *J. Phys. Conf. Ser.*, **2174**, 1, 12017, 2022.
67. Fröhlich, F., Hildebrand, J., Bergmann, J.P., Additive manufacturing with borosilicate glass and soda-lime glass. *2nd Int. Conf. Adv. Join. Process. (AJP 2021) Sel. Contrib. AJP*, vol. 2021, pp. 151–164, 2022.
68. de Moraes, E.G., Ferreira, I.M., Teixeira, L.B., Cartapati, L.H., Souza, M.T., de Oliveira, A.P.N., Additive manufacturing of cellular structures from recycled soda-lime glass printing inks by robocasting. *Ceram. Int.*, **49**, 4, 6554–6562, 2023.
69. Chen, H., Lu, T., Wang, Y., Liu, Y., Shi, T., Prashanth, K.G., Kosiba, K., Laser additive manufacturing of nano-TiC particles reinforced CoCrFeMnNi high-entropy alloy matrix composites with high strength and ductility. *Mater. Sci. Eng. A*, **833**, 142512, 2022.
70. Qiao, G., Zhang, B., Bai, Q., Gao, Y., Du, W., Zhang, Y., Machinability of TiC-reinforced titanium matrix composites fabricated by additive manufacturing. *J. Manuf. Processes*, **76**, 412–418, 2022.
71. Raj Mohan, R., Venkatraman, R., Raghuraman, S., Kumar, P.M., Rinawa, M.L., Subbiah, R., Arulmurugan, B., Rajkumar, S., Processing of aluminium-silicon

alloy with metal carbide as reinforcement through powder-based additive manufacturing: A critical study. *Scanning*, **2022**, 1–14, 2022.
72. Wang, Q.Z., Lin, X., Kang, N., Wen, X.L., Cao, Y., Lu, J.L., Peng, D.J., Bai, J., Zhou, Y.X., El Mansori, M., Effect of laser additive manufacturing on the microstructure and mechanical properties of TiB2 reinforced Al-Cu matrix composite. *Mater. Sci. Eng. A*, **840**, 142950, 2022.
73. Chen, W., Xu, L., Zhang, Y., Han, Y., Zhao, L., Jing, H., Additive manufacturing of high-performance 15-5PH stainless steel matrix composites. *Virtual Phys. Prototyping*, **17**, 2, 366–381, 2022.
74. Sivagnanamani, G.S., Begum, S.R., Siva, R., Kumar, M.S., Experimental investigation on influence of waste egg shell particles on polylactic acid matrix for additive manufacturing application. *J. Mater. Eng. Perform.*, **31**, 5, 3471–3480, 2022.
75. Karaş, B., Smith, P.J., Fairclough, J.P.A., Mumtaz, K., Additive manufacturing of high density carbon fibre reinforced polymer composites. *Addit. Manuf.*, **58**, 103044, 2022.
76. Deepa, C., Rajeshkumar, L., Ramesh, M., Preparation, synthesis, properties and characterization of graphene-based 2D nanomaterials for biosensors and bioelectronics. *J. Mater. Res. Technol.*, **19**, 2657–2694, 2022.
77. Ramesh, M., Rajeshkumar, L., Bhoopathi, R., Carbon substrates: A review on fabrication, properties and applications. *Carbon Lett.*, **31**, 4, 557–580, 2021.
78. Ganguly, A., Shankar, S., Das, A., Shukla, M., Swaroop, C., Bhardwaj, T., Natural fibre reinforced composites: A review based on additive manufacturing routes and biodegradability perspective. *Mater. Today Proc.*, 62, 1, 131–135, 2022.
79. Adeniran, O., Cong, W., Aremu, A., Material design factors in the additive manufacturing of carbon fiber reinforced plastic composites: A review. *Adv. Ind. Manuf. Eng.*, **5**, 100100, 2022.
80. Nguyen, H.D., Pramanik, A., Basak, A.K., Dong, Y., Prakash, C., Debnath, S., Shankar, S., Jawahir, I.S., Dixit, S., Buddhi, D., A critical review on additive manufacturing of Ti-6Al-4V alloy: Microstructure and mechanical properties. *J. Mater. Res. Technol.*, **18**, 4641, e4661, 2022.
81. Yu, L., Chen, K., Zhang, Y., Liu, J., Yang, L., Shi, Y., Microstructures and mechanical properties of NiTi shape memory alloys fabricated by wire arc additive manufacturing. *J. Alloys Compd.*, **892**, 162193, 2022.
82. Guo, X., Li, H., Pan, Z., Zhou, S., Microstructure and mechanical properties of ultra-high strength Al-Zn-Mg-Cu-Sc aluminum alloy fabricated by wire+arc additive manufacturing. *J. Manuf. Processes*, **79**, 576–586, 2022.
83. Sehhat, M.H., Mahdianikhotbesara, A., Yadegari, F., Impact of temperature and material variation on mechanical properties of parts fabricated with fused deposition modeling (FDM) additive manufacturing. *Int. J. Adv. Manuf. Technol.*, **120**, 7–8, 4791–4801, 2022.
84. Zheng, J., Chen, S., Jiang, L., Ye, X.-X., Xu, C., Li, Z., Effect of carbon content on the microstructure and mechanical properties of NiCrFe-7A alloys

synthesized by wire arc additive manufacturing. *Mater. Sci. Eng. A*, **842**, 142925, 2022.

85. Jiang, Q., Zhang, P., Tan, J., Yu, Z., Tian, Y., Ma, S., Wu, D., Influence of the microstructure on mechanical properties of SLM additive manufacturing Fe-based bulk metallic glasses. *J. Alloys Compd.*, **894**, 162525, 2022.
86. Jandaghi, M.R., Pouraliakbar, H., Shim, S.H., Fallah, V., Hong, S.I., Pavese, M., *In-situ* alloying of stainless steel 316L by co-inoculation of Ti and Mn using LPBF additive manufacturing: Microstructural evolution and mechanical properties. *Mater. Sci. Eng. A*, **857**, 144114, 2022.
87. Klein, T., Arnoldt, A., Lahnsteiner, R., Schnall, M., Microstructure and mechanical properties of a structurally refined Al–Mg–Si alloy for wire-arc additive manufacturing. *Mater. Sci. Eng. A*, **830**, 142318, 2022.
88. Shi, Y., Zhao, N., Chen, Y., Liu, J., Li, J., Hou, Z., Zhou, G., Spectroscopic analysis of Nd/Al co-doped silica glass fabricated by laser additive manufacturing. *J. Non. Cryst. Solids*, **604**, 122139, 2023.
89. Aktitiz, I., Aydın, K., Darıcık, F., Topcu, A., Production of different metal oxide nanoparticle embedded polymer matrix composite structures by the additive manufacturing technology and investigation of their properties. *Polym. Compos.*, **43**, 11, 7826–7835, 2022.
90. Xu, G., Nallappan, K., Cao, Y., Skorobogatiy, M., Infinity additive manufacturing of continuous microstructured fiber links for THz communications. *Sci. Rep.*, **12**, 1, 1–13, 2022.
91. Vidakis, N., Petousis, M., Mountakis, N., Kechagias, J.D., Optimization of friction stir welding parameters in hybrid additive manufacturing: Weldability of 3D-printed poly (methyl methacrylate) plates. *J. Manuf. Mater. Process.*, **6**, 4, 77, 2022.
92. Liu, F., He, H., Cheng, L., Li, H., Bian, H., Liu, S., Influence of layer thickness on microstructure and dielectric properties of Mg2TiO4 microwave ceramics fabricated by vat photopolymerization. *Addit. Manuf.*, **17**, 103413, 2023.
93. Goulas, A., McGhee, J.R., Whittaker, T., Ossai, D., Mistry, E., Whittow, W., Vaidhyanathan, B., Reaney, I.M., Vardaxoglou, J.Y.C., Engstrøm, D.S., Synthesis and dielectric characterisation of a low loss BaSrTiO3/ABS ceramic/polymer composite for fused filament fabrication additive manufacturing. *Addit. Manuf.*, **55**, 102844, 2022.
94. de Oliveira Neto, A.M., Justo, J.F., Beccaro, W., de Oliveira, A.M., Designing and building radio frequency devices with tailored dielectric properties using additive manufacturing. *Microw. Opt. Technol. Lett.*, **65**, 3, 777–784, 2023.
95. Spinelli, G., Kotsilkova, R., Ivanov, E., Georgiev, V., Naddeo, C., Romano, V., Thermal and dielectric properties of 3D printed parts based on polylactic acid filled with carbon nanostructures. *Macromol. Symp.*, **405**, 1, 2100244, 2022.

96. Banks, J.D. and Emami, A., Carbon-based piezoresistive polymer nanocomposites by extrusion additive manufacturing: Process, material design, and current progress. *3D Print. Addit. Manuf.*, 2022.
97. Ryan, K.R., Down, M.P., Hurst, N.J., Keefe, E.M., Banks, C.E., Additive manufacturing (3D printing) of electrically conductive polymers and polymer nanocomposites and their applications. *eScience*, **2**, 4, 365–381, 2022.
98. Singh, N., Edachery, V., Rajput, M., Chatterjee, K., Kailas, S.V., Prashanth, K.G., Ti6Al7Nb–TiB nanocomposites for ortho-implant applications. *J. Mater. Res.*, **37**, 16, 2525–2535, 2022.
99. Gowtham, S., Kumar, T.C., Devi, N., Chakravarthi, M.K., Pradeep Kumar, S., Karthik, R., Anandaram, H., Kumar, N.M., Ramaswamy, K., A survey on additively manufactured nanocomposite biomaterial for orthopaedic applications. *J. Nanomater.*, **2022**, 1–7, 2022.
100. Rouf, S., Malik, A., Raina, A., Haq, M.I.U., Naveed, N., Zolfagharian, A., Bodaghi, M., Functionally graded additive manufacturing for orthopedic applications. *J. Orthop.*, **33**, 70–80, 2022.
101. Groetsch, A., Stelzl, S., Nagel, Y., Kochetkova, T., Scherrer, N.C., Ovsianikov, A., Michler, J., Pethö, L., Siqueira, G., Nyström, G., Microscale 3D printing and tuning of cellulose nanocrystals reinforced polymer nanocomposites. *Small*, **19**, 3, 2202470, 2023.
102. Ghofrani, A., Taghavi, L., Vavdareh, B.K., Rohanishirvan, A., Nouri, A., Additive manufacturing and advanced functionalities of cardiac patches: A review. *Eur. Polym. J.*, **174**, 111332, 2022.
103. Vidakis, N., Petousis, M., Mangelis, P., Maravelakis, E., Mountakis, N., Papadakis, V., Neonaki, M., Thomadaki, G., Thermomechanical response of polycarbonate/aluminum nitride nanocomposites in material extrusion additive manufacturing. *Mater. (Basel)*, **15**, 24, 8806, 2022.
104. Kurapati, S.K., Mahendar Reddy, N., Sujithra, R., Kola, R., Ramesh, G.V., Saritha, D., Nanomaterials and nanostructures in additive manufacturing: Properties, applications, and technological challenges, in: *Nanotechnology-Based Addit. Manuf. Prod. Des. Prop. Appl.*, vol. **1**, pp. 53–102, 2023.
105. Muthukumar, C., *et al.*, (eds.), *Polymer based bio-nanocomposites: Properties, durability and applications*. Springer Nature, 2022.
106. Shanmugam, V. and Velu, R., 3D printing of nanocomposites, in: *Smart 3D Nanoprinting*, pp. 149–164, CRC Press, United States, 2023.
107. Ramesh, M., Rajeshkumar, L., Balaji, D., Bhuvaneswari, V., Green composite using agricultural waste reinforcement, in: *Green Composites for Automotive Applications*, T. Sabu, and B. Preetha, (Eds.), pp. 21–34, Springer Singapore, Singapore, 2021.
108. Sapuan, S.M., Norrrahim, M.N.F., Ilyas, R.A., *Industrial applications of nanocellulose and its nanocomposites*, Woodhead Publishing, United Kingdom, 2022.

109. Srikanth, G., Kailasa, S., Manikanta, D., Sadasivuni, K.K., Rao, K.V., Additive manufacturing of hydrogels, in: *Nanotechnology-Based Addit. Manuf. Prod. Des. Prop. Appl.*, vol. **1**, pp. 189–218, 2023.
110. Bashandeh, K., Amiri, A., Rafieerad, A., Rahman, S., Yan, W., Dhingra, S., Polycarpou, A.A., MXene-aromatic thermosetting copolyester nanocomposite as an extremely wear-resistant biocompatible implant material for osteoarthritis applications. *Appl. Surf. Sci.*, **600**, 154124, 2022.
111. Ramesh, M., Maniraj, J., Rajeshkumar, L., 11 - Thermomechanical characterization of cellulose fiber composites, in: *Woodhead Publishing Series in Composites Science and Engineering*, R. ArunRamnath, M.R. Sanjay, S. Siengchin, V.B.T. Fiore, C.F.R.C. (Eds.), pp. 183–201, Woodhead Publishing, United Kingdom, 2023.
112. Toong, D.W.Y., Ng, J.C.K., Cui, F., Leo, H.L., Zhong, L., Lian, S.S., Venkatraman, S., Tan, L.P., Huang, Y.Y., Ang, H.Y., Nanoparticles-reinforced poly-l-lactic acid composite materials as bioresorbable scaffold candidates for coronary stents: Insights from mechanical and finite element analysis. *J. Mech. Behav. Biomed. Mater.*, **125**, 104977, 2022.
113. Channa, I.A., Ashfaq, J., Gilani, S.J., Shah, A.A., Chandio, A.D., bin Jumah, M.N., UV blocking and oxygen barrier coatings based on polyvinyl alcohol and zinc oxide nanoparticles for packaging applications. *Coatings*, **12**, 7, 897, 2022.
114. Garimella, A., Ghosh, S.B., Bandyopadhyay-Ghosh, S., Bioactive fluorcanasite reinforced magnesium alloy based porous bio-nanocomposite bone scaffold with controlled degradation. *Mater. Technol.*, **37**, 13, 2795–2808, 2022.
115. Shah, J., Patel, V., Chavda, V., Patel, J., Sustainable nanobiocomposites, in: *Sustain. Nanotechnol. Strateg. Prod. Appl.*, pp. 217–236, 2022.
116. Sharma, N., Saxena, T., Alam, S.N., Ray, B.C., Biswas, K., Jha, S.K., Ceramic-based nanocomposites: A perspective from carbonaceous nanofillers. *Mater. Today Commun.*, **31**, 103764, 2022.
117. Siripongpreda, T., Hoven, V.P., Narupai, B., Rodthongku, N., Emerging 3D printing based on polymers and nanomaterial additives: Enhancement of properties and potential applications. *Eur. Polym. J.*, **184**, 111806, 2022.
118. Shah, M., Ullah, A., Azher, K., Rehman, A.U., Juan, W., Aktürk, N., Tüfekci, C.S., Salamci, M.U., Vat photopolymerization-based 3D printing of polymer nanocomposites: Current trends and applications. *RSC Adv.*, **13**, 2, 1456–1496, 2023.
119. Latif, S., Izhar, F., Imran, M., Hussain, N., Bilal, M., Polymer nanocomposites for dielectric and energy storage applications, in: *Smart Polym. Nanocomposites*, pp. 435–460, 2023.
120. Patel, A. and Taufik, M., Advances in nanocomposite material for fused filament fabrication. *Polym. Technol. Mater.*, **61**, 15, 1617–1661, 2022.
121. Tank, M., De Leon, A., Huang, W., Patadia, M., Degraff, J., Sweat, R., Manufacturing of stereolithographic 3D printed boron nitride

nanotube-reinforced ceramic composites with improved thermal and mechanical performance. *Funct. Compos. Struct.*, **5**, 1, 1–15, 2023.
122. Kallagunta, H. and Tate, J.S., Influence of surface functionalization of ceramic nanofibers on morphological, physical and mechanical properties of unsaturated polyester nanocomposites. *Compos. Adv. Mater.*, **31**, 26349833221087276, 2022.
123. Ahmadi, Y., Popalzai, N.R., Furmuly, M., Azizi, N., Fiber-reinforced nanocomposites, in: *Nanotechnology in the Automotive Industry*, pp. 215–227, Elsevier, United Kingdom, 2022.
124. Loginos, P., Patsidis, A., Vrettos, K., Sotiriadis, G., Psarras, G.C., Kostopoulos, V., Georgakilas, V., Thermomechanical properties of carbon nanocomposites PEGDA photopolymers. *Molecules*, **27**, 20, 6996, 2022.
125. Mahmood, T., Ullah, A., Ali, R., Improved nanocomposite materials and their applications, in: *Nanocomposite Materials*, IntechOpen, London, England, 2022.
126. Kaliyannan, G.V., Rathanasamy, R., Palaniappan, S.K., Mahesh Kumar, K.V., Chinnasamy, M., A review on the effect of carbon based nanofillers on the properties of elastomers. *Mater. Sci. Eng. Int. J.*, **3**, 3, 89–101, 2019.
127. Ali, M.A. and Gould, M.L., Innovativeness and sustainability of polymer nanocomposites, in: *Advanced Polymer Nanocomposites*, pp. 515–535, Elsevier, United Kingdom, 2022.
128. Chandran, G.U., Parappanal, A.S., Sambhudevan, S., Shankar, B., A critical review on cellulose nano structures based polymer nanocomposites for packaging applications. *Polym. Technol. Mater.*, **61**, 17, 1933–1958, 2022.
129. Meng, Q., Song, X., Han, S., Abbassi, F., Zhou, Z., Wu, B., Wang, X., Araby, S., Mechanical and functional properties of polyamide/graphene nanocomposite prepared by chemicals free-approach and selective laser sintering. *Compos. Commun.*, **36**, 101396, 2022.
130. Maturi, M., Spanu, C., Fernández-Delgado, N., Molina, S.I., Franchini, M.C., Locatelli, E., de León, A.S., Fatty acid–functionalized cellulose nanocomposites for vat photopolymerization. *Addit. Manuf.*, **61**, 103342, 2023.

9

Additive Manufacturing of Hydrogels: Process and Properties

V. Bhuvaneswari, Balaji Devarajan and L. Rajeshkumar*

Department of Mechanical Engineering, KPR Institute of Engineering and Technology, Coimbatore, Tamil Nadu, India

Abstract

Additive manufacturing (AM) or three-dimensional (3D) printing has emerged as an indispensable manufacturing technique for the direct manufacturing of parts having the most complicated shapes and intricacies. The control level offered by these AM technologies for the design and fabrication of tissues in tissue engineering and other biomedical applications may facilitate the understanding of governing factors in these areas. Hydrogels are well-known materials that can be manufactured using 3D or four-dimensional (4D) printing owing to their biocompatibility and low cost. Manufacturing of hydrogels through AM has become the recent trend owing to the potential application of hydrogels and their adaptability to any manufacturing method. Hydrogels possess commendable mechanical properties due to which they are used in biomedical and tissue engineering applications for the manufacturing of implants and other biocompatible devices. Soft robotics and organ-on-a-chip are some other promising areas where the utilization of hydrogels has penetrated. For such advanced applications, using AM methods can enhance the applicability of hydrogels. Parts made of hydrogels are tuneable with a wide range of diffusive and mechanical characteristics and they are controllable without a power supply even after their implantation. This chapter comprehensively deals with the various processes of the synthesis of hydrogels from various materials and the properties of hydrogels that make them suitable for various applications.

Keywords: 3D printing, robotics, hydrogels, tissue engineering, materials

**Corresponding author*: lrkln27@gmail.com

R. Rajasekar, C. Moganapriya and P. Sathish Kumar (eds.) Additive Manufacturing with Novel Materials: Processes, Properties and Applications, (295–318) © 2024 Scrivener Publishing LLC

9.1 Introduction

Prior to the advent of additive manufacturing (AM), also known as three-dimensional (3D) printing, significant progress was made in the knowledge and governed technology of hydrogels. In a dramatic fashion, hydrogels expand or shrink depending on the amount of water they contain. They can be designed to react to specific stimuli [1–3], for example, heat, cold, light, acidity, alkalinity, mechanical or electrical fields. Hydrogels, due to their observable behavior, can be programmed to respond to these types of ecological triggers [4, 5]. Hydrogels engineered as actuators may also respond to external stimulus factors as control signals. Before the advent of AM, many potential practical applications had already been proposed. There were early examples that could be used as synthetic muscular or microchannels stream-regulated machines [6, 7]. Hydrogels can perform a synthetically energetic feature in soluble remedy or by being functionalized at the polymer level. Hydrogels containing glucose delivered via permeation have been used in conjunction with amylase to produce a carbohydrate device [8]. The term "additive manufacturing" describes a method whereby components are built up from a digital model in successive layers. Hydrogels can be used in AM to create intricate 3D configurations, and the accurate indicator of the final product can be adjusted locally. When producing hydrogels additively, liquid precursors are typically crosslinked in successive layers. As a result, it is simple to work with a wide variety of materials at once during production. Stereolithography and digital light processing (DLP) are examples of the AM process categories used to print hydrogel materials [9]. Given the promising future of hydrogels and the exceptional capabilities of AM, hydrogel AM has emerged as an early and cutting-edge application of the technology. In an overview of tissue engineering in 1993, hydrogels were only briefly mentioned twice: once for their use in corneal tissue creation and again as microcapsules for chamber shift [10]. Tissue engineering using hydrogel structures produced using AM [11] was demonstrated by 1998. As a result, syringe-based 3D robotic distributing systems were developed and put to use in the production of hydrogel scaffolds for use in bioengineering [12–14]. Hydrogel scaffolds with living cells were generated using photolithography at the identical period [15]. In the last twenty years, the field of AM of hydrogels has made significant strides. Numerous insightful analyses of this development have been published, including works on the following subjects:

Detailed analyses of relevant materials, methods, and medical implementations [16–19]; biofabrication of synthetic tissues and tissue technology, with a special focus on the implementations and techniques of tissue engineering [20–22]; complicated craft constructions, shape-memory materials, and design field of robotics have all been recently explored [23–26]; solution of a biomaterial with nanobiomaterials for cell transplantation; hydrogels with synthesized polymeric materials; hydrogels with aggressive additives and embedded cells; hydrogel materials [27–30]; biodegradable electronic sector [31]; AM using various resources, with potential in bioengineering and design field of robotics [32].

Recently, intriguing applications for hydrogels have been developed, with help from AM techniques. Even with these improvements, AM is still not a good fit for many possible uses. The author suggests eligibility requirements for the use of instances that are a better match for 3D printers, drawing on experience from other application areas of AM. Design field of robotics and organ-on-a-chip engineering are two prominent fields where hydrogels may soon be incorporated into operational creations. The use of AM processes has facilitated many of these developments. The rate of progress in this area of study is accelerating. Hydrogel materials can be designed to break the traditional connections among structural rigidity and variables like stress behavior and healing responsiveness, resulting in significant improvements in dynamic performance. Significant progress has been made in the performance and dependability of hydrogel-based actuators, as well as in the combination manufacturing of hydrogel-polymer laminated composite materials.

Digital light processing (DLP) vat photopolymerization is an AM process that can be used to share and form electron pairs of highly stretchable hydrogel structures with ultraviolet (UV)-cured polymers [33]. This method is appealing because it allows designers to autonomously achieve the special characteristics of hydrogels for detecting, narcotic solubilization, or mechanical reaction to the surroundings, and to regulate 3D stiffness contours with great precision. When compared to previous research that produced similar hydrogel-polymer systems via material extrusion [34], this method is a significant improvement in terms of speed and resolution. Design field of robotics is another promising area for these covalently bonded hydrogel-polymer structures because restricted hydrogel bending can be constructed to contract when the hydrogel components grow [35]. Recent advances in hydrogel mechanical characteristics will undoubtedly be utilized by the AM sector. Stronger, more durable dual-hydrogel networks can be used to design self-healing hydrogel structures or ones that grow stronger in response to recurring pressure [36]. Incorporating zinc-peptide-based steel cooperation structures into the hydrogel channel has recently been shown to produce strong, tough, and quickly recovering hydrogels [37]. By incorporating multimaterial and slope fabrication, AM can make full use of the enhanced hydrogels' broader range of physical characteristics.

9.2 Application of 3D-Printed Hydrogels

9.2.1 Hydrogel Printed Sensors, Actuators, and Design Aspect of Robotics

For AM, scientists designed a hydrogel robot that can move autonomously using only light power and is propelled by oscillations; they also created a performance model that allows them to fine-tune the operating occurrence and automotive force for a range of robot dimensions [38]. Hydrogels containing nanomagnetic particles embedded in them are magnetically affected, supplying a second non-invasive type of controlling design field of robots [39]. CRISPR's Cas enzyme can be employed to purposefully separate hydrogels embedded with DNA sequences for use in cell and release of drug, detection, and motion [40]. Hydrogel-based robotics have been modified using synthetic biology techniques, with engineered cells converting physiological detection to fluorescence and then using the resulting optical signal to provide the control loop of a hydrogel oscillator [41]. By encasing steel cylinders in hydrogel, the conventional manufacturing method produced two-dimensional (2D) gradients in amplitude, which proved beneficial for ultrasonic detection [42]. Enhanced acoustic sensor performance is predicted as a result of AM's increased command through 3D acoustic impedance contours.

9.2.2 New Advances in the 3D Printing of Hydrogels for Use in Tissue Engineering

The difficulty of providing blood vessels to tissue developed on composites is a significant obstacle. In a recent study [43], researchers demonstrated that hydrogels with interconnected microchannels could be manufactured using DLP-based vat photopolymerization AM for use as synthetic skin grafts. Scientists are working to develop synthetic cells through engineering [44]. Additive manufacturing of development of "organs-on-a-chip" technology constructs will eventually expand to incorporate synthetic cells.

9.3 Hydrogel Materials for Additive Manufacturing

Additive manufacturing is an excellent aid for quick iterative processes in design and research due to its capacity to quickly synthesize intricate shapes. The AM of hydrogel components with gradients in composition is the most compelling application. Parts crafted by means of AM can often be manufactured through the development of manufacturing methods using standard mechanisms [45]. This is so despite the fact that AM is suitable for (internal) geometrically complex parts, can fabricate multiple materials, and generate versatile parts (for instance, with integrated semiconductors) [46]. After a design has been validated using quick redesign tools, it is typically transmitted to production using conventional processes to maximize efficiency. This is also true for uses involving hydrogels. Hydrogels have historically been made through either sheet lamination or casting [47, 48]. Cast hydrogels around complex structures to improve their features [49]. Multimaterial AM may be largely replaced by improvements in multi-step traditional methods, such as casting hydrogel components of mixtures in numerous investment casting methods or casting straight in an injection-molded plastic part [48].

With an aid of lab-on-a-chip modeling utilizing conventional polymeric resources, injection molding with laminated films is a cost-effective manufacturing technique [50]. Hydrogels cast in place add a similar level of complexity to these multimaterial designs. With automated fluid distribution and healing equipment, the boundaries become more porous. The robotic dispensing of hydrogel is more similar to the automatic vehicle injection of glues or valves in traditional manufacturing, as it does not involve building layers upon layers. Hydrogels are likely further along in their use in skin area and organ-on-a-chip frameworks than they are in soft robotics. These biomedical uses are excellent examples of where the most value can be added by AM beyond rapid prototyping. Inserting hydrogels with intricate intrinsic shapes into organ-on-a-chip gadgets is typically accomplished through a process called sacrificial molding [44, 48]. Hydrogels with lumens of 250 m in diameter may be micromolded to overcome the difficulty of creating long lumens with a small internal diameter for vascular transport [51]. Therefore, "vascularized" hydrogel tissue engineering constructs may be made using two-stage casting procedures. These techniques eliminate the need for AM.

Therefore, in most cases, the conventional manufacturing methods of hydrogels, which involve a multi-step fabrication process, offer a considerable low price over AM. It is difficult to debate for mass customizability of most hydrogel-centric layout implementations due to the hydrogels' malleable nature. For modeling that depends on the slope structure production to accomplish strongly localized behavior, AM will continue to have a long-term advantage over traditional processes, even if there are no utilization instances that recommend a need for "manufacture on demand" of hydrogel devices. This includes hydrogel designs for localized contraction, detection, and drug carriers [38–41], such as those already mentioned. It is reasonable to assume that the working model of the investigation test table and design applications will continue to be the main position for AM of hydrogels until implications are enhanced that make use of the localized effectiveness attained across contour structure production.

9.3.1 Conductive Hydrogels

Increased options for electromagnetic interference (EMI) shielding may arise from additively manufactured conductive polymer hydrogels with desirable conductivity and biocompatibility. However, it is still difficult to print conductive polymer hydrogels with elevated permeability, layout liberty, and shape adherence all at the same time. Here, we have a poly(3,4-ethylenedioxythiophene) functionalized with Ti3C2-MXene that is soluble in water. We propose a freeze-thaw protocol to directly transform the printed objects into strongly semiconducting and sturdy hydrogels with high shape fidelity on the macro- and microscales, using polystyrene sulfonate ink for extrusion printing to create –3D objects of arbitrary geometries. The as-obtained hydrogel has excellent mechanical properties, including versatility, durability, and fatigue life, and a high electrical conductivity of 1525.8 S m^{-1} at moisture content up to 96.6 wt%. In addition, we show how the printed hydrogel can be used in a variety of EMI-shielding layouts [52].

9.3.2 Shape Memory Hydrogels

There is potential for fast, volume-based AM in solids. Outer space missions (microgravity) and/or extreme conditions are ideal settings for this AM technology. In this case, UV crosslinking in a solid state is shown by using a specialized thermal gel. Hydrogels are produced, and their properties, including their responsiveness to heat and cold and their shape memory properties, are elucidated. In this final stage of AM, the shape-memory property is necessary to prevent the distortion caused by separating uncrosslinked and crosslinked components [53].

9.3.3 Chitosan-Based Hydrogels

Hydrogel's properties (geometry and hierarchical roughness) impact particle adsorptive and emission profiles in cultured cells. After pouring chitosan (CS) solutions into 3D-printed (3DP) molds depicting complex shapes, researchers submerged them in alkaline solutions of sodium hydroxide and sodium chloride. Hydrogels were successfully shaped and molded using 3DP molds. Adsorption and release profiles were studied using methyl orange (MO) along with paracetamol (PMOL). It was discovered that the presence of centralized roughness in the sample, in addition to the sample's topography, yielded a wide range of MO sorption/PMOL discharge attributes. The ability of CS samples to adsorb MO was found to depend on both the surface hierarchy and the geometry of the samples themselves. Therefore, cube specimens without a hierarchical surface had the highest dye extraction efficiency. Hydrogel surface area was also more relevant to PMOL release measurements, with semispherical samples exhibiting hierarchical roughness demonstrating the fastest (1.13 times) drug delivery profiles. The findings of this study suggest that indirect 3D printing could be an easy way to create hydrogels with novel adsorptive properties. Indirect 3D printing using fused filament fabrication (FFF) technology is a user-friendly and inexpensive method for creating CS hydrogels with a variety of geometries and hierarchical surfaces. Chitosan hydrogels' surface area and the existence of centralized roughness can be adjusted with this straightforward method. The release of drug features of nonionic atoms in bi-distilled water was most favorable for hydrogels with increased surface areas produced via indirect 3D printing. Because of local surface charge gradients, the centralized rough surface of polycationic hydrogels impeded their ability to remove anionic dye. These adjustments were stimulated by the removal of impurities, which is especially important for 3DP molds featuring complex geometries and hierarchical roughness [54].

9.3.4 PEGD-Based Hydrogels

The stereolithography apparatus (SLA) uses a laser light of precisely 405 nm in wavelength for its operations. To produce the 3D objects, we go with the default settings for the SLA when working with transparent liquid resins, which means we use a laser power at 55 mW, and the layer is 0.50 mm thick. Size reduction of the SLA is required before the design specimens can be made. When this happens, from a volume of roughly 90 ml, the chamber rapidly reduces to a mere 10 ml. As a result, we can conserve a great deal of useful substance without compromising the quality of the necessary samples. After the connector is changed, the platform can be securely erected. The chamber is also swapped out to drastically cut down on the amount of filling required. Additionally, a custom unit that was 3D printed replaces and repairs the chamber. When the chamber needs to be secured, a suspense clamp is used. The object is secured by means of a tension lock. Sample sizes can be kept low while searching for the optimal formulation. It has been decided that between 3 and 5 ml of liquid resin should be used to examine the specimens. It follows that the SLA must be adjusted initially for such inquiries. All 3D models are created independently to prevent manipulation. Since the laser beam's focus must be maintained throughout the manufacturing process, this rules out simultaneous production of multiple 3D objects.

Postprocessing involves rinsing the specimens in 2-propanol (99.5%, Ph.Eur. purest) for 9–15 sec to remove any traces of polymer resin from the 3D object. Next, they're placed in an UV chamber with a 405-nm wavelength for a post-curing time of 15–20 min. A curing chamber at 0°C is available. Sigma Aldrich, Inc., in the United States sources poly(ethylene glycol) diacrylate (PEGD) in three different formulations, with a mean molecular weight range of 700 Mn, 575 Mn, and 250 Mn, respectively. Omnirad 2022 (photoinitiator) from IGM Resins B.V., and 2,2'-dihydroxy-4,4'-dimethoxybenzophenone (UV radiation absorber) from TCI Deutschland GmbH are combined to form the final product. First, 5-ml samples are mixed at 40°C for 12 h with 1000 rpm to find the optimal mixing ratio for the SLA. After that, the formulations sit for an hour at 23°C. Three 270-ml formulations are performed after the resin has been successfully adjusted on the SLA. 168 h of stirring at 40°C and 900 rpm is necessary because of the increased mass. After being made, the formulas have a 5-h break [55].

9.3.5 Ion-Based Hydrogels

To print composites, sensors, and actuators via direct-ink-write (DIW), researchers are creating stimuli-responsive hydrogels, ion gels, and ion elastomers. This is the first work to demonstrate 3D printing of soft sensors with improved ductility due to the substance's mechanical characteristics and the usable energy absorption arrangement, and the first to do so using direct ink writing (DIW) printing of ion gel without the requirement for treatment among layer upon layer or the use of a solvent. It investigates the use of polymerizable metal ions in inkjet printing of shape memory atom elastomers capable of exhibiting multidirectional flexing on such push and discharge of the printed parts. Photocuring of the uncured hydrogel/particle composite materials after printing with triplet merging up transformation has been done to eliminate the restrictions of classical UV curing [56].

9.3.6 Magnetic Hydrogel

Next-generation biomimetic soft robotics have been fueled by hydrogel with magnetic property and composites for soft material applications due to their precise control and non-cytotoxic nature. While remote regulation of free magnetic nanoparticles is challenging, the system becomes a soft mussel-like artificial organism when the nanoparticles are arrested within polymeric matrices. Similarly, the existence of an electromagnet makes these polymeric magnetic soft materials sensitive (static or oscillatory). Additive manufacturing, which entails the geographic gathering of polymeric catalysts accompanied by actuation-like actions, is a comparatively new method of producing magnetic soft materials. We evaluated the clinical (hyperthermia) and morphing (shape) applications of magnetic nanoparticles that have been entrapped in polymeric matrices. Dual responsive soft microbots are the result of combining both behaviors. Methods and examples of their use are described in greater detail in Ganguly and Mangel [57].

9.3.7 Weldable Hydrogels

A major challenge to 3D-printed biomaterials with better resolution, diversified topology, and controllable mechanical performance are the decreased photo-reactivity of the enhanced containing catalyst and the decreased flexural strength of the photo-cured hydrogel. In this study, a dry gel of acryloyl morpholine, acrylic acid, and polyethylene glycol diacrylate was used to print at high resolution; the gel was then post-treated with a –water-soluble solution comprising particular atoms, which fully saturated the hydrophobicity connectivity and organized the carboxyl and atoms. The subsequent dual-crosslinked hydrogel was remarkably elastic and possessed a high degree of morphological ambiguity. This method separated the hydrogels' printability from their mechanical properties. This advantage was further demonstrated in comparison to conventional direct hydrogel 3D printing by demonstrating self-supported lattice structures made of ultrasoft hydrogels and ionic covalently bonded hydrogels with excellent self-weldability [58].

9.3.8 Hydrogels With Integrated Cameras

Repairing bone damage after surgical procedures such as amputation, bone fracture, or joint replacement remains a major clinical challenge. Although there have been significant advances in the fields of AM and bone tissue engineering, there is still a dearth of fully operational bone augmentations with the vascularization capacity, which is important to ensure bone rejuvenation and long-term physiological merging of cracks. In this study, an in-cultured cells vascularized bone implant was developed by fusing cell-laden hydrogels with porous titanium alloys using direct metal printing (DMP) (Ti–6Al–4V). To investigate the simultaneous osteogenic and vasculogenic performance of human mesenchymal stromal cells (hMSCs) and human umbilical vein endothelial cells (HUVECs), we co-encapsulated these cells with a 5 wt% allylatedgelatin (GelAGE). The potential of hybrid implants for cell delivery and the engineering of vascular-like networks were investigated by infusing/enriching direct metal-printed macroporous Ti-6Al-4V scaffolds with cell-laden GelAGE. The clinical feasibility of this strategy was further demonstrated by impregnating a large porous Ti-6Al-4V acetabular cup with cell-laden hydrogel.

The ability of microcapillary configuration and maturation/stabilization of the vascular system to transport blood was used to evaluate vasculogenic potential. Alkaline phosphatase activity, along with the expression of osteocalcin and osteopontin, were used to evaluate osteogenic differentiation. Based on our findings, GelAGE promoted both HUVECs' ability to migrate and form a vascular-like network, as well as hMSCs' ability to differentiate into osteoblasts. Titanium-hybrid hydrogel-loaded titanium constructs showed enhanced osteogenesis and comparable vasculogenic capacity compared to hydrogel-loaded titanium constructs. Full-scale implants with cell-laden hydrogel coatings showed related cell dispersion and spreading, indicating the technology may have clinical applications in the future. Our study shows that it is possible to combine biofunctional hydrogels with porous titanium implants to generate a vascularized hybrid construct with support structure and preferred physiological abilities (osteogenesis/vasculogenesis), which opens the door to more effective methods for promoting bone healing in complex, large bone defects and achieving long-term bone-implant fixation [59].

9.3.9 pNIPAM Hydrogels

This is possible with 4D printing because it allows for the creation of structures that can react to their environment. Poly(N-isopropylacrylamide) (pNIPAM) stands out among the many materials that can be used for both 4D printing and biological applications. The low critical solution temperature (LCST) of pNIPAM is similar to the human physiological temperature, and it is biocompatible with many different types of cells. The processing parameters, however, have a substantial impact on these features. To investigate the effects of changing the printing temperature from 5°C to 10°C to 15°C on the final properties of pNIPAM hydrogels, this study employed AM with DLP to create the hydrogels. Hydrogels were examined using compression tests, SEM images, scanning electron microscopy (SEM), and differential scanning calorimetry (DSC) to determine the effect of the stimulus. According to our findings, a rise of 10°C in the hydrogels' production temperature resulted in a variation of 6.5°C in the LCST, a 10% rise in water retention, and a 50% drop in maximum swelling capacity. Because of an increase in crosslinking rate and a thickening of the hydrogel's outer layer, water is unable to freely move within the material, which explains the observed behavior [60].

9.3.10 Polymer-Based Hydrogels

The rheological properties of polymer melts and polymer inks, the two different kinds of substances in use for extrusion 3D printing, can vary greatly from one another. For this type of processing, they need to be shear-thinning and quickly gain back their higher solubility upon deformation. Most polymer melts meet both criteria because of their viscoelastic nature. Most of the time, the higher the shear rate, the lower their viscosity becomes. They quickly return to a high viscosity state after printing, and then continue to harden as they cool and crystallize, preserving the original voxel's dimensions. However, 3D printing inks are typically made up of a variety of precursors, such as solvents, crosslinkers that serve a dual or even multiple purposes. This results in solutions with low viscosity that would deliquesce after printing and that typically lack shear-thinning properties. The addition of rheology modifiers is necessary because of this. These are polymers or particles that interact with the ink's other molecules in a way that can be reversed (e.g., van der Waals or hydrogen bonds, dipolar or ionic interactions). When shear forces are applied, numerous transient intermolecular interrelationships are temporarily disrupted, resulting in a reduction of local viscosity by directives of amplitude [61]. Figure 9.1 shows the 3D-printed hydrogel used for wearable electronics applications. Figure 9.1 A1 and A2 show the hydrogels in cylindrical shape, B1 and B2 in sheet structure, C shows the load bearing capacity of hydrogel, D1-3 shows the hydrogel employed in circuit connection.

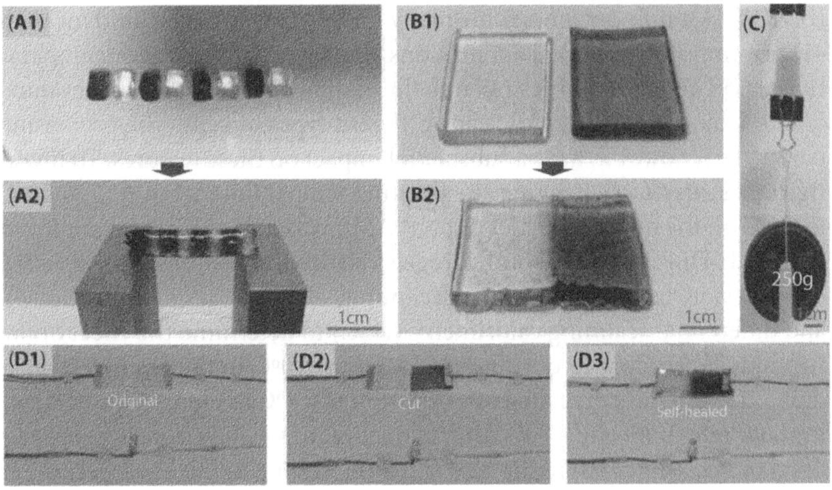

Figure 9.1 (A1, A2) Cylindrical shaped hydrogel, (B1, B2) sheet structured hydrogel, (C) hydrogel carrying 250 g load, (D1-D3) hydrogel employed in circuits [61].

Hydrophilic nanoclays (such as laponite) [63], polymeric microparticles (such as those derived from poly(acrylic acid)] 64], and triblock copolymers that usually contains hydrogen bond functionalization (such as Pluronic F127) [62], are all examples of rheology modifiers. These additives help printed objects keep their shape during post-treatment (like UV crosslinking) and allow for the production of covalently connected polymer hydrogels with excellent shape retention. The shear-thinning behavior and the shape fidelity of the ink are two crucial rheological properties that must be taken into account when optimizing the rheology modifier amount. Hydrogel inks with shear thinning properties are less viscous under higher shear stress, as was previously mentioned. When the shear stress is greater than the material's yield stress, the ink will flow viscoelastically through the nozzle (G"> G'). The rheology modifier's intermolecular interactions with the other ink components are disrupted in this environment.

When a rheology modifier is present in too high of a concentration, it forms strongly bound clusters, which cause the material to flow in a discontinuous fashion and are the primary cause of nozzle clogging. In order for hydrogels to undergo a transition from elastic (G'> G") to viscous (G"> G') behavior during printing, the crosslinks must break under the applied shear stress. By quickly returning to a high viscosity state after the shear stress has been removed, a high degree of shape fidelity can be achieved. This is contingent upon the specifics of the rheology modifier's intermolecular interactions to both these ink components [65, 66]. The optimum value of rheology modifier, like the optimal values for a whole viscosity and tensile stress thinning behavior, is experimentally determined and is additive-specific. According to the literature, the lowest effective concentrations of frequently used rheological properties modifiers are 18% w/v of Pluronic F-127, 5% w/v of nanoclay, and 0.5% w/v of carbomer microgels [64]. Alterations to rheological properties improve the 3D print quality of hydrogels, but also change the characteristics of the final product, so using the right amount is essential (as much as you need, as little as you can get away with). An excessive amount of charged laponite nanoparticles, for instance, can have a negative impact on the physical characteristics of hydrogels [67].

Polymer hydrogels are predicated on a vibrant network structure rather than a covalent crosslinking of the polymer chains. Reversible covalent bonds hold these networks together, and their chemically stable scaffolds can be printed directly from inks. In order to print via extrusion, shear forces are applied to the networks, breaking the reactive covalent bonds, and causing the network to change from an elastic to a viscous state. This is because the formation of dynamic covalent bonds is reversible, so the dynamic network structure can be reformed after printing has been completed. In order for the printed voxels to maintain their shape, the rate at which the vibrant bonds reform is critical in such systems [68]. In theory, any reversible and powerful enough interaction can produce such dynamic networks. In the perspective of 3D printing, dynamic covalent bonds involving heteroatoms are of particular interest. Carbon-heteroatom bonds, in contrast to the more static carbon-carbon bonds, go through continuous bonding-debonding transitions at a fast enough rate for instant bond restoration as once shear encountered all through printing are excluded [69].The derivatives (including Schiff's base), hydrazones, and Diels-Alder reaction adducts are all examples of such reactive functional groups [70]. By adding hydrazide and aldehyde residues to the raw materials, Wang *et al.* created shear-thinning 3D printable inks for hyaluronic acid hydrogels. Physical intermolecular interactions, such as hydrogen bonds [71], hydrophobic interactions [72], ionic interactions [73], and supramolecular host-guest interactions, can also be used to implement dynamic crosslinking. Because of the relative weakness of these interactions compared to covalent bonds, they are able to be broken temporarily under shear forces and undergo shear-thinning during the printing process. Complexation between polymers with opposite charges is one example of a physical interaction that can enhance printing resolution. In the presence of cationic CS, hydrogels of negative charges, hyaluronan methacrylate, showed this phenomenon [74].

The interaction strength between neighboring voxels that were not printed along the same filament strand is crucial to the stability of the target object, especially when subjected to sustained mechanical loads, given that 3D-printed objects are fabricated voxel by voxel. The printed layers will separate if the interaction between neighboring voxels is insufficient, whether it be adhesive or chemical. Due to the internal stress encountered by hydrogel actuators during swelling and actuation, this is especially pertinent in the sense of 3D printing. Hydrogel bilayers, for instance, can only be made to bend if their two layers are deeply linked at the interface, as this allows for the use of materials with various swelling ratios [75, 76]. Hydrogel actuators are capable of rapid and repeatable actuation due to the strong bonding at their interfaces [77].

For 3D-printed hydrogels, delamination between voxels or layers can be prevented using the same techniques that are used to connect and/or crosslink the material in its bulk form. Some of these include UV crosslinkers, nanofillers for reinforcement, polymers with opposite charges, and the formation of interconnected polymer networks (IPNs). Ultraviolet crosslinking for stabilization is mentioned earlier in this paragraph. Nanofillers are particles that have a high concentration of functional groups on their surface. Using these, you can create chemical or physical bonds between voxels of a hydrogel. Hydrogel bilayers containing laponite nanoclay, for instance, were found to be resistant to delamination during reversible actuation and bending stresses. Delamination was observed in the absence of nanoclay in these types of structures. The PNIPAm-PNAGA hydrogel bilayer, which lacked laponite nanoclays, was shown to delaminate easily after two bending cycles [78]. Tight interfacial adhesion can also be achieved through electrostatic interactions between polyelectrolyte building blocks with opposite charges. Hydrogels with opposite charges can adhere to one another through molecular-level polyelectrolyte complexes. Charged polysaccharides are a common component of natural actuators [76], and these devices use this strategy to ensure compatibility between different components at interfaces.

Different from the aforementioned methods is the formation of IPNs. Different hydrogel building blocks are not bonded to one another at their interfaces, rather, a second network is formed around them. The printed parts (which make up the first network, say a dynamic covalent one) are swollen with the designated monomer for the second network, and then crosslinked. Due to the high swelling ability of hydrogels in monomers with matching hydrophilicity, this is easily accomplished. When more polymer chains are added, the first polymer network is penetrated to create interpenetrating networks. As a result, the high mechanical strength typically observed in IPNs is due to the two networks becoming irreversibly entangled at the molecular level [79]. More targeted intermolecular interactions (like hydrogen bonds), for example, could potentially increase the effect [80].

9.3.11 Smart Hydrogel

Hydrogels have been studied from a variety of angles over the past few decades with the goal of better characterizing and understanding their properties for application as biomaterials. Smart hydrogels have garnered a lot of interest as of late because of how crucial they are in a variety of contexts due to their responsiveness. The sol-gel transformation, inflammation behavior, mechanical characteristics, network structure, etc., of smart hydrogels can, for example, undergo dramatic changes in response to changes in environmental conditions such as heat, pH, conductivity, and electromagnetic waves [81]. Hydrogels that are pH- and temperature-sensitive (thermoresponsive) are widely used for biological applications and as biomaterials despite their stability under *in vivo* and *in vitro* circumstances, and due to their physiological significance inside the human body. Controlled drug delivery to specific organ and tissue biomedical applications, biosensors, tissue regeneration, and tissue engineering are just some of the many areas where smart hydrogels have been put to use [82, 83]. Among the many types of smart hydrogels, thermoresponsive hydrogels have proven to be the most effective for cornea tissue engineering. For tissue engineering purposes, the ideal thermoresponsive hydrogel is one that is liquid at room temperature but solidifies when brought to 37°C (the human body's average temperature). Because of this property, this category of hydrogels is the most important class of intelligent hydrogels for use in biomedicine. Hydrogels that react to temperature changes can be used for a wide variety of applications, including injectable gels, drug delivery, and controlling cell adhesion and detachment. Although scientists have found thermoresponsive hydrogels to be helpful for biological applications, they have worked to address two major concerns: the hydrogels' mechanical characteristics and their toxicity [84, 85].

The unique properties of the cornea have led to the development of thermoresponsive hydrogels for use in tissue engineering. Examples of such serious forms on the surface of the eye include corneal neovascularization. By 2020, Xu *et al.* had developed a poloxamer-based thermoresponsive hydrogel for sustained release of bone morphogenetic protein 4 (BMP4). To further enhance its temperature-sensitive qualities, Polylysine (EPL) was incorporated. The study aimed to determine the optimal

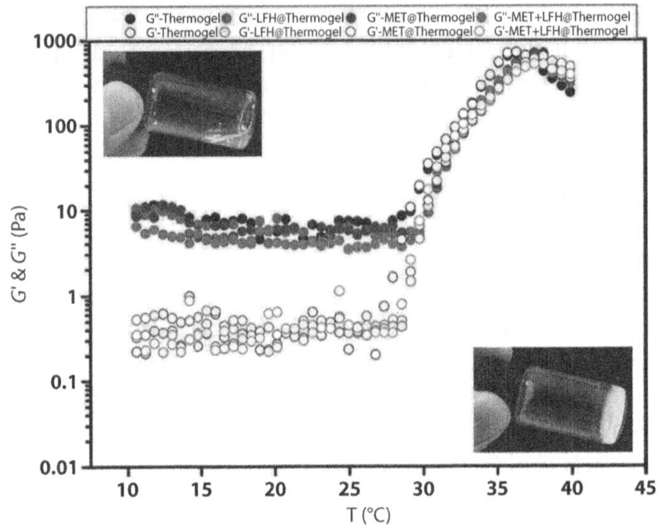

Figure 9.2 Storage and loss modulus curve for MET- and LFH-based smart hydrogel [86].

hydrogel phase transition temperature to prevent corneal neovascularization following ocular surgery. Therefore, the thermoresponsive hydrogel transforms from its liquid form to its gelled state very quickly at normal physiological temperature. To get rid of corneal neovascularization, another study utilized a thermoresponsive hydrogel composed of a triblock copolymer of poly(D, L-lactic-co-glycolic acid)-block-poly(ethylene glycol)-block-poly(D, L-lactic-co-glycolic acid) (PLGA-PEG-PLGA) [86]. Metformin (MET) and levofloxacin hydrochloride (LFH) were chosen as drugs by Liu *et al.* because they, like the polymers, were soluble in water at room temperature. This type of hydrogel was found to be beneficial for anti-angiogenic therapy and to release drugs sustainably for 30 d. Figure 9.2 shows the storage and loss modulus of MET- and LFH-based smart hydrogel used for drug delivery applications.

9.4 Conclusion

This chapter comprehensively discussed the synthesis of hydrogels from various materials and their properties, and which suits them better for various technological applications including the biomedical field. Smart and responsive hydrogels were recently developed for application in the field of tissue engineering and sometime after that, the preparation of high-strength hydrogels was developed, which led to the development of high-strength smart responsive hydrogels that can soon be used in advanced applications. Hydrogel applications can be diversified through the development of smart responsive hydrogels with thermal, optical, and electrical characteristics which can be the potential prospect of the hydrogel materials. Multifunctional properties and multiple response mechanisms are to be made an integral part of the smart responsive hydrogels which develop material attributes like self-healing, self-adaptation, and self-deformation. Characterization techniques like scanning electron microscopy, IR spectroscopy, and X-ray diffraction, and mechanical properties testing methods like tensile tests were used to explore the mechanical characteristics of smart hydrogels which expand the utilization of smart hydrogel in dynamic applications. Apart from the mechanical properties such as self-recovery and fatigue resistance, very few research focus on the various other properties of the hydrogels. As a direction for future research, if some other properties of smart responsive hydrogels such as electrical conductivity and self-healing attributes are measured other than the mechanical properties, the application spectrum of hydrogels may widen.

References

1. Tanaka, T., Nishio, I., Sun, S.T., Ueno-Nishio, S., Collapse of gels in an electric field. *Science, 218*, 4571, 467–469, 1982.
2. Ramesh, M., Rajeshkumar, L., Balaji, D., Influence of process parameters on the properties of additively manufactured fiber-reinforced polymer composite materials: A review. *J. Mater. Eng. Perform., 30*, 7, 4792–4807, 2021.
3. Priyadharshini, M., Balaji, D., Bhuvaneswari, V., Rajeshkumar, L., Sanjay, M.R., Siengchin, S., Fiber reinforced composite manufacturing with the aid of artificial intelligence–a state-of-the-art review. *Arch. Comput. Methods Eng.*, 29, 5511–5524, 2022.
4. Michie, W.C., Culshaw, B., Konstantaki, M., McKenzie, I., Kelly, S., Graham, N.B., Moran, C., Distributed pH and water detection using fiber-optic sensors and hydrogels. *J. Lightwave Technol., 13*, 7, 1415–1420, 1995.
5. MacLean, A., Michie, W.C., Pierce, S.G., Thursby, G., Culshaw, B., Moran, C., Graham, N.B., Hydrogel/fiber optic sensor for distributed measurement of humidity and pH value, in: *Smart Structures and Materials 1998: Sensory Phenomena and Measurement Instrumentation for Smart Structures and Materials*, vol. 3330, SPIE, pp. 134–144, 1998, July.

6. Calvert, P.D. and Liu, Z., Electrically stimulated bilayer hydrogels as muscles, in: *Smart Structures and Materials 1999: Electroactive Polymer Actuators and Devices*, vol. 3669, SPIE, pp. 236–241, 1999, May.
7. Ramesh, M., Rajeshkumar, L., Balaji, D., Bhuvaneswari, V., Polymer composites for enzyme sensors, in: *Polymeric Nanocomposite Materials for Sensor Applications*, pp. 343–366, Woodhead Publishing, USA, 2023.
8. Ramesh, M., Rajeshkumar, L., Balaji, D., Bhuvaneswari, V., Sustainable and renewable nano-biocomposites for sensors and actuators: A review on preparation and performance. *Curr. Anal. Chem.*, *19*, 1, 38–69, 2023.
9. Astm, I., ASTM52900-15 standard terminology for additive manufacturing—General principles—Terminology, vol. 3(4), ASTM International, West Conshohocken, PA, p. 5, 2015.
10. Langer, R. and Vacanti, J.P., Tissue engineering. *Science*, *260*, 5110, 920–926, May 14, 1993.
11. Kim, S.S., Utsunomiya, H., Koski, J.A., Wu, B.M., Cima, M.J., Sohn, J., Mukai, K., Griffith, L.G., Vacanti, J.P., Survival and function of hepatocytes on a novel three-dimensional synthetic biodegradable polymer scaffold with an intrinsic network of channels. *Ann. Surg.*, *228*, 1, 8, 1998.
12. Landers, R. and Mülhaupt, R., Desktop manufacturing of complex objects, prototypes and biomedical scaffolds by means of computer-assisted design combined with computer-guided 3D plotting of polymers and reactive oligomers. *Macromol. Mater. Eng.*, *282*, 1, 17–21, 2000.
13. Ramesh, M., Janani, R., Deepa, C., Rajeshkumar, L., Nanotechnology-enabled biosensors: A review of fundamentals, design principles, materials, and applications. *Biosensors*, *13*, 1, 40, 2023.
14. Ramesh, M., Bhuvaneswari, V., Balaji, D., Rajeshkumar, L., Self-healable conductive and polymeric composite materials, in: *Aerospace Polymeric Materials*, pp. 231–258, 2022.
15. Liu, V.A. and Bhatia, S.N., Three-dimensional photopatterning of hydrogels containing living cells. *Biomed. Microdevices*, *4*, 4, 257–266, 2002.
16. Li, J., Wu, C., Chu, P.K., Gelinsky, M., 3D printing of hydrogels: Rational design strategies and emerging biomedical applications. *Mater. Sci. Eng.: R: Rep.*, *140*, 100543, 2020.
17. Deepa, C., Rajeshkumar, L., Ramesh, M., Thermal properties of kenaf fiber-based hybrid composites, in: *Natural Fiber-Reinforced Composites: Thermal Properties and Applications*, pp. 167–182, 2022.
18. Ramesh, M., Rajeshkumar, L., Balaji, D., Bhuvaneswari, V., Keratin-based biofibers and their composites, in: *Advances in Bio-Based Fiber*, pp. 315–334, Woodhead Publishing, USA, 2022.
19. Ramesh, M., Rajeshkumar, L., Bhuvaneswari, V., Leaf fibres as reinforcements in green composites: A review on processing, properties and applications. *Emergent Mater.*, 5, 3, 833–857, 2021.

20. Ong, C.S., Yesantharao, P., Huang, C.Y., Mattson, G., Boktor, J., Fukunishi, T., Zhang, H., Hibino, N., 3D bioprinting using stem cells. *Pediatr. Res.*, *83*, 1, 223–231, 2018.
21. Deepa, C., Balaji, D., Bhuvaneswari, V., Rajeshkumar, L., Ramesh, M., Priyadharshini, M., Deep learning for the selection of multiple analogs, in: *Drug Design Using Machine Learning*, pp. 117–142, 2022.
22. Bhuvaneswari, V., Priyadharshini, M., Deepa, C., Balaji, D., Rajeshkumar, L., Ramesh, M., Deep learning for material synthesis and manufacturing systems: A review. *Mater. Today: Proc.*, *46*, 3263–3269, 2021.
23. Hann, S.Y., Cui, H., Nowicki, M., Zhang, L.G., 4D printing soft robotics for biomedical applications. *Addit. Manuf.*, *36*, 101567, 2020.
24. Ramesh, M., Kumar, L.R., Khan, A., Asiri, A.M., Self-healing polymer composites and its chemistry, in: *Self-Healing Composite Materials*, pp. 415–427, Woodhead Publishing, USA, 2020, https://doi.org/10.1016/B978-0-12-817354-1.00022-3.
25. Ramesh, M., Rajeshkumar, L., Saravanakumar, R., Mechanically-induced self-healable materials, in: *Self-Healing Smart Materials and Allied Applications*, pp. 379–403, 2021, https://doi.org/10.1002/9781119710219.ch15.
26. Ramesh, M., Rajeshkumar, L., Balaji, D., Bhuvaneswari, V., Sivalingam, S., Self-healable conductive materials, in: *Self-Healing Smart Materials and Allied Applications*, pp. 297–319, 2021, https://doi.org/10.1002/9781119710219.ch11.
27. Du, X., Zhou, J., Shi, J., Xu, B., Supramolecular hydrogelators and hydrogels: From soft matter to molecular biomaterials. *Chem. Rev.*, *115*, 24, 13165–13307, 2015.
28. Ramesh, M., Rajeshkumar, L., Balaji, D., Mechanical and dynamic properties of ramie fiber-reinforced composites, in: *Mechanical and Dynamic Properties of Biocomposites*, pp. 275–291, 2021, https://doi.org/10.1002/9783527822331.ch15.
29. Ramesh, M., Balaji, D., Rajeshkumar, L., Bhuvaneswari, V., Saravanakumar, R., Khan, A., Asiri, A.M., Tribological behavior of glass/sisal fiber reinforced polyester composites, in: *Vegetable Fiber Composites and their Technological Applications*, pp. 445–459, 2021, https://doi.org/10.1007/978-981-16-1854-3_20.
30. Ramesh, M., Rajeshkumar, L., Balaji, D., Priyadharshini, M., Properties and characterization techniques for waterborne polyurethanes, in: *Sustainable Production and Applications of Waterborne Polyurethanes*, pp. 109–123, 2021, https://doi.org/10.1007/978-3-030-72869-4_6.
31. Huo, W., Li, J., Ren, M., Ling, W., Xu, H., Tee, C.A.T., Huang, X., Recent development of bioresorbable electronics using additive manufacturing. *Curr. Opin. Chem. Eng.*, *28*, 118–126, 2020.

32. Han, D. and Lee, H., Recent advances in multi-material additive manufacturing: Methods and applications. *Curr. Opin. Chem. Eng.*, *28*, 158–166, 2020.
33. Ge, Q., Chen, Z., Cheng, J., Zhang, B., Zhang, Y.F., Li, H., He, X., Yuan, C., Liu, J., Magdassi, S., Qu, S., 3D printing of highly stretchable hydrogel with diverse UV curable polymers. *Sci. Adv.*, *7*, 2, eaba4261, 2021.
34. Bakarich, S.E., Gorkin III, R., Gately, R., Naficy, S., in het Panhuis, M., Spinks, G.M., 3D printing of tough hydrogel composites with spatially varying materials properties. *Addit. Manuf.*, *14*, 24–30, 2017.
35. Zhang, H., Guo, X., Wu, J., Fang, D., Zhang, Y., Soft mechanical metamaterials with unusual swelling behavior and tunable stress-strain curves. *Sci. Adv.*, *4*, 6, eaar8535, 2018.
36. Matsuda, T., Kawakami, R., Namba, R., Nakajima, T., Gong, J.P., Mechanoresponsive self-growing hydrogels inspired by muscle training. *Science*, *363*, 6426, 504–508, 2019.
37. Sun, W., Xue, B., Fan, Q., Tao, R., Wang, C., Wang, X., Li, Y., Qin, M., Wang, W., Chen, B., Cao, Y., Molecular engineering of metal coordination interactions for strong, tough, and fast-recovery hydrogels. *Sci. Adv.*, *6*, 16, eaaz9531, 2020.
38. Zhao, Y., Xuan, C., Qian, X., Alsaid, Y., Hua, M., Jin, L., He, X., Soft phototactic swimmer based on self-sustained hydrogel oscillator. *Sci. Rob.*, *4*, 33, eaax7112, 2019.
39. Podstawczyk, D., Nizioł, M., Szymczyk, P., Wiśniewski, P., Guiseppi-Elie, A., 3D-printed stimuli-responsive magnetic nanoparticle embedded alginate-methylcellulose hydrogel actuators. *Addit. Manuf.*, *34*, 101275, 2020.
40. English, M.A., Soenksen, L.R., Gayet, R.V., de Puig, H., Angenent-Mari, N.M., Mao, A.S., Nguyen, P.Q., Collins, J.J., Programmable CRISPR-responsive smart materials. *Science*, *365*, 6455, 780–785, 2019.
41. Justus, K.B., Hellebrekers, T., Lewis, D.D., Wood, A., Ingham, C., Majidi, C., LeDuc, P.R., Tan, C., A biosensing soft robot: Autonomous parsing of chemical signals through integrated organic and inorganic interfaces. *Sci. Rob.*, *4*, 31, eaax0765, 2019.
42. Dong, E., Song, Z., Zhang, Y., Ghaffari Mosanenzadeh, S., He, Q., Zhao, X., Fang, N.X., Bioinspired metagel with broadband tunable impedance matching. *Sci. Adv.*, *6*, 44, eabb3641, 2020.
43. Zhou, F., Hong, Y., Liang, R., Zhang, X., Liao, Y., Jiang, D., Zhang, J., Sheng, Z., Xie, C., Peng, Z., Zhuang, X., Rapid printing of bio-inspired 3D tissue constructs for skin regeneration. *Biomaterials*, *258*, 120287, 2020.
44. Hindley, J.W., Law, R.V., Ces, O., Membrane functionalization in artificial cell engineering. *SN Appl. Sci.*, *2*, 4, 1–10, 2020.
45. Campbell, S.B., Wu, Q., Yazbeck, J., Liu, C., Okhovatian, S., Radisic, M., Beyond polydimethylsiloxane: Alternative materials for fabrication of organ-on-a-chip devices and microphysiological systems. *ACS Biomater. Sci. Eng.*, *7*, 7, 2880–2899, 2020.

46. MacDonald, E. and Wicker, R., Multiprocess 3D printing for increasing component functionality. *Science*, *353*, 6307, aaf2093, 2016.
47. Wirthl, D., Pichler, R., Drack, M., Kettlguber, G., Moser, R., Gerstmayr, R., Hartmann, F., Bradt, E., Kaltseis, R., Siket, C.M., Schausberger, S.E., Instant tough bonding of hydrogels for soft machines and electronics. *Sci. Adv.*, *3*, 6, e1700053, 2017.
48. Najaf Zadeh, H., Huber, T., Nock, V., Fee, C., Clucas, D., Complex geometry cellulose hydrogels using a direct casting method. *Bioengineering*, *7*, 2, 58, 2020.
49. McNulty, J.D., Marti-Figueroa, C., Seipel, F., Plantz, J.Z., Ellingham, T., Duddleston, L.J., Goris, S., Cox, B.L., Osswald, T.A., Turng, L.S., Ashton, R.S., Micro-injection molded, poly (vinyl alcohol)-calcium salt templates for precise customization of 3D hydrogel internal architecture. *Acta Biomater.*, *95*, 258–268, 2019.
50. Temiz, Y., Lovchik, R.D., Kaigala, G.V., Delamarche, E., Lab-on-a-chip devices: How to close and plug the lab? *Microelectron. Eng.*, *132*, 156–175, 2015.
51. Dumont, C.M., Carlson, M.A., Munsell, M.K., Ciciriello, A.J., Strnadova, K., Park, J., Cummings, B.J., Anderson, A.J., Shea, L.D., Aligned hydrogel tubes guide regeneration following spinal cord injury. *Acta Biomater.*, *86*, 312–322, 2019.
52. Liu, J., Mckeon, L., Garcia, J., Pinilla, S., Barwich, S., Möbius, M., Stamenov, P., Coleman, J.N., Nicolosi, V., Additive manufacturing of Ti3C2-MXene-functionalized conductive polymer hydrogels for electromagnetic-interference shielding. *Adv. Mater.*, *34*, 5, 2106253, 2022.
53. Salvekar, A.V., Nasir, F.H.B.A., Chen, Y.H., Maiti, S., Ranjan, V.D., Chen, H.M., Wang, H., Huang, W.M., Rapid volumetric additive manufacturing in solid state: A demonstration to produce water-content-dependent cooling/heating/water-responsive shape memory hydrogels. *3D Printing and Additive Manufacturing*, 2022.
54. Moreira, M.F., Kanaan, A.F., Piedade, A.P., Indirect additive manufacturing: A valid approach to modulate sorption/release profile of molecules from chitosan hydrogels. *Polymers*, 14, 13, 2530, 2022.
55. Klimaschewski, S.F., Küpperbusch, J., Kunze, A., Vehse, M., Material investigations on poly (ethylene glycol) diacrylate-based hydrogels for additive manufacturing considering different molecular weights. *J. Mech. Energy Eng.*, *6*, 1, 33–42, 2022.
56. Wong, J., The design of functional ion gels and hydrogels for additive manufacturing of devices and composites (Doctoral dissertation), 2022.
57. Ganguly, S. and Margel, S., 3D printed magnetic polymer composite hydrogels for hyperthermia and magnetic field driven structural manipulation. *Prog. Polym. Sci.*, 131, 101574, 2022.

58. Kumar, P.S., Pal, S.K., Chinnasamy, M., Rajasekar., R., Organic/silica nanocomposite membranes, in: *Organic-Inorganic Composite Polymer Electrolyte Membranes: Preparation, Properties, and Fuel Cell Applications*, pp. 47–72, Springer International Publishing AG, Switzerland, 2017, https://doi.org/10.1007/978-3-319-52739-0_3.
59. Li, J., Cui, X., Lindberg, G.C., Alcala-Orozco, C.R., Hooper, G.J., Lim, K.S., Woodfield, T.B., Hybrid fabrication of photo-clickable vascular hydrogels with additive manufactured titanium implants for enhanced osseointegration and vascularized bone formation. *Biofabrication, 14*, 3, 034103, 2022.
60. Solis, D.M. and Czekanski, A., The effect of the printing temperature on 4D DLP printed pNIPAM hydrogels. *Soft Matter, 18*, 17, 3422–3429, 2022.
61. Heidarian, P., Kouzani, A.Z., Kaynak, A., Paulino, M., Nasri-Nasrabadi, B., Dynamic hydrogels and polymers as inks for three-dimensional printing. *ACS Biomater. Sci. Eng.*, *5*, 6, 2688–2707, 2019.
62. Arslan, H., Nojoomi, A., Jeon, J., Yum, K., 3D printing of anisotropic hydrogels with bioinspired motion. *Adv. Sci.*, *6*, 2, 1800703, 2019.
63. Jin, Y., Liu, C., Chai, W., Compaan, A., Huang, Y., Self-supporting nanoclay as internal scaffold material for direct printing of soft hydrogel composite structures in air. *ACS Appl. Mater. Interfaces, 9*, 20, 17456–17465, 2017.
64. Chen, Z., Zhao, D., Liu, B., Nian, G., Li, X., Yin, J., Qu, S., Yang, W., 3D printing of multifunctional hydrogels. *Adv. Funct. Mater.*, *29*, 20, 1900971, 2019.
65. Xue, X., Hu, Y., Deng, Y., Su, J., Recent advances in design of functional biocompatible hydrogels for bone tissue engineering. *Adv. Funct. Mater.*, *31*, 19, 2009432, 2021.
66. Rajeshkumar, L., Biodegradable polymer blends and composites from renewable resources, in: *Biodegradable Polymers, Blends and Composites*, pp. 527–549, Woodhead Publishing, USA, 2022, https://doi.org/10.1016/B978-0-12-823791-5.00015-6.
67. Zhao, D., Liu, Y., Liu, B., Chen, Z., Nian, G., Qu, S., Yang, W., 3D printing method for tough multifunctional particle-based double-network hydrogels. *ACS Appl. Mater. Interfaces, 13*, 11, 13714–13723, 2021.
68. Morgan, F.L., Moroni, L., Baker, M.B., Dynamic bioinks to advance bioprinting. *Adv. Healthcare Mater.*, *9*, 15, 1901798, 2020.
69. Jungst, T., Smolan, W., Schacht, K., Scheibel, T., Groll, J., Strategies and molecular design criteria for 3D printable hydrogels. *Chem. Rev.*, *116*, 3, 1496–1539, 2016.
70. Xu, C., Dai, G., Hong, Y., Recent advances in high-strength and elastic hydrogels for 3D printing in biomedical applications. *Acta Biomater.*, *95*, 50–59, 2019.
71. Wang, L.L., Highley, C.B., Yeh, Y.C., Galarraga, J.H., Uman, S., Burdick, J.A., Three-dimensional extrusion bioprinting of single- and double-network hydrogels containing dynamic covalent crosslinks. *J. Biomed. Mater. Res. Part A*, *106*, 4, 865–875, 2018.

72. Lei, Z., Wang, Q., Wu, P., A multifunctional skin-like sensor based on a 3D printed thermo-responsive hydrogel. *Mater. Horiz.*, *4*, 4, 694–700, 2017.
73. Zhu, F., Cheng, L., Yin, J., Wu, Z.L., Qian, J., Fu, J., Zheng, Q., 3D printing of ultratough polyion complex hydrogels. *ACS Appl. Mater. Interfaces*, *8*, 45, 31304–31310, 2016.
74. Gong, J., Schuurmans, C.C., Genderen, A.M.V., Cao, X., Li, W., Cheng, F., He, J.J., López, A., Huerta, V., Manríquez, J., Li, R., Complexation-induced resolution enhancement of 3D-printed hydrogel constructs. *Nat. Commun.*, *11*, 1, 1267, 2020.
75. Li, H., Tan, C., Li, L., Review of 3D printable hydrogels and constructs. *Mater. Des.*, *159*, 20–38, 2018.
76. Duan, J., Liang, X., Zhu, K., Guo, J., Zhang, L., Bilayer hydrogel actuators with tight interfacial adhesion fully constructed from natural polysaccharides. *Soft Matter*, *13*, 2, 345–354, 2017.
77. Zheng, S.Y., Shen, Y., Zhu, F., Yin, J., Qian, J., Fu, J., Wu, Z.L., Zheng, Q., Programmed deformations of 3D-printed tough physical hydrogels with high response speed and large output force. *Adv. Funct. Mater.*, *28*, 1803366, 2018.
78. Li, J., Ma, Q., Xu, Y., Yang, M., Wu, Q., Wang, F., Sun, P., Highly bidirectional bendable actuator engineered by LCST–UCST bilayer hydrogel with enhanced interface. *ACS Appl. Mater. Interfaces*, *12*, 49, 55290–55298, 2020.
79. Sun, P., Zhang, H., Xu, D., Wang, Z., Wang, L., Gao, G., Hossain, G., Wu, J., Wang, R., Fu, J., Super tough bilayer actuators based on multi-responsive hydrogels crosslinked by functional triblock copolymer micelle macro-crosslinkers. *J. Mater. Chem. B*, *7*, 16, 2619–2625, 2019.
80. Liang, X., Ding, H., Wang, Q., Wang, M., Yin, B., Sun, G., Nature-inspired semi-IPN hydrogels with tunable mechanical properties and multi-responsiveness. *New J. Chem.*, *45*, 2, 861–871, 2021.
81. Bacelar, A.H., Cengiz, I.F., Silva-Correia, J., Sousa, R.A., Oliveira, J.M., Reis, R.L., "Smart" hydrogels in tissue engineering and regenerative medicine applications, in: *Handbook of Intelligent Scaffolds for Tissue Engineering and Regenerative Medicine*, pp. 333–367, Jenny Stanford Publishing, Singapore, 2017.
82. Kumar, P.S., Senthil, S.M., Pal, S.K., Rajasekar, R., Organic/montmorillonite nanocomposite membranes, in: *Organic-Inorganic Composite Polymer Electrolyte Membranes: Preparation, Properties, and Fuel Cell Applications*, pp. 133–164, Springer International Publishing AG, Switzerland, 2017, https://doi.org/10.1007/978-3-319-52739-0_6.
83. Tahrir, F.G., Ganji, F., Ahooyi, T.M., Injectable thermosensitive chitosan/glycerophosphate-based hydrogels for tissue engineering and drug delivery applications: A review. *Recent Pat. Drug Deliv. Formul.*, *9*, 2, 107–120, 2015.

84. Zhang, X.Z., Xu, X.D., Cheng, S.X., Zhuo, R.X., Strategies to improve the response rate of thermosensitive PNIPAAm hydrogels. *Soft Matter*, *4*, 3, 385–391, 2008.
85. Ramesh, M., Rajeshkumar, L., Balaji, D., Aerogels for insulation applications. *Mater. Res. Found*, *98*, 57–76, 2021, https://doi.org/10.21741/9781644901298-4.
86. Liu, D., Wu, Q., Zhu, Y., Liu, Y., Xie, X., Li, S., Lin, H., Chen, W., Zhu, F., Co-delivery of metformin and levofloxacin hydrochloride using biodegradable thermosensitive hydrogel for the treatment of corneal neovascularization. *Drug Delivery*, *26*, 1, 522–531, 2019.

10

Additive Manufacturing of Bulk Metallic Glasses

**Ashutosh Sahu[1] and Anil Kumar[2]*

[1]*Department of Mechanical Engineering, Chaitanya Bharathi Institute of Technology, Hyderabad, India*

[2]*Department of Metallurgical and Materials Engineering, Visvesvaraya National Institute of Technology, Nagpur, India*

Abstract

After the discovery of bulk metallic glasses (BMGs), these materials have gained great attention due to their wider industrial applications and scientific importance, ascribed to their superior properties compared to conventional alloys and composites. However, synthesis of these materials with large dimensions by conventional methods (like rapid solidification and powder metallurgy routes) is difficult due to the inherent brittleness at room temperature and necessity to attain high cooling rates for restricting crystallization. On the other hand, additive manufacturing (AM) or 3D printing is an advanced technique to synthesize large-dimension intricate designs starting from the materials in powder or ribbon form. In this chapter, various AM routes are discussed in detail for fabricating BMG components. A broad perspective on mechanical behavior of additively manufactured BMGs is also presented. Finally, imperative aspects of AM of BMGs have been summarized for the development of such materials.

***Keywords*:** Additive manufacturing, bulk metallic glass, glass forming ability, defects, crystallization, mechanical properties

**Corresponding author*: kumar.anilsah14@mme.vnit.ac.in

R. Rajasekar, C. Moganapriya and P. Sathish Kumar (eds.) Additive Manufacturing with Novel Materials: Processes, Properties and Applications, (319–342) © 2024 Scrivener Publishing LLC

10.1 Introduction

10.1.1 Bulk Metallic Glass

Metallic glasses (MGs) are noncrystalline solid materials that are produced by continuous cooling from the liquid state and exhibit glass transition [1–4]. Generally, MGs having a diameter or section thickness of 1 mm or more are known as bulk metallic glasses (BMGs) [1]. Owing to the lack of long-range atomic periodicity and crystalline defects (slip systems, dislocations, stacking faults, and grain boundaries), MGs show excellent combination of properties compared to their crystalline counterparts [1–4]. Tensile strength and Vickers hardness vs Young's modulus graphs of different conventional metallic alloys and various BMGs shown in Figure 10.1 reveal that MGs are the strongest known engineering materials. In addition to the mechanical properties, MGs possess soft magnetic, good catalytic, excellent wear-resistant, and superior anti-corrosion properties [1–4]. Because of the combination of such properties, MGs embrace the great potential for various structural and functional applications. Therefore, MGs are regarded as superior materials for applications in aerospace, manufacturing, biomedical, and sports industries [1].

To retain the noncrystalline structure, metallic liquids should be quenched at an adequately higher cooling rate. The minimum cooling rate that can restrict the development of crystalline phases during solidification is termed as critical cooling rate. Value of critical cooling rate to form a glassy structure (i.e., to freeze the liquid structure) usually varies in the

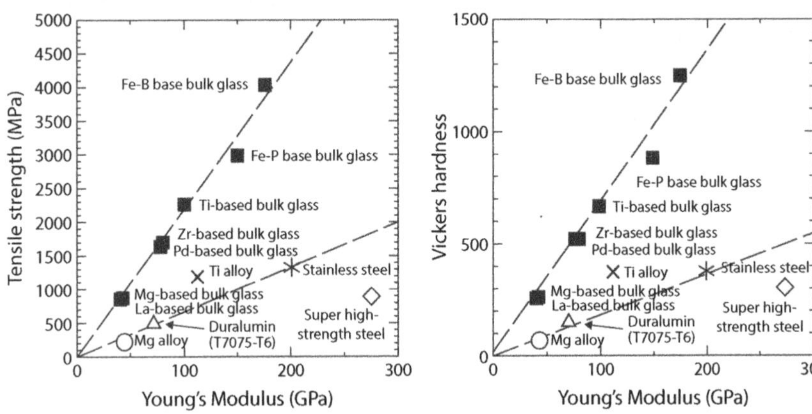

Figure 10.1 Property comparison of metallic glass and other materials. Reproduced from [2], with permission from Elsevier, 2023.

range of 10^4–10^6 k/s, depending on the glassy alloy composition [1]. The first metallic glass was experimentally developed by Pol Duwez's research group by chilling metallic liquid of $Au_{75}Si_{25}$ (at.%) composition at a critical cooling rate of about 10^6 K/s in 1960 at Caltech University, USA [5]. Afterwards, various research works were carried out in the search for new metallic glasses by using different processing techniques. It has been found that the formation of MGs is mainly affected by its alloy composition, i.e., glass forming ability (GFA) as well as cooling rate. Also, industrial applications of MGs are rather restricted due to the difficulty encountered in bulk quantity production of BMGs ascribed to their limited GFA.

Powder metallurgy (PM) synthesis technique can be adopted to improve the dimensions of MGs in place of rapid quenching methods. Higher amounts of glassy powders can be synthesized via either gas atomization or mechanical alloying and then consolidated by using advanced sintering techniques like vacuum hot pressing, spark plasma sintering, and extrusion, to synthesize large-dimension BMGs [6]. However, it is still difficult to further process these BMGs by either machining or plastically deforming them owing to their high brittleness at room temperature. Thus, BMGs of intricate designs cannot be synthesized by either rapid solidification or PM routes. On the other hand, additive manufacturing (AM), which is also known as 3D printing, is considered to be very an effective manufacturing process for developing complex designs of BMG components. In this process, glassy parts can be synthesized by depositing feedstock in the form of powders, thin ribbons or filaments in a layer-by-layer manner according to 3D data models [7–11]. The synthesized parts or components usually do not require further machining or plastic deformation to produce the final product. Different AM processes to synthesize BMGs with their advantages and disadvantages have been mentioned in detail in the following sections.

10.1.2 Additive Manufacturing Techniques to Synthesize Bulk Metallic Glass

Additive manufacturing was developed by Charles Hull in 1986 in a process known as stereolithography [7–11]. In this technique, materials are joined in a layered manner to produce 3D objects from computer-controlled 3D model data. It is an excellent technique to fabricate intricately designed 3D metallic glass parts because cooling rates in the range of 10^3–10^8 K/s can be achieved during layer-by-layer deposition of material in the process. The cooling rates obtained in this range are usually higher compared to the critical cooling rates for most of the metallic alloy compositions. Thus, it is possible to retain the amorphous nature of alloy after solidification.

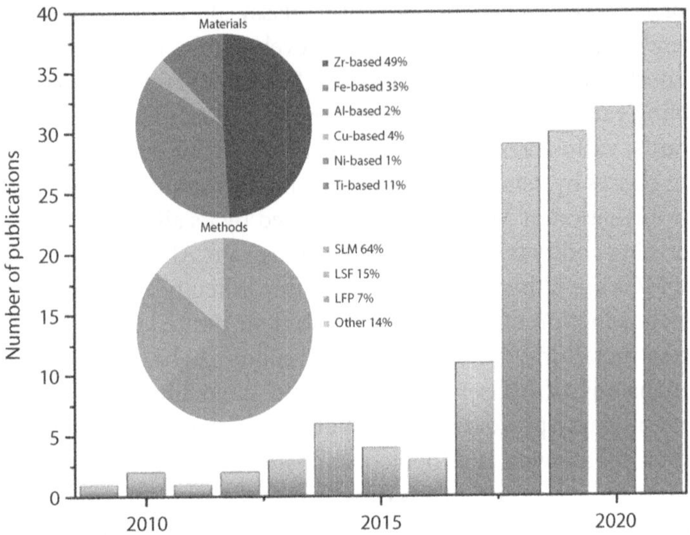

Figure 10.2 Statistics of development in metallic glasses via AM route [9].

The main advantages associated with the AM processes are lower wastage, cost-effectiveness, productivity, possibility to synthesize engineering parts of complex geometries, etc. Additive manufacturing processes are generally categorized based on (i) base material like metal, ceramic, and polymer, which is used to synthesize the final product, (ii) indirect and direct methods subjected to bonding types, and (iii) based on state of raw material (i.e., liquid, molten, powder, and solid layer). Apart from the above-mentioned categorization, the AM process can also be categorized into seven different types based on adhesion and bonding methods, such as (i) vat photopolymerization, (ii) material jetting, (iii) material extrusion, (iv) powder bed fusion, (v) binder jetting, (vi) direct energy deposition, and (vii) sheet lamination. Research related to AM processes for developing MGs has increased dramatically in the last decade, which can be easily seen by the increasing number of research papers related to this field, as shown in Figure 10.2. Various commonly used AM techniques to synthesize MGs are discussed in the next section.

Selective Laser Melting (SLM)

In this AM technique, high-energy beam is used for selective melting of metallic powder particles to fuse locally in layer-by-layer mode based on computer-aided design (CAD) data to deposit 3D solid samples, as shown in Figure 10.3. Once powders get fused with a layer beneath, then the powder bed is lowered and a fresh powder layer is deposited, which is locally exposed to the laser beam. With this repetitive process of powder melting layer by layer, any type of components/parts can be built [8, 12]. The chamber used in this technique must be flushed with nitrogen or argon to minimize the possibility of oxidation of materials. Laser spot in focal plane is kept in the range of 50–80 μm, while scanning speed is maintained at approximately 15 m/s [8, 12]. Cooling rate of fused powder particles are generally found in the range of 10^4–10^6 K/s, ascribed to the high laser beam power, high scanning speed, and small beam diameter. This level of cooling rate is higher than the critical cooling rate required for most of the glass forming alloy compositions. However, synthesized metallic glass can be subjected to certain defects such as pores and microcracks due to such high cooling rates. Thus, optimization of processing parameters like laser power, scanning speed, scanning hatch, and scanning layer thickness can be very useful, depending on the raw material composition, to take care of these defects.

Many researchers have synthesized BMGs and their composites of various compositions by using SLM technique [13–23]. In 2018, Mahbooba *et al.* synthesized Fe-based (FeCrMoCB) cylindrical-shaped BMG, of 45-mm diameter and thickness of 30 mm using SLM [14]. This particular work inspired researchers to synthesize completely glassy monolithic

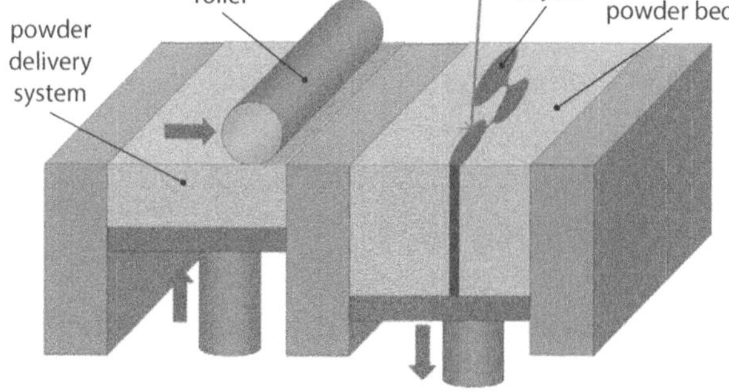

Figure 10.3 Schematic showing the principle of selective laser melting technique. Modified from [12], with permission from Elsevier, 2023.

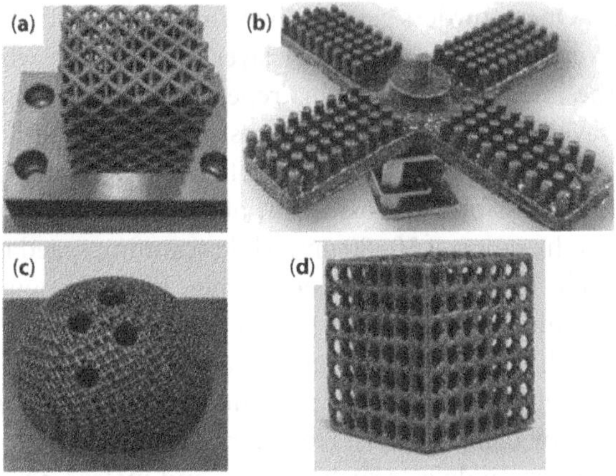

Figure 10.4 (a–d) Different shapes of Zr-based BMG samples synthesized by SLM. Modified from [8], with permission from Elsevier, 2023.

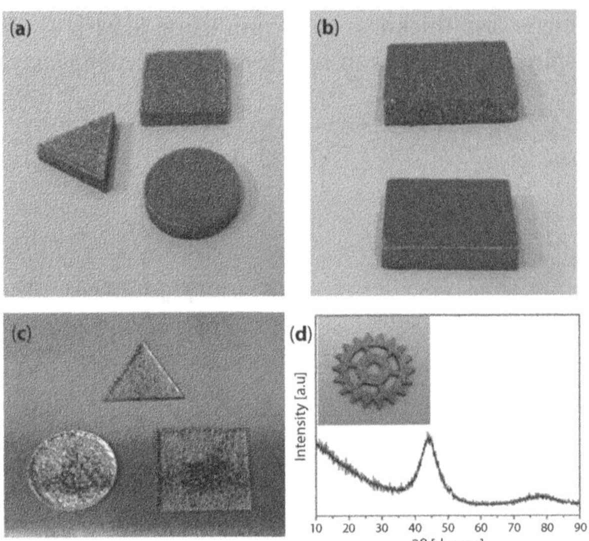

Figure 10.5 (a–c) Different shapes of Fe-based glassy parts synthesized by SLM process at different laser powers and (d) gear-shaped component along with the related X-ray diffraction (XRD) pattern revealing its amorphous nature. Modified from [17], with permission from Elsevier, 2023.

BMGs of any dimension out of any composition with even marginal GFA. Researchers have successfully synthesized different intricate designs of glassy materials by using the SLM process, which have been presented in Figure 10.4. Fe-based glassy components of different shapes fabricated by Nong *et al.* using SLM is shown in Figure 10.5 [17]. These complicated designs are not possible to be synthesized via rapid solidification process (RSP) or PM routes. This is because machining or plastic deformation of BMGs to form any shape is extremely difficult owing to the inherent brittleness of metallic glasses.

The major advantages and disadvantages associated with the selective laser melting method are mentioned below.

Major advantages of SLM process:

i. It can print intrinsically designed parts.
ii. Unused powder can be recycled.
iii. Components with refined microstructure and minimum microsegregation can be produced due to high solidification rates.
iv. Components with high densification and better mechanical properties can be obtained.

Major disadvantages of SLM process:

i. Residual thermal stresses are generated, which can lead to distortion, delamination, and cracking of components.
ii. Microstructural anisotropy may occur due to preferred growth of solidifying grains and phase transformations.
iii. Stress relieving, hot isostatic pressing, etc., may be required to reduce porosity.
iv. Component size may be limited due to the build plate.
v. Various defects such as variation in desired atomic ratio, micro-voids, and oxidation can be generated.
vi. Although a rapid solidification rate can be achieved, thermal oscillation of high laser power may alter the chemical homogeneity, which can deteriorate GFA of materials.
vii. Higher thermal fluctuation and thermal stresses can be produced because of rapid cooling in small regions which will lead to formation of microcracks and thus, premature failure of BMG components can occur.

Laser-Engineered Net Shaping (LENS)

This is also a laser-based AM technique for synthesizing BMGs, which is also termed as laser-directed energy deposition (LDED) or laser solid forming (LSF) or direct metal deposition (DMD). The setup for this process is schematically shown in Figure 10.6. Like the SLM process, powder particles are used as raw materials in laser-engineered net shaping (LENS). Also, a high-power laser melts the powders which is supplied through nozzles by injecting gases. A molten pool is created on a substrate during melting of the powder particles. Laser gun is pulled up after the deposition of the first layer and successive layers are deposited until the predesigned portion is synthesized [24]. Amorphous structure in synthesized components/parts is obtained by rapidly solidifying the molten pool. This technique is generally used in high-tech fields (e.g., aerospace, automobiles, medical treatment, and manufacturing industries) because high-density metallic glass parts with complex geometries can be fabricated. Processing parameters that mainly affect the molten pool are temperature gradient and solidification rate. Sun *et al.* were the first to synthesize BMG using LENS in 2008 [25]. Afterwards, Lin *et al.* reported Zr-based ($Zr_{55}Cu_{30}Al_{10}Ni_{5}$) BMG part with a diameter of 54 mm and height of 10 mm by using pulsed-laser LENS system, as shown in Figure 10.7 [26].

The advantages and disadvantages of laser engineered net shaping process are listed below.

Advantages of the LENS process:

i. This process allows the independent deposition of different composition powders through different nozzles.

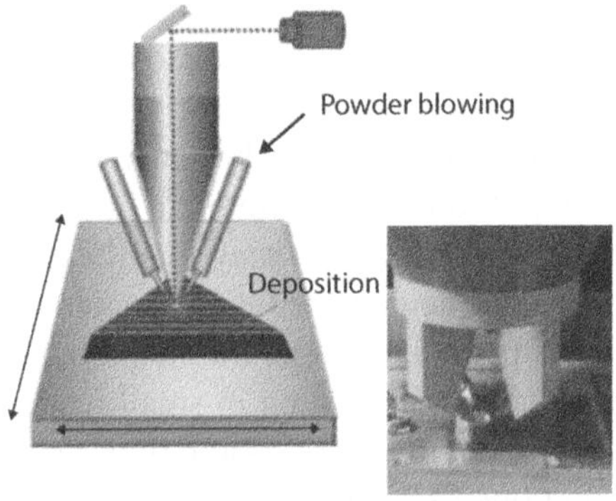

Figure 10.6 Schematic showing the overall process of laser engineered net shaping. Modified from [8], with permission from Elsevier, 2023.

Figure 10.7 (a) Top view and (b) side view of $Zr_{55}Cu_{30}Al_{10}Ni_5$ glassy specimen synthesized via LENS technique. Reproduced from [26], with permission from Elsevier, 2023.

ii. LENS devices usually have beam sizes in the range of 0.5–5 mm, which is apt for synthesizing large metallic material parts.
iii. LENS can be used to print metallic materials onto curved surfaces.

Disadvantages of LENS process are:

i. Surface finish of components produced by this method is not as smooth as those produced by SLM.
ii. Post-processing like machining is required for final finishing.
iii. Irregular heating and cooling cycles can lead to accumulation of residual thermal stress during synthesis process, which deteriorates the mechanical properties of fabricated parts.
iv. Heat accumulated in the heat affected zone (HAZ) hampers formation of fully amorphous structure.

Laser Foil 3D Printing (LFP)

Laser foil 3D printing was developed by Chen *et al.* in 2016 [8, 27]. A schematic diagram of the LFP process and various steps are presented in Figure 10.8. Unlike SLM and LENS techniques, glassy ribbons/foils are used as raw materials in place of powders. This technique consists of two different processes: welding and cutting. Welding of new foils to substrates is done by continuous wave lasers and cutting of additional regions is done by ultraviolet (UV) lasers. In LFP system, rapid cooling is obtained by a cooling chamber with liquid Ar and liquid N as cooling media. The purpose of the cooling chamber is to prevent formation of HAZ during the synthesis of BMGs. A glassy foil is spot welded on a substrate at first and then, as per the design, selective laser welding is done. After that, the deposited part is cut with the help of UV laser beam. Again, a fresh glassy foil is brought in for the next cycle of welding and cutting. Completely glassy parts with low porosity can be synthesized by this technique. In fact, laminate structures with gradient can be produced by stacking and welding metal strips in an ordered manner. Researchers have synthesized Zr-based MG/Ti-6Al-4V and MG/Zr/Ti/V materials exhibiting superior properties [28, 29]. Some of the complicated structures formed by using LFP are shown in Figure 10.9. The major drawback of this process is that good quality glassy ribbons

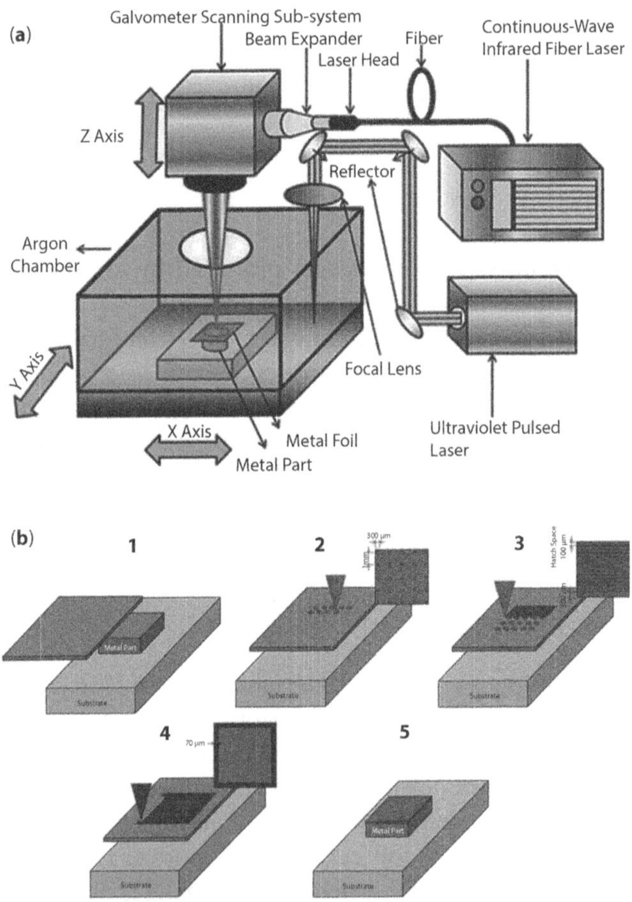

Figure 10.8 Schematic images of (a) LFP system and (b) various process steps. Modified from [27], with permission from Elsevier, 2023.

Figure 10.9 (a–d) Various BMG parts fabricated by LFP with different geometries. Reproduced from [4], with permission from Elsevier, 2023.

are difficult to produce, and the process is a bit complicated. However, the cost of powder production can be reduced by using ribbons.

Thermal Spray 3D Printing (TS3DP)

Thermal spray by high-velocity oxygen fuel (HVOF) is generally used for producing coatings on metals. This technique was developed by Liu *et al.* based on the technique to make thick coatings in the millimeter range on metal substrates [8]. Metal powders are used as raw materials in this process. High-velocity oxygen fuel gun is used to preheat the glassy powders to a semi-molten state and then it is deposited on the substrate. Metallic glasses are obtained in a layered manner due to the rapid deposition and cooling of the molten metal. Intricate glassy structures can be obtained by using a mask between the HVOF gun and the substrate [8].

The main advantages of TS3DP are:

i. Thermal stresses can be reduced and hence, the formation of cracks.
ii. High amount of amorphous phase can be retained.

Researchers have synthesized large-sized, crack-free, nearly fully amorphous BMGs and their composites by this technique. For example, Zhang *et al.* have synthesized nearly completely amorphous $Fe_{48}Cr_{15}Mo_{14}C_{15}B_6Y_2$

BMGs with dimension of $100 \times 100 \times 5$ mm^3 and $20 \times 20 \times 20$ mm^3 [30]. Similarly, Hofmann *et al.* also reported the synthesis of FeCrMoBC BMG of dimension $50 \times 50 \times 2$ mm^3 by this technique [19]. The alloy composition can be modified by using mixtures of different composition powders in this process, which leads to synthesis of glassy composites with desirable mechanical properties. For example, Zhang *et al.* have manufactured glassy composites with good mechanical properties by reinforcing stainless steel and high entropy alloy [31]. The main disadvantage of TS3DP is that alloy systems are limited to Fe-based glassy systems with a low tendency to oxidation, ascribed to carrying out the operation in the air. Hence, alloy systems which are prone to oxidation, such as Zr- and Ti-based alloy compositions, cannot be synthesized using TS3DP. Cold spray AM can be adopted in these cases.

Fused Filament Fabrication (FFF)

Fused filament fabrication was developed by Schroers *et al.* from Yale, MIT in collaboration with Desktop Metal to synthesize glassy alloys [8, 12]. The main aim was to avoid devitrification and segregation of composition in AM of glassy alloys. In this technique, glassy rods are taken as raw materials and are extruded in the supercooled region where viscous flow occurs. A pulsed Joule heater can be used to make multiple layers. The previously printed layer below the nozzle is heated to the extrusion temperature by the heater. Adjacent layers are bonded by thermoplastic forming. Although FFF can yield high quality amorphous alloys of a nearly completely glassy nature, the essential requirement is that the glass forming alloy compositions must have a wide supercooled liquid region, low viscosity, and high oxidation resistance. Bordeenithikasem *et al.* fabricated $Zr_{65}Cu_{17.5}Ni_{10}Al_{7.5}$ amorphous sheet metal by using this process [32].

Laser Forward Transfer 3D Printing (LFT3D)

Laser forward transfer 3D printing was developed by Gorodesky *et al.* to fabricate metallic glass components of dimension in the micrometer range [33]. In this technique, a transparent substrate coated with a thin layer of metallic glass called a donor is melted by a laser pulse. Then droplets of the glassy alloy of micrometer size are deposited on a substrate (acceptor). The liquid droplet vitrifies under the effect of extremely high cooling rates ranging from 10^8 to 10^{11} K/s. ZrPd amorphous pillars of high aspect ratio and nearly amorphous microbridges were fabricated by this technique [33].

High resolution along with the capability of printing multimaterials make LFT3D an important technique to synthesize 3D glassy microdevices.

Direct Metal Writing (DIW)

Direct metal writing is also called direct ink writing (DIW). In this process, samples are made in a layered manner by depositing inks with micrometer- or nanometer-size particles. The carrier fluid evaporates after deposition and the amorphous particles are bound by a polymer matrix in 3D shapes [34]. This technique is used to synthesize a variety of materials such as Bi-Sn alloy, AuAg alloy and CoCrFeNi high-entropy alloys [35–37]. The same technique can also be used to synthesize metallic glasses [8]. Researchers have used glassy ink consisting of FeSiB glassy particles along with liquid suspension and polymer binders. They extruded the ink on a substrate from a micro nozzle and a computer program was controlling the movement of the nozzle. The glassy particles were bonded by the polymer in the desired shape. Then, spark plasma sintering was carried out to remove the polymer binder and consolidate the glassy particles. FeSiB metallic glass component with porous structure was fabricated exhibiting high strength by the same technique [8].

10.2 Overview of Various Bulk Metallic Glasses Synthesized by Additive Manufacturing

Among various additively manufactured BMGs, Zr-based ones have gained much more attention (as presented in Figure 10.2) compared to other BMG systems, especially because of their attractively high GFA and relatively low material cost [11]. Li *et al.* synthesized the first AM Zr-based BMG in 2016 [49]. Afterwards, a lot of efforts have been made to produce Zr-based BMGs of different compositions via 3D printing. Interestingly, most of the investigations in AM of Zr-based BMGs are related to $Zr_{59.3}Cu_{28.8}Al_{10.4}Nb_{1.5}$, which is an industrial-grade alloy, and it contains more impurities compared to lab-grade alloys. Therefore, this is considered an economical alloy and is widely known for its industrial name AMZ4. The second element which has been largely used to synthesize AM BMGs is Fe-based ones. The first additively manufactured Fe-based BMG ($Fe_{74}Mo_4P_{10}C_{7.5}B_{2.5}Si_2$) was reported in 2013 [13]. Although the BMG was partially crystallized, it showed the way to synthesize BMGs of any shape via 3D-printing route. Synthesis of Fe-based BMGs via the conventional casting method restricts their applications to small dimension devices (<2 mm), ascribed to insufficient cooling rates associated with this method to produce large amorphous parts. However, it has been observed that the main advantage associated with the Fe-based system is the relatively low affinity towards oxygen, which allows it to be handled easily compared to Ti-based or Zr-based alloys. Also, various investigations have shown that AM method can be a potential technique for the fabrication of larger sized Fe-based BMGs [11].

From the perspective of engineering applications, Al-based alloys have also gained significant attention especially due to their high strength-to-weight ratio and corrosion resistance. However, Al-based glassy alloy compositions exhibit generally low GFA and thus, synthesizing BMGs with diameter >1 mm is a very challenging task using conventional techniques. On the other hand, the AM route paves the way to synthesize high dimension Al-based BMGs. The glass forming compositions in Al-based systems that are mostly used to synthesize BMGs via AM route are $Al_{86}Ni_6Y_{4.5}Co_2La_{1.5}$, $Al_{85}Nd_8Ni_5Co_2$, and $Al_{85}Ni_5Y_6Co_2Fe_2$ [11]. Apart from the above-mentioned alloy systems, Ti-based alloys have also been considered as potential engineering materials for applications in aerospace and biomedical fields because of their high strength-to-weight ratio as well as biocompatibility. In 2018, the first additively manufactured fully amorphous with delicate geometries of the miniature skull and hand $Ti_{47}Cu_{38}Zr_{7.5}Fe_{2.5}Sn_2Si_1Ag_2$ BMG was synthesized [8]. It was observed that even very flimsy details can be obtained without any macroscopic defects in such products by using AM methods.

10.3 Mechanical Properties of Bulk Metallic Glasses Synthesized Via Additive Manufacturing

Investigating the mechanical properties of BMGs synthesized via AM processes is very important for their practical application in various fields. Metallic glasses exhibit superior mechanical properties compared to crystalline metallic alloys and composites, ascribed to the absence of long-range atomic periodicity and absence of crystalline defects [1–4]. However, these materials are subjected to catastrophic failure as soon as the elastic limit is crossed, which is associated with the rapid propagation of shear bands (i.e., plastic instability arising in glassy alloys) under deformation [1]. The propagation of these shear bands can be prohibited by incorporating crystalline phases in the amorphous matrix (i.e., synthesizing metallic glass composites) [1, 38]. Although the AM route provides the flexibility to synthesize components of complicated shapes, structural defects and heterogeneities arise in regions near HAZ during the synthesis process. These defects decrease the mechanical properties such as tensile/compressive strength and plasticity, compared to the glassy alloys of the same composition synthesized via RSP or PM routes [8, 12]. Mechanical properties such as hardness, compressive strength, tensile strength, plasticity, hardness, and fatigue of additively manufactured metallic glasses and their composites are discussed in the following sections.

10.3.1 Strength and Plasticity

Strength and plasticity of MGs synthesized via AM differ from that of MGs of the same composition synthesized via rapid solidification or PM routes, ascribed to the presence of defects such as porosity and micro cracks. For example, Ouyang *et al.* have synthesized $Zr_{55}Cu_{30}Ni_5Al_{10}$ metallic glass via SLM technique which exhibited compressive strength of ~1504 MPa, whereas the same glassy alloy synthesized via casting exhibited compressive strength of ~1870 MPa, as shown in Figure 10.10 [39]. Similarly, Deng *et al.* observed reduced yield strength and plastic strain in additively manufactured BMGs compared to the as-cast sample of the same composition. The main reason was the presence of large pores which act as

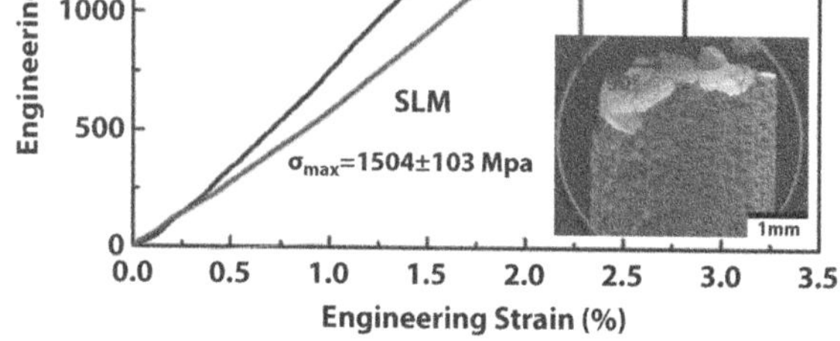

Figure 10.10 Compressive engineering stress-strain curves of the selective laser-melted $Zr_{55}Cu_{30}Ni_5Al_{10}$ glassy alloy compared to the same alloy synthesized via casting. Reproduced from [39], with permission from Elsevier, 2023.

stress concentrators and are responsible for the reduction in the plasticity of 3D-printed samples [40].

The 3D-printed MGs can also show size effects. Plasticity of additively manufactured glassy alloys increases with a decrease in the diameter of components. For example, Huang *et al.* reported approximately three times increase in plasticity of $Zr_{60.14}Cu_{22.31}Fe_{4.85}Al_{9.7}Ag_3$ BMG when the diameter decreased from 3 mm to 1 mm [41], ascribed to the lesser amount of defects in small samples. Apart from the pore and size effect, oxygen content in 3D-printed BMGs also affects plasticity. Presence of oxygen can alter the short-range order of BMGs and hence their plasticity [42]. Best *et al.* found that oxygen content in a Zr-based BMG synthesized by SLM was 1271 ppm which was 7.5 times higher than that of BMG synthesized via the casting of the same composition [42]. The increase in oxygen content led to significant reduction of plasticity and fracture toughness of the additively manufactured sample. Using a mixture of Ar and H_2 gases in SLM process can avoid the possibility of oxygen contamination.

Under tensile loading, additively manufactured BMGs usually exhibit quasi-brittle behavior without much plastic deformation. Best *et al.* in their investigation on additively manufactured BMG observed that catastrophic failure of such materials is related to the lack of fusion (LoF) defects which resulted due to low input energy, where feedstock does not melt completely [43]. Another study by Shi *et al.* revealed that the ultimate tensile strength (UTS) of BMG is more sensitive to the porosity level than that of Young's modulus [44]. Sohrabi *et al.* compared the effects of defects at the surface region and inside the samples [45]. The tensile strength increased up to 27% by improving the parameters related to the surface region, whereas the same increased only up to 5% by increasing the core power. In every investigation reported so far, it has been observed that the removal of defects can lead to improvement in strength and plasticity.

10.3.2 Hardness

Although BMGs exhibit much higher hardness than that of conventional crystalline alloys, it can be significantly affected by the presence of defects (i.e., porosity and secondary crystalline phases) arising during synthesis. Porosity has always had a negative influence on the hardness of any material including glassy alloys. Apart from the porosity, formation of various crystalline phases in the glassy matrix also has a significant impact on the hardness of additively manufactured BMGs. For example, intermetallic phases in the matrix increase the hardness and make the material more brittle. Such observations are reported by Marattukalam *et al.*, who found that the formation of intermetallic crystalline phases in BMG made it more brittle [46].

10.3.3 Fatigue Strength

Different mechanical parts and components of machinery are subjected to fatigue failure due to the application of fluctuating loads. Usually, fatigue ratio (fatigue limit/ultimate tensile strength) of BMGs lies below 20%, which is significantly lower compared to conventional engineering crystalline materials (i.e., low carbon steel or aluminum alloys), where the ratio varies from 30% to 40% [47]. Fatigue strength of materials depends on loading mode, sample geometry, testing environment, and history of thermal treatment. Some measures like relaxation heat treatment, laser shock peening, and changing the build orientation can be taken to improve fatigue strength of BMGs synthesized via AM [48]. Presence of oxygen and porosity as impurities in the BMGs synthesized by AM processes have also had a huge negative impact on their fatigue life. Such defects at the surface or core regions of BMGs can be reduced by increasing laser power during the synthesis process. Table 10.1 lists the mechanical properties of different metal-based BMGs synthesized via various 3D-printing techniques.

10.4 Summary and Future Development Perspective

Bulk metallic glasses (BMGs) have gained great attention after the emergence of additive manufacturing (AM) techniques, as these opened new avenues to fabricate BMGs with complex geometries as well as dimensions larger than their critical casting diameter. Additive manufacturing has shown great potential for accelerating the development of BMGs with excellent mechanical properties for promising future technology to synthesize bulk engineering components. The main challenges associated with the synthesis of BMGs via AM are low ductility, parameter optimization, and controlling the defects and the crystallization, all of which are interconnected. From the synthesis viewpoint, it is desirable to minimize/avoid the formation of defects and crystalline phases in the metallic glass component. Mechanical properties of BMGs fabricated via AM techniques are not only influenced by the degree of crystallization and defects, but also by the level of oxygen content. Both chemical composition and defects have a significant influence on the thermal stability and glass transition temperature of such materials. Selection of optimized manufacturing parameters can assist in generating the critical cooling rate required to circumvent

Table 10.1 Mechanical properties of different metal-based BMGs synthesized via different additive manufacturing processes.

Composition	Synthesis route	Hardness	Tensile strength (MPa)	Compressive strength (MPa)	Reference
$Zr_{52.5}Cu_{17.9}Ni_{14.6}Al_{10}Ti_{5}$	SLM	6.5 GPa	-	500–1710 depending on processing parameters	[49–51]
$Zr_{55}Cu_{30}Al_{10}Ni_{5}$	SLM	-	-	1504	[39]
$Zr_{60.14}Cu_{22.31}Fe_{4.85}Al_{9.7}Ag_{3}$	SLM	4.17 GPa	-	1734–1770 depending on processing parameters	[52]
$Zr_{59.3}Cu_{28.8}Nb_{1.5}Al_{10.4}$	SLM	4.3 GPa	-	~1300 in bending and 2500 in compression	[53]
$Zr_{39.6}Ti_{33.9}Nb_{7.6}Cu_{6.4}Be_{12.5}$	LENS	4.33 to 4.42 GPa	-	880–1300 depending on processing parameters	[54, 55]
$Zr_{55}Cu_{30}Al_{10}Ni_{5}$	LENS	-	<600	1450	[26]
$Zr_{65}Cu_{17.5}Ni_{10}Al_{7.5}$	LFP	4.07 tp 4.27 GPa	-	2200 in bending	[56]
$Zr_{44}Ti_{11}Cu_{10}Ni_{10}Be_{25}$	FFF	-	-	1200	[57]

(Continued)

Table 10.1 Mechanical properties of different metal-based BMGs synthesized via different additive manufacturing processes. (*Continued*)

Composition	Synthesis route	Hardness	Tensile strength (MPa)	Compressive strength (MPa)	Reference
$Ti_{47}Cu_{38}Zr_{7.5}Fe_{2.5}Sn_2Si_1Ag_2$	SLM	-	-	1690	[58]
$Cu_{50}Zr_{43}Al_7$	SLM	9.4 GPa		1044	[59]
$Fe_{43.7}Co_{7.3}C_{14.7}B_{4.3}Mo_{12.6}Cr_{15.5}Y_{1.9}$	SLM	-	-	100	[16]
$Fe_{55}C_2B_2Mo_{16}Cr_{25}$	SLM	12.7	-	6000	[15]
$Fe_{48}C_{15}B_6Mo_{14}Cr_{15}Y_2$	TS3DP	-	-	2000	[30]

crystallization, and therefore, defect-free fully glassy structure can be obtained. In addition, presence of defects and crystalline phases in the BMGs can also be eliminated/minimized with proper control of alloying composition.

References

1. Suryanarayana, C. and Inoue, A., *Bulk metallic glasses*, pp. 1–525, CRC Press, Taylor and Francis Group, Boca Raton, Florida, USA, 2011.
2. Inoue, A., Stabilization of metallic supercooled liquid and bulk amorphous alloys. *Acta Mater.*, 48, 279, 2000.
3. Wang, W.H., Dong, C., Shek, C.H., Bulk metallic glasses. *Mater. Sci. Eng. R*, 44, 45, 2004.
4. Shen, Y., Li, Y., Chen, C., Tsai, H.L., 3D printing of large, complex metallic glass structures. *Mater. Des.*, 117, 213, 2017.
5. Jun, W.K., Willens, R.H., Duwez, P., Non-crystalline structure in solidified gold-silicon alloys. *Nature*, 187, 869, 1960.
6. Suryanarayana, C., Mechanical alloying and milling. *Prog. Mater. Sci.*, 46, 1, 2001.
7. Ngo, T.D., Kashani, A., Imbalzano, G., Nguyen, K.T.Q., Hui, D., Additive manufacturing (3D printing): A review of materials, methods, applications and challenges. *Compos. Part B*, 143, 172, 2018.
8. Zhang, C., Ouyang, D., Pauly, S., Liu, L., 3D printing of bulk metallic glasses. *Mater. Sci. Eng. R*, 145, 100625, 2021.
9. Wu, W., Li, X., Liu, Q., Fuh, J.Y.H., Zheng, A., Zhou, Y., Ren, L., Li, G., Additive manufacturing of bulk metallic glass: Principles, materials and prospects. *Mater. Today Adv.*, 16, 1, 2022.
10. Lashgari, H.R., Ferry, M., Li, S., Additive manufacturing of bulk metallic glasses: Fundamental principle, current/future developments and applications. *J. Mater. Sci. Technol.*, 119, 131, 2022.
11. Sohrabi, N., Jhabvala, J., Loge, R.E., Additive manufacturing of bulk metallic glasses - process, challenges and properties: A review. *Metals*, 11, 1, 2021.
12. Leitz, K.H., Singer, P., Plankensteiner, A., Tabernig, B., Kestler, H., Sigl, L.S., Multi-physical simulation of selective laser melting. *Met. Powder Rep.*, 72, 331, 2017.
13. Pauly, S., Ber, L.L., Petters, R., Stoica, M., Scudino, S., Kuhn, U., Eckert, J., Processing metallic glasses by selective laser melting. *Mater. Today*, 16, 37, 2013.
14. Mahbooba, Z., Thorsson, L., Unosson, M., Skoglund, P., West, H., Horn, T., Rock, C., Vogli, E., Harrysson, O., Additive manufacturing of an iron-based bulk metallic glass larger than the critical casting thickness. *Appl. Mater. Today*, 11, 264, 2018.

15. Wang, L., Wang, H., Zhou, S., Yan, M., Wang, G., Dai, Y., Selective laser melting helps fabricate record-large bulk metallic glass: Experiments, simulation and demonstrative part. *J. Alloys Compd.*, 808, 151731, 2019.
16. Li, N., Zhang, J., Xing, W., Ouyang, D., Liu, L., 3D printing of Fe-based bulk metallic glass composites with combined high strength and fracture toughness. *Mater. Des.*, 143, 285, 2018.
17. Nong, X.D., Zhou, X.L., Ren, Y.X., Fabrication and characterization of Fe-based metallic glasses by selective laser melting. *Opt. Laser Technol.*, 109, 20, 2019.
18. Nam, Y.G., Koo, B., Chang, M.S., Yang, S., Yu, J., Park, Y.H., Jeong, J.W., Selective laser melting vitrification of amorphous soft magnetic alloys with help of double-scanning-induced compositional homogeneity. *Mater. Lett.*, 261, 127068, 2020.
19. Hofmann, D.C., Bordeenithikasem, P., Pate, A., Roberts, S.N., Vogli, E., Developing processing parameters and characterizing microstructure and properties of an additively manufactured FeCrMoBC metallic glass forming alloy. *Adv. Eng. Mater.*, 20, 1800433, 2018.
20. Natarajan, S., Rathanasamy, R., Palaniappan, S.K., Velayudham, S., Subburamamurthy, H.B., Pal, K., Comparison of MA-g-PP effectiveness through mechanical performance of functionalised graphene reinforced polypropylene. *Polimeros: Ciênc. e Tecnol.*, 30, 3, 1–7, Article No. e2020035, 2020.
21. Xing, W., Ouyang, D., Li, N., Liu, L., Insight into micro-cracking in 3D-printed Fe-based BMGs by selective laser melting. *Intermetallics*, 103, 101, 2018.
22. Luo, N., Scheitler, C., Ciftci, N., Galgon, F., Fud, Z., Uhlenwinkel, V., Schmidt, M., Korner, C., Preparation of Fe-Co-B-Si-Nb bulk metallic glasses by laser powder bed fusion: Microstructure and properties. *Mater. Charact.*, 162, 110206, 2020.
23. Zou, Y.M., Wu, Y.S., Li, K.F., Tan, C.L., Qiu, Z.G., Zeng, D.C., Selective laser melting of crack-free Fe-based bulk metallic glass via chessboard scanning strategy. *Mater. Lett.*, 272, 12782, 2020.
24. Izadi, M., Farzaneh, A., Mohammed, M., Gibson, I., Rolfe, B., A review of laser engineered net shaping (LENS) build and process parameters of metallic parts. *Rapid Prototyp. J.*, 26, 1059, 2020.
25. Sun, H. and Flores, K.M., Laser deposition of a Cu-based metallic glass powder on a Zr-based glass substrate. *J. Mater. Res.*, 23, 2692, 2008.
26. Lin, X., Zhang, Y., Yang, G., Gao, X., Hu, Q., Yu, J., Wei, L., Huang, W., Microstructure and compressive/tensile characteristic of large size Zr-based bulk metallic glass prepared by laser solid forming. *J. Mater. Sci. Technol.*, 35, 328, 2019.
27. Hung, C.H., Sutton, A., Li, Y., Shen, Y., Tsai, H.L., Leu, M.C., Enhanced mechanical properties for 304L stainless steel parts fabricated by laser-foil-printing additive manufacturing. *J. Manuf. Process.*, 45, 438, 2019.

28. Li, Y., Shen, Y., Leu, M.C., Tsai, H.L., Building Zr-based metallic glass part on Ti-6Al-4V substrate by laser-foil-printing additive manufacturing. *Acta Mater.*, 144, 810, 2018.
29. Li, Y., Shen, Y., Hung, C.H., Leu, M.C., Tsai, H.L., Additive manufacturing of Zr-based metallic glass structures on 304 stainless steel substrates via V/Ti/Zr intermediate layers. *Mater. Sci. Eng. A*, 729, 185, 2018.
30. Zhang, C., Wang, W., Li, Y.C., Yang, Y.G., Wu, Y., Liu, L., 3D printing of Fe-based bulk metallic glasses and composites with large dimension and enhanced toughness by thermal spraying. *J. Mater. Chem. A*, 6, 6800, 2018.
31. Zhang, C., Wang, W., Xing, W., Liu, L., Understanding on toughening mechanism of bioinspired bulk metallic glassy composites by thermal spray additive manufacturing. *Scr. Mater.*, 177, 112, 2020.
32. Bordeenithikasem, P., Shen, Y., Tsai, H.L., Hofmann, D.C., Enhanced mechanical properties of additively manufactured bulk metallic glasses produced through laser foil printing from continuous sheet metal feedstock. *Addit. Manuf.*, 19, 95, 2018.
33. Gorodesky, N., Cohen, S.S., Altman, M., Fogel, O., Taguri, G.C., Fleger, Y., Kotler, Z., Zalevsky, Z., Concurrent formation of metallic glass during laser forward transfer 3D printing. *Adv. Funct. Mater.*, 30, 2001260, 2020.
34. Lewis, J.A., Direct ink writing of 3D functional materials. *Adv. Funct. Mater.*, 16, 2193, 2006.
35. Chen, W., Thornley, L., Coe, H.G., Tonneslan, S.J., Vericella, J.J., Zhu, C., Duoss, E.B., Hunt, R.M., Wight, M.J., Apelian, D., Pascall, A.J., Kuntz, J.D., Spadaccini, C.M., Direct metal writing: Controlling the rheology through microstructure. *Appl. Phys. Lett.*, 110, 094104, 2017.
36. Zhu, C., Qi, Z., Beck, V.A., Luneau, M., Lattimer, J., Chen, W., Worsley, M.A., Ye, J., Duoss, E.B., Spadaccini, C.M., Friend, C.M., Biener, J., Toward digitally controlled catalyst architectures: Hierarchical nanoporous gold via 3D printing. *Appl. Sci. Eng.*, 4, 1, 2018.
37. Kenel, C., Casati, N.P.M., Dunand, D.C., 3D ink-extrusion additive manufacturing of CoCrFeNi high-entropy alloy micro-lattices. *Nat. Commun.*, 10, 1, 2019.
38. Kaliyannan, G.V., Kumar, P.S., Kumar, S.M., Deivasigamani, R., Rajasekar, R., Mechanical and tribological behavior of SiC and fly ash reinforced Al 7075 composites compared to SAE 65 bronze. *Mater. Test.*, 60, 12, 1225–1231, 2018.
39. Ouyang, D., Li, N., Xing, W., Zhang, J., Liu, L., 3D printing of crack-free high strength Zr-based bulk metallic glass composite by selective laser melting. *Intermetallic*, 90, 128, 2017.
40. Deng, L., Kosiba, K., Limbach, R., Wondraczek, L., Kuhn, U., Pauly, S., Plastic deformation of a Zr-based bulk metallic glass fabricated by selective laser melting. *J. Mater. Sci. Technol.*, 60, 139, 2021.
41. Huang, Y.J., Shen, J., Sun, J.F., Bulk metallic glasses: Smaller is softer. *Appl. Phys. Lett.*, 90, 081919, 2007.

42. Best, J.P., Evenson, Z., Yang, F., Dippel, A.C., Stolpe, M., Gutowski, O., Hasib, M.T., Li, X., Kruzic, J.J., Structural periodicity in laser additive manufactured Zr-based bulk metallic glass. *Appl. Phys. Lett.*, 115, 031902, 2019.
43. Best, J.P., Ostergaard, H.E., Li, B., Stolpe, M., Yang, F., Nomoto, K., Hasib, M.T., Muránsky, O., Busch, R., Li, X., Kruzic, J.J., Fracture and fatigue behaviour of a laser additive manufactured Zr-based bulk metallic glass. *Addit. Manuf.*, 36, 101416, 2020.
44. Shi, J., Ma, S., Wei, S., Best, J.P., Stolpe, M., Markert, B., Connecting structural defects to tensile failure in a 3D-printed fully-amorphous bulk metallic glass. *Mater. Sci. Eng. A*, 813, 141106, 2021.
45. Sohrabi, N., Parrilli, A., Jhabvala, J., Neels, A., Loge, R.E., Tensile and impact toughness properties of a Zr-based bulk metallic glass fabricated via laser powder-bed fusion. *Mater. (Basel)*, 14, 5627, 2021.
46. Marattukalam, J.J., Pacheco, V., Karlsson, D., Riekehr, L., Lindwall, J., Forsberg, F., Jansson, U., Sahlberg, M., Hjorvarsson, B., Development of process parameters for selective laser melting of a Zr-based bulk metallic glass. *Addit. Manuf.*, 33, 101124, 2020.
47. Wang, G.Y., Liaw, P.K., Yokoyama, Y., Freels, M., Buchanan, R.A., Inoue, A., Brooks, C.R., Effects of partial crystallization on compression and fatigue behavior of Zr-based bulk metallic glasses. *J. Mater. Res.*, 22, 493, 2007.
48. Sohrabi, N., Nasab, M.H., Rouxel, B., Jhabvala, J., Parrilli, A., Vedani, M., Loge, R.E., Fatigue performance of an additively manufactured Zr-based bulk metallic glass and the effect of post-processing. *Metals*, 11, 1064, 2021.
49. Li, X.P., Roberts, M.P., O'Keeffe, S., Sercombe, T.B., Selective laser melting of Zr-based bulk metallic glasses: Processing, microstructure and mechanical properties. *Mater. Des.*, 112, 217, 2016.
50. Pauly, S., Schricker, C., Scudino, S., Deng, L., Kühn, U., Processing a glass-forming Zr-based alloy by selective laser melting. *Mater. Des.*, 135, 133, 2017.
51. Deng, L., Gebert, A., Zhang, L., Chen, H.Y., Gu, D.D., Kühn, U., Zimmermann, M., Kosiba, K., Pauly, S., Mechanical performance and corrosion behaviour of Zr-based bulk metallic glass produced by selective laser melting. *Mater. Des.*, 189, 108532, 2020.
52. Zhang, C., Li, X., Liu, S.Q., Liu, H., Yu, L.J., Liu, L., 3D printing of Zr-based bulk metallic glasses and components for potential biomedical applications. *J. Alloys Compd.*, 790, 963, 2019.
53. Bordeenithikasem, P., Stolpe, M., Elsen, A., Hofmann, D.C., Glass forming ability, flexural strength, and wear properties of additively manufactured Zr-based bulk metallic glasses produced through laser powder bed fusion. *Addit. Manuf.*, 21, 312, 2018.
54. Su, S. and Lu, Y., Laser directed energy deposition of Zr-based bulk metallic glass composite with tensile strength. *Mater. Lett.*, 247, 79, 2019.

55. Lu, Y., Su, S., Zhang, S., Huang, Y., Qin, Z., Lu, X., Chen, W., Controllable additive manufacturing of gradient bulk metallic glass composite with high strength and tensile ductility. *Acta Mater.*, 206, 116632, 2021.
56. Punnathat, B., Shen, Y., Tsai, H.L., Hofmann, D.C., Enhanced mechanical properties of additively manufactured bulk metallic glasses produced through laser foil printing from continuous sheetmetal feedstock. *Addit. Manuf.*, 19, 95, 2018.
57. Gibson, M.A., Mykulowycz, N.M., Shim, J., Fontana, R., Schmitt, P., Roberts, A., Ketkaew, J., Shao, L., Chen, W., Bordeenithikasem, P., Myerberg, J.S., Fulop, R., Verminski, M.D., Sachs, E.M., Chiang, Y.M., Schuh, C.A., Hart, A.J., Schroers, J., 3D printing metals like thermoplastics - fused filament fabrication of metallic glasses. *Mater. Today*, 21, 697, 2018.
58. Palaniappan, S.K., Chinnasamy, M., Rathanasamy, R., Pal, S.K., Synthetic binders for polymer division, in: *Green Adhesives: Preparation, Properties and Applications*, pp. 227–272, Wiley Scrivener Publishing LLC, River street, Hoboken, USA, 2020.
59. Lu, X., Nursulton, M., Du, Y., Liao, W., Structural and mechanical characteristics of $Cu_{50}Zr_{43}Al_7$ bulk metallic glass fabricated by selective laser melting. *Mater.*, 12, 775, 2019.

11

Additive Manufacturing of Tools and Dies for Metal Forming Applications

Yogeshwaran K. and Shubhajit Das*

Department of Mechanical Engineering, National Institute of Technology, Arunachal Pradesh, India

Abstract

The method of creating a product through the addition of metal is known as additive manufacturing (AM). In recent days, this technique is recommended for manufacturing tools and dies. The tools and dies for forging, extrusion, and sheet metal processes must be produced using a precise manufacturing method. In the AM process, the parts were accurately constructed layer by layer at the micro level. As a result, production time has decreased compared to the traditional approach and AM does not require part machining. Die production utilized AM techniques such as fused filament fabrication (FFF), selective laser sintering (SLS), wire arc additive manufacturing (WAAM), and laminated object manufacturing (LOM). This chapter explores the potential and application of AM in the creation of tools and dies for metal forming operations. In order to create complex geometries and get a superior surface polish, some hybrid production techniques have also been devised. Furthermore, the distinction between the direct and indirect quick tooling processes has been investigated. Also explored are the problems with metals in AM and solutions to them.

Keywords: Additive manufacturing, metal forming, tool and die, forging, extrusion and sheet metal process

11.1 Introduction

One of the typical industrial methods used to create a metal part is metal forming. Metal forming is the process of applying an external force on

**Corresponding author*: 06shubhajit@gmail.com; shubhajit@nitap.ac.in

R. Rajasekar, C. Moganapriya and P. Sathish Kumar (eds.) Additive Manufacturing with Novel Materials: Processes, Properties and Applications, (343–372) © 2024 Scrivener Publishing LLC

a solid material to change its geometry to a desired state. The numerous metal forming processes include rolling, extrusion, forging, and sheet metal forming (SMF). Forging and extrusion are two examples of metal forming processes that combine heat and external force. By exerting force on the workpiece over the stiff permanent die, the desired form can be achieved in the workpiece. According to their intended use, rigid dies are constructed from various metals and special alloys. The SMF processes are used to create automotive and aerospace components. The creation of the necessary tools and dies is the most important step in the production process. Geometry-specific and general tooling are two categories for dies used in metal forming. Geometry-specific tooling refers to the dies used in closed-die forging, extrusion, and stamping. Generic tooling includes dies that used for rolling, blanking, and open-die forging.

Cao *et al.* [1] provided an explanation of how metal forming tool manufacture works. The numerous essential elements and processes that must be used in the tool and die manufacturing process are briefly covered in various parts of their study. The essential functional requirements and various fabrication procedures for tooling in metal forming operations are shown in Figures 11.1–11.3. Next, he described the various tool materials, fabrication techniques, and process control systems used in the production of tools and dies. For the most part, computer numerical control (CNC) machines have been utilized to machine the tool and die for sheet metal processing. A skilled CNC programmer is then required for programming, and skilled labor is required to operate the CNC. Therefore, creating the

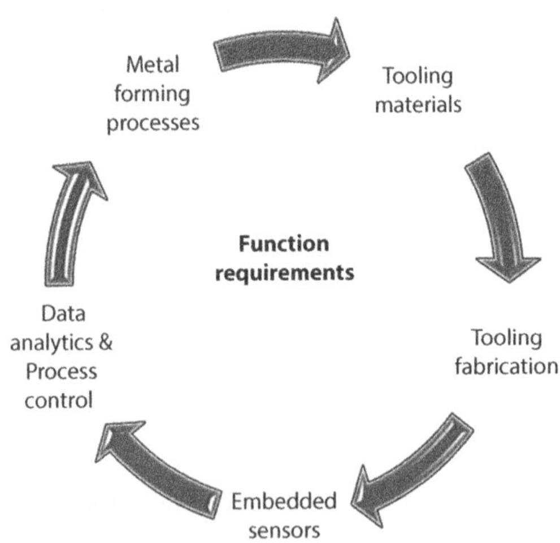

Figure 11.1 Key factors in fabricating the tools and die for metal forming.

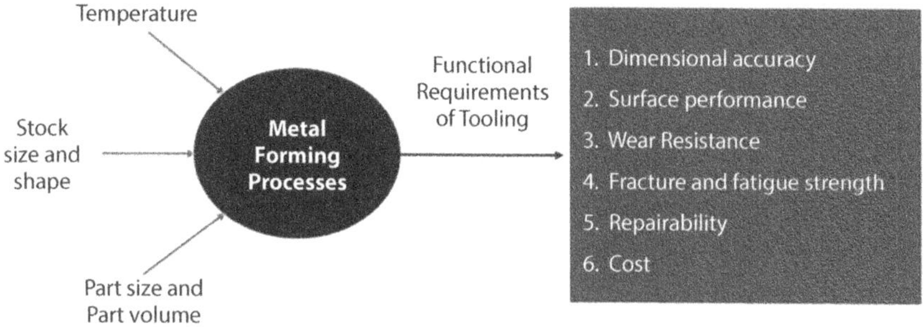

Figure 11.2 Functional requirements of tooling.

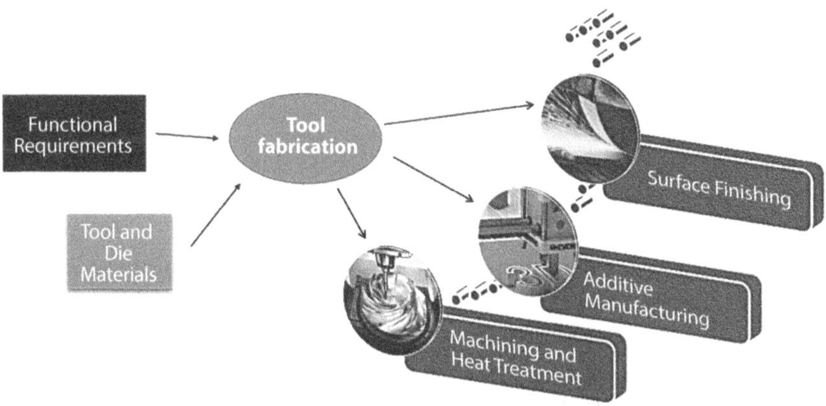

Figure 11.3 Fabrication methods of tool and die.

tool and die requires a lot of time. And the biggest drawback of conventional production methods is the lengthier time required to switch the production line to a new product. However, the short lead time and part customization are both possible with the additive manufacturing (AM) technique. However, the AM method makes it simple to switch quickly to producing new parts. For the manufacture of the tools and dies for the metal forming process, numerous AM types were used. In several circumstances, the production time and cost were decreased by using AM.

This chapter gives a detailed review and also discusses the numerous research projects on AM techniques that are used to create the tools and dies used in metal forming procedures. The fused filament fabrication (FFF) and selective laser sintering (SLS) were chosen for producing the tool and die for the SMF process in the majority of research efforts. Fused deposition modeling (FDM) is another name for FFF that is used commonly.

The various innovation techniques, hybrid manufacturing methods which are introduced to manufacture the tools and dies for the metal forming processes are also discussed. For locating the expedient tooling operations, Chung *et al.* [2] developed a tool guidance system. The build time, overall time, building cost, and overall cost of the tool and die may all be forecasted utilizing this designed tool advisory system for the FDM, SLS, and stereolithography (SLA) processes. This advisory system was constructed with the use of Java programming. The input data—material qualities and geometry details—were programmed using Java from earlier efforts. The final output product is made using direct rapid tooling. In indirect rapid tooling, the master tool is created, and the final tool is created via conventional techniques.

11.2 Extrusion-Based Additive Manufacturing

Extrusion-based additive manufacturing (EAM) is a method for creating parts out of polymer-based components. Fused deposition modeling is an extrusion-based kind of AM. The substance used for fabrication in this procedure is called filament. This approach is particularly successful when compared to other AM processes. For quickly and easily constructing the item, this FDM procedure is very cost-effective and effective. A thorough analysis of the potential applications of extrusion-based additive manufacturing (EAM) for producing tools for metal forming operations was carried out by Strano *et al.* [3]. The process of fused deposition modeling is also known as EAM. He explored the various materials and three-dimensional (3D) printing methods for metals using the FDM machine in this review. It is more cost-effective than the direct metal laser sintering method (DMLS). Stainless steel, ceramics, and thermoplastics with fiber reinforcement can all be employed in this procedure. Thermal and geometric qualities must be attained by the made tool. The tool and dies for metal shaping can be produced using either a polymer- or metal-based EAM method. The filament in metal-based EAM contains 60% metals. The filament must be preheated before printing in order to minimize the thermal deficit problem. The created component must then be sintered to permanently bond the layers of materials. Sintering and preheating are not required in polymer-based EAM. Metal forming tools and dies made of polymer can be utilized with sheets of 0.6 to 2.0 mm thickness. When creating dies out of polymer materials, the raster angle of 45° works well.

According to Tondini *et al.* [4], the FDM process could take the role of the traditional way of producing tools for sheet metal shaping. This cuts down on the price of raw materials, manufacturing time, and energy use. The component produced by the tool created utilizing AM has been analyzed. The tool was created using a filament made of polylactic acid (PLA). Compared to standard manufacturing methods, the cost of the tool was extremely low. Each tool's printing time, weight, and associated manufacturing costs were compared to those of a similar tool created utilizing a milling technique and wire electric discharging machine. The final product of the sheet metal technique used to shape the part using the die is shown in Figure 11.4. Vertical orientation has been chosen as a criterion for constructing the tool because the die working perpendicular to the created layers produces a compressive force that is resisted by this orientation. After that, he offered a plan for introducing flexible production in SMF processes.

In their investigation of the effectiveness of punch and die made using FDM methods for SMF, Nakamura *et al.* [5] utilizing PA, the punch and die for the v-bending operations were created. First, tests were done on the punch and die material's compressive strength. The punch and die were then modeled in computer-aided design (CAD) software and translated to a stereolithography file (STL file) format. The punch and die manufacturing process in FDM took 3 h. Figure 11.5 depicts the punch and die produced using FDM techniques.

The spring back effect of the bending material must be taken into account while designing the punch and die. Steel sheets composed of A1100-H14, A5052-H32, and 440 Mpa were bent using a PA punch and die. In order to

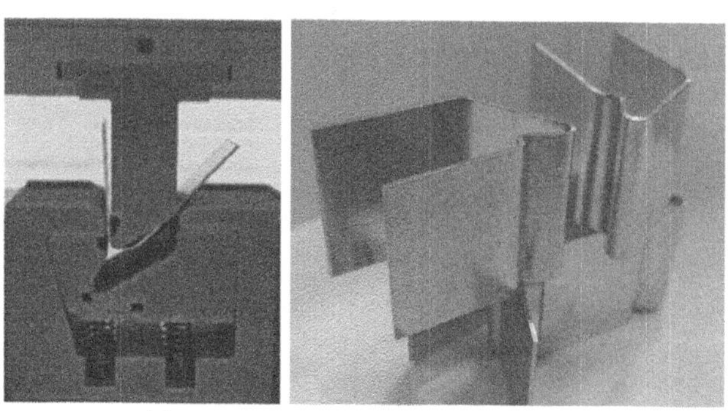

(a) (b)

Figure 11.4 (a) Die made by FDM and (b) finished component [4].

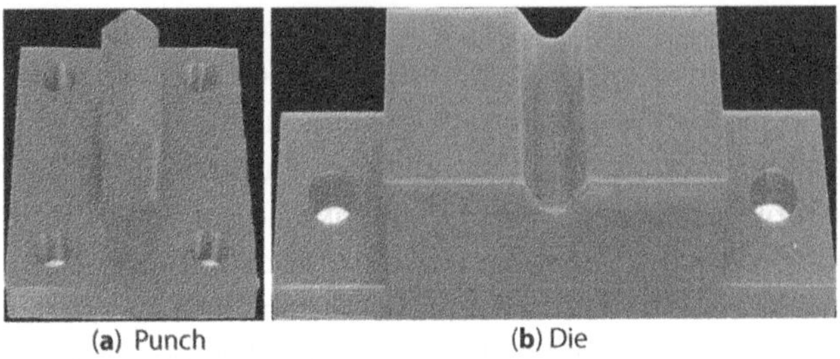

Figure 11.5 Polylactic acid tools manufactured by 3D printer [5].

measure the elastic deformation of the tool, the punch and die angles were changed. The v-bending procedure is shown in Figure 11.6. The bending angle of plastic tools is comparable to that of steel tools. The scratching of the bending sheets is avoided with plastic tools. Plastic tools perform similarly to steel tools in comparison. The use of acrylonitrile butadiene styrene (ABS) material as a die for the hydroforming process was looked into by Prithvirajan *et al.* [6]. A fluid was used to exert pressure on thin metal sheets during the metal forming process known as hydroforming in order to shape the metal. In this project, the bellow-shaped element must be manufactured utilizing a polymer die and a hydroforming technique that incorporates FDM. The die was created using an existing design, after which it underwent finite element analysis. For analysis, the die was divided into two parts: the top die and the bottom die. This static analysis was done to determine how much contact pressure will be applied to the die during the hydroforming operation. The ABS material property model was then put under the same amount of pressure to see if it can survive the load situation or if it had to be redesigned. The investigation revealed that the bellow's root region experiences the most stress. A die was then created using an FDM machine with ABS-P430. Then, utilizing SS304 metal and

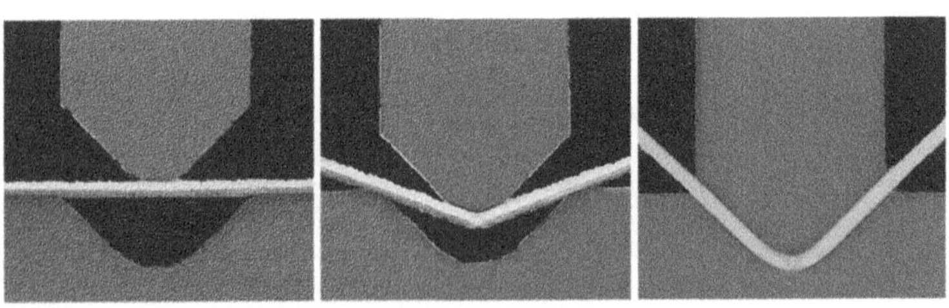

Figure 11.6 V-bending process for A1100-H14 sheet with plastic tools [5].

a 25-ton load, the die was used in a hydroforming procedure to create a bellow shape. The polymer die successfully produced a product more than 20 times.

Aksenov *et al.* [7] investigated the use of a plastic die for metal forming operations on American National Standards Institute (ANSI) 3003 aluminium sheet. As a plastic to create a die out of, polyethylene terephthalate was chosen. To prevent the spring back effect, the process was carried out using three sets of dies. The geometry shape of the dies and workpiece were simulated using the Simufact forming simulation software. It depicted the stress distribution on the tool and blank as well as the force needed to complete the operation. It was demonstrated in simulation that tools were created with elastic properties following simulation. This indicated that the tool was not deformed during this procedure, that it was safely designed, and that it may be used in real-world situations. The effectiveness of a carbon fiber reinforced polyphenylene sulphide material as a compression mold for the metal forming process was examined by Yeole *et al.* [8]. The schematic diagram of the extruder is depicted in Figure 11.7. The extruder was changed to create a huge shape of a product. This extruder was used to create the mold for the manufacturing process, and a sample model of the mold is shown in Figure 11.8.

For analyzing the sheet metal component and converting the file format needed by the rapid prototype machine, specialized software is available on the market. In order to transform the format of a part model into a rapid prototype format, Jirathearanat *et al.* [9] explored the virtual processing methodology. The surface geometry of the component to be made was translated utilizing the PAM-STL, PAM-STAMP, and FEM-STL software package to the format required by the rapid tooling machine. The model of a part was created using PAM-STL software and then converted to finite element model format. The sheet metal model was then simulated using PAM-STAMP software, then converted to an STL file for rapid prototyping. The PAM-STAMP meshed model was also converted with FEM-STL to an STL file for rapid prototyping. An examination of the use of 3D printing in sheet metal punching operations was conducted by Klimyuk *et al.* [10]. For small-scale batch production, the traditional method of making tools and dies can be replaced by employing the FDM 3D printing technology. The tool for the punching process was made of polyethylene terephthalate, and it must withstand the geometry and compression stress tests. The tensile and compression strength of 3D printed objects was tested in accordance with American Society for Testing and Materials (ASTM) D695 standards. The specimen was created using this standard and had a cylinder shape. After that, it was examined, and the results are depicted in Figure 11.9.

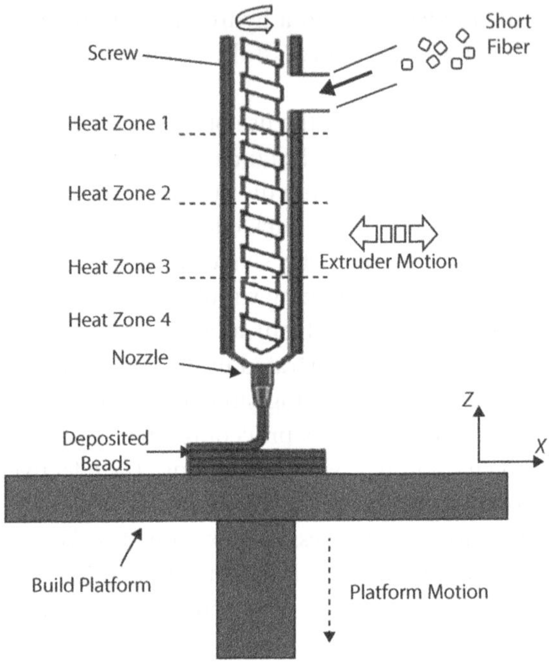

Figure 11.7 Schematic diagram of EAM [8].

The tool was constructed after the tested specimens were successfully tested. A hollow cylinder-shaped object was created utilizing a polyethylene terephthalate mold and die, as shown in Figure 11.10.

The punch and die used for deep drawing operations in big or small quantity production were explored by Schuh *et al.* [11]. This inquiry was done to find a cost-effective way to do a deep drawing procedure. For the creation of a punch and die, PLA was chosen as the material. For the deep drawing operation, the punch and die were made using the FDM technique. According to DIN EN ISO 20582, a cupping test was performed on 3D specimens to determine the material's ductility. When constructing the punch and die, the shrinkage of the polymer material during cooling was taken into account. Parameter-like infill pattern had more influence on compression strength of the printed parts. Tri-Hexagon and triangle shape infill pattern had a high compression strength. The steel screw was fastened with a punch and die to increase strength. It strengthened the

Figure 11.8 CAD model of the Mold.

internal structure of the printed pieces and lowered material usage. The designed part was virtually tested using the Abaqus simulation software. The results of the simulation test demonstrated that deep drawing operations may be performed using punch and die made of PLA.

The v-bending and groove pressing processes used with the 3D printed punch and die were investigated by Tondini *et al.* [12]. Fused deposition modeling and vat photopolymerization additive manufacturing (VPAM) were two processes used in the production of the punch and die. The resin used by VPAM was cured using ultraviolet lights. The substance for the FDM process was PLA, and the material for the VPAM process was an acrylate-based photopolymer with phosphine oxide photo initiator. Items for FDM must withstand compressive force. Therefore, pieces

Figure 11.9 Deformed shapes of samples [10].

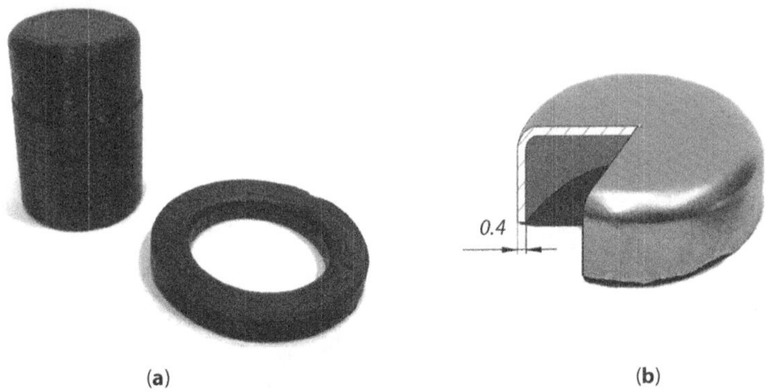

Figure 11.10 (a) Punch and die manufactured by using 3D printing: (b) workpiece manufactured by using the 3D printing tool [10].

were produced perpendicular to the vertical force. When parts were created with FDM, parameters had a significant impact on their qualities. The void in the manufactured part was decreased by utilizing a thin layer while printing the item in FDM. Layer thickness also had a significant impact on polymerization in VPAM. Infill density for polymerization and providing bonding between layers was 50% in FDM and 100% in VPAM. Punch and die wear and tear on sharp edges was greater. Therefore, various angles and radii were employed to examine the performance. When compared to the traditional process, 3D printing offered a punch and die bend radius that was correct. In comparison to the traditional punch-and-die manufacturing process, 3D printed items produced better results. Additionally, employing it in production was advised.

The structural and frictional performances of ABS P430 were examined by Leacock *et al.* [13]. Fused deposition modeling allowed for the 3D printing of parts that had been approved for use in SMF. The test specimens were produced and analyzed by adjusting the parameters in order to confirm the compression strength. When compared to others, the 45° raster angle and edge construction orientation offer good strength. To assess the material's ability to reduce friction, a strip pull friction test was performed. The pull strips were produced in the following dimensions: 315 × 50 × 1.01. The ABS P430 had a good coefficient of friction (both static and dynamic). The specimens' friction varied depending on how the orientations were combined. The ABS P430 had strong compression strength, making it suitable for rapid tooling.

11.3 Laser-Based Additive Manufacturing

Powder bed-based manufacturing is one method of AM. In this method, powdered raw materials are employed, and a laser is used to heat and mold the powder to create the desired shape. The cladding method is used to deposit the particles one at a time. The pieces are then subjected to postprocessing procedures like sintering or heat treatment to improve the bonding of the particles. Kain *et al.* [14] investigated the concept of integrating the process chain for the extrusion process. Critical forms and twist channels could not be created using the usual production procedure in the past. However, it is now possible to fabricate utilizing AM processes. The tool's design can be optimized by testing it using finite element (FE) software.

Figures 11.11 and 11.12 depict the technique and advantages of digitalizing manufacturing operations. In digitally enabled manufacturing procedures, the tool was created in three weeks, reducing the lead time and cost of the tool. The tool will go through postprocessing such as sintering and polishing. The dimensional accuracy will then be assessed using the coordinate measuring machine (CMM). Deviations may arise differently in some postprocessing approaches, and they should be permitted by employing tolerance applied throughout the design phase._Metal forming tools and dies produced by utilizing the SLS process were investigated by Cheah *et al.* [15]. The punch and die for the embossing process, were modeled and fabricated using SLS. For embossing operations, aluminium sheet blanks were employed. Comparisons were made between the conventional and rapid tooling processes' manufacturing costs and lead times. It was discovered through comparison that the rapid tooling process had lower costs and shorter lead times than the traditional approach. Following that, the pieces' hardness was evaluated and their dimensions were assessed. The blank sheet was precisely shaped throughout the embossing process to match the punch's exact shape. The component's surface roughness was depicted as having the texture of a punch and die. The punch and die also showed no signs of deterioration or wear after stamping. Tools and dies

Figure 11.11 Digital enabled manufacturing processes.

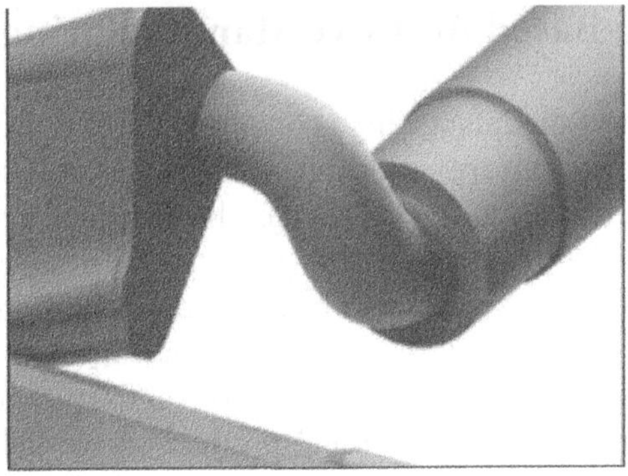

Figure 11.12 Inverted flow channel design for AM die manufacturing [14].

made with SLS are more practical and efficient than those made using the traditional method.

Hentschel *et al.* [16] examined laser metal deposition (LMD) processing of tool steel for forging tools. Laser metal deposition is also known as direct energy deposition (DED). Metal powder is fed via a gas stream and heated by a laser to produce metal in this procedure. This approach is already used in the production of die for injection molding. In this work, high carbon cold work tool steel was employed to create a forging tool. In this procedure, an inert gas, such as argon or helium, was used to produce an oxidation over a settled metal. The particle size of 5–190 μm was used to test the mass flow of the metal powder. The experimental study was carried out by adjusting the mass flow, laser power, and feed rate. The evaluation of the combination of process parameters was done utilizing the single weld track tests. The microhardness of the produced samples was assessed in this study using preheated and non-heated metal powders. The particles were preheated, which somewhat increased the microhardness. Preheated particles exhibited a dendritic microstructure, which was determined by examining the samples under a microscope. Particles in a size range of 20 to 60 μm provided an excellent mass flow rate for production and were advised for tool fabrication. Du *et al.* [17] investigated rapid sheet metal manufacture (RSMM), which is a computer-controlled closed loop method. RSMM consists of three steps: virtual and physical domains creation, and functional shaped parts. In the virtual domain, the part is modeled with a CAD package and evaluated for production feasibility. The tool for the sheet metal process is then created in the physical world utilizing rapid prototyping techniques and traditional machining processes such as milling. The fabricated tool is utilized for making the sample part for quality inspection in the third step in functional shape parts. If errors occur, the computer-aided verification (CAV) system automatically reports to the database and the design is automatically adjusted to correct the faults. The alignment of the punch and die should be prioritized when designing. The alignment can be matched by giving the pin holes, but adequate tolerances must also be allowed. The dimensional accuracy, manufacturing time, and cost of the SLS, SLA, CNC milling, and traditional tool making processes are then investigated. SLS, SLA, and CNC milling provide precision dimensioning in the tool. Except for traditional methods, all processes used less time in terms of production time.

Reggiani *et al.* [18] examined the selective laser melting (SLM) technique for creating conformal cooling channels (CCC) for extrusion dies. The liquid nitrogen flows in CCC during the extrusion operation to maintain a uniform temperature in the dies. Using SLM, the CCC was manufactured with AISI H13 steel to maximize cooling efficiency in dies. Conformal cooling channels was also used after the operation to cool the extruded product in the die. It was employed in the injection molding and blow molding processes. The tool for the extrusion process was built in two sections in this study. The first was an insert with CCC created using SLM, and the housing for the tool was made using a traditional machining method. The thermal efficiency of the planned CCC was tested using the COMSOL finite element simulation software at the peak temperature in both cooled and uncooled conditions. Lagrangian equations were also employed to assess the thermal efficiency of the SLM-based CCC. There was no difference in the values of the two approaches, proving that the design was secure. The die insert with CCC was then created using SLM. Yesildaga *et al.* [19] explored the difficulties associated with using SLM to make extrusion dies. In SLM dies, surface finish is a critical difficulty. If the tool's surface is rough, it will show in the end product. The polishing procedure was carried out on the die manufactured using SLM in this work. After machining, the rough surface was removed; this material must be compensated for in the design. As a result, the topology of the die must be optimized. The safest die topology must first be discovered. The prospective design was established using FE software (Polyflow tool) in Ansys. The die design was then optimized with Abaqus FE software before being produced with SLM.

Bambach *et al.* [20] studied postprocessing approaches on SLM components and compared them to forging-produced parts. Parts made with SLM cannot be used directly in the application because the component lacks strength in comparison to other production procedures such as forging. To address this shortcoming, academics developed a number of postprocessing approaches in recent decades. In this work, the sample part was manufactured with titanium alloy using SLM, as illustrated in Figure 11.13.

Following SLM fabrication, the pieces were evaluated for hot compression strength as two separate parts (without heat treatment and heat treated). The item was heated to 950°C for 30 min before being cooled using air. They were heated in induction by argon gas and tested to improve the hot compression strength and resistivity of SLM parts. Before and after testing, the microstructure of the SLM (without and with heat treatment) and forging samples was investigated. The porosity in the pieces was eliminated as a result of the hot compression. All samples had identical stress behavior, and SLM parts sustained more compression force than forging. Bi *et al.* [21] studied the porosity of a nickel-based alloy die insert produced using SLM for liquid forging. The most serious flaw in parts produced by the manufacturing process is porosity. In this study, samples were manufactured in SLM using various combinations of process parameters such as power, powder feed rate, shield gas timing, and layer thickness. The samples were then tested for microhardness and microstructure. When compared to other process factors, the powder feed rate parameter had a deficiency in fusion and porosity. Using the design of experiment (DOE), it was discovered that the 0.12-mm layer thickness resulted in less porosity. Oter *et al.* [22] investigated the advantages of laser-based AM for the fabrication of an extrusion die out of maraging steel. The experimental study was divided into two sections. During operation, the die should be able to withstand temperature stress, wear, and fatigue. The die design was designed and improved in the initial step for die preparation utilizing DMLS. The mechanical properties of the produced die were next evaluated. The die was heat treated after it had been produced in DMLS to improve its mechanical qualities. The hardness of the fabricated part was measured after it had been annealed at 550°C for 3–12 h and cooled by air. The die profile quality was good when manufactured with DMLS. The die's lifetime was then tested using a thermomechanical fatigue test, which yielded a positive result.

Foster *et al.* [23] studied the use of laser-based AM for hot forging die repair. The die, which was made in the traditional way, became worn after

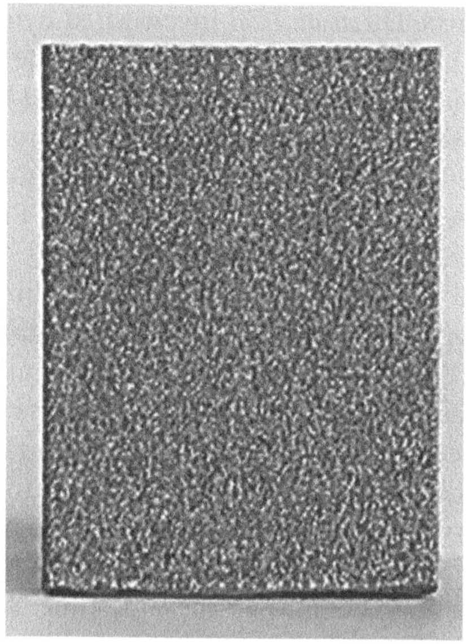

Figure 11.13 Sample made by SLM process [20].

multiple uses. The permanent die was fairly expensive. As a result, in order to cut manufacturing costs and reuse worn-out dies, the advanced forming research centre (AFRC) developed a low-cost test method to measure abrasive wear. The worn-out die was repaired and reused for production using the CNC machine and LMD. Computer numerical control was used to machine the worn-out section of the die. Then, utilizing LMD, the worn-out portion was recreated as an insert on die, and the surplus section was machined. Following that, the machined profile was measured for accuracy using a CMM.

Khalifa *et al.* [24] investigated a novel concept of AM extrusion die in order to improve the mechanical properties of the extruded part. This concept was investigated in order to obtain anisotropy properties in an extruded part manufactured using an AM die. When designing an extrusion die, several plastic deformation (SPD) techniques such as equal channel angular pressing or extrusion and asymmetric extrusion were used to change the strain path. The material flowed through the entrance and exit was modified in the design and achieved by using the SLM in this concept. The FE simulation software was used to analyze the performance of the billet in the die during extrusion. The inner part was then fabricated using SLM, and the outer part was made by hot working to reduce the cost of making the die. Figure 11.14 depicts the assembled inner and outer parts. Du *et al.* [25] investigated hybrid manufacturing methods for 18Ni maraging steel metallic parts. The precision and surface finish of the manufactured parts were improved by combining SLM and milling processes. This hybrid manufacturing process was developed to avoid the stair effect in the AM process. Metal addition and subtraction (A/SM) had been used in this hybrid process. Figure 11.15 depicts the A/SM procedure.

When compared to 3D welding and milling, combining SLM and milling is very effective because the machining operation performed after fabricating the part results in a greater amount of material collected as scrap. However, machining is frequently used in this concept. As a result, waste will be reduced, as will the time required to fabricate the tool or die. In this process, the initial part was created using SLM, and milling was used to form the shape. The process was then repeated until the final shape was achieved. Both additive and subtractive processes were carried out in separate machines at the same time. As a result, fabrication time was reduced when compared to other hybrid manufacturing processes. The process parameters that must be maintained in AM for this concept are laser energy, voltage, heat conductivity of the bed, and scanning speed. After the manufacturing process was completed, the tool or die was heated in a furnace for 3 h at 500°C for heat treatment. The microstructure was then examined in the cross section of the tool in the perpendicular and parallel directions of the metal settled. The fabricated tool's surface was smoother than the AM, and porosity was discovered to have been eliminated while examining the microstructure. Because of the heat treatment, the grains were properly aligned with fine cellular and

Figure 11.14 Assemble of die made by AM [24].

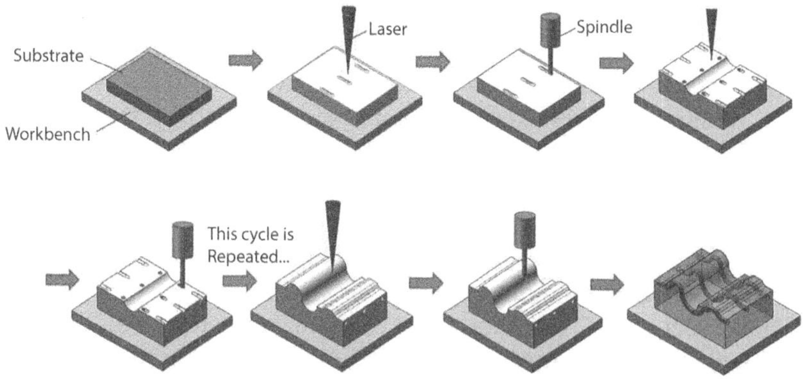

Figure 11.15 Schematic description of the hybrid A/SM process [25].

dendritic structures. Otherwise, the microstructure would have become martensite as it cools. The tool's hardness was measured on the top and side surfaces. However, there was no difference in hardness between the two surfaces.

11.4 Layer-Laminated Additive Manufacturing

Layer laminated AM is a process of fabricating the part or specimen with layer by layer of metal sheets and bonded together by using laser. It is also called laminated object manufacturing (LOM). At the time of discovery, the parts were manufactured by using paper. The paper was cut by laser according to the design and set layer by layer to make a prototype model. It was later developed, and nowadays metal sheets are implemented for producing tools and dies for manufacturing process. Bryden *et al.* [26] analyzed the operations of the Warwick Manufacturing Group's large scale tooling for rapid manufacture. This research was carried out in order to solve concerns that arose during the mass production of a tool designed using LOM. The problems are insufficient strength during mass production, automation limitations, and the impossibility to produce thicker materials. They looked into the equipment utilized in superplastic forming, hot creep forming, resin transfer molding, and injection molding. They first preferred the austenitic stainless-steel alloy for low-temperature manufacturing during the brainstorming session. They then determined whether the material should be constructed horizontally or vertically. It is best to make shallow parts horizontally and large parts vertically. Waste materials from horizontal build orientation can be reused as a core for another production; nevertheless, vertical build orientation gives excellent compression strength in large volume manufacturing in tool. To evaluate the bonding strength of the manufactured tool, the following tests were performed: single lap shear, lap shear (at higher temperatures and after coolant application), and flexural test. Horizontal build orientation tools are built faster and have a better sealing effect on injection molding.

Holkera *et al.* [27] used the LOM to evaluate the extrusion dies with conformal cooling channels. The excess heat produced in the die surface during the extrusion process is caused by the continuous operation and speed of extrusion in the die. It results in poor surface quality and other flaws. The cooling channel can be installed to maintain a consistent temperature in the die. However, in traditional production processes, the multidirectional cooling channel and cooling efficiency are inadequate. Additive manufacturing can be used to provide a better multidirectional cooling channel for the die. In this study, an efficient multidirectional cooling channel for the extrusion die was achieved by using the LOM and SLM. The design for heat dissipation in the cooling channel was simulated using FE software. The LOM was used to create the cooling channel, and the SLM was used to create the die's mandrel. The material flow and surface were examined using viscoplastic analysis, and the results were positive. Himmer *et al.* [28] explored metal laminated tooling (MELATO), a process chain that combines metal additive and subtractive processes to create a tool. Laser-based welding, cutting, and depositing are all used in this study. Initially, the CAD model of the part was created and converted to an STL file for further processing. The central section of the instrument was constructed utilizing metal sheet lamination in this MELATO method. Additional tool parts were manufactured with inserts using the traditional manner. Adaptive slicing approaches were used to determine the optimum way to develop the tool for improved slicing. The sliced sheets were then joined together by screws known as casting anchors. A low melting alloy was used to fill the space between the sheets. This alloy expanded during the solidification process, making the tool stiff and pulling back the casting screw to minimise shrinkage defects. The sheets were then connected and produced as a tool using laser welding, which provided good strength. The laser cutting or milling procedures were undertaken for finishing and dimensional correctness.

11.5 Wire Arc Additive Manufacturing

Wire arc additive manufacturing (WAAM) is a 3D printing process that is used to produce or repair parts. It is a method of DED. It combines gas welding with AM technology. It can manufacture large size objects with good mechanical qualities when compared to other AM processes. Metal deposition rate is higher when compared to other 3D printing methods. In WAAM, hybrid manufacturing is achievable by integrating with other processes. The surface finishing of this process is not nearly up to par. As a result, some finishing process must be involved in order to finish the produced object. Following printing, heat treatments must be used to remove residual tension and stress in part. It improves the quality and life of the manufactured part.

Hong *et al.* [29] researched WAAM in order to create a big die for the hot forging process. Because of the high temperature application, the tool or die used in hot forging has a shorter lifetime. The die producing process includes casting, metal removal processing, and heat treatment. As a result, the cost of producing dies is high. Wire arc additive manufacturing allows for the production of low-cost, high-precision, and accurate dies in an expedient manner. Tool path moment planning is critical in AM. There are two approaches to tool path planning: vertical additive manufacturing (VAM) and normal additive manufacturing (NAM). Vertical additive manufacturing is appropriate for large form parts, while NAM is suitable for coating processes. As a result, VAM is chosen to create the die for large-scale hot forging. However, NAM is occasionally chosen for large size manufacturing since VAM cannot make parts with smaller thickness. More heat is generated during the hot forging process. As a result, the heat dissipation test must be performed in the intended die for safety, utilizing FE software. The wear resistance of Fe-based, Ni-based, and Co-based alloys was compared. Cobalt-based alloys are used to create dies because they have a high wear resistance. The process parameters and the relationship between welding are assessed using an artificial neural network (ANN) for better die manufacturing utilizing WAAM. Welding process factors include voltage, welding speed, feeding speed, and workpiece size. The die is then created using WAAM, and the same procedure is also recommended for rebuilding worn-out hot forging dies. Duartea *et al.* [30] studied a new type of hot forging-wire arc additive manufacturing (HF-WAAM) that combines forging with AM. Following printing, the produced item is immediately forged, and viscoplastic deformation occurs at high temperatures.

Figure 11.16 depicts a comprehensive schematic diagram of hybrid AM. The fundamental goal of this approach is to achieve viscoplastic behavior of the process at high temperatures. Viscoplastic behavior is a material's inelastic behavior at high stress and temperature. Due to the continuous operation of the hot forging process, significant heat will rise in the die. It may cause the die's lifespan to be reduced. The creep strength of the die will be improved and the tool's life will be extended by employing this HF-WAAM. The quality of the manufactured item will vary as process factors such as layer count, forging frequency, distance to arc center, wire feed speed, voltage, shield gas, and gas flow rate are optimized. The created portion solidified from bottom to top in conventional WAAM. As a result, the microstructure at the bottom is fine, while the microstructure at the top is coarse. The portion does not have a homogeneous dendritic structure. To give good mechanical properties, the HF-WAAM technique produces consistent solidification and dendritic structure throughout the part. The yield and tensile strength were measured and found to be positive.

Hirtler *et al.* [31] investigated the hardness and microstructure of hybrid parts produced through AM, milling, and forging. In comparison to SLM, the WAAM investment cost is lower and easier to manage. This concept yields cold metal technology (CMT) welding technology. WAAM achieves a good feed rate, build rate, and splash-free droplet in the metal pool by utilizing this technology. As a result, the mechanical properties of the fabricated part will be high. For this study, EN AW-6082 was chosen as the base metal and ER 4047 as the filler material, and testing is depicted in Figure 11.17. First, the T shape part was created (via forging and milling) and heat treated. The balance of the part was then fabricated using AM. The forged and milled specimen was preheated at 190°C before performing WAAM. Tensile strength, hardness, microstructure, and porosity were all tested in the manufactured samples. Binding errors occurred in some samples of both concepts. The microstructure test of both concepts established a fine eutectic structure. Milled and forged AM samples had uniform hardness and less than 1% porosity. The tensile strength of the forged sample exceeded that of the milled sample.

Meiners *et al.* [32] investigated the hybrid manufacturing process developed by WAAM and LMD to fabricate high performance parts out of titanium alloy. Two different manufacturing concepts are available by combining AM and forging. The first method, depicted in Figure 11.18, is forging the pre-shape of the part and additional shapes are created using AM. Second, the part's pre-shape was created by AM, and the balance elements were created by forging. Both concepts are examined in this study

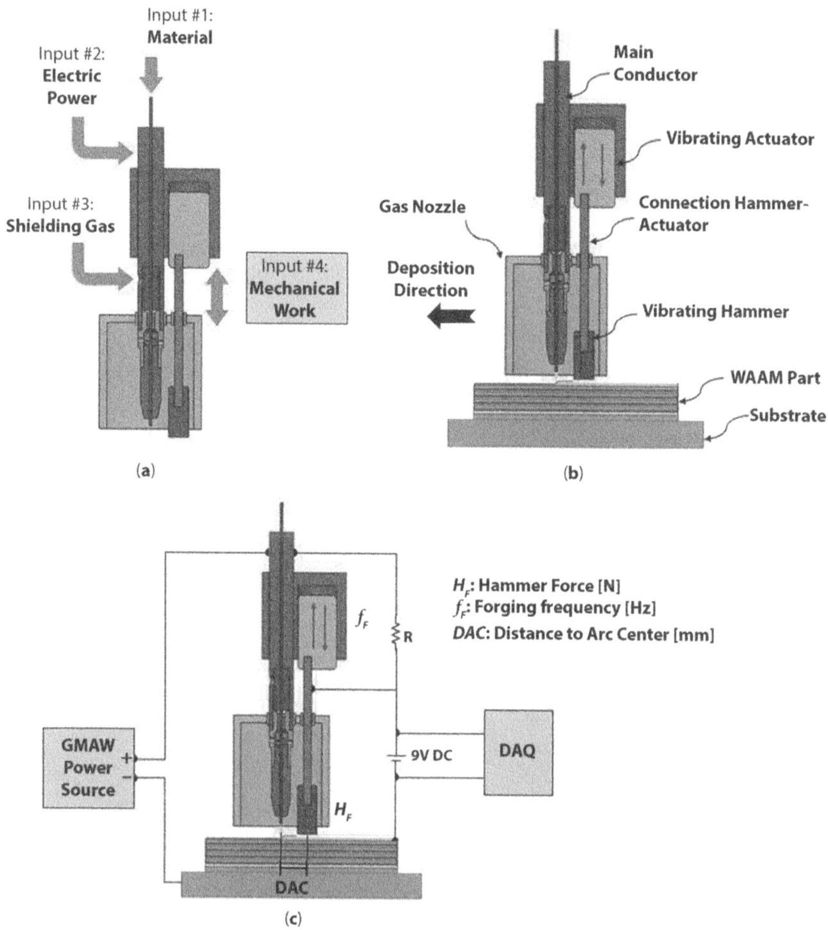

Figure 11.16 Schematic diagram of new type WAAM with forging (a) HF-WAAM inputs, (b) HF-WAAM components, and (c) process parameters and electrical scheme used to avoid arc between the vibrating hammer and welding torch [30].

using LMD and WAAM. Prior to that, heat treatment was used to improve the mechanical properties and relieve stress in fabricated pre-shape parts. Stress-relief annealing (at 710°C for 6 h) and annealing (at 1050°C for 3 h) were performed on forged pre-shape parts. The microstructure was studied in samples collected at three stages: before heat treatment, during heat treatment, and after heat treatment. Before and after heat treatment, the micro- and macrostructures were similar. The properties were the same due to the similar microstructure. The balance part was then created using

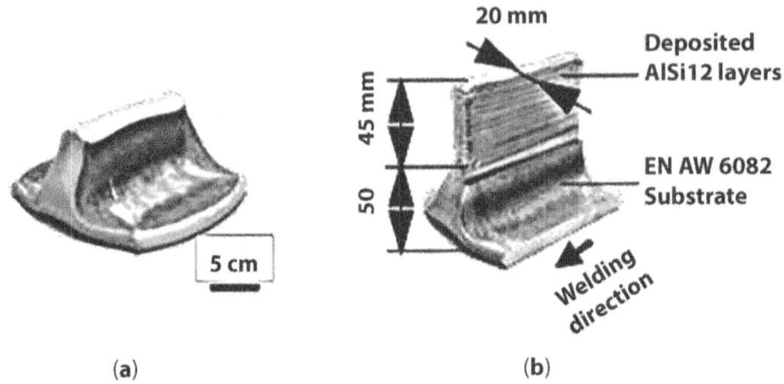

Figure 11.17 (a) Geometry of a forged sample; (b) shape of an AM deposited sample [31].

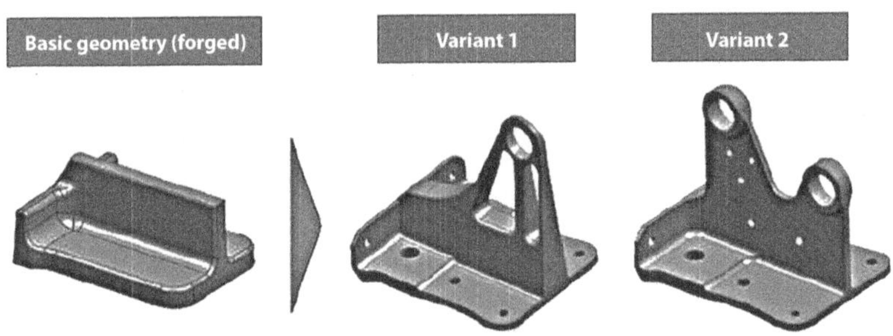

Figure 11.18 The process begins with basic forged geometry and progresses to AM [32].

the LMD and WAAM. Finally, LMD-produced parts performed better in fabrication than WAAM-produced parts. As a result, it was determined that hybrid manufacturing with LMD can be used to create tools or components for titanium alloys.

11.6 Discussions

The process of creating a tool or die is extremely precise and critical because the product's output is entirely dependent on the die. As a result, the die should be designed efficiently and manufactured at the lowest possible cost. In industries, tools and dies are made using the traditional machining process. However, completing the tool or die for manufacturing the part takes more time. Researchers recommend the AM process to shorten the time and lower the cost. Dies for SMF can be manufactured in both low- and high-volume production using the FDM process [3, 4, 8]. Polylactic Acid and ABS are the materials used to create the tool and die for SMF. The mechanical quality of the part manufactured is determined by the FDM process parameters [41]. According to research, raster angle 45° and vertical orientation provide good mechanical behavior in FDM-fabricated parts [13]. The extruder must be modified for good material feed rate and reduced time when manufacturing large size dies [8]. Metal dies for hot forging and extrusion can be created using WAAM, LOM, and SLS. A CCC is designed to maintain a uniform temperature in the extrusion die [18]. It can only be accomplished by employing AM techniques such as LOM and SLS. This channel is used to cool the die surface by passing coolants through it. It enhances the surface finish of the part produced by the extrusion process. In AM techniques, new concepts such as RSMM and MELATO are introduced to fabricate the tool and die for hot forging with automation [17, 24, 30].

The topology of the design can be analyzed and optimized for performance using FE software such as Ansys, Abaqus, COMSOL, PAM-STL, PAM-STAMP, and FEM-STL [14, 18, 19, 27, 29]. To fabricate dies for hot forging and extrusion, some hybrid manufacturing processes combining AM and conventional production processes are used [24, 25, 28, 30–32]. Additive manufacturing techniques such as LOM, WAAM, and SLS are combined with traditional manufacturing processes such as forging and milling. Selective laser sintering with milling produces a higher output in die manufacturing in this hybrid manufacturing [31, 32] because the porosity in the fabricated part is lower than in the other concept. The HF-WAAM concept, on the other hand, produces the part in less time and with viscoplastic behavior. However, when compared to others, the properties of the SLS with milling fabricated part are quite high.

11.7 Challenges of Metals in Additive Manufacturing

Metals are used as raw materials in the SLS, WAAM, and LOM processes. The metals in powder or sheet form are set to the desired shape of the model using a laser and heat energy. Metals in AM present a number of challenges [46–57] because some materials are not suitable for the AM process. When metals are employed in the AM process, required characteristics cannot be obtained due to its varying melting point temperature and incorrect bonding of incompatible materials [33–35]. This will occur particularly when choosing bi-materials like metal with ceramics and plastics. The items will be manufactured with residual stress, porosity, and voids as a result of the temperature mismatch. It shortens the life of the manufactured item and develops cracks in the part [36–45]. The production's scalability comes next as a challenge. When compared to traditional manufacturing methods, AM has a lower production rate and higher facility investment costs. In order to increase the scalability of AM processes, it must overcome this challenge by creating new production systems. Production time can be cut down and productivity can be increased by optimizing the component topology.

A greater number of trials must be conducted to obtain the desired physical qualities in multimaterial applications for good composition and optimum process parameters [51]. The cost of production and investment go up as a result. The optimization technique models in tools like Minitab and MATLAB can be used to get around this. Simulation software can then be used to assess the mechanical qualities. The metal should be preheated before manufacture and given postproduction heat treatments in order to improve the mechanical qualities. It will enhance the grain size of the metal and remove the defects (voids and residual stress) in the manufactured component and improve its attributes. Alloys of titanium, copper (Inconel), aluminum, stainless steel, silver, and platinum are suitable metals for AM processes. In AM, only a few alloys are used. For the development of novel alloys for AM, further research must be done. The main factors influencing metal implementation in AM are metallurgical defects such as porosity, residual stress formation, low quality, finishing, and cost. Metal particles are melted and solidified during the manufacturing process. During the solidification process, gas or air can become trapped in the part, causing it to fail, as well as the formation of residual stress in metals during AM processes due to the uneven thermal temperature of melting metals in contact with the environment temperature. Both challenges were overcome by manufacturing the part in a closed environment and protecting the melting zone with argon gas. Another minor concern in AM is low quality and surface finishing. Insufficient heat energy sources for melting the metals in AM processes result in low-quality SLS, WAAM, and LOM products. It can be avoided by keeping the metals at the proper temperature to melt while making the component. The surface finishing of AM-produced parts is not excellent. By applying postprocessing methods like grinding and polishing, it can be resolved.

11.8 Conclusion

Rapid tooling was in high demand in the metal forming process in order to reduce lead time, tool and die costs. The various concepts of creating a tool and die for metal forming using AM were investigated. Dies for SMF can be made efficiently, cheaply, and quickly using the FDM process with a 45° raster angle and vertical orientation to withstand compression loads. Extrusion and forging use a greater amount of heat. As a result, metals with good thermal properties can be used to create tools and dies. Metals in high temperatures can be handled using various AM techniques such as SLS, LOM, and WAAM. When compared to the other two methods, the WAAM process produces tools and dies with less lead time and cost because of the high material feed rate and deposition rate. However, the mechanical properties of the part manufactured by the above three methods are similar and will vary depending on the materials used. The performance of the designed part can be digitally evaluated and optimized using FE software simulation (Ansys, COMSOL, Abaqus, PAM-STL, PAM-STAMP, FEM-STL). Hybrid manufacturing techniques that combine AM (such as SLS and WAAM) with traditional machining processes (such as forging and milling) are introduced to produce tools and dies with better mechanical properties and in a more efficient manner. The HF-WAAM concept, reduces production time and cost of the die. Finally, the FDM process is appropriate for creating tools and dies for SMF. Selective laser sintering with milling is the best way to create a tool and die for extrusion and hot forging with an optimized topology using FE software. However, scalability, multimaterial usage, investment expense, and metallurgical flaws are issues for metals in AM. By enhancing the material composition, appropriate thermal energy, safe atmosphere, pre- and postprocessing procedures, these issues can be resolved.

References

1. Cao, J., Brinksmeier, E., Fu, M., Gao, R.X., Liang, B., Merklein, M., Schmidt, M., Yanagimoto, J., Manufacturing of advanced smart tooling for metal forming. *CIRP Ann.*, 68, 2, 605–628, 2019.
2. Chung, C.W., Hong, J., Lee, A., Ramani, K., Tomovic, M., Methodology for selection of rapid tooling process for manufacturing applications. *Innovations and Applied Research in Mechanical Engineering Technology*, 2003.

3. Strano, M., Rane, K., Farid, M.A., Mussi, V., Zaragoza, V., Monno, M., Extrusion-based additive manufacturing of forming and molding tools. *Int. J. Adv. Manuf. Technol.*, 117, 7–8, 2059–2071, 2021.
4. Tondini, F., Arinbjarnar, U., Basso, A., Nielsen, C.V., 3D printing to facilitate flexible sheet metal forming production. *Proc. CIRP*, 103, 91–96, 2021.
5. Nakamura, N., Mori, K.I., Abe, F., Abe, Y., Bending of sheet metals using plastic tools made with 3D printer. *Proc. Manuf.*, 15, 737–742, 2018.
6. R., P., M., S., N., S., G., A., Metal bellow hydroforming using additive manufactured die: A case study. *Rapid Prototyping J.*, 25, 4, 765–774, 2019.
7. Aksenov, L.B. and Kononov, I.Y., 3D printed plastic tool for Al thin-sheet forming. *IOP Conf. Ser.: Earth Environ. Sci.*, 337, 1, 012053, 2019.
8. Yeole, P., Kim, S., Hassen, A.A., Kumar, V., Kunc, V., Vaidya, U., Large-scale additive manufacturing tooling for extrusion-compression molds. *Addit. Manuf. Lett.*, 1, 100007, 2021.
9. Jirathearanat, S., Vazquez, V., Rodríguez, C.A., Altan, T., Virtual processing application of rapid prototyping for visualization of metal forming processes. *J. Mater. Process. Technol.*, 98, 1, 116–124, 2000.
10. Klimyuk, D., Serezhkin, M., Plokhikh, A., Application of 3D printing in sheet metal forming. *Mater. Today: Proc.*, 38, 1579–1583, 2021.
11. Schuh, G., Bergweiler, G., Bickendorf, P., Fiedler, F., Colag, C., Sheet metal forming using additively manufactured polymer tools. *Proc. CIRP*, 93, 20–25, 2020.
12. Tondini, F., Basso, A., Arinbjarnar, U., Nielsen, C.V., The performance of 3D printed polymer tools in sheet metal forming. *Metals*, 11, 8, 1256, 2021.
13. Leacock, A.G., Cowan, G., Cosby, M., Volk, G., McCracken, D., Brown, D., Structural and frictional performance of fused deposition modelled acrylonitrile butadiene styrene (P430) with a view to use as rapid tooling material in sheet metal forming. *Key Eng. Mater.*, 639, 325–332, 2015.
14. Kain, M., Calaon, M., Pedersen, D., Tosello, G., On the implementation of metal additive manufacturing in the tooling process chain for polymer profile extrusion. *Proc. CIRP*, 93, 26–31, 2020.
15. Cheah, C., Chua, C., Lee, C., Lim, S., Eu, K., Lin, L., Rapid sheet metal manufacturing. Part 2: Direct rapid tooling. *Int. J. Adv. Manuf. Technol.*, 19, 7, 510–515, 2002.
16. Hentschel, O., Scheitler, C., Fedorov, A., Junker, D., Gorunov, A., Haimerl, A., Merklein, M., Schmidt, M., Experimental investigations of processing the high carbon cold-work tool steel 1.2358 by laser metal deposition for the additive manufacturing of cold forging tools. *J. Laser Appl.*, 29, 2, 022307, 2011.
17. Du, Z., Chua, C., Chua, Y., Loh-Lee, K., Lim, S., Rapid sheet metal manufacturing. Part 1: Indirect rapid tooling. *Int. J. Adv. Manuf. Technol.*, 19, 6, 411–417, 2002.

18. Reggiani, B. and Todaro, I., Investigation on the design of a novel selective laser melted insert for extrusion dies with conformal cooling channels. *Int. J. Adv. Manuf. Technol.*, 104, 1–4, 815–830, 2019.
19. Yesildag, N., Hopmann, C., Windeck, C., Bremen, S., Wissenbach, K., Merkt, S., Opportunities and challenges of profile extrusion dies produced by additive manufacturing processes. *AIP Conf. Proc.*, 2011.
20. Bambach, M., Sizova, I., Emdadi, A., Development of a processing route for Ti-6Al-4V forgings based on preforms made by selective laser melting. *J. Manuf. Processes*, 37, 150–158, 2019.
21. Bi, G., Ng, G.K.L., Teh, K.M., Jarfors, A.E., Feasibility study on the laser aided additive manufacturing of die inserts for liquid forging. *Mater. Des.*, 31, S112–S116, 2010.
22. Oter, Z.C., Coskun, M., Akca, Y., Surmen, M., Yilmaz, M.S., Ozer, G., Tarakci, G., Koc, E., Benefits of laser beam based additive manufacturing in die production. *Optik*, 176, 175–184, 2019.
23. Foster, J., Cullen, C., Fitzpatrick, S., Payne, G., Hall, L., Marashi, J., Remanufacture of hot forging tools and dies using laser metal deposition with powder and a hard-facing alloy Stellite 21®. *J. Remanuf.*, 9, 3, 189–203, 2018.
24. Khalifa, N.B., Isakovic, J., Bohlen, J., New concepts of extrusion dies to reduce the anisotropy of extruded profiles by means of additive manufacturing. *CIRP Ann.*, 70, 1, 231–234, 2021.
25. Du, W., Bai, Q., Zhang, B., A novel method for additive/subtractive hybrid manufacturing of metallic parts. *Proc. Manuf.*, 5, 1018–1030, 2016.
26. Bryden, B., Pashby, I., Wimpenny, D., Adams, C., Laminated steel tooling in the aerospace industry. *Mater. Des.*, 21, 4, 403–408, 2000.
27. Holker, R., Haase, M., Khalifa, N.B., Tekkaya, A.E., Hot extrusion dies with conformal cooling channels produced by additive manufacturing. *Mater. Today: Proc.*, 2, 10, 4838–4846, 2015.
28. Himmer, T., Techel, A., Nowotny, S., Beyer, E., Recent developments in metal laminated tooling by multiple laser processing. *Rapid Prototyping J.*, 9, 1, 24–29, 2003.
29. Hong, X., Xiao, G., Zhang, Y., Zhou, J., Research on gradient additive remanufacturing of ultra-large hot forging die based on automatic wire arc additive manufacturing technology. *Int. J. Adv. Manuf. Technol.*, 116, 7–8, 2243–2254, 2021.
30. Duarte, V.R., Rodrigues, T.A., Schell, N., Miranda, R., Oliveira, J., Santos, T.G., Hot forging wire and arc additive manufacturing (HF-WAAM). *Addit. Manuf.*, 35, 101193, 2020.
31. Hirtler, M., Jedynak, A., Sydow, B., Sviridov, A., Bambach, M., A study on the mechanical properties of hybrid parts manufactured by forging and wire arc additive manufacturing. *Proc. Manuf.*, 47, 1141–1148, 2020.
32. Meiners, F., Ihne, J., Jürgens, P., Hemes, S., Mathes, M., Sizova, I., Bambach, M., Hama-Saleh, R., Weisheit, A., New hybrid manufacturing routes

combining forging and additive manufacturing to efficiently produce high performance components from Ti-6Al-4V. *Proc. Manuf.*, 47, 261–267, 2020.

33. Junker, D., Hentschel, O., Schmidt, M., Merklein, M., Qualification of laser based additive production for manufacturing of forging tools. *MATEC Web Conf.*, 21, 08010, 2015.
34. Morrow, W., Qi, H., Kim, I., Mazumder, J., Skerlos, S., Environmental aspects of laser-based and conventional tool and die manufacturing. *J. Cleaner Prod.*, 15, 10, 932–943, 2007.
35. Junker, D., Hentschel, O., Schmidt, M., Merklein, M., Tailor-made forging tools by laser metal deposition. *Key Eng. Mater.*, 651–653, 707–712, 2015.
36. Krol, M., Musiorski, M., Pagac, M., Design and selective laser melting manufacturing of TPE extrusion die. *Solid State Phenom.*, 308, 51–63, 2020.
37. Mertens, R., Vrancken, B., Holmstock, N., Kinds, Y., Kruth, J.P., Van Humbeeck, J., Influence of powder bed preheating on microstructure and mechanical properties of H13 tool steel SLM parts. *Phys. Proc.*, 83, 882–890, 2016.
38. Abe, F., Osakada, K., Shiomi, M., Uematsu, K., Matsumoto, M., The manufacturing of hard tools from metallic powders by selective laser melting. *J. Mater. Process. Technol.*, 111, 1–3, 210–213, 2001.
39. Almangour, B., Grzesiak, D., Yang, J.M., Selective laser melting of TiB2/H13 steel bulk nanocomposites: Influence of nanoscale reinforcment. *TMS 2016 145th Annual Meeting & Exhibition*, pp. 171–176, 2016.
40. Hölker, R., Jäger, A., Ben Khalifa, N., Tekkaya, A.E., Controlling heat balance in hot aluminum extrusion by additive manufactured extrusion dies with conformal cooling channels. *Int. J. Precis. Eng. Manuf.*, 14, 8, 1487–1493, 2013.
41. Yogeshwaran, K. and Das, S., Mechanical properties of fused deposition modelling processed parts: A review. *Mater. Today: Proc.*, 2022.
42. Pruncu, C.I., Hopper, C., Hooper, P.A., Tan, Z., Zhu, H., Lin, J., Jiang, J., Study of the effects of hot forging on the additively manufactured stainless steel preforms. *J. Manuf. Processes*, 57, 668–676, 2020.
43. Kuhnhen, C., Knoche, J., Reuter, J., Hassan Al-Maeeni, S.S., Engel, B., Hybrid tool design for a bending machine. *Proc. CIRP*, 99, 370–374, 2021.
44. Bambach, M., Sizova, I., Sydow, B., Hemes, S., Meiners, F., Hybrid manufacturing of components from Ti-6Al-4V by metal forming and wire-arc additive manufacturing. *J. Mater. Process. Technol.*, 282, 116689, 2020.
45. Turazza, A., Davoudinejad, A., Calaon, M., Pedersen, D., Tosello, G., Towards the integration of additively manufactured photopolymer dies in the polymer profile extrusion process chain. *Proc. CIRP*, 93, 3–8, 2020.
46. Nazir, A., Gokcekaya, O., Billah, K.M.M., Ertugrul, O., Jiang, J., Sun, J., Hussain, S., Multi-material additive manufacturing: A systematic review of design, properties, applications, challenges, and 3D printing of materials and cellular metamaterials. *Mater. Des.*, 226, 111661, 2023.

47. Tshephe, T.S., Akinwamide, S.O., Olevsky, E.A., Olubambi, P.A., Additive manufacturing of titanium-based alloys- A review of methods, properties, challenges, and prospects. *Heliyon*, 8, 3, e09041, 2022.
48. Altiparmak, S., Yardley, V.A., Shi, Z., Lin, J., Challenges in additive manufacturing of high-strength aluminium alloys and current developments in hybrid additive manufacturing. *Int. J. Lightweight Mater. Manuf.*, 4, 2, 246–261, 2021.
49. Maleki, E., Bagherifard, S., Bandini, M., Guagliano, M., Surface post-treatments for metal additive manufacturing: Progress, challenges, and opportunities. *Addit. Manuf.*, 37, 101619, 2021.
50. Nirish, M. and Rajendra, R., Suitability of metal additive manufacturing processes for part topology optimization – a comparative study. *Mater. Today: Proc.*, 27, 1601–1607, 2020.
51. Vafadar, A., Guzzomi, F., Rassau, A., Hayward, K., Advances in metal additive manufacturing: A review of common processes, industrial applications, and current challenges. *Appl. Sci.*, 11, 3, 1213, 2021.
52. Ansari, M.A., Jabari, E., Toyserkani, E., Opportunities and challenges in additive manufacturing of functionally graded metallic materials via powder-fed laser directed energy deposition: A review. *J. Mater. Process. Technol.*, 294, 117117, 2021.
53. Gao, J., Chen, S., Liu, T., Ye, J., Liu, J., Additive manufacture of low melting point metal porous materials: Capabilities, potential applications and challenges. *Mater. Today*, 49, 201–230, 2021.
54. Magar, S.P., Khedkar, N.K., Kumar, S., Review of the effect of built orientation on mechanical Properties of metal-plastic composite parts fabricated by additive manufacturing technique. *Mater. Today: Proc.*, 5, 3926–3935, 20182018.
55. Ngo, T., Kashani, A., Imbalzano, G., Nguyen, K., Hui, D.S., Additive manufacturing (3D printing): A review of materials, methods, applications and challenges. *Compos. Part B-Eng.*, 143, 172–196, 2018.
56. Kumar, S., Elangovan, S., Mohanraj, R., Ramakrishna, J., Review on the evolution and technology of state-of-the-art metal additive manufacturing processes. *Mater. Today: Proc.*, 46, 7907–7920, 2021.
57. Schloder, M., Overcoming the challenges with metal additive manufacturing, 2022, https://www.alphaprecisionpm.com/blog/common-mam-issues-to-avoid.

12

Environmental Aspects of 3D Printing Metal and Alloys

Palivela Bhargav Chandan[1], Pranav G. Bhat[2], Sangeeth Purushothaman[1], Devara Venkata Krishna[1], Tadi Siva Prasad[1], Aye Aye Thet[3] and Mamilla Ravi Sankar[1*]

[1]*Department of Mechanical Engineering, Indian Institute of Technology Tirupati, Andhra Pradesh, India*
[2]*Department of Mechanical Engineering, Vidyavardhaka College of Engineering, Mysuru, Karnataka, India*
[3]*Department of Mechanical Engineering, Technological University Lashio, Northern Shan, Myanmar*

Abstract

3D printing is a revolutionary manufacturing technique with several advantages over traditional manufacturing processes. This technology has vast applications from aerospace to the food industry, construction of solid architectural buildings, brittle glass, and rugged titanium products. It also reduces specific energy demand during processes and lowers carbon footprints. During 3D printing, various metals like titanium, zinc, cobalt, and iron, result in hazardous waste emissions during their metal printing process. These emissions include volatile organic compounds (VOCs), nanoparticles of metals, particulate matter, fumes, and oxides of metals. Exposure to these waste products leads to serious health problems related to the respiratory system, dermal and ocular issues. Emissions penetrate the skin and reach the tissue level, leading to pulmonary inflammation, allergies, and other related health issues for nearby workers. So, proper safety precautions must be taken while using 3D printing technology. This review addresses the environmental aspects of 3D printing metals and alloys as metal 3D printing has become very popular in schools, colleges, and research institutions.

***Keywords*:** Additive manufacturing, 3D metal printing, sustainable technology, carbon footprint, emissions and wastes, health issues and hazards

**Corresponding author*: evmrs@iittp.ac.in

R. Rajasekar, C. Moganapriya and P. Sathish Kumar (eds.) Additive Manufacturing with Novel Materials: Processes, Properties and Applications, (373–408) © 2024 Scrivener Publishing LLC

12.1 Introduction

3D printing is a ground-breaking manufacturing mechanization. This method has raised the bar for various rigid copper or tungsten alloy components, brittle glass components, and soft, delicate cakes. It has a vast application in the medical, aerospace, food and craft, military, architecture, and fashion industries [1]. Additive manufacturing, also referred to as 3D printing, is creating 3D models from their digital representations in computer-aided design (CAD) files and depositing materials layer by layer to precisely and accurately construct an entire structure. The basic principles behind 3D printing are the preprinting process, i.e., modeling, printing, and postprinting. Initially, a vertical 3D replica of the model is created using CAD design packages or 3D scanners. This process involves examining and acquiring data on the shape and design of the required model. Then comes the printing stage. The printers use the material necessary to print the thing from the virtual model, the stereolithography (STL) file. They use software that virtually breaks the model into thin layers, later replicated by the printer. Additive manufacturing involves depositing material layer upon layer to create the final product. Although most products are ready to use, in other circumstances, such as metals, we need to sinter the mold to improve material properties such as strength and hardness. The postprinting operation sometimes involves superfinishing processes for high dimensional accuracy, as shown in Figure 12.1 [2].

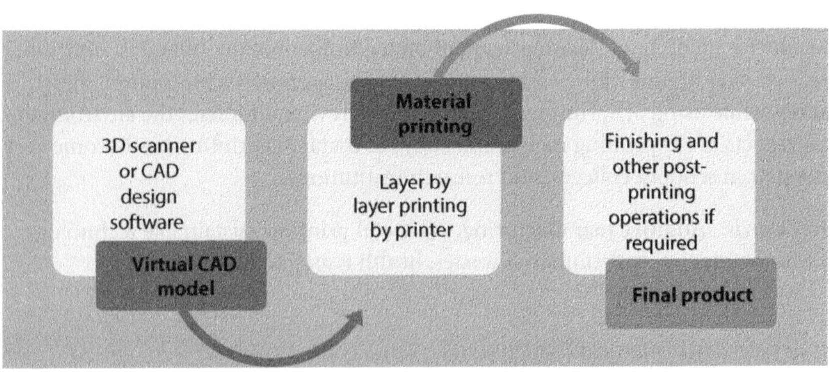

Figure 12.1 Flow chart for manufacturing objects by 3D printing [2].

12.2 Additive Manufacturing Technologies

A variety of techniques are used in material 3D printing. While certain materials are liquid, others have softened materials. An overview of the various additive manufacturing technologies is presented in this section. A detailed discussion of each additive manufacturing technology is beyond this chapter's scope.

12.2.1 Selective Laser Sintering

This process uses a laser power source for sintering powdered material. It directs a laser in space resulting in the binding of the material to form a solidified area, while in selective laser melting (SLM), the material is fully melted. These are generally used for metals [3].

12.2.2 Fused Deposition Modeling

This technique uses thermoplastic wire or filament as a coil to feed material to the extrusion nozzle head. Tiny beads of material are extruded to form a layer when hardened. They are generally used for thermoplastic materials [4].

12.2.3 Stereolithography

Stereolithography uses photopolymerization to produce models in a layer-by-layer trend. The connection of molecules is carried out using a light source with photosensitive resins. This is used for polymers [5].

12.2.4 Laminated Object Manufacturing

This 3D printing technology combines layers of materials like paper, plastics, or metals with an adhesive coating that is integrated and trimmed to the required shape by a laser cutter [6]. One more technique is to use ultrasonic vibrations to join metals by scribing them with each other and applying low continuous force [7].

12.2.5 Binder Jetting

In this technology, binder material moves across the powder layer to devise the product layer by layer using printheads. In the end, the product's final shape is sintered for metals to get desired properties and strength. It is sometimes called a powder bed and inkjet. This process is used for metal, ceramics, or plastic [8].

12.2.6 Directed Energy Deposition

In order to liquefy metal granules into a male-shaped product in an inert environment where the molten metals would not react with the oxygen in the air, this procedure requires a powerful laser as an energy source. This process is commonly used for metals like titanium, aluminum, and stainless steel [9]. These are less common as they have low precision and postprinting processing of manufactured features [10].

12.3 Metal and Alloy Materials

Additive manufacturing of high-purity metals is challenging due to their poor mechanical characteristics. Titanium has proven to be an excellent material for the additive manufacturing industry out of its enhanced strength-to-weight ratio [10]. Titanium can be alloyed with Mo, V, Ta, Nb, Al, Sn, N, and O to enhance its characteristics [11]. Several pre-alloyed materials used in additive manufacturing are alloys based on Fe, Co, Al, Ni and Ti. Commonly used engineering material steel [12] is an Fe-alloy. For corrosion-resistant, high-yield strength applications, Ni-alloy such as Inconel is used. Cobalt-based alloys are used for wear resistance and high temperature. Aluminum-alloys are known for their lightweight characteristics. Ti-6Al-4V is an extensively used Ti alloy for aerospace and medical applications. Traces of other metals are also added to get the desired properties in the manufactured product [13]. Some of the materials used in the powder bed fusion technique are presented in Table 12.1, and the material and the energy flow are shown in Figure 12.2.

12.4 Positive Environmental Impact of Additive Manufacturing

3D printing has expanded tremendously in recent years. It has had a sizable positive influence on the environment. As a material addition technique, 3D printing has the benefit over traditional production processes of lowering scrap. 3D printing produces objects by a layer-by-layer approach with more than 90% accuracy and efficiency [14]. The estimated efficiency potential of

Table 12.1 Powder bed fusion technology and materials [10].

Type	Technology	Materials
Powder bed	Inkjet 3D printing (3DP) and powder bed	powdered polymers, almost metal alloy
	Electron beam melting (EBM)	Almost metal alloy, including titanium alloys
	Selective laser melting of alloys (SLM)	Superalloys, stainless steel, titanium alloys, aluminum
	Aluminum selective heat sintering (SHS)	Thermoplastic powder
	Selective laser sintering (SLS)	Ceramic powder, thermoplastic powder, metal powder
	Direct metal laser sintering (DMLS)	Almost metal alloy

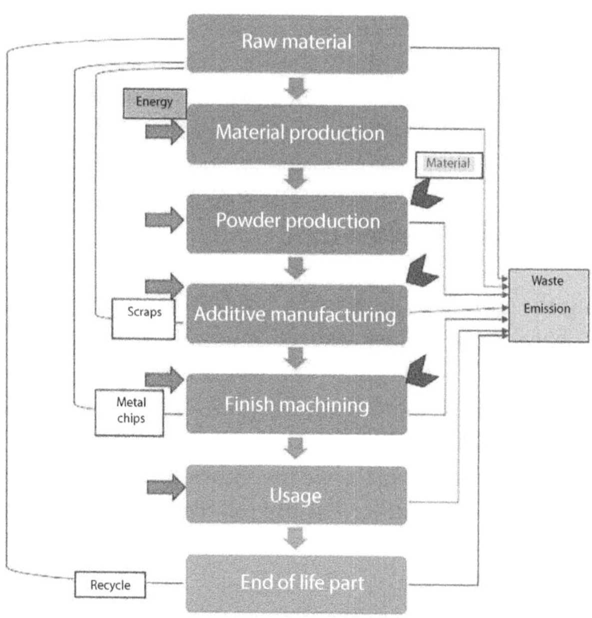

Figure 12.2 The material and energy flow in additive manufacturing [10].

the process is over 97%. The life assessment cycle of additive manufacturing has shown that it reduces carbon dioxide emissions leading to curbing greenhouse gases [15]. A study was conducted using GaBi software-based life cycle assessment of inkjet-printed perovskite-based solar cells. This study concluded that the global warming potential and cumulative energy demand of inkjet-printed perovskite solar cells of 7.54 $kg\,CO_2eq/m^2$ and 200.18 MJ/m^2, respectively. On the other hand, spin-coating had reported a median value of global warming potential and cumulative energy demand of 74.5 $kg\,CO_2eq/m^2$ and 1204 MJ/m^2, respectively. These outcomes suggest an environmental advantage for inkjet-printed perovskite solar cells. This study also used novel green solvent-developed precursor ink, eliminating toxic and hazardous solvent materials. The developed green solvent-based precursor inks demonstrated up to six orders of magnitude lower impacts than the conventional inks, as shown in Figure 12.3 [16]. Another parameter that is not recommended in any production process is specific energy consumption. Specific energy consumption is the term used to describe the entire amount of energy used in the manufacturing process to create the model. Typically, it is expressed in units of kJ/kg of material. Low specific energy consumption is preferred because energy directly affects the carbon footprint of the process [17, 18]. Energy consumption in various methods is shown in Table 12.2. 3D printing of sand molds provides a new path for exploring ways to reduce CO_2 emissions from the traditional casting of metal components. A study showed that mold design optimization reduces energy and metal, resulting in a two-thirds reduction in CO_2 emission compared to the conventional casting process [19].

12.5 Life Cycle Assessment

Determining the ultimate impact of a procedure and model on habitat at the design phase is a regularly utilized method. It assesses its risks using a variety of eco-parameters [20]. The ReCiPe is a process for life cycle impact assessment. It resulted from the coordination of RINM, Radboud University Nijmegen, Leiden University, and PRe sustainability in 2008. Its main aim was to transform life cycle inventories into numbered indicative scores. Its indicative score describes the relative severity of environmental impact. It uses two levels to determine indicators (18 midpoint and three endpoint indicators) [21].

Life cycle assessment (LCA) breakdown of impact by printing via inkjet machine printing and CNC machining to make one pair of models per week and remaining in idle span for the remaining time [22]. Comparing

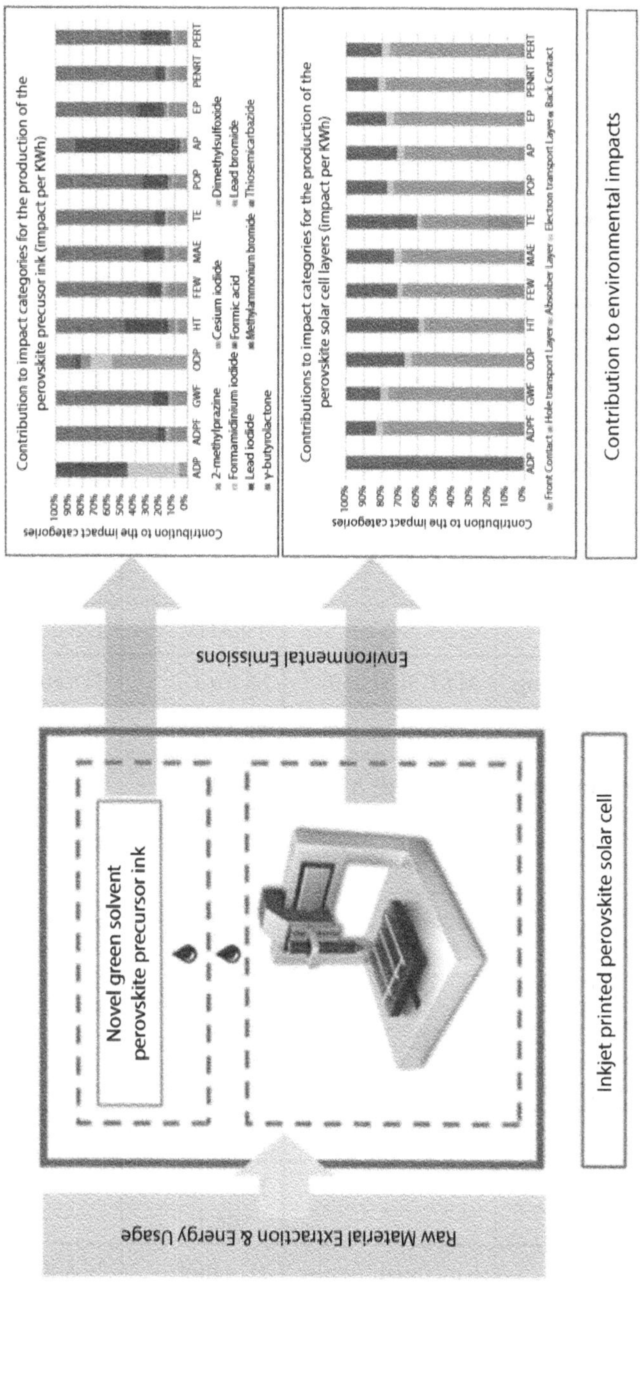

Figure 12.3 Carbon dioxide emission during various manufacturing processes [16].

Table 12.2 Energy utilization during fabrication using various processes [17].

Process	Machine	Material	Energy usage (MJ/kg)
Directed energy deposition	-	H13 tool steel	7708
Directed energy deposition	Optomec LENS 750	Nistelle 625	1052
Finishing	-	H13 tool steel	~600
Selective laser melting	Renishaw AM250	AlSi10Mg	566.2
Selective laser melting	Concept Laser M3 Linear	SS 316L	423–588
Selective laser melting	Concept Laser Mlab	Aluminium	309.1–533.0
Directed energy deposition	EOSINT M 270	SS 17-4 PH	241-339
Finishing	-	Ti6Al4V	178.0
Selective laser melting	MTT SLM 250	SS 316L	111.60–139.50
Selective laser melting	MTT SLM 250	SS 316L	83–106
Electron beam melting	Arcam AB A1	Ti-6Al-4V	61.20–176.67
Selective laser melting	Concept Laser M3 Linear	SS 316L	96.8
Electron beam melting	Arcam	Ti-6Al-4V	94
Directed energy deposition	-	Carbon steel	61.1
Electron beam melting	Arcam AB A1	Ti-6Al-4V	60.0
Gas metal arc welding	-	Carbon steel	32.3
Roughening	-	H13 tool steel	~24
Semi-finishing	Cincinnati Arrow 500	H13 tool steel	16.8–68.4

the given charts, the ReCiPe endpoint H pts is high for the CNC machining process for electricity consumption compared to inkjet machine printing. This shows that inkjet machine printing utilizes less energy than CNC machining. It indirectly also reduced gaseous emissions from the primary source of energy. A review of life cycle models for enhancing the

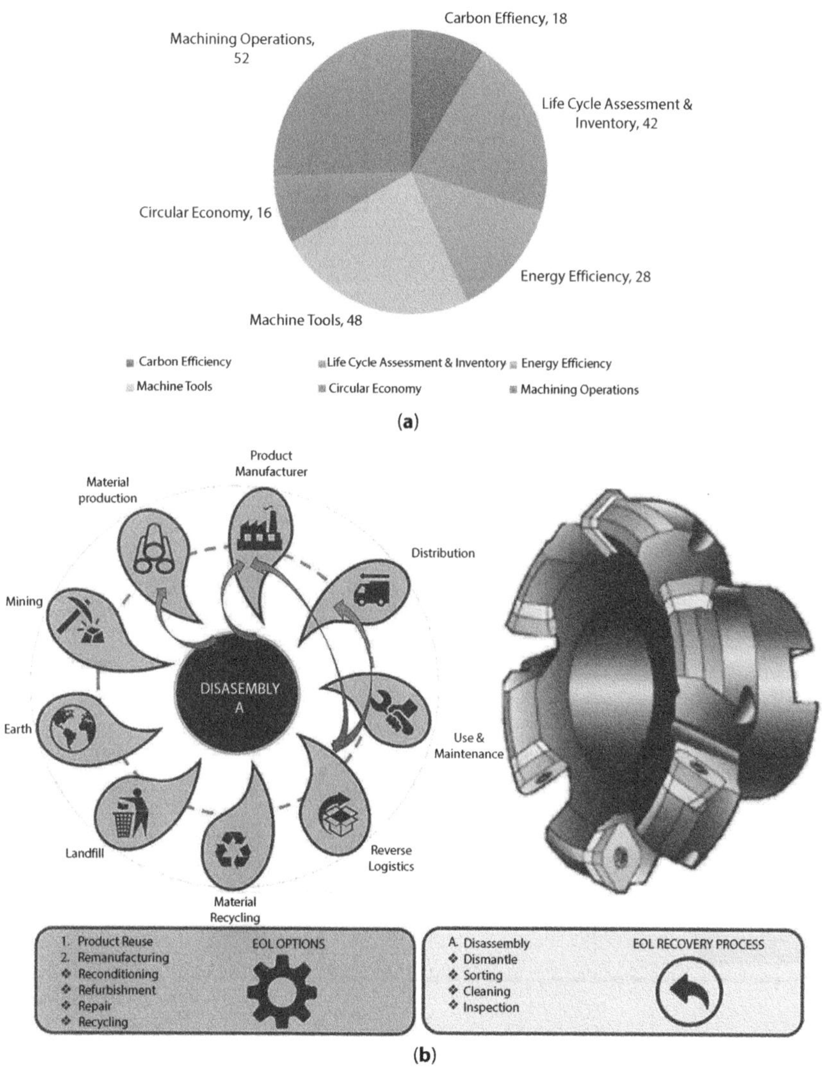

Figure 12.4 (a) Life cycle of a machine tool conceptual view, (b) product recovery sustainability model. *(Continued)*

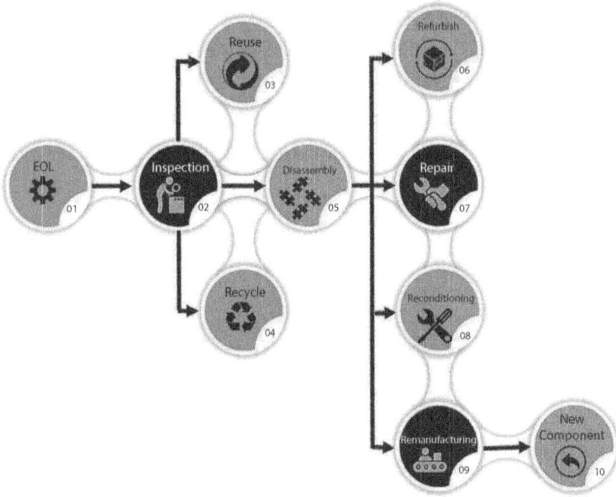

(c)

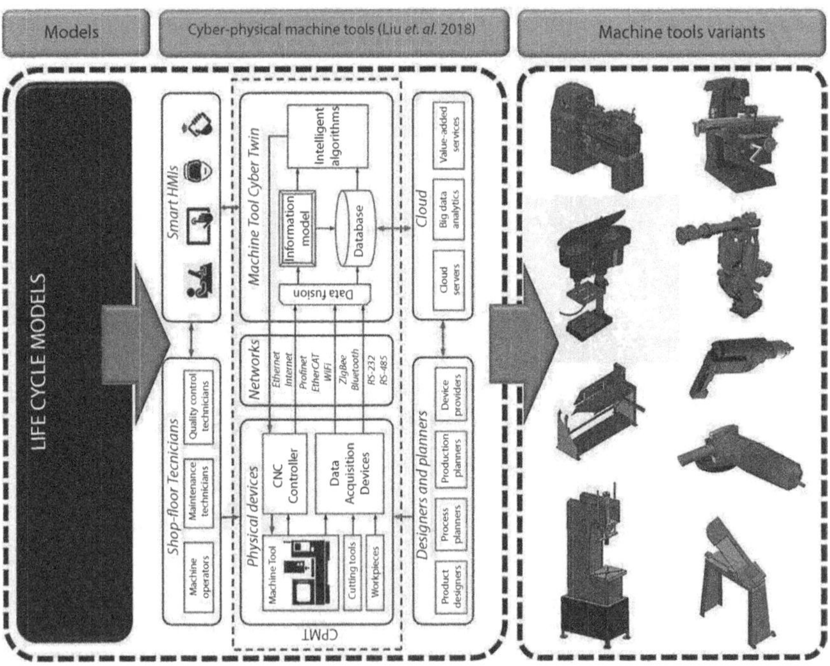

(d)

Figure 12.4 (Continued) (c) integration of life cycle and cyber-physical model of the machine tool, and (d) an integrated cyber physical machine tool model and life cycle [23].

LCA Results

Midpoint Categories	Damage Categories
1. Carcinogens	1. Human Health
2. non-Carcinogens	
3. Respiratory Inorganic	
4. Ionizing radiation	
5. Ozone layer depletion	
6. Respiratory Organic	
7. Aquatic ecotoxicity	2. Ecosystem Quality
8. Terrestrial ecotoxicity	
9. Terrestrial acidification	
10. Land occupation	
11. Aquatic acidification	
12. Eutrophication	
13. Global Warming	3. Climate Change
14. Non-Renewable energy	4. Resources
15. Mineral Extraction	

Figure 12.5 Impact categories during life cycle assessment [34].

sustainability of machine tool models states that 60% of the articles apply LCA methodology, and 40% of the papers used other assessment tools to improve environmental sustainability and reduce energy consumption. The classification of literature review based on the research focus on LCA states that most of the literature is focused upon the machining operations followed by machine tools, whereas the least being circular economy, as shown in Figure 12.4(a). The conceptual view of LCA for a machine tool model and product recovery sustainability model is shown in Figure 12.4(b-c). The integration of the life cycle machine tool model with modern cyber-physical systems is shown in Figure 12.4(d) [23]. The impact of carcinogens produced during the LCA cycle is depicted in Figure 12.5. More information on the LCA of metals can be found in the available literature [21, 22, 24–26].

12.6 Sustainable Manufacturing

3D printing is a modern way to work in sustainable manufacturing, reducing habitat threats. Conventional machining involves subtractive manufacturing resulting in many scraps, leading to energy and resource waste. Additive manufacturing produces less debris, and most of the trash produced can be reused. This process also eliminates the possibility of wastage from defects [27]. 3D printing has a wide range of environmental implementations. Systems for measuring air quality at the nanoscale are being created to monitor indoor air quality. Along with lighting conditions, radiant temperature, air speed, relative humidity, and air temperature, it also describes CO_2 concentration [28]. A wide variety of membranes used in wastewater treatment are supported by 3D printing.

Materials based on ceramic membranes used in wastewater filtering are suitable for 3D printing [29, 30]. Equipment for renewable energy sources like wind turbine blades is made using additive manufacturing [31]. The manufacturing industry is a significant contributor to air pollution [32]. About 19% of greenhouse gases are associated with manufacturing processes [33]. We have to use sustainable manufacturing technology for a clean manufacturing process. The sustainability of 3D-printed products could be enhanced for reducing the impact of metals on the environmental ecology by adapting the practices like material recycling, LCA (discussed in the earlier section), designing 3D-printed models with the intention of repair, refurbishment or remanufacturing, local production for reducing transportation-related emissions and minimization of carbon footprint, by using renewable energy resources for operating 3D printers, and by creating awareness for the users about the prominence of sustainable 3D printing practices.

12.7 Negative Environmental Impact of Additive Manufacturing

Every technology has two aspects, valuable and harmful, but when its beneficial aspect outnumbers its dangerous part, it becomes a revolutionary technology. There are several gases and fine particles emissions and several other non-recyclable wastes.

We have to measure these parameters, for which we use various measurement devices. Nanoparticles are measured by a Nanotracer.

This apparatus discovers and computes airborne particles ranging from 10 nm to 300 nm that are ultrafine. Both particle concentration (particle per cubic centimeter) and average particle diameters are continuously monitored by Nanotracer. A Lighthouse Handheld 3016 IAQ is used to measure airborne ultrafine particles. It can identify airborne particles as small as 300 nm and as large as 10 m. It is an optical particle-counting device. It offers a number of facts on counting, temperature, and relative humidity. Laser diffraction analysis can determine the particle size dispersion by measuring the rotational shift in the brightness of scattered light as a beam of laser travels through a scattered collection of particulate matter. Compared to bigger particles, which scatter laser light at fewer angles, more minor flecks or particles scatter light at greater angles. The analysis of the rotational scattering intensity data is then used to identify the particles size that is accountable for the breakdown or dispersion of laser beams [35]. Few studies also have used robust carbon quantum dots as nanotracer for groundwater studies, as shown in Figure 12.6 [36].

A carbon footprint is not advisable in any process. This process reduces the carbon footprint as differentiated from conventional manufacturing processes. Several gas emissions include carbon dioxide, VOCs, nitrogen oxides, and other harmful gases. Many emissions occur during the manufacturing of titanium components using binder jetting technology. This is done by making a layer of titanium powder as a base. Binder material polyvinyl alcohol (PVA) or acrylonitrile butadiene styrene (ABS) is used to hold metal together. After settling for some time, they become a green mold of the product. They are then further sintered in a furnace to gain strength [37]. These polymeric binders burn away during the sintering process, producing fumes of VOCs, metal oxides, and other harmful

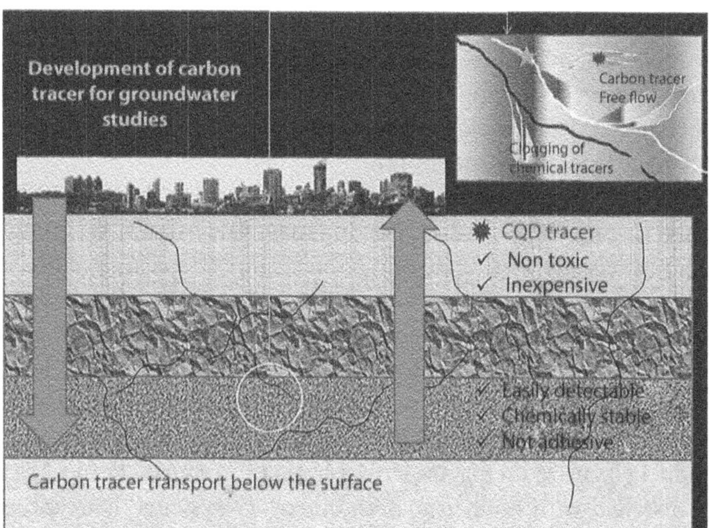

Figure 12.6 Carbon tracer for groundwater studies [36].

emissions. Emissions from 3D printers are hazardous for indoor use in less-ventilated spaces [38]. There is a chance of dust explosion from carbon nanoparticle emission using metal powders as they are highly combustible [39]. Based on the study of five batches of material X5CrNi18-10, it was found that waste constitutes 20.4% of the weight of material through the SLM process. 204g of trash for 1 kg of product is generated [40]. High-energy additive manufacturing, like direct energy deposition, is done in a closed chamber to reduce its exposure to the surrounding. These chambers are filled with inert gases to make a static environment around the final product to reduce the chances of oxidation of metals or their alloys. After completing the process, the chamber is opened to take the product out and clean unused materials. During the manufacturing process, a lot of nanoparticles of metals are formed. When cleaning, bigger particles are recovered and reused in the following manufacturing loop. Still, ventilation of the chamber releases metal particulate matter into the environment, which poses a threat to the health of the workers exposed to it [41]. 3D printing processes with high energy consumption contribute to a large amount of minor particulate matter. Many ultrafine particulate particles are found during metal recycling after the SLM process [42]. The selective laser sintering process also contributes to ultrafine particulate matter of metals and nanoparticles as by-products harmful to living beings [43]. The settling of particles in the air depends on their size. Coarser particles (100 μm) formed from processes like direct metal deposition take less time to settle than finer particles (20 μm or less) formed during SLM. A test was conducted on particles ranging from 20 μm to 80 μm in a closed one-square-meter chamber box. It was found that 2.5 h are needed to reduce particles ten times, starting from 105 to 104 particles/cm^3 in the encompassed chamber. Particulate matters float in the air during manufacturing hours in a closed environment [44].

A few practices that can be adopted for overcoming the negative impact of unused raw materials and supporting materials from 3D printing technologies are self-supporting printing objects or designs at specified angles for reducing the need for support and minimizing material waste and usage. Also, implementing proper recycling practices and setting up systems for collecting and recycling unused materials could help to reduce waste and minimize the negative impact of supporting materials. Opting for biodegradable or compostable materials in place of conventional plastics could help in reducing the environmental impacts. Some 3D printers are installed with built-in material recovery systems that capture and recycle unused materials thereby reducing waste. Educating users about the importance of waste reduction and proper management of unused raw materials and support materials from 3D printing could be effective. Collaborating with other stakeholders in the community of 3D printing like manufacturers of material, printers, and recycling companies could help in the reduction of waste and finding sustainable solutions.

12.8 Direct Sampling Devices

Direct sampling devices can assess particulate matter and nanoparticles of metals and alloys.

12.8.1 Condensation Particle Counters (CPC)

The particle in the air stream that needs to be detected is first magnified and compressed. The particles penetrate via a "cloud" of flowing liquid (water or N-butanol) before being counted; the liquid condenses and binds to the particles. As a result, they are larger and more straightforward to find and disturb. Total condensation particle counters (CPCs) based on N-butanol offer high sensitivity but emit odor and safely deal with problems [44]. Water-based total particle counters have an extremely complex arrangement to provide comparable detection sensitivity to butanol-based CPCs due to the different voltages on the floor of the condensed droplets [45].

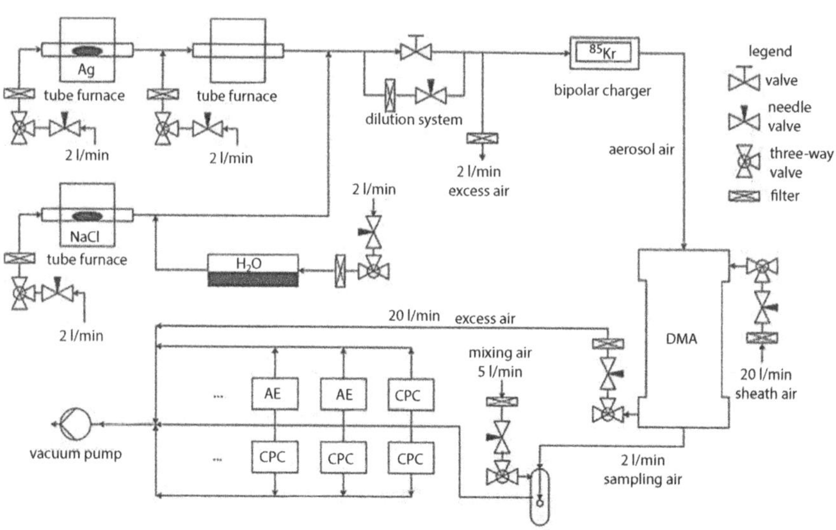

Figure 12.7 Condensation particle counter setup [46].

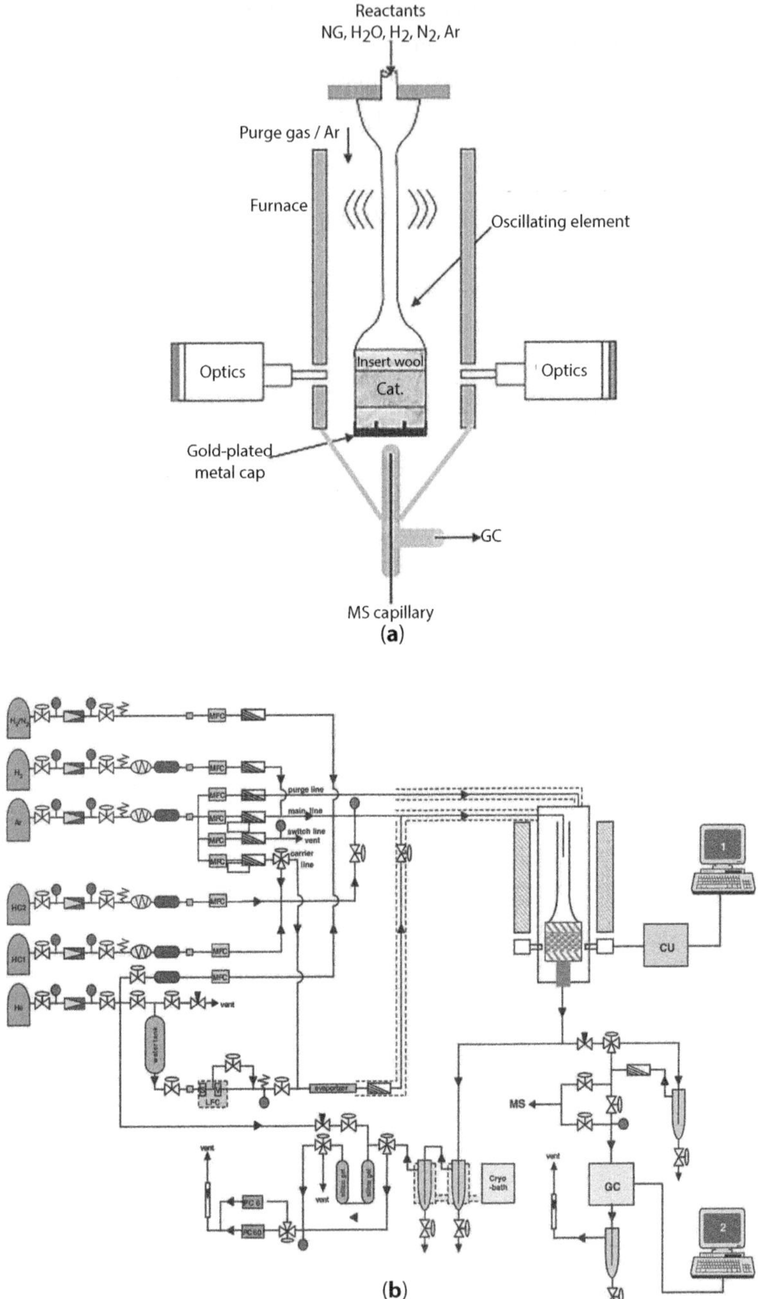

Figure 12.8 (a) Tapered element oscillating microbalance (TEOM) and (b) TEOM apparatus for high-pressure operations [49].

Small particles multiply to a diameter of 10–12 μm; up to a particle size of approximately 1 μm can be counted using a CPC. The CPC calibration setup is shown in Figure 12.7, and more information can be found in Reference [46].

12.8.2 Tapered Element Oscillating Microbalance (TEOM)

Instantaneously determining the mass distribution of solid particles is possible with a cone-element oscillating microbalance. In a conical oscillating microbalance, an air sample is taken using a filter that is attached to the tip of a tubing-made glass that is vibrating. The vibration frequency identifies the accumulated mass of the particle. Another improvement would be size-based quantification [47]. Figure 12.8(a) shows the TEOM, and Figure 12.8(b) shows the TEOM apparatus for high-pressure applications [48, 49].

12.8.3 Nanoparticle Surface Area Monitoring (NSAM)

When considering the risks to human health from exposure to fine particulate matter (respirable suspended particulate matter), the surface area measure offers more information than mass (Figure 12.9). Since a substance's surface area grows as it degrades, particles other than nanoparticles often have larger surfaces than facets for the same mass.

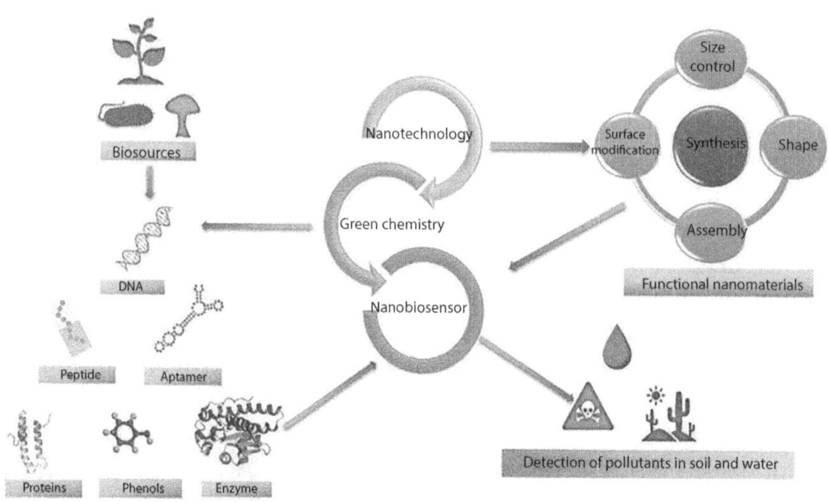

Figure 12.9 Nanoparticle surface area monitoring [51].

The concentrations of the surface area of particles deposited in the lungs can be readily identified by nanoparticle surface area monitoring [50].

12.9 Health Damage Due to Exposure to Emissions

Emissions from additive manufacturing severely damage the health of living beings in the vicinity. Emissions can penetrate the open pores of the body, skin, eyes, and nose and solidify in the lungs or tissues, which can be a nightmare in everyone's life [52]. Copper (Cu) oxide nanoparticles are specified as one among the biologically reactive metal oxide nanoparticles. Further, it is crucial to look into any potential toxicity mechanisms for this chemical in order to produce and use it safely. The three key elements may be the cause of the toxicological stimulation of CuO nanoparticles to pulmonary tissues, according to the research that have already been conducted.

They include oxidative damage, absorption of Cu ions from the surface of CuO nanoparticles, and autophagic cell death brought on by CuO nanoparticle exposure [53]. Widespread applications of CuO nanoparticles increase human vulnerability and thus, the menace of short-term and long-term toxicity. Andujar [54] furnished the first corroboration of an association between human vulnerability to metal nanoparticles and everlasting pulmonary effects in exposed welders occupationally. However, the literature does not report pulmonary toxicity in humans or metal fume fever following inhalation subjection to CuO nanoparticles.

Metal fume fever is a state where the affected person has flu-like symptoms. It is due to the exposure to fumes of specific metals such as cadmium, copper, and zinc [55]. The Finnish Institute of Occupational Health recommended an occupational exposure limit below 20,000 nanoparticles/cm^3. According to a test, the geometric mean of cadmium, nickel, and cobalt in the worker's urine samples was higher than the reference mass of office workers [56]. Metal oxides bond with particulate matter and biomolecules of living beings more efficiently than pure metals. Hence, metal oxides are more harmful than pure metals. Titanium powder is used as feedstock material for various additive manufacturing products.

Studies that examined this resulted in multiple lung diseases from titanium oxide particles. An exposure experiment was done on rats for 12 wk and pulmonary inflammation was observed on them at nanometer size but not on a sub-micron level due to TiO_2 [57]. Figure 12.10 shows

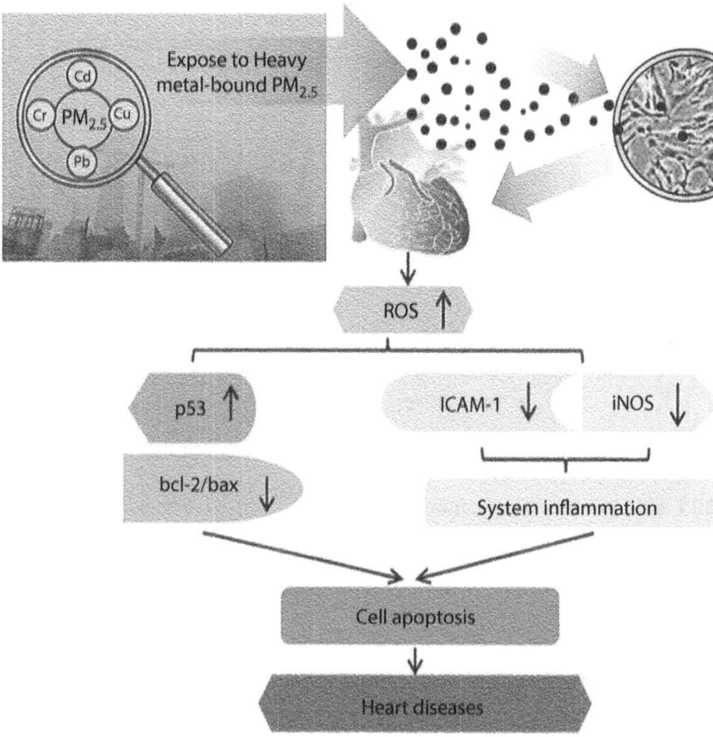

Figure 12.10 Health risks of exposure to heavy metal ions [58].

season-dependent health risks from heavy metals bound to fine particulate matter from northern China, like myocardial toxicity [58].

In general, adequately coated rutile TiO_2 nanoparticles added to sunscreens are claimed to be excellent inorganic UV filters. But then, prolonged exposure to nanoparticles leads to dermal penetration. Nano-TiO_2 (4 nm to 60 nm) penetrates the skin with prolonged exposure and reaches the deep places of the epidermis, as shown in Figure 12.11.

An experiment on mice exposed to TiO_2 nanoparticles for 60 d showed that the particles reached organ tissues. They have the capability of causing oxidative damage to the skin, as shown in Figure 12.12 [60].

Fine particles have higher adsorption per unit mass than larger particles. Due to the larger area, they are deposited more in the irregularly shaped dendrites at the end of the lungs, as shown in Figure 12.13 [62].

Due to the large facet area of the small particles, there is a high chance of surface oxidation (phenomenon shown in Figure 12.14). Titanium, chromium, and cobalt have poor solubility and are unstable.

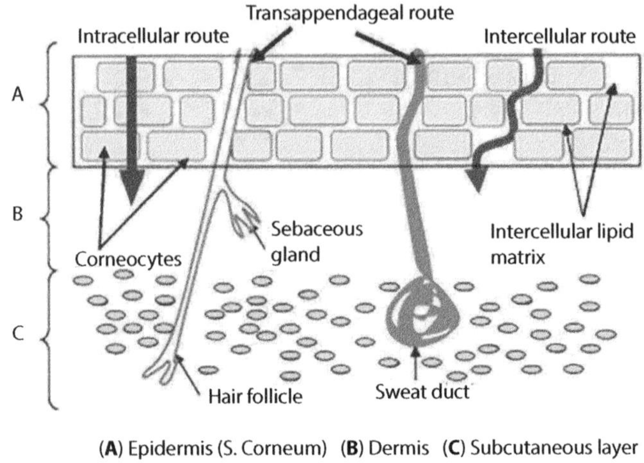

Figure 12.11 Skin penetration of nanoparticles [59].

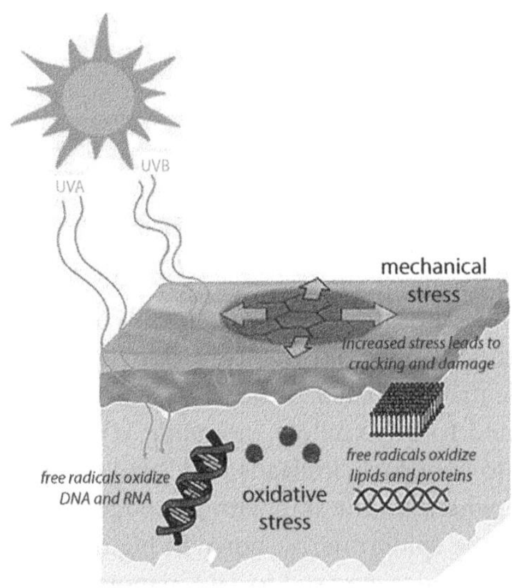

Figure 12.12 Oxidative stress damage the skin due to exposure to metal particles [61].

The insoluble oxides' toxicity, like inflammation and oxidative stress, agrees with the cellular redox potential between -4.12 and -4.84 eV. The conduction band energy levels (Ec) overlap with this range. According to this field of study, the hazardous metal oxide particles include Co_3O_4, Cr_2O_3, Ni_2O_3, Mn_2O_3, CoO, and TiO_2, which sadly overlap with oxides

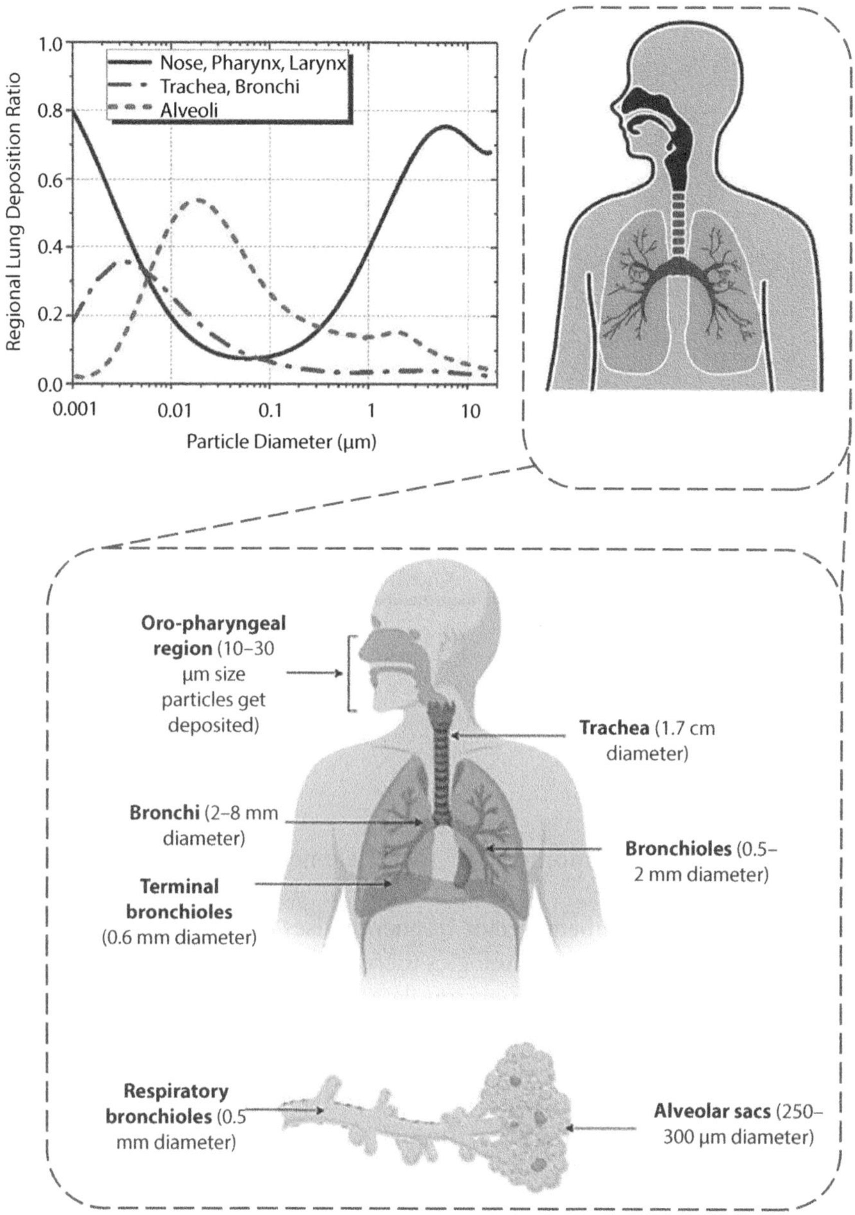

Figure 12.13 Regional particle deposition in the lungs based on the particle diameter [63] and different parts of the respiratory tract along with the diameter of the deposited particles [64].

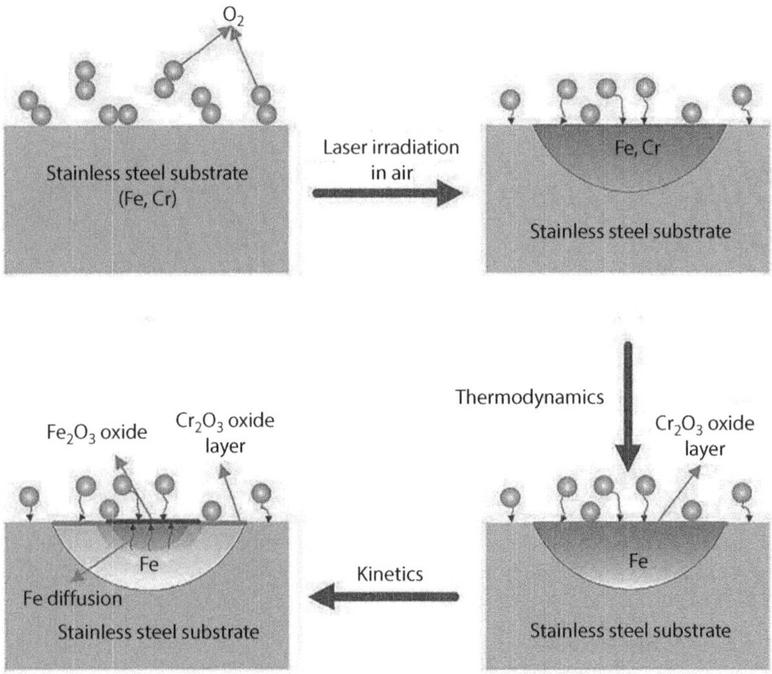

Figure 12.14 Surface oxidation phenomenon of stainless steel under laser irradiation in the air [66].

produced from commonly used metal elements utilized in additive manufacturing [65].

The conventional metal sector frequently uses aluminum powder, and additive manufacturing technologies increasingly use it. Excess aluminum and lead (Figure 12.15) can accumulate in biological tissues after prolonged work exposure, leading to various problems, notably neurological illnesses [67].

Other metal nanoparticles, such as iridium-192, titanium dioxide, copper, and gold, will enter the brain through this track, producing unpleasant outcomes (Figure 12.16).

Additionally, evidence supports a link between subclinical human functional impairment (Figure 12.17) and constant contact with metallic PM (particulate matter).

A study on human exposure to metals in fused filament fabrication or 3D printing processes states that research is necessary on the

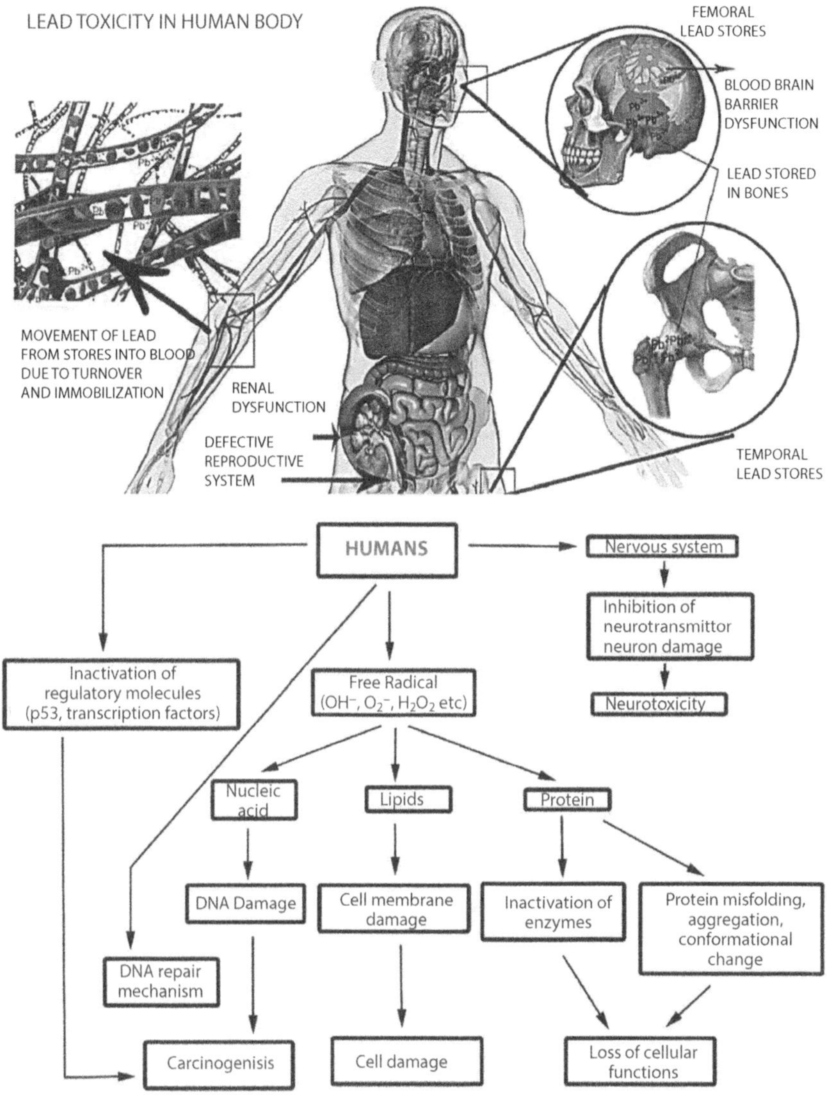

Figure 12.15 Diagrammatical representation and bioaccumulation of lead-heavy metal and its effect on the human body [68].

sustainability aspects of using metal-containing filaments in schools and homes [71].

Figure 12.18 depicts the consumer exposure to emissions in the fused filament fabrication process. Pushing the 3D-printed filament from the heated nozzle releases the emissions. Emissions consist of free and

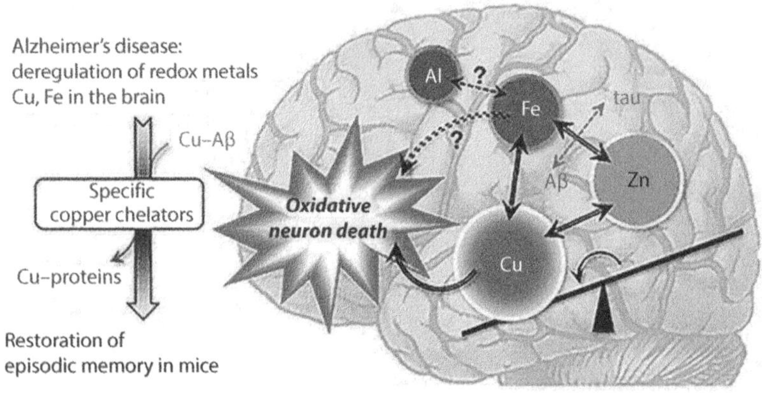

Figure 12.16 Metal ions in Alzheimer's disease [69].

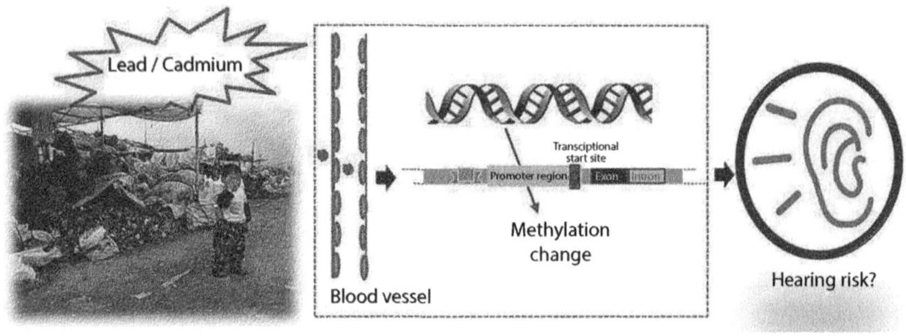

Figure 12.17 Hearing loss risk and DNA methylation due to lead and cadmium exposure [70].

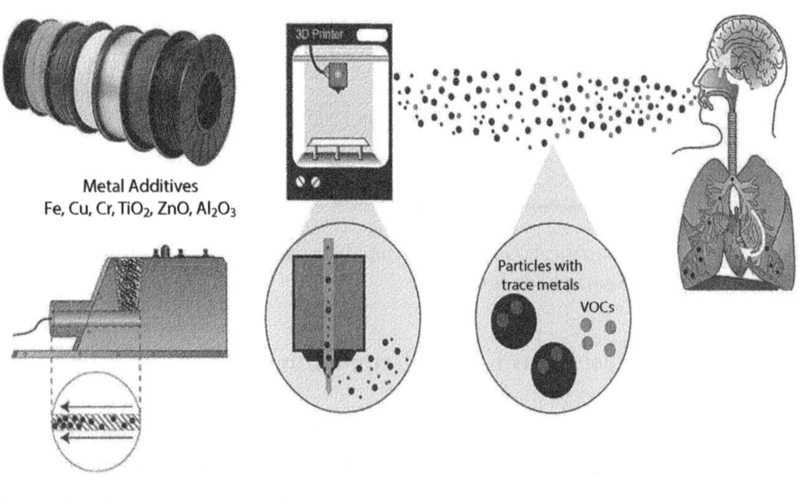

Figure 12.18 Human exposure to metals in fused filament fabrication 3D printing [71].

particle-bound vapors of VOCs, particles, and trace metals. The emissions are prone to deposit along the respiratory tract, including in the pulmonary regions, tracheobronchial regions, and extrathoracic or nasal regions. Ultrafine metal particulate matter may transfer directly to the brain or from the respiratory system to circulation. Material from the tracheobronchial area may be swallowed into the gut and distributed systematically. Even though the transport of fused filament emission particles through gastrointestinal and nasal pathways is credible, this process is yet to be explored and demonstrated further [71].

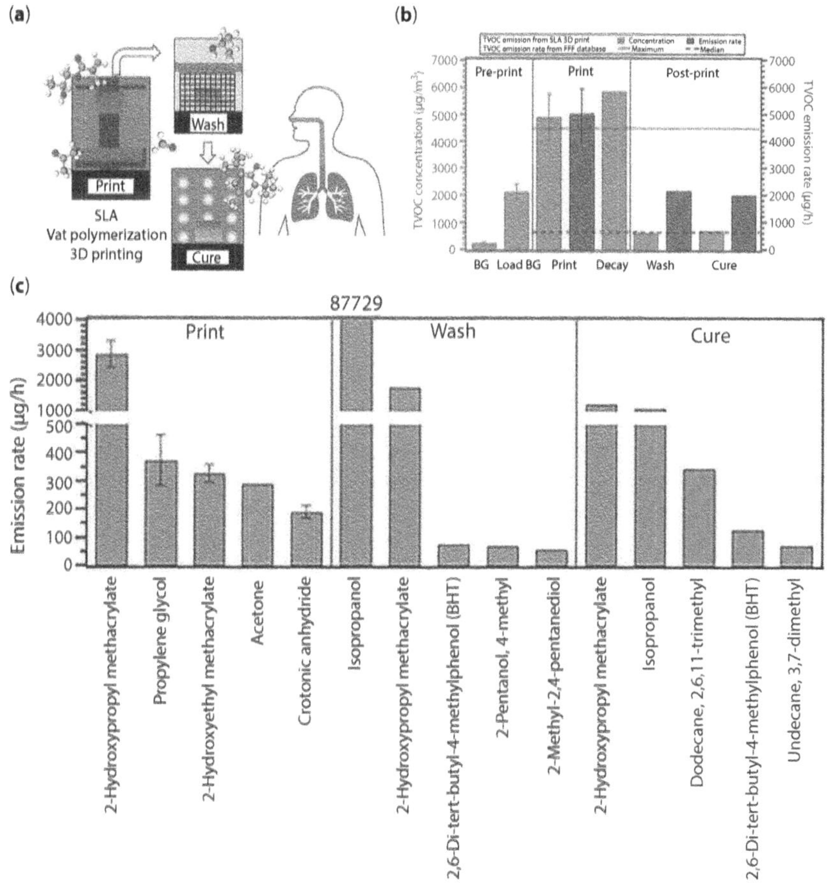

Figure 12.19 (a) Emissions and chemical exposure potential from stereolithography (STL) 3D printing and postprocessing units, (b) volatile organic compounds (VOC) concentrations, and (c) emission rates from each SLA operation phase, and top five highly emitting volatile organic compounds from each process (print, wash, and cure) with their corresponding emission rates (Error bar shows one standard deviation) [74].

12.10 Safety Measures to be Followed During 3D Printing

Workers should use proper protection to reduce exposure to emissions in their bodies. They should use personal protective equipment (PPE) kits and specs while 3D printing. They should use printers that produce less waste and emissions and reuse and recycle materials left from the previous

Table 12.3 Potential hazards from different technologies used in 3D printers with their respective materials [29].

Category	Feedstock material	Feedstock form	Binding/ fusing	Potential hazards
Binder jetting	Metal, ceramic, plastic, or sand	Powder	Adhesive	Inhalation/ dermal exposure to powder, explosion, inhalation of VOCs, dermal exposure to binders
Directed energy deposition	Metal	Powder or wire	Laser/ electron beam heating	Inhalation or dermal exposure to powder, fumes, explosion, laser radiation exposure
Metal extrusion	Thermoplastics	Spooled filament, pellet, or granulates	Electrical heating element-induced melting/ cooling	Inhalation exposure to Volatile organic compounds (VOCs), particulate, additives, burns

(*Continued*)

Table 12.3 Potential hazards from different technologies used in 3D printers with their respective materials [29]. (*Continued*)

Category	Feedstock material	Feedstock form	Binding/ fusing	Potential hazards
Material jetting	Photopolymer or wax	Liquid ink	Ultraviolet light-induced curing	Inhalation of VOCs, ultraviolet exposure, dermal exposure to resins and solvents
Powder bed fusion	Metal, ceramics or plastic	Powder	High-powered laser or electron beam heating	Inhalation or dermal exposure to the powder, fumes, explosion, laser/ radiation exposure
Sheet lamination	Metal, ceramic, plastic	Rolled film or sheet	Adhesive or ultrasonic welding	Inhalation of fumes, VOCs, shock, laser/ radiation exposure
Vat photo polymerization	Photopolymer	Liquid resin	Ultraviolet laser-induced curing	Inhalation of VOCs, ultraviolet exposure, dermal exposure to resins and solvents

manufacturing cycle. Proper ventilation at the manufacturing site should be maintained [72].

The hierarchy of hazard control states that elimination or substitution of hazardous material, engineering and administrative management, and PPE kits are the ways to prevent hazards from nanoparticle emissions [73].

The toxicity of products should be minimized by choosing less toxic materials and postprinting processes.

For example, intense exposure to ultraviolet rays on resin-based products reduces their toxicity affecting aquatic life and human life [Figure 12.19 (a-c)] [75]. Table 12.3 shows the potential hazards from different additive manufacturing processes with materials and technology used for the production.

12.11 Conclusion

A gist is presented in the lines of sustainability aspects on several merits and demerits of additive manufacturing or 3D printing. It is practical and has given us a new way of manufacturing products. Various technologies are used in the 3D printing of metals like binder jetting, SLM, and direct energy deposition, which are used as our requirement. 3D printing technology has an advantage over conventional manufacturing in social, economic, and environmental aspects. 3D printing, an additive manufacturing process, reduces scrap wastage and post-manufacturing costs and resources. It reduces carbon footprint and lowers greenhouse gases in the environment during production. It generally has low energy consumption as compared to conventional subtractive processes. It is used to manufacture environmental applications like air and water purification technologies that are very difficult using traditional manufacturing. Concerning demerits, 3D printing of metal releases fumes, nanoparticles, and particulate matter during melting or printing materials using titanium, zinc, magnesium, cobalt, iron, and copper powders. These hazardous wastes directly affect the environment, and exposure to these wastes leads to severe respiratory diseases, nervous disorders, allergies, inflammation of organs, irritation of the eyes, etc. These wastes penetrate through the skin, causing severe illness in human beings. However, these effects can be reduced to a permissible level by ensuring safety measures like using PPE kits for workers directly involved in processing, regular cleaning, and maintenance of the equipment. 3D printing is a revolutionary technology and a sustainable manufacturing process. It will have more advancement soon and should be encouraged.

Acknowledgment

The authors are grateful to the Department of Science and Technology (DST)-Innovation of Science Pursuit for Inspire Research (INSPIRE) for fellowship support (IF190716). The authors also thank Elsevier for granting permissions for figures and tables. License numbers are as follows: 5461860206784 (Table 12.2) [16], open access under creative commons license (Table 3) [28], open access under creative commons license (Figure 12.3) [16], 5463440326527 (Figure 12.4) [22], 5461870523250 (Figure 12.6) [35], 5461870744475 (Figure 12.7) [45], 5461871275868 (Figure 12.8) [48], open access under creative commons license (Figure 12.9) [50], 5461960087107 (Figure 12.10) [57], 5461960456648 (Figure 12.11) [58], 5461961015533 (Figure 12.12) [60], open access under creative commons license (Figure 12.13) [62], 5461970445126 (Figure 12.13) [63], 5461970760529 (Figure 12.14) [65], open access under creative commons license (Figure 12.15) [67], 5461971411965 (Figure 12.17) [69], 5461980086455 (Figure 12.18) [70], and open access under creative commons license (Figure 12.19) [73].

References

1. Mitsouras, D. *et al.*, Medical 3D printing for the radiologist. *Radiographics*, 35, 7, 1965–1988, 2015, doi: 10.1148/rg.2015140320.
2. Gokhare, V.G., Raut, D.N., Shinde, D.K., A review paper on 3D-printing and various processes used in the 3D-printing. *Int. J. Eng. Res. Technol.*, 6, 6, 953–958, 2017, doi: 10.55041/ijsrem13278.
3. Trapp, J., Rubenchik, A.M., Guss, G., Matthews, M.J., *In situ* absorptivity measurements of metallic powders during laser powder-bed fusion additive manufacturing. *Appl. Mater. Today*, 9, 341–349, 2017, doi: 10.1016/j.apmt.2017.08.006.
4. Hamzah, H.H., Shafiee, S.A., Abdalla, A., Patel, B.A., 3D printable conductive materials for the fabrication of electrochemical sensors: A mini review. *Electrochem. Commun.*, 96, 27–31, 2018, doi: 10.1016/j.elecom.2018.09.006.
5. Bártolo, P.J. and Gibson, I., History of stereolithographic processes, in: *Stereolithography*, pp. 37–56, Springer, Boston, MA, 2011, doi: 10.1007/978-0-387-92904-0_2.
6. Molitch-Hou, M., *Additive manufacturing: Materials, processes, quantifications and applications*, Elsevier, Amsterdam, 2018.

7. White, D.R., Ultrasonic consolidation of aluminum tooling. *Adv. Mater. Process.*, 161, 1, 64, 2003, doi: https://link.gale.com/apps/doc/A96991341/AONE?u=anon~9bd3ae2a&sid=googleScholar&xid=394589e4.
8. Gokuldoss, P.K., Kolla, S., Eckert, J., Additive manufacturing processes: Selective laser melting, electron beam melting and binder jetting-selection guidelines. *Mater. (Basel)*, 10, 6, 672, 2017, doi: 10.3390/ma10060672.
9. Beese, A.M. and Carroll, B.E., Review of mechanical properties of Ti-6Al-4V made by laser-based additive manufacturing using powder feedstock. *JOM*, 68, 3, 724–734, 2016, doi: 10.1007/s11837-015-1759-z.
10. Duda, T. and Raghavan, L.V., 3D metal printing technology. *IFAC-PapersOnLine*, 49, 29, 103–110, 2016, doi: 10.1016/j.ifacol.2016.11.111.
11. Santos, E.C., Osakada, K., Shiomi, M., Kitamura, Y., Abe, F., Microstructure and mechanical properties of pure titanium models fabricated by selective laser melting. *Proc. Inst. Mech. Eng. Part C J. Mech. Eng. Sci.*, 218, 7, 711–719, 2004, doi: 10.1243/0954406041319545.
12. Zadi-Maad, A., Rohib, R., Irawan, A., Additive manufacturing for steels: A review. *IOP Conf. Ser. Mater. Sci. Eng.*, 285, 1, 012028, 2018, doi: 10.1088/1757-899X/285/1/012028.
13. Cooke, S., Ahmadi, K., Willerth, S., Herring, R., Metal additive manufacturing: Technology, metallurgy and modelling. *J. Manuf. Process.*, 57, 978–1003, 2020, doi: 10.1016/j.jmapro.2020.07.025.
14. Peng, T., Kellens, K., Tang, R., Chen, C., Chen, G., Sustainability of additive manufacturing: An overview on its energy demand and environmental impact. *Addit. Manuf.*, 21, 694–704, 2018, doi: 10.1016/j.addma.2018.04.022.
15. Gebler, M., Uiterkamp, A.J.M.S., Visser, C., A global sustainability perspective on 3D printing technologies. *Energy Policy*, 74, 158–167, 2014, doi: 10.1016/j.enpol.2014.08.033.
16. Okoroafor, T., Maalouf, A., Oez, S., Babu, V., Wilk, B., Resalati, S., Life cycle assessment of inkjet printed perovskite solar cells. *J. Cleaner Prod.*, 373, 133665, 2022, doi: 10.1016/j.jclepro.2022.133665.
17. Society, Z.Y., Santana, A., Afonso, P., Zanin, A., Wernke, R., Energy consumption in additive manufacturing of metal parts energy consumption in additive manufacturing of metal parts costing models for capacity optimization in Industry 4.0: Trade-off capacity and operational efficiency. *Proc. Manuf.*, 26, 834–845, 2018, doi: 10.1016/j.promfg.2018.07.104.
18. Palivela, B.C., Manoj, A., Gajrani, K.K., Dhaka, S., Sankar, M.R., Sustainability issues in advanced machining processes, in: *Adv. Mach. Sci.*, pp. 425–468, CRC Press, Boca Raton, Florida, USA, 2022, doi: 10.1201/9780429160011-15.
19. Sivarupan, T., Upadhyay, M., Ali, Y., El, M., Dargusch, M.S., Reduced consumption of materials and hazardous chemicals for energy efficient production of metal parts through 3D printing of sand molds. *J. Cleaner Prod.*, 224, 411–420, 2019, doi: 10.1016/j.jclepro.2019.03.158.
20. Allenby, B.R., Environmental constraints and the evolution of the private firm, in: *Ind. Green Game*, 1997.

21. Goedkoop, M., Heijungs, R., Huijbregts, M., De Schryver, A., Struijs, J., Van Zelm, R., ReCiPe 2008, a life cycle impact assess. method which comprises harmon. categ. indic. midpoint endpoint lev., vol. 1, pp. 1–126, 2009, [Online]. Available: http://www.pre-sustainability.com/download/misc/ReCiPe_main_report_final_27-02-2009_web.pdf.
22. Faludi, J., Bayley, C., Bhogal, S., Iribarne, M., Comparing environmental impacts of additive manufacturing vs. traditional machining via life-cycle assessment introduction. *Rapid Prototyp. J.*, 21, 1, 14–33, 2015, doi: 10.1108/RPJ-07-2013-0067.
23. Daniyan, I., Mpofu, K., Ramatsetse, B., Gupta, M., Review of life cycle models for enhancing machine tools sustainability: Lessons, trends and future directions. *Heliyon*, 7, 4, e06790, 2021, doi: 10.1016/j.heliyon.2021.e06790.
24. Iacopo, B., Valerio, D.P., Tommaso, M., Massimiliano, P., Alessio, V., Environmental impacts assessment of bound metal deposition 3D printing process for stainless steel. *Proc. CIRP*, 105, 386–391, 2022, doi: 10.1016/j.procir.2022.02.064.
25. Peng, T., Wang, Y., Zhu, Y., Yang, Y., Yang, Y., Tang, R., Life cycle assessment of selective-laser-melting-produced hydraulic valve body with integrated design and manufacturing optimization: A cradle-to-gate study. *Addit. Manuf.*, 36, 101530, 2020, doi: 10.1016/j.addma.2020.101530.
26. Nuss, P. and Eckelman, M.J., Life cycle assessment of metals: A scientific synthesis. *PLoS One*, 9, 7, e101298, 2014, doi: 10.1371/journal.pone.0101298.
27. Ghobadian, A., Talavera, I., Bhattacharya, A., Kumar, V., Garza-reyes, J.A., Regan, N.O., Examining legitimatisation of additive manufacturing in the interplay between innovation, lean manufacturing and sustainability. *Int. J. Prod. Econ.*, 219, 457–468, 2020, doi: 10.1016/j.ijpe.2018.06.001.
28. Salamone, F., Belussi, L., Danza, L., Ghellere, M., Meroni, I., Design and development of nEMoS, an all-in-one, low-cost, web-connected and 3D-printed device for environmental analysis. *Sensors*, 15, 6, 13012–13027, 2015, doi: 10.3390/s150613012.
29. Dommati, H., Ray, S., Wang, J., Chen, S., 3D printing technique for ceramic membrane fabrication for water purification. *RSC Adv.*, 9, 29, 16869–16883, 2019, doi: 10.1039/c9ra00872a.
30. Palivela, B.C., Bandari, S.D., Mamilla, R.S., Extrusion-based 3D printing of bioactive glass scaffolds-process parameters and mechanical properties: A review. *Bioprinting*, 27, e00219, 2022, doi: 10.1016/j.bprint.2022.e00219.
31. Nadagouda, M.N., Ginn, M., Rastogi, V., A review of 3D printing techniques for environmental applications. *Curr. Opin. Chem. Eng.*, 28, 173–178, 2020, doi: 10.1016/j.coche.2020.08.002.
32. Vallero, D., *Fundamentals of air pollution*, vol. 5, izd, San Diego, California, USA, 2014.
33. Diaz, N., Redelsheimer, E., Dornfeld, D., Energy consumption characterization and reduction strategies for milling machine tool use. *Glocalized Solut. Sustain. Manuf.*, 53, 263–267, 2011, doi: 10.1007/978-3-642-19692-8.

34. da Barros, S.K.Z.P. and Mansur, A.I., Where do the environmental impacts of Additive Manufacturing come from? Case study of the use of 3d-printing to print orthotic insoles. *12ème Congrès Int. Génie Ind.*, 2017.
35. Graff, P., Ståhlbom, B., Nordenberg, E., Graichen, A., Johansson, P., Karlsson, H., Evaluating measuring techniques for occupational exposure during additive manufacturing of metals: A pilot study. *J. Ind. Ecol.*, 21, S1, S120–S129, 2017, doi: 10.1111/jiec.12498.
36. Bhattacharjee, L. *et al.*, Emergence of robust carbon quantum dots as nanotracer for groundwater studies. *Diam. Relat. Mater.*, 103, 107701, 2020, doi: 10.1016/j.diamond.2020.107701.
37. Wheat, E., Vlasea, M., Hinebaugh, J., Metcalfe, C., Sinter structure analysis of titanium structures fabricated via binder jetting additive manufacturing. *Mater. Des.*, 156, 167–183, 2018, doi: 10.1016/j.matdes.2018.06.038.
38. Steinle, P., Characterization of emissions from a desktop 3D printer and indoor air measurements in office settings measurements in office settings. *J. Occup. Environ. Hyg.*, 13, 2, 121–132, 2016, doi: 10.1080/15459624.2015.1091957.
39. Turkevich, L.A., Fernback, J., Dastidar, A.G., Osterberg, P., Potential explosion hazard of carbonaceous nanoparticles: Screening of allotropes. *Combust. Flame*, 167, 218–227, 2016, doi: 10.1016/j.combustflame.2016.02.010.
40. Kellens, K. and Yasa, E., Environmental assessment of selective laser melting and selective laser sintering. *Methodology*, 4, 5, 5, 2010.
41. Chen, R. *et al.*, Chemosphere exposure , assessment and health hazards of particulate matter in metal additive manufacturing: A review. *Chemosphere*, 259, 55, 127452, 2020, doi: 10.1016/j.chemosphere.2020.127452.
42. Åkermo, M., Fernberg, P., Nordenberg, E., Brodin, H., Strondl, A., Nanosized by-products from metal 3D printing , composite manufacturing and fabric production. *J. Cleaner Prod.*, 139, 1224–1233, 2016, doi: 10.1016/j.jclepro.2016.08.141.
43. Fonseca, A.S. *et al.*, Science of the total environment process-generated nanoparticles from ceramic tile sintering: Emissions , exposure and environmental release. *Sci. Total Environ.*, 565, 922–932, 2016, doi: 10.1016/j.scitotenv.2016.01.106.
44. Shi, X. *et al.*, Evaluation of nanoparticles emitted from printers in a clean chamber , a copy center and office rooms: Health risks of indoor air quality. *J. Nanosci. Nanotechnol.*, 15, 12, 9554–9564, 2015, doi: 10.1166/jnn.2015.10314.
45. Taylor, P. *et al.*, Use of a condensation particle counter and an optical particle counter to assess the number concentration of engineered nanoparticles. *J. Occup. Environ. Hyg.*, 7, 9, 535–545, 2010, doi: 10.1080/15459624.2010.496072.
46. Hermann, M. *et al.*, Particle counting efficiencies of new TSI condensation particle counters. *J. Aerosol Sci.*, 38, 6, 674–682, 2007, doi: 10.1016/j.jaerosci.2007.05.001.

47. Ruppecht, E., Meyer, M., Patashnick, H., The tapered element oscillating microbalance as a tool for measuring ambient particulate concentrations in real time. *J. Aerosol Sci.*, 23, 635–638, 1992, doi: 10.1016/0021-8502(92)90492-E.
48. Giaya, A. and Thompson, R.W., Single-component gas phase adsorption and desorption studies using a tapered element oscillating microbalance. *Microporous Mesoporous Mater.*, 55, 3, 265–274, 2002, doi: 10.1016/S1387-1811(02)00428-6.
49. Chen, D., Bjorgum, E., Christensen, K.O., Holmen, A., Lodeng, R., Characterization of catalysts under working conditions with an oscillating microbalance reactor. *Adv. Catal.*, 51, 351–382, 2007, doi: 10.1016/S0360-0564(06)51007-7.
50. Ding, Y. and Koutrakis, P., Development of a dichotomous slit nozzle virtual impactor. *J. Aerosol Sci.*, 31, 12, 1421–1431, 2000, doi: 10.1016/S0021-8502(00)00048-3.
51. Heya, M.S., Parra-saldívar, R., Iqbal, M.N., Nano-biomaterials in-focus as sensing/detection cues for environmental pollutants. *Case Stud. Chem. Environ. Eng.*, 2, 100055, 2020, doi: 10.1016/j.cscee.2020.100055.
52. Kellens, K., Baumers, M., Gutowski, T.G., Flanagan, W., Lifset, R., Duflou, J.R., Environmental dimensions of additive manufacturing: Mapping application domains and their environmental implications. *J. Ind. Ecol.*, 21, S1, S49–S68, 2017, doi: 10.1111/jiec.12629.
53. Basak, A. and Das, S., Epitaxy and microstructure evolution in metal additive manufacturing. *Annu. Rev. Mater. Res.*, 46, 125–149, 2016, doi: 10.1146/annurev-matsci-070115-031728.
54. Andujar, P. *et al.*, Role of metal oxide nanoparticles in histopathological changes observed in the lung of welders. *Part. Fibre Toxicol.*, 11, 1, 1–13, 2014, doi: 10.1186/1743-8977-11-23.
55. Ahamed, M., Akhtar, M.J., Alhadlaq, H.A., Alrokayan, S.A., Assessment of the lung toxicity of copper oxide nanoparticles: Current status. *Nanomedicine*, 10, 15, 2365–2377, 2015, doi: 10.2217/nnm.15.72 View Article.
56. Ljunggren, S.A. *et al.*, Biomonitoring of metal exposure during additive manufacturing (3D Printing). *Saf. Health Work*, 10, 4, 518–526, 2019, doi: 10.1016/j.shaw.2019.07.006.
57. Renwick, L.C., Brown, D., Clouter, A., Donaldson, K., Increased inflammation and altered macrophage chemotactic responses caused by two ultrafine particle types. *Occup. Environ. Med.*, 61, 5, 442–447, 2004, doi: 10.1136/oem.2003.008227.
58. Zhang, Y., Ji, X., Ku, T., Li, G., Sang, N., Heavy metals bound to fine particulate matter from northern China induce season-dependent health risks: A study based on myocardial toxicity. *Environ. Pollut.*, 216, 380–390, 2016, doi: 10.1016/j.envpol.2016.05.072.
59. Nafisi, S. and Maibach, H.I., *Skin penetration of nanoparticles*, Elsevier Inc., Boston, Massachusetts, USA, 2018.

60. Wu, J. *et al.*, Toxicity and penetration of TiO2 nanoparticles in hairless mice and porcine skin after subchronic dermal exposure. *Toxicol. Lett.*, 191, 1, 1–8, 2009, doi: 10.1016/j.toxlet.2009.05.020.
61. Levi, K. and damage, U.V., and sun care: Deciphering mechanics of skin to develop next generation therapies. *J. Mech. Behav. Biomed. Mater.*, 28, 471–473, 2013, doi: 10.1016/j.jmbbm.2013.02.008.
62. Delfino, R.J., Sioutas, C., Malik, S., Potential role of ultrafine particles in associations between airborne particle mass and cardiovascular health. *Environ. Health Perspect.*, 113, 8, 934–946, 2005, doi: 10.1289/ehp.7938.
63. Lee, M., Yang, W., Chae, N., Choi, S., High resolution size characterization of particulate contaminants for radioactive metal waste treatment. *Nucl. Eng. Technol.*, 53, 7, 2277–2288, 2021, doi: 10.1016/j.net.2021.01.029.
64. KumarThakur, A., Kaundle, B., Singh, I., Mucoadhesive drug delivery systems in respiratory diseases, in: *Targeting Chronic Inflammatory Lung Diseases Using Advanced Drug Delivery Systems*, pp. 475–491, Academic Press, Amsterdam, Netherlands, 2020.
65. Zhang, H. *et al.*, Use of metal oxide nanoparticle band gap to develop a predictive paradigm for oxidative stress and acute pulmonary inflammation. *ACS Nano*, 6, 5, 4349–4368, 2012, doi: 10.1021/nn3010087.
66. Cui, C.Y., Cui, X.G., Ren, X.D., Qi, M.J., Hu, J.D., Wang, Y.M., Surface oxidation phenomenon and mechanism of AISI 304 stainless steel induced by Nd: YAG pulsed laser. *Appl. Surf. Sci.*, 305, 817–824, 2014, doi: 10.1016/j.apsusc.2014.04.025.
67. Taylor, P. *et al.*, Human health risk assessment for aluminium, aluminium oxide, and aluminium hydroxide. *J. Toxicol. Environ. Health Part B*, 10, S1, 1–269, 2007, doi: 10.1080/10937400701597766.
68. Collin, M.S. *et al.*, Bioaccumulation of lead (Pb) and its effects on human: A review. *J. Hazard. Mater. Adv.*, 7, 100094, 2022, doi: 10.1016/j.hazadv.2022.100094.
69. Liu, Y., Nguyen, M., Robert, A., Meunier, B., Metal ions in Alzheimer's disease: A key role or not? *Acc. Chem. Res.*, 52, 7, 2026–2035, 2019, doi: 10.1021/acs.accounts.9b00248.
70. Xu, L., Huo, X., Liu, Y., Zhang, Y., Qin, Q., Xu, X., Hearing loss risk and DNA methylation signatures in preschool children following lead and cadmium exposure from an electronic waste recycling area. *Chemosphere*, 246, 125829, 2020, doi: 10.1016/j.chemosphere.2020.125829.
71. Tedla, G., Jarabek, A.M., Byrley, P., Boyes, W., Rogers, K., Human exposure to metals in consumer-focused fused filament fabrication (FFF)/3D printing processes. *Sci. Total Environ.*, 814, 152622, 2021, doi: 10.1016/j.scitotenv.2021.152622.
72. Khosravani, M.R. and Reinicke, T., On the environmental impacts of 3D printing technology. *Appl. Mater. Today*, 20, 100689, 2020, doi: 10.1016/j.apmt.2020.100689.

73. Centers for Disease Control and Prevention (CDC), Hierarchy of Controls US National Institute for Occupational Safety and Health, Atlanta, Georgia, USA, 2017.
74. Zhang, Q., Davis, A.Y., Black, M.S., Emissions and chemical exposure potentials from stereolithography vat polymerization 3D printing and post-processing units. *ACS Chem. Health Saf.*, 29, 2, 184–191, 2022, doi: 10.1021/acs.chas.2c00002.
75. Oskui, S.M. *et al.*, Assessing and reducing the toxicity of 3d-printed parts. *Environ. Sci. Technol. Lett.*, 3, 1, 1–6, 2016, doi: 10.1021/acs.estlett.5b00249.

13

Current Aspects of Additive Manufacturing in the Aerospace Industry

Dheeraj Lal Soni[1] and Jagadish[1,2*]

[1]National Institute of Technology Raipur, Chhattisgarh, India
[2]SQC & OR Unit, Indian Statistical Institute Bangalore Centre, Bangalore, India

Abstract
Additive manufacturing (AM) is a layer-by-layer process of manufacturing that enables a product to be generated using a simple 3D model or 3D scanner. Additive manufacturing technology is utilized in automotive, aerospace, medical, and manufacturing industries due to its diversity in part size, material composition, and production time. The aircraft industry was one of the early users of AM prototyping technology, which is now utilized to make both final and replacement parts. Furthermore, thin-walled aircraft engine components, complicated geometries, and material processing problems are propelling the aviation industry's use of AM. It is poised to revolutionize the aerospace industry's production of intricate, lightweight, and virtually waste-free components. This chapter examines current aspects and development efforts in the aerospace industries in the field of AM and provides a literature review. From the literature, AM technology has shown enhanced printing speed due to multiple nozzle applications. Significant output characteristics of products such as ultimate tensile strength, surface finish, and skewness has improved for different alloys in liquid- and powder-based AM. Titanium powder-based alloys AM has shown difficulties due to high temperature characteristics. Researchers are currently working on the possibilities of AM-attached subtractive manufacturing process.

Keywords: Additive manufacturing (AM), aerospace industry, AM material

**Corresponding author*: jagadishbaridabad.s@gmail.com

R. Rajasekar, C. Moganapriya and P. Sathish Kumar (eds.) Additive Manufacturing with Novel Materials: Processes, Properties and Applications, (409–428) © 2024 Scrivener Publishing LLC

13.1 Introduction

Additive manufacturing (AM) is a layer-over-layer process of manufacturing used to construct different structures and complex geometries from model data by superimposing successive layers of material [1]. The AM process has attracted academics and benefited several sectors over the past three decades. The AM technology is extensively applied in European countries and the United States of America for design of AM component structure, biomaterials-forming using laser, selective laser melting of aerospace materials, and research of innovative laser sintering materials [2]. The AM process has altered the method and fundamentals of conventional manufacturing [3], hence subverting the conventional industrial model. It has revolutionized traditional manufacturing, which has turned the traditional industrial model on its head. In addition, the aerospace industrial application of expensive materials like titanium and other alloy necessitates more effort and time to recycle conventionally produced scrap. In the aerospace industries, AM process decreases the rate of skewness from 80–90% to 10–20%, hence decreasing the cost of time and effort, because maintenance, repair, and regeneration are important components of the aircraft industry. Aerospace comprises civil as well as military aircraft, space launch and in-space systems, satellites, missiles, and also civil aviation. In this situation, AM is not only extremely useful, but the industry can build the standard tool for each component that cannot be purchased from other industries. Additionally, AM is gaining ground in the commercial production and fabrication sector. Metal fabrication companies throughout the globe that have historically relied on subtractive processes like computer numerical control (CNC) turning and milling have begun offering AM alternatives. The AM process is explained in Section 13.2. The current challenges involved in AM technologies up to nanoscale fabrication using different materials with technological advancement were discussed in the previous literatures [3, 4]. In this direction, the present chapter aims to study the current aspects of AM for aerospace industry applications. The chapter focuses on three major aspects such as development in AM techniques, process parameters of AM, and different AM materials. The rest of the chapter is organized as follows: the different AM technologies for aerospace industries are discussed in Section 13.2. Section 13.3 explains the summary of the literature study. Finally, conclusion and recommendation are presented in Section 13.4.

13.2 AM Technologies for the Aerospace Industries

Every user has distinct requirements for the AM technique. Each AM technology has its own distinct benefits, drawbacks, and applications. Additive manufacturing machines differ in terms of application, quality, cost, precision, size, utilized material, etc. The technology used in AM is classified into three major groups (Figure 13.1). Most of the AM consumers have options to select an AM technology that meets their demands (Figure 13.2).

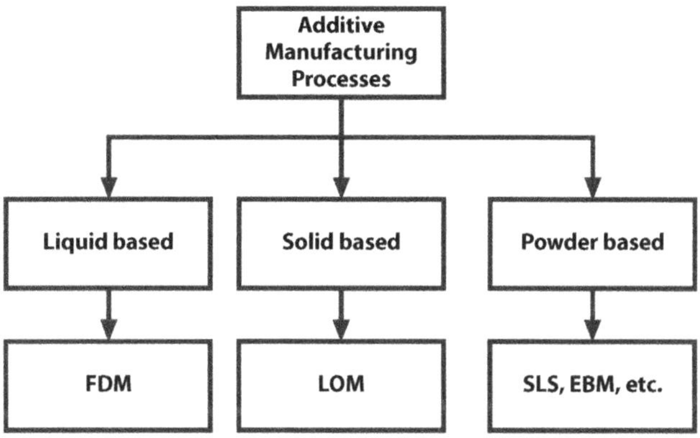

Figure 13.1 Classification of AM technology for aerospace applications.

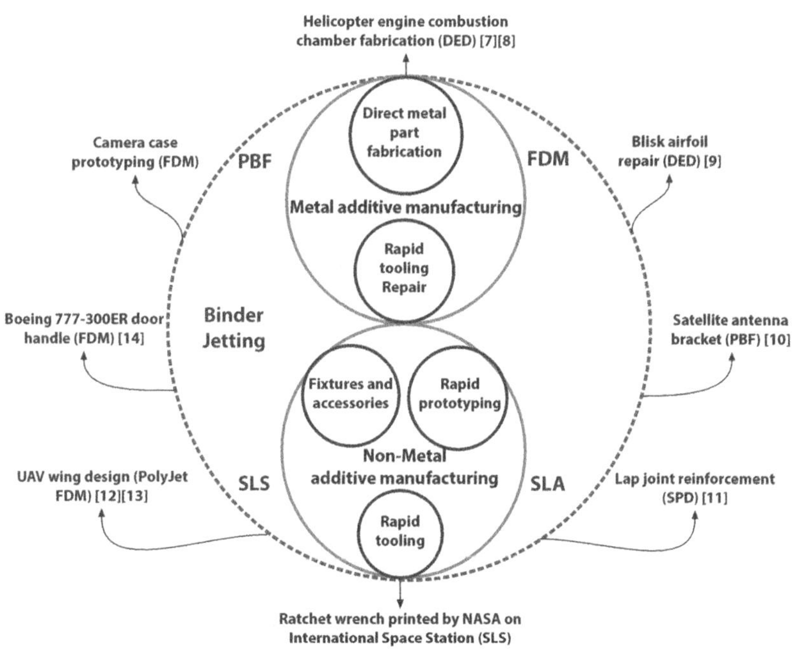

Figure 13.2 Metal and nonmetal AM technologies for aerospace applications with examples [6].

First is the powder AM technique, i.e., selective laser sintering (SLS), in which a focused laser moves on the bed to sinter a particular area of powder material layer after layer, creating a solid object. Surrounding powder near the sintering area acts as a support for the next layer. In SLS, the materials used are nylon and polyamide powder. Selective laser sintering is used in the medical field to impart bone and aerospace hardware. In electron beam melting (EBM), parts are produced in layers with a high vacuum electron beam by melting powder. The material used in EBM is conductive metal instead of a thermoplastic polymer. Second is the liquid-based AM technique: stereolithography (SLA), which is a technology used to manufacture layer-by-layer models using photochemical processes that bind chemical monomers into polymers. Stereolithography is an AM method used to build models, prototypes, and layer-by-layer patterns through photo polymerization (Figure 13.3), in this process, light links molecular chains and forms polymers [5]. The third solid AM technique is laminated object manufacturing (LOM), in which plastic or metal laminates are glued together and formed with a laser cutter. Fused deposition modeling (FDM) is accomplished by loading the extrusion head with filament material and heating it to a semi-liquid condition. The head is controlled by a computer to define the printed shape. Polylactic acid (PLA),

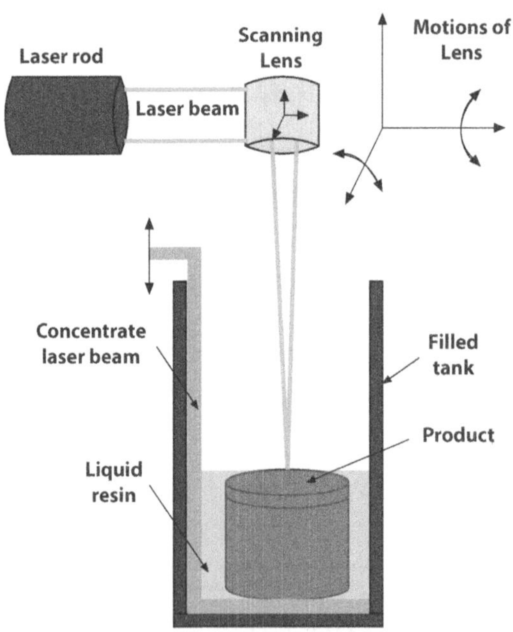

Figure 13.3 Schematic representation of SLA process.

high-impact polystyrene (HIPS), and thermoplastic polyurethane (TPU) are the materials utilized in FDM. There is a large number of unique users for the various AM methods. Moreover, each year, a variety of new technologies are introduced, e.g., friction stir-involved AM, and CNC-attached AM. Automobiles, aircraft, medicine, and construction are just a few of the industries that utilize AM technology.

Metal AM is a process where metal powders are melted and fused layer by layer to create a three-dimensional (3D) object. This process has many advantages over traditional manufacturing methods like ability to create complex geometries, reduce material waste, and produce fully dense, highly precise, and accurate components. One of the most common metal AM techniques is powder bed fusion (PBF), which uses either a laser or an electron beam to melt and fuse the metal powder. This technique has been widely adopted in the aerospace, medical, and automotive industries, where high-performance and customized parts are required [15].

Nonmetal AM, on the other hand, applies various materials, such as plastics, ceramics, and composites, to create 3D objects. This technique is commonly used in product design, architecture, and the production of consumer goods, where the cost effectiveness and flexibility of the process are highly valued [16]. Some of the most popular nonmetal AM techniques include different processes. First, SLA this technique is commonly used in product design and prototyping [17]. Second, FDM, this technique is widely used in the production of consumer goods and architectural models [18]. Third, SLS, this technique is commonly used in the production of aerospace components and medical devices [19]. While metal AM is generally more expensive than nonmetal AM, it offers unique advantages in terms of precision, accuracy, and strength. Nonmetal AM, on the other hand, is more cost effective and flexible, making it ideal for applications where these factors are more important.

Both metal and nonmetal AM processes have been paid significant attention in the aerospace industry because both have complex component production abilities with high precision with less weight. Various metal AM applications are as follows:

i. Turbine components: turbine blades and other components in aircraft engines require high strength, corrosion resistance, and complex geometries. Metal AM can produce these components with higher precision and fewer defects than traditional manufacturing methods.
ii. Spacecraft components: AM has been used to produce lightweight, high-strength parts for spacecraft such as satellites,

rockets, and space probes. For example, NASA has used AM to produce rocket engine components, and Boeing has used it to produce satellite components.

iii. Aircraft structural parts: Metal AM has the potential to reduce the weight of aircraft components while maintaining strength and durability. Airbus has used metal AM to produce parts for the A350 XWB, including structural components and brackets.

A few metal AM applications are as follows:

i. Interiors: Nonmetal AM has been used to produce interior components such as seat parts, cabin air grilles, and overhead bins. This technology allows for the production of parts with intricate designs and lightweight materials.
ii. Prototyping: Nonmetal AM has been used to produce prototypes of aircraft components, such as wings and landing gear. This allows for the testing of new designs and modifications to existing components without the need for expensive tooling.
iii. UAV Components: Nonmetal AM has been used to produce components for unmanned aerial vehicles (UAVs) such as drones. This technology allows for the production of lightweight, high-strength components that are essential for UAVs.

The detailed schematic of aerospace application is shown in Figure 13.3 with different applications, where ii describes the metal and nonmetal applications in various fields using different AM processes for specific applications. Manufacturing of all the above components involves different AM techniques as follows:

13.2.1 Stereolithography (SLA)

Stereolithography method employs (Figure 13.3) polymer liquid, which solidifies and forms the modal when ultraviolet photopolymer light strikes [20]. Liquid photopolymer resin (thermosetting plastic) is illuminated and cured layer over layer using a light emitting device (digital light processing or laser) to create 3D objects using the selective laser annealing (SLA) technique, also known as vat photo polymerization [6]. These polymers then create a solid entity with three dimensions. The region was studied in the 1970s, but Charles (Chuck) W. Hull coined this appellation in 1986 when he copyrighted the procedures [21]. This technique is used to generate the aerospace parts such as fully transparent light full-size panels and cabin accessories [22]. Stereolithography is used to generate seamless, high-detail scale models of aircraft designs, as well as to prototype aerospace components. A wide range of cabin components, from functional knobs and full-sized panels to seat components and main door can be created using standard SLA resin [6, 23]. Clear SLA resins are utilized to manufacture prototypes of completely transparent, highly detailed aircraft headlamps (Table 13.2).

Stereolithography can generate fine features and offer a smooth surface with minimal stair-stepping [24]. Rapid prototypes with high model similarity were created using low-stiffness resins for design, testing, and verification of aero elastic aerofoils [25]. Mold patterns and injection molds are made respectively with castable and high temperature SLA resins [6]. Stereolithography clear resins are used to create prototypes of entirely transparent, extremely detailed aircraft headlights [26]. Polystyrene (PS), Polyamide (PA), thermoplastic-elastomer (TPE), and polyaryletherketone (PAEK) are generally used in SLA. Process parameters used in the SLA process are extremely important for the strength of the product. In this way, according to Schuab *et al.* [27], the main selective laser (SL) process parameters that affect accuracy and strength are layer thickness, part orientation, and over-cure depth. According to Jacobs [20], the mechanical properties and part strength of SL prototypes are primarily determined by the layer thickness, laser power, scanning velocity, and orientation of the process. Jawahar *et al.* [28] and Ghazy [29] conducted research on the mechanical strength of SL-processed prototypes. They discovered that the most critical SL process factors to have the greatest influence on the prototype's ultimate tensile strength (UTS) are orientation, layer thickness, and post-curing time. Layer thickness was found as one of the most influential process parameters by Chockalingam *et al.* [30]. To maximize precision and reduce build time, Cheng *et al.* [31] proposed a multi-objective method for determining the optimal orientation of part construction. The study found that the orientation of the fabricated component during the SLA manufacturing showed significance, and that the orientation of the part affected the strength of the generated model.

13.2.2 Electron Beam Melting (EBM) or Sublimation

Electron beam melting is based on the fusion of the powder used that generate layers by concentrating heat to a specific area of metal powder on a substrate [32, 33]. Instead of ultraviolet light, electron beams are utilized by 3D printers [28]. As shown in Figure 13.4, the EBM method creates dense metal components layer by layer from metal powder using a strong electron beam as a sculptor [34]. In each layer, the appropriate geometrical (3D) CAD model is used. Large parts and cross sections are warped or deformed due to the high thermal residual stress produced by fusion AM [35]. Due to the cooling of the molten pool, the structure may crack [36]. A substantial amount of energy can be utilized for maximum fusion capability and output. Components produced in a vacuum at high temperatures are superior to cast components and comparable to enfolded material, resulting in stress relieving components [36]. To maintain the correct high temperature for each alloy, the electron beam scans the bed of powder for each layer of powder. To maintain the correct high temperature for each alloy, the electron beam scans the bed of powder for each layer of powder. Thereafter, the electron beam is fusing the perimeter and mass of the component. This technology is utilized in the aerospace and automotive industries to produce metal components with complex geometries. Parts generated by either wire-fed lasers or arc beam deposition are suitable for use in aerospace applications, with similar mechanical properties

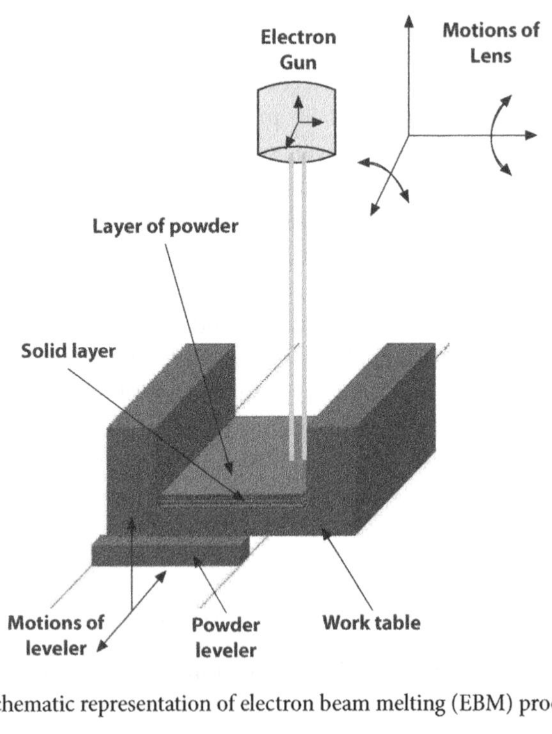

Figure 13.4 Schematic representation of electron beam melting (EBM) process.

depending on load direction, process parameters, and post-build heat treatment [37].

The experimental parameters generally used to generate a product in EBM are beam current, scan speed, line offset, etc. [38]. In particular, it has been shown that different factors, such as electron beam current, scanning rate, layer thickness, powder particle size, and hatching method affect the microstructural evolution and the generation of defects [39, 40]. The commercial materials used in the EBM process are titanium and titanium alloys, cobalt chrome, copper, nickel alloy, tool steel, stainless steel, tantalum and titanium tantalum alloy, etc.

13.2.3 Fused Deposition Modeling (FDM)

A filament roller feeds filament to both rollers, which is then turned into a semi-fluid object using heated nozzles in FDM. Today, FDM, also referred to as material extrusion, is the most common type of AM technology [21]. The technique is used to build the sculpture by extruding tiny pearls of material, which then solidify in layers (see Figure 13.5). The head of an extruder is fed by a coil of thermoplastic filament or wire. Before discharging the material, the dust head commences the flow and warms it to a specified temperature. Step motors are commonly used to adjust the extrusion's flow rate and to move the extrusion head in the Z-axis. A microcontroller-based computer aided manufacturing (CAM) software application can remove the sensor heads from both vertical and horizontal sensors and operate the mechanism.

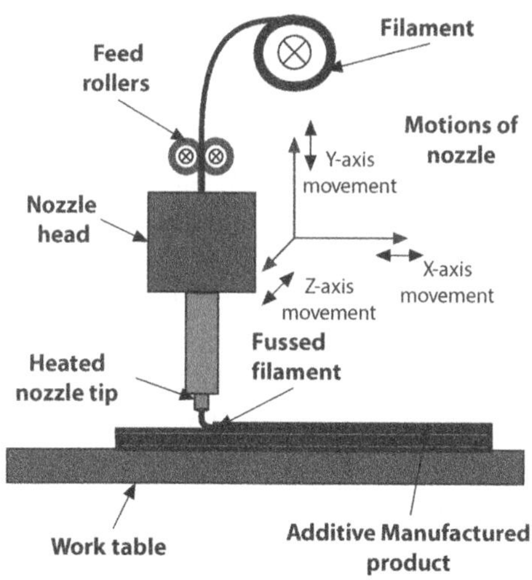

Figure 13.5 Schematic representation of fused deposition modeling (FDM).

Fused deposition modeling enables the production of high-strength thermoplastic components such as Ultem, polycarbonate [41], polyphenylsulfone [6], PLA, and acrylonitrile butadiene styrene (ABS). Fused deposition modeling systems have a wide range of applications, from quick and inexpensive rapid prototyping to robust and stiff end-use products (Table 13.2). The aerospace industry has been replacing old metal components with suitably robust FDM-produced ones in order to reduce weight and turnaround time for part repairs. Because they are lightweight and sturdy enough to resist the rigors of space, over 70% of 'NASA's Mars rover is made up of production-grade thermoplastic parts. Fused deposition modeling technology is appropriate for prototyping application that do not need high resolution or surface finish because it is cost effective and does not require chemical postprocessing [42]. The significant parameters in this process are laser beam width, sheet thickness, orientation angle, laser scan velocity, and laser Power [39]. Effects of the parameters are illustrated in Table 13.1.

13.2.4 Selective Laser Sintering

Selective laser sintering process (Figure 13.6) involves PBF that typically involves melting polymer powders with a laser energy source [44]. It is capable of creating huge components with good mechanical strength at a reasonable investment [24]. In aerospace applications, SLS is typically employed for rapid prototyping of non-functional parts and direct-digital manufacture (DDM) of non-critical components.

In the fabrication of SLS engine chambers requiring heat protection, such as tarmac nozzle bezels, glass-filled nylon is utilized [45]. Nylon-12 is utilized in SLS-generated components that need flexibility like bellow airflow and ducts directors (i.e., flow controllers). Degree of the particulate melt significantly affects the SLS parts and percent crystalline, such that sufficiently varied volumes of parts with varying percent crystalline can be handled as distinct materials [46]. This enables the production of functionally graded materials (FGM) components with SLS (Table 13.2). The mechanical properties of an SLS component made with acrylic and nylon are virtually identical to those of plastic injection-molded parts [47]. Selective laser sintering is cost effective for production of modest quantities of aircraft parts in the vast majority of applications [44]. Also, SLS has been used to make aerospace parts out of steel, Ti-alloys, and Ni-based super alloys, among other things. Selective laser sintering is the first AM technology that is being used to make important parts for a jet engine. How the SLS process parameters affected Ti47Al2Cr2Nb were studied. High scanning speeds made the melt pool less stable and led to

Table 13.1 FDM process parameters and their descriptions for application [43].

S.N.	Process parameters for FDM	Description
1	Thickness of layers	It is the measurement along the Z-axis of the height (or thickness) of layers deposited following extrusion from the nozzle tip (or vertical direction of the FDM machine). Typically, it is less than the diameter of the 'extruder's nozzle.
2	Orientation of building	An FDM 'machine's build orientation refers to the 'part's alignment (orientation) on the build platform in respect to the X, Y, and Z axes and the 'part's angle of production.
3	Orientation of raster	It refers to the angle (direction) at which extruded material is deposited in relation to the X-axis of the build platform. It is the angle between the X-axis and the raster pattern. Typically, raster angle ranges between 0° and 90°.
4	Air gap	Spacing between tool paths (or raster) on a single layer of an FDM-printed part.
5	Extrusion temperature	Prior to extrusion during the FDM process, the temperature to which thermoplastic filaments were heated within the nozzle. This parameter is dependent on the print speed and thermoplastic material type.
6	Printing speed	It is the speed with which the nozzle tip moves along the XY plane of the construction platform.
7	Infill density/ interior infill percentage	Most FDM-printed objects have solid outer layers, but their internal structure can be porous and made up of a wide variety of infill patterns, sizes, and shapes. It follows that the higher the infill density, the more robust the FDM-printed 'part's internal structure.
8	Nozzle diameter	Diameter of the 'extruder's nozzle tip
9	Width of contour	The thickness of the contour layers that surround the internal structure

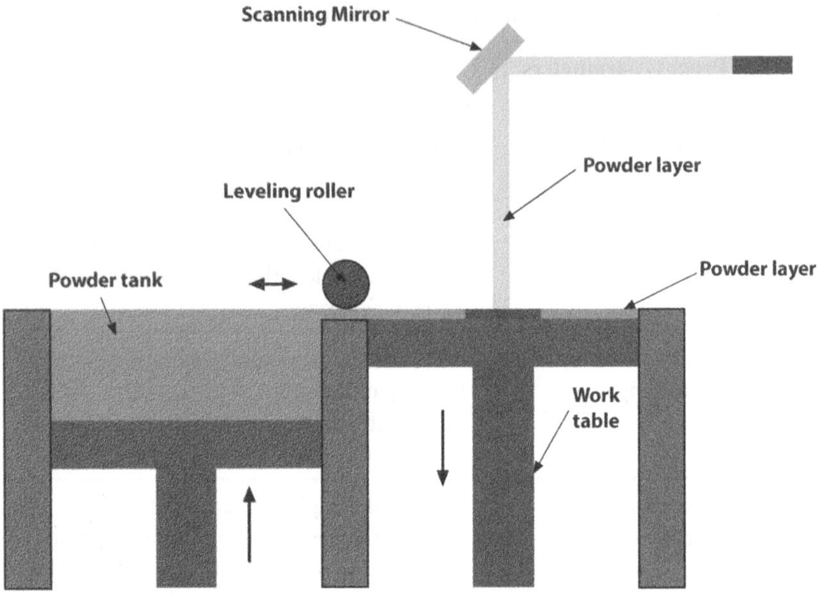

Figure 13.6 Working schematic of selective laser sintering.

more cracks after it solidified. The process has been optimized, but the range of parameters used in the study did not completely stop the cracking effect, which was caused by the high cooling rate of the SLS process [48].

13.2.5 Laminated Object Manufacturing (LOM)

Laminated object manufacturing (Figure 13.7) is an AM process in which plastic, paper, or metal is fused and laser cut into the required shape in layers of adhesive material. Laser cutters slice and create paper, plastic, and metal laminates into the appropriate shape. After the printing process, the printed objects can be modified by machining.

During the printing process, printed objects are suitable for machining. Depending on the material feedstock, the typical layer resolution of this technique ranges from one to several thick sheets. Laminated object manufacturing is mainly used to generate the emergency exit doors, ceilings, sidewalls, etc., of the aircraft. Few assembled parts of the seating, e.g., seating shells, can also be generated by the LOM technique.

13.2.6 Fusion Metal Additive Manufacturing (FMAM)

Fusion Metal AM is the process of adding materials layer over layer through the directional, programmable molten pool to create a near/

Table 13.2. Additive nonmetal technologies in the aerospace industry additive [43].

Application	Part fabrication	Requirement	AM process	Material used
UAV	Wings and fuselage	High strength light weight durability	FDM	ULTEM
Accessories parts of the cabin	Door handle covers	Easily modifiable, durable handles	FDM	ULTEM
Full-size panels	Main doors and parts of seat	Reasonably large, perfectly finished parts	SLA	Standard resin
Metal parts (casted)	Door handles and brackets	Casted metal parts	SLA	Castable resin or wax
Lights and its arrangements	Prototypes of the headlight	Detailed model with required transparency	SLA	Clear resin
Engine component	Termec bezel of nozzles	Heat resisting functional part	SLS	Glass-filled nylon
Air duct	Airflow ducting	Flexible-duct	SLS	Nylon 12

net form product (Figure 13.8). Consequently, material waste, lead time, design limitations, and potentially, the price of the completed product, are decreased [49]. Fusion AM can deliver cost-effective, small-batch production by reducing environmental impacts, material waste, and energy consumption compared to standard subtractive techniques. Distortion and cracking problem are very common in the FMAM process [50]. Some distortion can be stopped by putting in place physical limits and reducing the severity of thermal gradients by heating the body while it is being made.

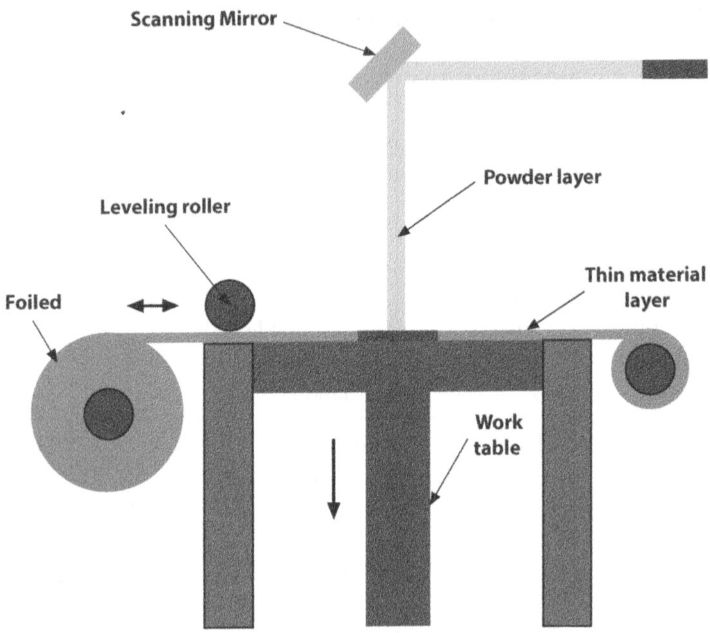

Figure 13.7 Working schematic of laminated object manufacturing.

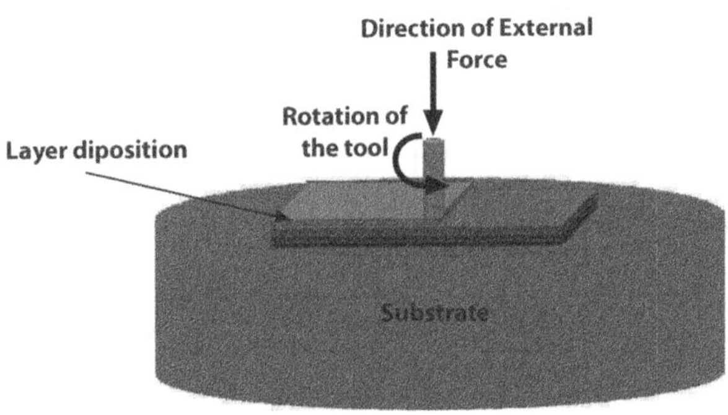

Figure 13.8 Schematic of additive friction stir deposition [49].

Crack modeling is more complicated and needs more information about the material's properties than distortion modeling. Some cracks and pores can be fixed with post-hot isostatic processing, which, when combined with annealing and quenching, can also relieve stress and improve the microstructure. Forging after the fact can sometimes close pores and cracks and grid of distortion, but it costs more [35]. Friction stir welding (FSW) and its variations, called friction stir processing (FSP) that can be adopted as FSAM to make materials that are mostly without limitations of fusion AM process. Some naval applications use FSAM for small structure applications [35].

13.3 Summary of the Literature Study

The AM industry is in the development stage and expanding rapidly. Some technologies exist to produce the product, while others are in the development stage. Compared to metallic products, the use of AM for nonmetallic materials is expanding exponentially. Aerospace and automotive industries use both nonmetallic and metallic items manufactured by AM, which has led to numerous improvements in AM techniques. Both of these sectors have difficulties associated with the generation of defective and complex metallic parts. Titanium alloys experience the same problems or challenges as AM in general. The primary issue in the AM of metals is alloy applications. In addition, the relationship between AM and subtractive manufacturing has attracted the focus of researchers. Consequently, certain CNC machines also offer the AM attachment, and advancements such as dual-nozzle FDM applications are now available for use. Few AM techniques including laser beam technology and FSP are in the developmental stages and require refinement. At the same time for research and development, these technologies are implemented in some application areas of aerospace industries and naval industries. The aerospace industry is most profited by employing AM technology.

13.4 Conclusion & Recommendation for Future Work

Various aircraft components, including the cabin, seating shells, ceilings, turbine blade, and engine cover, are manufactured using AM technology. In the aerospace sector, AM is also utilized to produce safety-related components such as emergency doors and windows. Upgradation of AM technology to minimize the fabrication time and design metal additively made products appropriately. By analyzing the uses and aspects of the aerospace industry, it is feasible to draw the following conclusions and recommend future work.

- To expedite the aerospace industry's use of AM metal components, present quality control requirements require further development.

- It has been determined that the SLS approach is the most suited and accurate way of AM, although greater component production and cost reduction are limitations.
- Most uses of metal AM in aircraft industries have demonstrated the benefits of cost and scheduled reduction, while some applications have demonstrated improvements in mass reduction and multiple part consolidation compared to earlier designs.
- Novel alloy composite generation for AM and aeronautical applications is recommended.
- The precision of the AM process can be improved using optimization methodologies such as topological optimization (TO) and hybrid analytical thermal optimization (HAATO)
- Components with many functions, like integrated electronic circuits and sensors, and incorporating image processing technologies have a broad field of work in the future for AM process.

References

1. Wong, K.V. and Hernandez, A., A review of additive manufacturing. *ISRN Mech. Eng.*, 2012, 1–10, 2012, https://doi.org/10.5402/2012/208760.
2. Shi, Y., Yan, C., Zhou, Y., *Polymer materials for additive manufacturing: Liquid materials*, Huazhong University of Science and Technology Press, Published by Elsevier Ltd. All rights reserved, China, 2021.
3. Adapa, S.K. and Jagadish, Prospects of natural fiber-reinforced polymer composites for additive manufacturing applications: A review. *JOM*, 75, 920–940, 2023, https://doi.org/10.1007/s11837-022-05670-w.
4. Tiwari, S. and Jagadish, Recent advancement in sustainable hybrid fiber-reinforced biocomposites: A state of the art, in: *Emerging Trends in Mechanical and Industrial Engineering: Select Proceedings of ICETMIE 2022*, X. Li, M.M. Rashidi, R.S. Lather, R. Raman (Eds.), pp. 699–711, Springer Nature Singapore, Singapore, 2023.
5. Xu, X., Awad, A., Robles-Martinez, P., Vat photopolymerization 3D printing for advanced drug delivery and medical device applications. *J. Control. Release*, 329, 743–757, 2021, https://doi.org/10.1016/j.jconrel.2020.10.008.
6. Najmon, J.C., Raeisi, S., Tovar, A., *Review of additive manufacturing technologies and applications in the aerospace industry*, Elsevier Inc., China, 2019.
7. Broad, United States Patent date of patent: Syst method program a weigh scale usinga key signal to enter a program mode 1:14, 12, 2009.

8. Dong, Y.H., Wang, Z.L., Peng, Z.L., Design and fabrication of the micro turbine for micro gas turbine engine. *Adv. Mater. Res.*, 69–70, 530–534, 2009, https://doi.org/10.4028/www.scientific.net/AMR.69-70.530.
9. Berger, R., Häfele, J., Hofmeister, B., Rolfes, R., Blend repair shape optimization for damaged compressor blisks, in: *Advances in Structural and Multidisciplinary Optimization*, A. Schumacher, T. Vietor, S. Fiebig, *et al.*, (Eds.), pp. 1631–1642, Springer International Publishing, Cham, 2018.
10. E.O.S. Gmb, H., Facts lightweight and highly stable: The topologically optimized antenna bracket for sentinel satellites with EOS aluminium AlSi10Mg using additive manufacturing certified for universal success: Additive manufacturing of satellite components, 2018.
11. Garcia, R. and Prabhakar, P., Bond interface design for single lap joints using polymeric additive manufacturing. *Compos. Struct.*, 176, 547–555, 2017, https://doi.org/10.1016/j.compstruct.2017.05.060.
12. Goetzendorfer, B., Mohr, T., Hellmann, R., Hybrid approaches for selective laser sintering by building on dissimilar materials. *Mater. (Basel)*, 13, 1–11, 2020, https://doi.org/10.3390/ma13225285.
13. Koc, E., Çalişkan, C., Coskun, M., Khan, H.M., Unmanned aerial vehicle production with additive manufacturing. *J. Aviat.*, 4, 22–30, 2020, https://doi.org/10.30518/jav.681037.
14. Continental Airlines 777 flight manual. 1913, 1–827, 2002.
15. Gibson, I., Rosen, D.W., Stucker, B., *Additive manufacturing technologies*, Springer, Sheffield, United Kingdom, 2015.
16. Wu, J. and An, J., The research status and development trend of non-metallic additive manufacturing technology. *J. Mech. Eng.*, 51, 7, 1–10, 2015.
17. C.W. Hull, Apparatus for production of three-dimensional objects by stereolithography. US Patent No. 4575330, 1986.
18. S.S. Crump, Apparatus and method for creating three-dimensional objects. US Patent No. 5121329, 1992.
19. C.R. Deckard, Method and apparatus for producing parts by selective sintering. US Patent No. 4863538, 1989.
20. Jacobs, P.F., Rapid prototyping & manufacturing—Fundamentals of stereolithography. *J. Manuf. Syst.*, 12, 430–433, 1993, https://doi.org/10.1016/0278-6125(93)90311-g.
21. Thakar, C.M., Parkhe, S.S., Jain, A., 3d Printing: Basic principles and applications. *Mater. Today Proc.*, 51, 842–849, 2021, https://doi.org/10.1016/j.matpr.2021.06.272.
22. Gasman, L., *Additive aerospace considered as a business*, Elsevier Inc., WA, United States, 2019.
23. Froes, F., Combining additive manufacturing with conventional casting and reduced density materials to greatly reduce the weight of airplane components such as passenger seat frames, Elsevier Inc., USA, 2019.

24. Goh, G.D., Agarwala, S., Goh, G.L., Additive manufacturing in unmanned aerial vehicles (UAVs): Challenges and potential. *Aerosp. Sci. Technol.*, 63, 140–151, 2017, https://doi.org/10.1016/j.ast.2016.12.019.
25. Zhu, W., Li, D., Zhang, Z., Design and fabrication of stereolithography-based aeroelastic wing models. *Rapid Prototyp. J.*, 17, 298–307, 2011, https://doi.org/10.1108/13552541111138423.
26. Gulati, M.S., Kumar, M.S., Arora, J., Stereolithography in implant placement: A review. *Int. J. Health Sci. (Qassim)*, 5, 23–27, 2021, https://doi.org/10.53730/ijhs.v5nS2.5344.
27. Chockalingam, K., Jawahar, N., Ramanathan, K.N., Banerjee, P.S., Optimization of stereolithography process parameters for part strength using design of experiments. *Int. J. Adv. Manuf. Technol.*, 29, 79–88, 2006, https://doi.org/10.1007/s00170-004-2307-0.
28. Chockalingam, K., Jawahar, N., Ramanathan, K.N., Banerjee, P.S., Optimization of stereolithography process parameters for part strength using design of experiments. *Int. J. Adv. Manuf. Technol.*, 29, 79–88, 2006, https://doi.org/10.1007/s00170-004-2307-0.
29. Ghazy, M. and Hossam, M., Effect of SLA process parameters on part build-time, in: *2015 25th International Conference on Computer Theory and Applications (ICCTA)*, pp. 40–45, 2015.
30. Chockalingam, K., Jawahar, N., Ramanathan, K.N., Banerjee, P.S., Optimization of stereolithography process parameters for part strength using design of experiments. *Int. J. Adv. Manuf. Technol.*, 29, 79–88, 2005, https://doi.org/10.1007/s00170-004-2307-0.
31. Cheng, W., Fuh, J.Y.H., Nee, A.Y.C., Multi-objective optimization of part-building orientation in stereolithography. *Rapid Prototyp. J.*, 1, 12–23, 1995, https://doi.org/10.1108/13552549510104429.
32. Liu, R., Wang, Z., Sparks, T., *Aerospace applications of laser additive manufacturing*, Elsevier Ltd, Germany, 2017.
33. Günther, J., Brenne, F., Droste, M., Design of novel materials for additive manufacturing-isotropic microstructure and high defect tolerance. *Sci. Rep.*, 8, 1–14, 2018, https://doi.org/10.1038/s41598-018-19376-0.
34. Kovalchuk, D. and Ivasishin, O., *Profile electron beam 3D metal printing*, Elsevier Inc., USA, 2019.
35. Withers, J.C., *Fusion and/or solid state additive manufacturing for aerospace applications*, Elsevier Inc., Netherlands, 2019.
36. Abdi, F., Yavari, P., Harik, V., Godines, C., Material allowable generation and AM process parameters effect on porosity. *Coatings*, 10, 625–644, 2020, https://doi.org/10.3390/coatings10070625.
37. Brandl, E., Baufeld, B., Leyens, C., Gault, R., Additive manufactured Ti-6A1-4V using welding wire: Comparison of laser and arc beam deposition and evaluation with respect to aerospace material specifications. *Phys. Proc.*, 5, 595–606, 2010, https://doi.org/10.1016/j.phpro.2010.08.087.

38. Silvestri, A.T., Foglia, S., Borrelli, R., Electron beam melting of Ti6Al4V: Role of the process parameters under the same energy density. *J. Manuf. Process.*, 60, 162–179, 2020, https://doi.org/10.1016/j.jmapro.2020.10.065.
39. Murr, L.E., Metallurgy of additive manufacturing: Examples from electron beam melting. *Addit. Manuf.*, 5, 40–53, 2015, https://doi.org/https://doi.org/10.1016/j.addma.2014.12.002.
40. Béraud, N., Vignat, F., Villeneuve, F., Dendievel, R., Improving dimensional accuracy in EBM using beam characterization and trajectory optimization. *Addit. Manuf.*, 14, 1–6, 2017, https://doi.org/https://doi.org/10.1016/j.addma.2016.12.002.
41. Shi, Y., Yan, C., Zhou, Y., Polymer materials for additive manufacturing—Powder materials, 2021.
42. Klippstein, H., De Cerio Sanchez, A.D., Hassanin, H., Fused deposition modeling for unmanned aerial vehicles (UAVs): A review. *Adv. Eng. Mater.*, 20, 1700552, 2018, https://doi.org/https://doi.org/10.1002/adem.201700552.
43. Jaisingh Sheoran, A. and Kumar, H., Fused deposition modeling process parameters optimization and effect on mechanical properties and part quality: Review and reflection on present research. *Mater. Today Proc.*, 21, 1659–1672, 2020, https://doi.org/10.1016/j.matpr.2019.11.296.
44. Wong, K., Yap, B., Fung, B.K.-P., Treatment of neck pain with collateral meridian acupressure therapy: A randomised, sham-intervention controlled trial. *Aust. J. Acupunct Chin. Med.*, 7, 10–15, 2012.
45. Quincieu, J., Robinson, C., Stucker, B., Mosher, T., Case study: Selective laser sintering of the USUSat II small satellite structure. *Assem. Autom.*, 25, 267–272, 2005, https://doi.org/10.1108/01445150510626389.
46. Majewski, C., Zarringhalam, H., Hopkinson, N., Effect of the degree of particle melt on mechanical properties in selective laser-sintered Nylon-12 parts. *Proc. Inst. Mech. Eng. Part B J. Eng. Manuf.*, 222, 1055–1064, 2008, https://doi.org/10.1243/09544054JEM1122.
47. Tang, H.H., Chiu, M.L., Yen, H.C., Slurry-based selective laser sintering of polymer-coated ceramic powders to fabricate high strength alumina parts. *J. Eur. Ceram. Soc.*, 31, 1383–1388, 2011, https://doi.org/10.1016/j.jeurceramsoc.2011.02.020.
48. Chen, W. and Li., Z., *Additive manufacturing of titanium aluminides*, Elsevier Inc., USA, 2019.
49. Mishra, R.S., Haridas, R.S., Agrawal, P., Friction stir-based additive manufacturing. *Sci. Technol. Weld. Join.*, 27, 141–165, 2022, https://doi.org/10.1080/13621718.2022.2027663.
50. Gouge, M. and Michaleris., P., Distortion in metal additive manufacturing: Modelling and mitigation. *Met Add Manuf.*, vol. 3, pp. 61–71, 2017.

38. Silvestri, A.T., Foglia, S., Borrelli, R., [illegible]

[illegible]

14

Four-Dimensional (4D) Microprinting: Materials, Processes, Challenges and Applications

Roja Rani Korrayi[1], Anita Jena[2], Bharat C. G. Marupalli[3], Bijaya Bikram Samal[2*], Shailendra Kumar Varshney[4] and Cheruvu Siva Kumar [5]

[1]Materials Science Centre, Indian Institute of Technology Kharagpur, West Bengal, India
[2]Advanced Technology Development Centre, Indian Institute of Technology Kharagpur, West Bengal, India
[3]Department of Metallurgical and Materials Engineering, Indian Institute of Technology Kharagpur, West Bengal, India
[4]Department of Electronics and Electrical Communication, Indian Institute of Technology Kharagpur, West Bengal, India
[5]Department of Mechanical Engineering, Indian Institute of Technology Kharagpur, West Bengal, India

Abstract
Four-dimensional (4D) printing technology has enabled researchers and engineers to print structures that can change their properties when an external fillip is imparted to them. 4D printing on micro- and nanoscale is still challenging and requires a lot of innovation for its widespread applications. Novel technologies like 4D microprinting can be an important enabler of industry 4.0 and industry 5.0 with broad applications starting from smart electronics and photonics devices to smart micro- and nanorobotics. Therefore, the present chapter reviews state-of-the-art and important developments in the field of 4D microprinting. It provides important micro AM processes and smart materials applicable for 4D microprinting. Further, major challenges and future outlooks of the novel technology are discussed in detail. Micro 4D printing will revolutionize the micro/nano fabrication methods currently used.

**Corresponding author*: bijaybikram@iitkgp.ac.in

R. Rajasekar, C. Moganapriya and P. Sathish Kumar (eds.) Additive Manufacturing with Novel Materials: Processes, Properties and Applications, (429–458) © 2024 Scrivener Publishing LLC

Keywords: 4D printing, micro 4D printing, 4D microprinting, smart materials, micro additive manufacturing

14.1 Introduction

Three-dimensional (3D) printing technology, also known as additive manufacturing (AM), is a process of creating 3D objects by laying down successive layers of material. The technology has come a long way since its invention in the 1980s, and it is now being used in a wide range of industries, including manufacturing, healthcare, and aerospace. One of the main advantages of 3D printing (3DP) is that it allows for the creation of complex geometries and shapes that would be difficult or impossible to produce using traditional manufacturing methods [1]. Additionally, it enables rapid prototyping, which allows for the rapid testing and iteration of designs. Direct ink writing (DIW), digital light processing (DLP), fused filament fabrication (FFF), selective laser sintering and melting (SLS/SLM), and stereolithography (SLA) are all common 3D printing/AM techniques [2]. Due to material and AM difficulties, the smallest features achievable in printed structures using most 3D printers are in the order of 50μm. Manufacturing at even smaller sizes is in high demand because, despite impressive advances at the macroscale, modern electronics are constantly being shrunk down in size [3]. Microscale 3D printing, also known as micro AM, is a subcategory of 3DP that focuses on creating small, highly detailed, and precise parts and structures. These structures can be as small as a few micrometers and require specialized equipment, techniques, and materials. Micro AM can have a wide range of applications in microelectromechanical systems (MEMS) devices, 3D-printed photonics, micro additively manufactured microfluidics, and microstents [4].

Four-dimensional (4D) printing has grown in popularity in recent years and is emerging as a useful technology for producing smart, dynamic, and adaptive structures, which was first introduced in 2013 [5]. Four-dimensional printing combines the versatility of 3DP with the capability of smart materials to respond to external stimuli [6, 7]. Controlled changes in the characteristics of 4D printed items, such as shape, size, color, or functionality, occur over time [8]. The fourth dimension, time, permits previously inconceivable possibilities by adopting adaptive materials, which can intentionally change their features in reaction to a stimulus [9]. Structures with a substantially small feature size of an order of magnitude of a few microns to nanometer have not yet been carefully investigated in 4D printing [10]. Three-dimensional/four-dimensional printing of accurate micro- and nanostructures, whose characteristics can be adjusted on demand, is greatly sought after for several applications, such as those in biomedicine, micro-optics, microrobotics, and microactuators. Two photon lithography/polymerization (TPP), which is also known as direct laser writing (DLW), and 3D laser is one such technology that has evolved in recent years to facilitate 3D/4D printing at ever smaller dimensions not possible before [11]. Fabricating 3D objects with sub-micrometer precision is made possible by DLW's reliance on a two-photon or multiphoton polymerization method employing a femtosecond-pulsed laser that is finely focused into a photoresist, thus curing it [12].

Macro 4D printing refers to the use of materials that can change shape or properties at a scale visible to the naked eye, typically on the order of millimeters to centimeters. The discipline of architecture and construction represents one of the most significant applications of macro 4D printing [13]. By utilizing 4D-printed materials, it is possible to design buildings that can adapt to altering environmental conditions, such as temperature, humidity, and wind. A 4D printed facade, for instance, can be programmed to open and close its apertures in response to changes in temperature or humidity, thereby modulating the passage of air and light into the building [13]. Micro 4D printing refers to the use of materials that can change shape or properties at a scale too small to be visible to the naked eye, typically on the order of micrometers to millimeters [14]. The discipline of -MEMS- is one of the most significant applications of micro 4D printing [9]. It is possible to construct MEMS devices that can alter their shape or properties in response to external stimuli using 4D-printed materials. For instance, a 4D-printed MEMS device can be programmed to respond to changes in temperature or pressure by altering its shape or rigidity. Nano 4D printing refers to the use of materials that can change shape or properties at a scale too small to be visible even with a microscope, typically on the order of nanometers. In the field of nanoelectronics and photonics, nano 4D printing has one of its most important applications. It is possible to construct nanoelectronic and photonic devices that can alter their shape or properties in response to external stimuli using 4D-printed materials. For instance, a 4D-printed nanoelectronic switch can be programmed to respond to changes in temperature or electric fields by altering its conductivity. Nano 4D printing can be utilized to create 4D photonic metamaterials for stimulus-driven properties due to shape change. As technology continues to develop, we can expect to see even more exciting applications in the future. Though 4D printing of macroscopic items has been accomplished, microscale 4D printing is still in its infancy. Therefore, in this chapter, we are reporting the recent developments with regard to the area of 4D microprinting, with an emphasis on smart materials, methods, and applications.

14.2 4D Printing Technologies

Four-dimensional (4D) printing, often called "dynamic printing" or "smart printing," is a cutting-edge technology with game-changing potential across several sectors. The process involves printing with particular materials and inks to make parts and components. Their shape or properties alter over time in response to environmental factors like temperature and humidity [15]. This innovation is still in its infancy, but it is already the subject of study in various disciplines, including architecture, manufacturing, agriculture, medicine, and robotics, and it has the potential to significantly alter the way we fabricate things [7, 16, 17]. The idea behind 4D printing is to make items whose form or qualities can evolve over time [18]. This is accomplished by printing a certain pattern or configuration into a substance, such as a shape memory alloy (SMA) or polymers (SMP) capable of undergoing shape alterations [19]. Smart materials, which can change their behavior in response to environmental factors like temperature, humidity, or electrical current, are often employed in 4D printing [20].

Four-dimensional printing can be classified broadly into three types, as shown in Figure 14.1 [21]. The first is 4D printing by embedding shape memory material during the process of 4D printing [22], the second is direct printing of smart material [23], and the third is multimaterial 4D printing [24]. The most common types of smart materials used in 4D printing are shape memory alloys, polymers with shape memory, and electroactive polymers. Metals known as SMA can be heated to a certain temperature and then reshaped into a previously stored pattern. This property makes them ideal for fabricating temperature-responsive shapes [25].

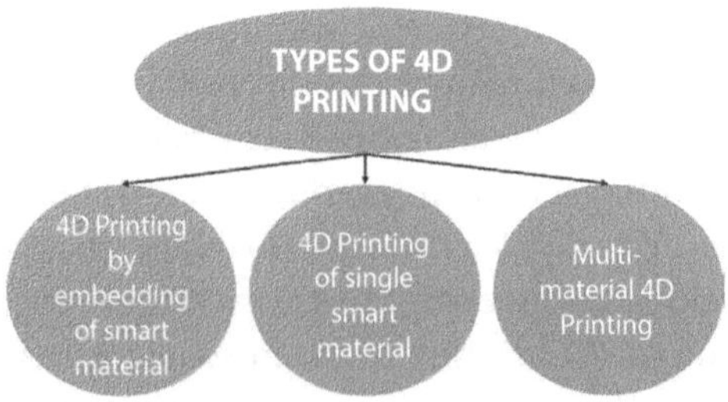

Figure 14.1 Various types of 4D printing.

Most 3D printers employ an extrusion-based printing approach to generate an economically viable 4D object, in which the material is forced through a nozzle to form a desired shape or pattern [26]. The potential for 4D printing to disrupt numerous markets remains high despite the technology's infancy. Research in architecture using 4D printing to produce resilient structures that can withstand extreme weather or natural disasters is one of the most exciting avenues of study [27]. A 4D-printed structure may, for instance, be made to contract or expand in reaction to temperature variations, making it resistant to the effects of wind and earthquakes. Medical implants that can undergo shape or property changes over time are another exciting prospect for 4D printing research. A 4D-printed stent, for instance, may be customized to expand or contract in response to variations in blood flow [28]. Additionally, 4D printing may cause a seismic shift in robotics by making it possible to build robots with dynamically adaptable forms and functions. To adapt to new activities or settings, a 4D-printed robot, for instance, may be programmed to alter its physical form in reaction to external stimuli [29].

14.3 Micro 4D Printing: State of the Art

Over the past few years, 4D printing at the microscale has gained the attention of researchers across the globe. Significant efforts are also being made with the fabrication of microscale devices through various routes like micro stereolithography (μSLA) and DLW. Direct laser writing is a powerful technique to obtain complex structures at the micro/nanoscale. The high-resolution structures of poly(N-isopropylacrylamide) (PNIPAm) and acrylamide were 4D printed by Han *et al.* [30] using the μSLA technique. The hydrogel's ability to expand at low and contract at high temperatures was independently controlled by adjusting the material and process parameters. Since the polymer chain length was shorter due to the high molar ratio of the crosslinker to the monomer, the degree of swelling was lower. The concentration of monomer affected the volume of the shrinking hydrogel above lower critical solution temperature (LCST). Grayscale patterns were used to control the curing light intensity, allowing spatial control of the hydrogel's swelling behavior within printed layers. An anisotropic swelling behavior can be produced by adjusting the layer thickness and maintaining a gradient of crosslinking density throughout the thickness.

A novel method for producing the temperature-sensitive polymer PNIPAm was proposed by Hippler *et al.* [31]. The PNIPAm block showed a response to increased heat generated due to the absorption of focused light. They also included a typical non-responsive material [pentaerythritol triacrylate (PETA)] and a responsive photoresist (PNIPAm) to demonstrate considerable 3D actuation responses. Huang *et al.* [32] presented a self-morphing programmable modular design modeled after Lego-like construction components. These modular structures actively showed deformation in a preprogrammed manner. This shape transformation behavior was activated by programming the exposure of femtosecond laser pulses to control the crosslink density of the material.

A novel technique for creating dynamic, shape-shifting, pH-responsive microstructures were developed by Lee *et al.* by tuning the crosslinking density of protein-based resist [33]. The control over the actuation mechanism was achieved by controlling the crosslinking density at the nanoscale level. Photo-responsive hydrogels of polyvinyl alcohol (PVA) were presented by Qin *et al.* [34]. These hydrogels provided fully synthetic matrices, multicellular patterning, and cell invasion. Hu *et al.* [35] described a pH-responsive hydrogel with many degrees of freedom shape change that was achieved by utilizing pH-triggered expansion, contraction, and torsion in micro botanical-inspired 4D printing. They showed how this 4D-printed microarchitecture might be used for medicine administration, single-cell analysis, and selective micro-object cargo release. A gelatin-methacryloyl (GelMA)-based hydrogel-based highly integrated multifunctional soft helical microrobot was introduced by Dong and colleagues. In addition to being impregnated with composite multiferroic nanoparticles to react to magnetoelectric stimuli, the helical body also fostered the formation of brain cells. It was broken down by enzymatic processes [36]. As Nishiguchi's study [37] demonstrated, spiral 4D-printed microstructures can serve as microactuators. The combined soft actuators of thermoresponsive PNIPAm and gold nanorods (AuNR) were 3D printed. The resulting composite used the photothermal effect to react to nearby heat. It was produced through AuNR's light absorption. As a result, these systems are susceptible to non-equilibrium and predetermined fast actuation, consisting of noticeable changes in form to the anterior and posterior. The creation of 3D angles is made possible by 4D direct laser lighting, which combines 3D manufacturing techniques with intelligent materials. The resulting 4D structures can carry out challenging tasks like limited foldable, bending, and twisting that are within control. These sophisticated 3D morphing micromachines can convert from one 3D shape to another quickly, precisely, and irreversibly in response to several environmental stimuli, opening a gateway for numerous engineering applications. Recently, Carbonell and Braunschweig [38] proposed the 4D nanoprinting process. It is anticipated that the bottom-up 4D nanoprinting method will be a successful substitute for the traditional top-down methods used to fabricate electronics, bioarrays, and functional substrates. Future advancements in this field are dependent on improvements in instrumentation and surface chemistry.

14.4 Smart Materials for Micro 4D Printing

Smart materials are materials that can respond to external stimuli. Thermoresponsive materials respond to temperature [39, 40], and magnetoresponsive materials show a response toward the magnetic field. Hence they are termed thermal and magnetic shape memory materials, respectively. There are other smart materials that respond to light, like pH of the solution, moisture, and pressure. This chapter discusses the various smart materials developed by micro or nano 4D printing for functional applications at the micro/nanoscale.

14.4.1 Polymer-Based Materials

Polymers are widely used materials for 4D printing at the micro or nanoscale owing to their significant properties such as excellent fluidity, curing performance, low cost, flexibility, transparency, moisture responsiveness, easy processing, and biocompatibility. Photopolymers are often used in vat photopolymerization. The reason for polymerization to take place fast is because of the absorption of ultraviolet (UV) light by the photoinitiators at a specific wavelength [41]. Hydrophilic polymers or hydrogels help produce moisture-responsive materials using natural or synthetic polymers [42]. Cellulose and polyacrylamide are common hydrophilic polymers. Thermoplastic polymers such as PLA and polycaprolactone (PCL) are highly used in 4D printing. As the demand for functional components and parts emerged, the properties provided by a single material filament was not sufficient and therefore, composite materials were introduced in AM. Apart from the basic thermoplastic materials utilized for 3DP, other high end materials like nylon and polyethylene glycol (PEG) have been used to produce micro structures for triboelectric nanogenerators (TENG) [43]. To summarize, thermoplastic polymers with medium viscosities and low melting temperatures have become crucial building blocks for high-performance composites.

14.4.2 Metallic-Based Materials

Metallic materials display excellent mechanical and chemical properties [44]. Producing metal-associated composites can enhance the performance of the metallic material. These composites are produced by combining metals with hydrophilic polymers and thermoplastic polymers. Titanium and titanium alloy powders are widely used in AM because of their low density, corrosion resistance, and high specific strength [45]. The processing time of titanium to fabricate 3D microscale structures decreased greatly with the usage of micro-SLM and its methods. However, optimizing the process parameters for each metal delays the materials' development and increases the cost of production [46]. Hydrogel-metal composites provide an alternative by incorporating metallic nanoparticles in hydrogels for designing low-cost materials with good biocompatibility and mechanical properties [47]. Metal-filled thermoplastic composites are produced by incorporating metal filler materials in the polymer matrix. Sang *et al.* fabricated a composite with metal filler showing improved mechanical properties and heat-responsive behavior. Metallic-based composites with 10% iron and copper were developed to improve the material's thermal and electrical conductivity and efficiency [48].

14.4.3 Ceramic-Based Materials

Ceramic-based materials possess high hardness, temperature, and corrosion resistance. The ceramic-based materials include oxides, carbides, nitrides, and a combination. Tricalcium phosphate (TCP) is one of the ceramic materials whose characteristics are similar to that of human bone [49]. Hence, they are useful in producing tissue scaffolds by 4D printing. Several studies have reported the fabrication of silica-based materials by micro 4D printing. Kotz *et al.* printed high-resolution molten glass structures using projection micro stereolithography (PμSL) [50]. The printed structures have shown properties equivalent to that of commercially produced glass.

14.4.4 Biomaterials

Biomaterials have great potential in tissue engineering, promoting cell growth and drug delivery applications. Biocomposite hydrogels and photosensitive biocomposites are primarily produced by combining bioactive ingredients with hydrophilic and photosensitive polymers. Hydrogel-based biocomposites can be utilized for getting tailored properties in 3D cell/tissue bioprinting. Photosensitive biocomposites capable of sensing light have the advantage of rapid curing and advanced biocompatibility and biodegradability. Photosensitive biocomposites can be classified into UV photosensitive and visible light photosensitive based on the absorbed range of light. Researchers have mixed photoinitiators with biomaterials to fabricate photosensitive biocomposites [51]. After composite preparation, they used bio printers to 4D print biomedical scaffolds which were later used to form a human ear structure. In the comparison between visible and UV absorbance monomers, it was seen that the visible absorbance photoinitiators and monomers are better for biomedical applications as it improves the curing and also the tissues and organs will not get damaged due to UV rays.

Biopolymers are another important classification under biomaterials. Biopolymers can promote tissue cell formation and regulate drug release, and they have improved mechanical and biodegradable characteristics comparable to those of human soft tissues. Stereolithography, DIW, and FFF are the procedures most frequently employed for printing biopolymers. The two main categories of biopolymers are natural and synthetic biopolymers. Common examples of natural biopolymers are gelatin, fibrinogen, and collagen. generally derived from animal tissue proteins and natural polysaccharides [52]. Some of the widely used synthetic biopolymers are PCL, PLA, and PEG.

14.5 Micro Additive Manufacturing Processes

Due to the recent application in various fields requiring the fabrication of micro- and nanoscale structures, it is highly necessary to develop an AM process capable of producing parts at that scale for AM's inherent advantages over other fabrication methods. Micro AM methods are a subset of AM used to construct microscale objects with micrometer-sized features. These techniques involve sophisticated technologies to attain great precision and accuracy. They are used to construct various microscale structures, including MEMS, bio-inspired structures, and functional materials.

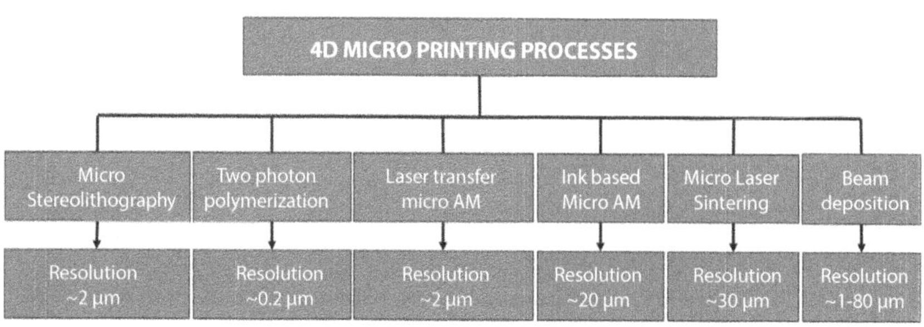

Figure 14.2 Various AM processes involved in 4D microprinting.

The various micro AM processes applicable for 4D microprinting along with a range of resolution are represented in Figure 14.2.

14.5.1 Micro Stereolithography

Stereolithography is an AM process was the first to be commercialized among all other processes. The process starts with a vat of photosensitive liquid resin, which on interaction with a UV laser moving in a CAD-driven predefined path, gets solidified. The platform moves down precisely to one layer, and this process continues to form the 3D object. As the spot size of a laser can be controlled well and minimized in the X-Y plane, this technology can be used for creating microstructures. The movement of the platform has to be modified to create layers of micrometer thickness, and the other mechanism of material addition remains the same, which is called micro sterolithography (μSLA) [53]. It is a technique that combines micro engineering with fast prototyping. Because the resolution of the μSLA approach is significantly superior to that of rapid prototyping technologies, it is of particular relevance in the micro engineering area, where its 3D capacity enables the manufacturing of components that no

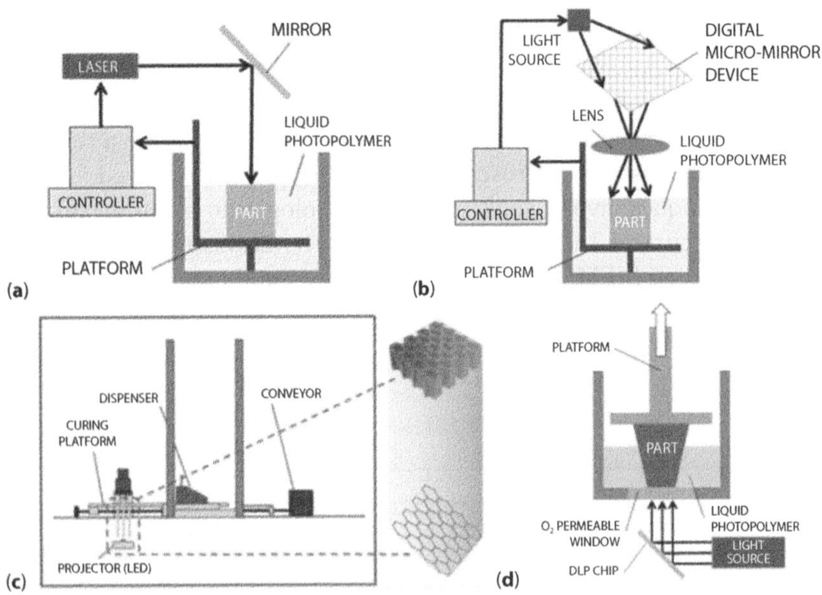

Figure 14.3 Various μSLA setups (a) scanning type μSLA and (b) projection type μSLA, (c) mask image projection type μSLA, and (d) continuous liquid interphase type μSLA [55, 56].

other microfabrication technology can produce. The production concept of μSLA is similar to that of SLA, but it uses refinements to achieve a much higher resolution. The earliest improvements in the μSLA process began in 1993, and since then, many research teams have built machines employing various techniques [54]. These techniques are demonstrated in Figure 14.3.

14.5.2 Two-Photon Polymerization

Two-photo polymerization (TPP) allows for the direct manufacture of arbitrary 3D prototypes from CAD models with resolutions as low as 100 nm or less. This method has been widely used for manufacturing bio-micro devices and bio-nanodevices, such as microneedles and microfluidic devices, because of its exceptional flexibility and great spatial resolution [57]. The simultaneous absorption of two photons in a photosensitive material which is a photoresist, results in TPP, a non-linear optical process. This method alters the photosensitive material, causing polymerization by activating photoinitiators in the resist. These react to form radicals, which polymerize the resist selectively. The non-polymerized material is then rinsed away to reveal that the structure is one of the most flexible

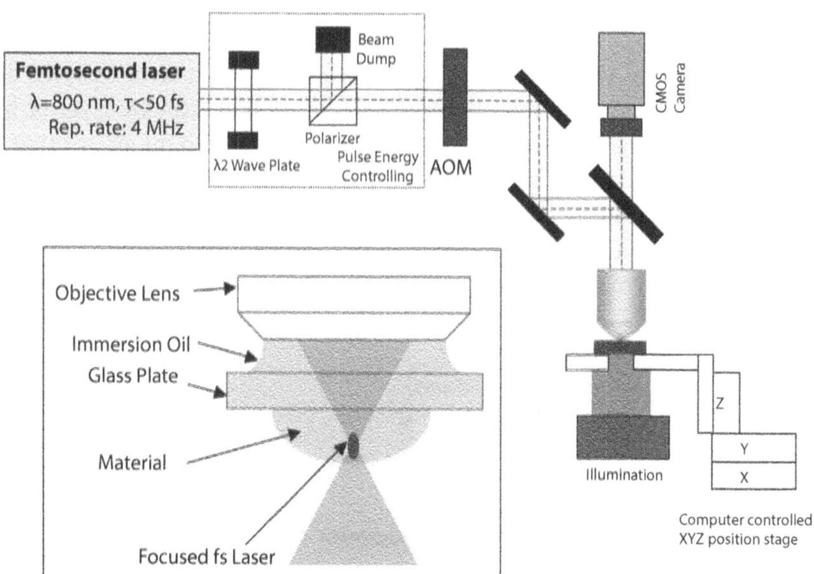

Figure 14.4 Schematic illustration of the two-photon polymerization setup [59].

and accurate AM technologies [58]. The mechanism of TPP is illustrated through a schematic in Figure 14.4.

14.5.3 Laser Transfer Micro AM

Laser transfer micro AM is a micro patterning-specific additive laser direct-writing technology. It uses lasers with great focusing strength to deposit microstructures of various materials. This process is an excellent example of additive laser direct writing (LDW), which can produce high-resolution patterns without requiring additional lithographic stages [60]. The same equipment configuration needed by laser-induced forward transfer (LIFT) may also be utilized for subtractive (laser micromachining) and altering LDW techniques (laser sintering, laser annealing, etc.). The schematic of LIFT process is given in Figure 14.5. LIFT's capacity to laser transfer a broad range of materials with reasonably high resolution conformal to the surface distinguishes it from other processes. The LIFT method, for example, has no negative impact on the electrical, chemical, or even biological aspects of the material voxels that comprise the digital pattern [61]. One of the key benefits of laser-based microfabrication methods is the ability to transfer functional materials and then treat or change them to produce the desired characteristics and behavior [62].

14.5.4 Ink-Based Micro AM

Ink-based micro AM is the process of fabricating a part layer by layer by continuously depositing ink on top of each other in a layer-by-layer fashion. One of the methods of ink-based micro AM is micro inkjet printing.

Because of its simple processing, low cost, and higher adaptability for large-scale fabrication of electronic devices, sensors, and light-emitting diodes, inkjet printing is recognized as an efficient method for directly depositing functional materials on flexible substrates in predesigned patterns.

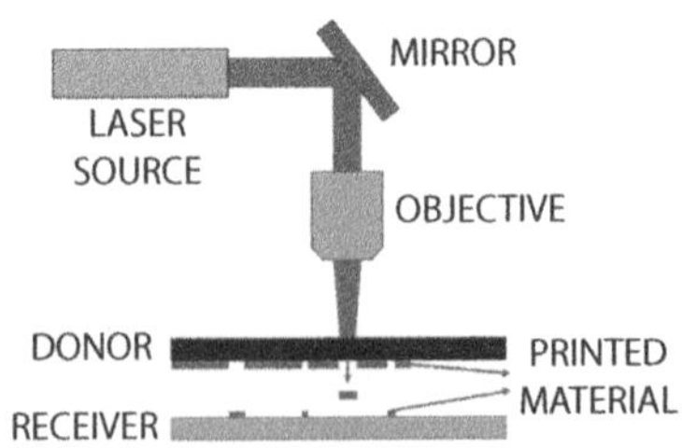

Figure 14.5 Process of laser transfer micro AM.

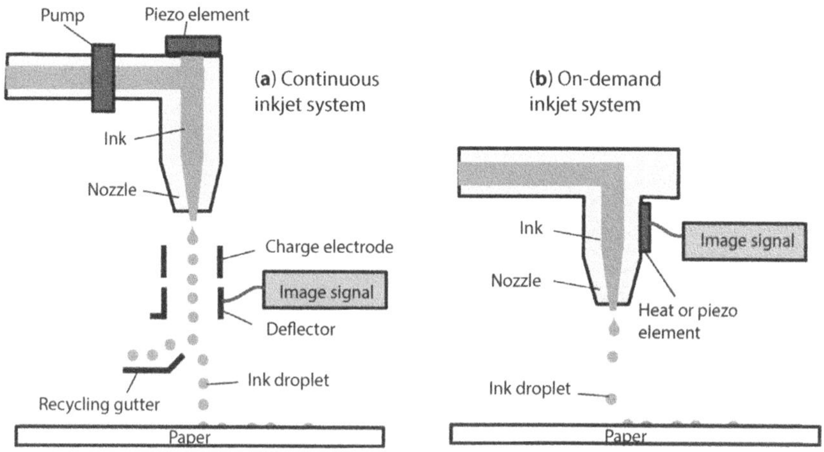

Figure 14.6 Schematics showing: (a) a continuous inkjet printer, (b) an on-demand inkjet printer [63].

Organic polymers, metal nanoparticles, and carbon-based nanoparticles are micro inkjet printing inks [64]. Another ink-based micro AM technology is 3DP. This is based on jetting the liquid/binder on the deposited powder to create the structure layer by layer, as shown in Figure 14.6. Still, it suffers from challenges like maintaining the resolution of the deposited powders to micron size, microparts fabricated have a very rough surface, the powder handling also becomes difficult as the powder size goes low [65].

14.5.5 Micro Laser Sintering

It combines the benefits of additive printing with micromachining to create sophisticated micro metal components with amazing precision, detail resolution, and surface quality. The technology provides maximum production flexibility and minimal operating costs. Micro SLS can process nonreactive and reactive materials, such as gold, stainless steel, titanium, and tungsten. The method of micro SLS can be used for actuators, sensors, and MEMS packaging applications. Micro SLS AM technique can be used for the fabrication of complex MEMS devices, micro optoelectronic components, and medical devices. The advantage of very thin laser beam spot size and micro-sized metal powders enable the production of micro-sized components [66]. Due to these features, it is possible to produce moving components and their assemblies in a single step. A detailed description of various parts and mechanism of micro SLS are presented in Figure 14.7.

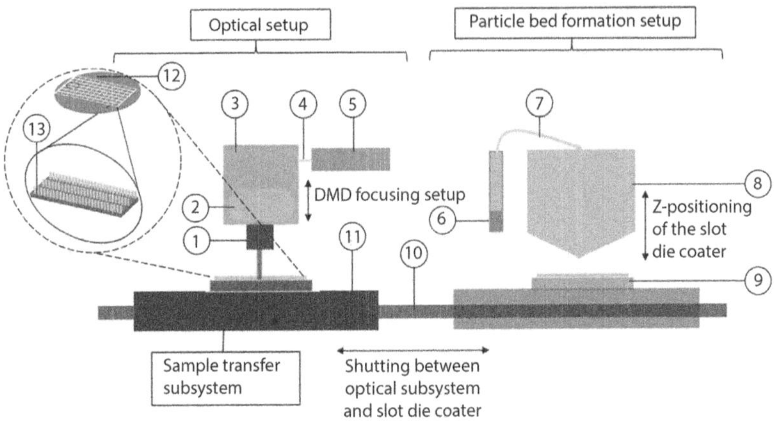

Figure 14.7 Micro laser sintering system: 1-secondary focusing lens/optics, 2-beam shaping primary optics/lens, 3-micro mirror array casing and regulator, 4-optical fiber coupling, 5-continues wave laser and particle bed formation setup, 6-syringe hydraulic delivery system of nano ink, 7-tube for nano ink, 8-nano ink deposition system head with Z axis positioning, 9-substrate holder with heating and vacuum condition, 10-shuttling system for transfer of substrate between optical setup and coating, 11-XY axis nanopositioning, 12-micro SLS 3D/ 4D printed part, and 13-enlarged view of micro SLS 3D part [67].

Micro SLS is a perfect technique for producing micro metal components for the medical device, semiconductor, and aerospace industries. It can also be used for producing intricate jewelry design and minute components for producing watches [67].

14.5.6 Beam Deposition

Beam deposition (BD) techniques allow the fabrication of components by melting and depositing material from its powdered form or wire feedstock. Beam deposition techniques allow the fabrication of components by melting and depositing material from powder or wire feedstock. Beam deposition techniques utilize energy concentrated into a confined area to heat a material being deposited. Every run of the BD head forms a path of solidified material and subsequent lines of material form layers. Support material or a multi-axis deposition head is required for complex 3D geometry [68]. Unlike powder bed fusion procedures where it is utilized to melt materials as they are deposited, in BD, they are set in a powder bed. Beam deposition methods, like extrusion-based processes, employ a focused heat source (laser, electron beam, or plasma arc) to melt the input material

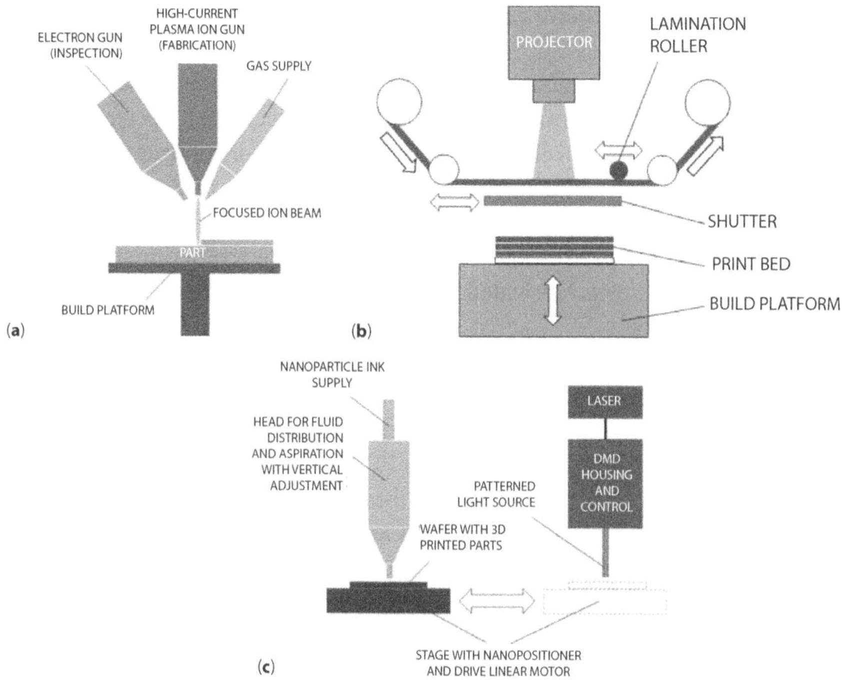

Figure 14.8 Beam deposition AM: (a) focused electron/ion beam-induced deposition, (b) laminated resin printing, (c) inkjet selective laser sintering [55].

and develop 3D objects [69]. Various kinds of BD AM processes and their mechanisms are shown in Figure 14.8.

14.6 Applications of Micro 4D-Printed Structures

Micro 4D printing can fabricate the structures utilizing smart materials that can change their characteristic properties like electromagnetic metamaterials (EMM). It can be used for absorbing radar waves and also for tuning the electromagnetic waves. These materials are useful in aerospace and communication fields to manufacture stealth aircraft [70]. Thermal metamaterials can be fabricated with complex geometries and multilayers with micro 4D printing to act as a thermal cloak [71]. Magnetic-responsive actuators show much faster responses in comparison to thermoresponsive actuators. Schmidt *et al.* fabricated a magnetic responsive micromotor used for targeted drug delivery upon application of a magnetic field [72].

Photo-responsive actuators are easy to control in micro- or nanoscale manipulation.

The fabrication of electronic devices at a small scale is easier through micro 4D printing. A microscale transparent electronic device was 3D printed by Byun *et al.* for light transmittance [73]. Self-folding and flexible electronic devices are hugely important in soft robotics and medical applications. Zeng *et al.* designed a microscopic walker using 3D laser lithography, which has shown a response to laser light [74]. A temperature-sensitive PLA microgripper was fabricated by 4D printing [75]. It has shown a 3-mm actuation distance.

A corkscrew propeller made from the TPP technique shows potential applications to absorb and transport microparticles in a specific manner. Due to its high resolution, this complex structure can be fabricated precisely through the TPP method. The TPP enables magnetic nanoparticles to be dispersed in smart material devices while patterning to control the device remotely with a magnetic field application. A crawling or rolling device can be fabricated at the microscale using TPP. A cylindrical structure that can roll with surface-mediated propulsion in response to a magnetic field was made by TPP [76]. Additionally, TPP-made helix and star-shaped structures have shown an unusual combination of rolling and tumbling motion triggered by a rotating magnetic field [77]. These smart structures are capable of both motion and transportation. Li *et al.* [78] developed spherical microrobots for *in-vivo* cell transportation. A robot with a size of 90 μm was operated inside the live zebrafish embryos. A photosensitive micromotor was designed by Vizsnyiczai *et al.* to capture and transport bacteria. Microchambers were designed inside the device to accommodate the captured bacteria [79].

Sun *et al.* designed a hydrogel-based microrobot that is magnetically operated by an oscillating magnetic field and is composed of stiff cylinders connected by compliant springs [80]. This microrobot is comprised of biocompatible materials such as poly(ethylene glycol) diacrylate (PEGDA) and PETA, and Fe_3O_4 nanoparticles are incorporated to enable the magnetic actuation mechanism. This microrobot's components, which include PEGDA, PETA, and Fe_3O_4 magnetic nanoparticles (MNPs), provide magnetic actuation and make it biocompatible and biodegradable. The PETA was added to improve the overall structure's structural integrity. Targeted drug release is one of the microrobot's key applications. The structural integrity of the device is the key objective for such applications. A lab-made microoperation system is used to conduct the structural integrity test of the microswimmer. A microneedle applied force on the middle section of the microswimmer to cause the deformation, as shown in Figure 14.9.

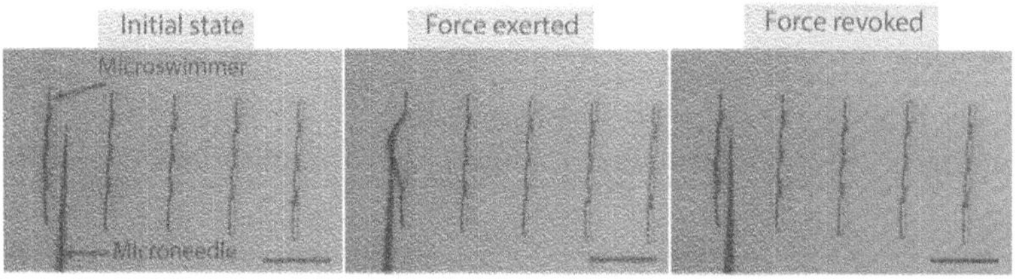

Figure 14.9 Structural integrity test of microswimmer. Reprinted with permission from [80] copyright@2020, MDPI, Creative Commons CC BY 4.0 license.

Upon releasing the load, the microswimmer regained its original shape, confirming that the fabricated microswimmer is structurally stable.

The functional behavior of this microswimmer was confirmed by conducting free-swimming experiments. A magnetic actuation was employed to control the motion, and a direct current (DC) motor was used to control the frequency of the magnetic field and to provide the propulsion. Figure 14.10 (a) shows the initial position of the microswimmer, and Figure 14.10 (b) shows the change in the position of the microswimmer over time with the application of the magnetic field. This 'device's velocity with the magnetic 'field's change in frequency is represented in Figure 14.10 (c). Figure 14.10 (d-f) represents the path and controlled motion of the microswimmer

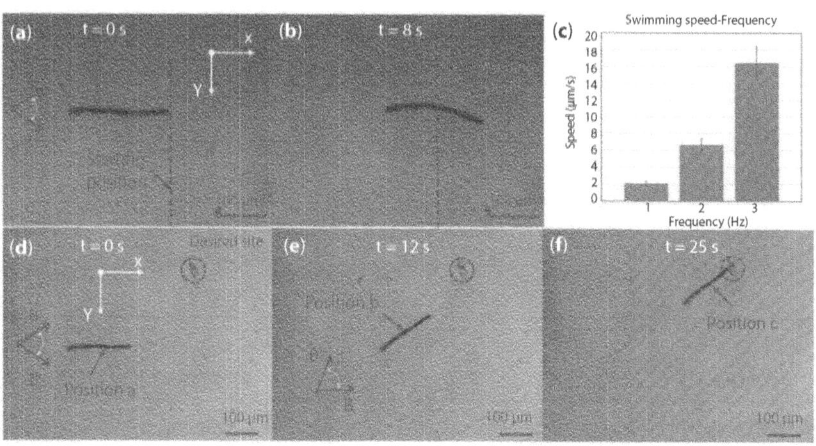

Figure 14.10 Functional test of microswimmer with magnetic actuation. (a) Initial position (b) Displacement with the application of magnetic field after t=8 s. (c) The velocity of the microswimmer with variation in the frequency of the magnetic field. (d-f) Controlled motion of microswimmer to reach the desired location. Reprinted with permission from [80], copyright@2020, MDPI, Creative Commons CC BY 4.0 license.

Table 14.1 Various applications based on materials and structures from the literature.

Material	The shape of the structure	Dimensions (µm)	Application	Ref
Ni/Ti	Helices	4-65	Transportation	[81]
Fe or Fe/Ti	Helix	15.7	biocompatible	[82]
Ni/TiO_2	Helical scaffold with needle	300	Scaffold architecture	[83]
Ni/Ti	Cylindrical	144–160	Rolling motion	[76]
MNPs	Concave and tapered	35–75	Dragging and tumbling	[84]
Ni/Pt	Serrated trapezoid	65	Multichannel moving jets in H_2O_2	[85]
Au	Triple-barreled tube	45	Thermophoretic propulsion	[86]
MNPs	Helix	20	Encapsulation of biomolecules	[87]
Ni/Ti	Hollow conical helices	45–75	Scaffold for NSC cells	[88]
Cr/Ni	Helix with two pointed heads	45–50	Microfluidic sensor	[89]
PVA	Helix	20	Cell instructive hydrogel	[34]
AuNRs	Scaffold	30–130	Self-folding in response to light	[90]
PNIPAm–AAc–PVP	Scaffold	45-150	Self-folding in response to pH	[91]
PNIPAm/AuNRs	Pillar and helix	20	Self-folding in response to temperature	[37]

to reach the desired location. The direction and speed of this device were controlled by varying the magnetic field.

Similarly, Ceylan *et al.* designed a microswimmer for drug delivery application by incorporating Fe_3O_4 nanoparticles [92]. This microswimmer was manipulated using a magnetic field. Tottori *et al.* employed a helical microrobot with a modified head structure to manipulate micro-objects utilizing a mix of creative design and magnetic actuation by using superparamagnetic iron oxide nanoparticles (SPIONs) for propulsion [81]. Bozuyuk *et al.* also presented a magnetic-driven microswimmer for drug delivery application [93]. This device comprises photosensitive material, so it can also be actuated with light application. Controlling a microscale device with a multi-stimuli mechanism improves functional accuracy and performance. Since magnetic actuation presents the most reliable and good control for *in-vivo* applications, most of the earlier presented microdevices are responsive toward the magnetic field. It is important to optimize the printing parameters to obtain the structures with the best functional parameters to improve the functional performance of these micro/nano devices. Some of the important applications of micro 4D printing utilized for creating smart devices are provided briefly in Table 14.1.

14.7 Challenges and Future Outlook

Micro 4D printing, like any other technology, must overcome various obstacles to realize its full potential. Among the most significant obstacles in micro 4D printing are:

- Material development: it is one of the most significant obstacles in micro 4D printing since it requires the creation of innovative materials that can alter their form or function in response to certain environmental circumstances. These substances must be biocompatible, biodegradable, and simple to manipulate. Microprinting techniques such as SLA, DLP, and LIFT can print complex shaped parts. However, they are limited by the materials that can be used for printing.
- Printing resolution: a further obstacle in micro 4D printing is printing microscale structures with excellent resolution. The printing resolution must be good enough to build structures with feature sizes in the micrometer range, which demands sophisticated printing procedures and equipment.

- Control over the process of changing shape: micro 4D printing demands precise control over the process of changing shape, which remains a formidable obstacle. More profound knowledge of smart materials' shape-changing mechanism is required to gain control over the transformation behavior.
- Cyclic stability: the functional performance of microdevices over time or number of cycles reduced. To overcome this issue, the printing parameters and material composition must be optimized for individual applications.
- Scalability: micro 4D printing is a relatively new technology, and the scalability of the process is still being developed. There is a need to create ways to cost-effectively and efficiently print many microscale structures.
- Residual elements: although many 4D-printed polymers are biocompatible, the same is not true for the monomers and photoinitiators employed in their synthesis. It is necessary to conduct more research to understand the influence of these residual elements on the living systems and their general effects.

Despite these obstacles, micro 4D printing can modernize several industries, including healthcare, robotics, and fashion. Micro 4D printing has the potential to be used in the healthcare industry to create implantable devices like drug-delivery systems, scaffolds for tissue engineering, and other bio-inspired structures. To confirm their functional performance, the testing methods must be improved to achieve *in-situ* inspection of these microdevices. In robotics, micro 4D printing can be used to produce tiny-scale robots capable of changing form and adapting to their surroundings. A genuinely spatially controlled response will be possible with the development of sophisticated materials with an independent multi-response capability or multimaterial 4D-printed microrobots, leading to greater adaptability and degree of freedoms (DoFs) in design and actuation.

Additionally, more research must be done on robotic performance measures, including positioning precision, repeatability, usable life, and generated force, to satisfy important microrobotic needs. In the fashion industry, micro 4D printing may produce fabrics that change color, form, or texture according to environmental variables. In the aerospace industry, micro 4D printing may be used to construct microscale structures that can change form in reaction to changes in temperature, allowing them to be used on space missions where materials are subjected to high temperatures.

Micro 4D printing has enormous promise for constructing new and creative microscale structures in various industries, but it must first overcome several obstacles to realize its full potential. As technology advances, it is anticipated that these obstacles will be resolved, and new applications will be created.

14.8 Conclusions

During the last decade, promising applications of 4D printing technology have been illustrated by various researchers. This includes utilizing materials like hydrogels, shape-memory alloys, shape-memory polymers, liquid crystals, and composite materials for creating actuating structures. Although a lot of research has been carried out in the field of 4D printing, cases of microscale use are very limited. The following conclusion is deduced from the present chapter:

- PNIPAm-based thermo-sensitive hydrogel has been widely utilized for micro 4D printing, showing very good reversible shape change at lower critical solution temperatures.
- Besides hydrogels, SMA, hydrogel-metal composites, and ceramic materials like TCP and glass can also be used for 4D microprinting.
- Micro 4D bioprinting has shown its applicability for creating smart microstents and shape-changing micro scaffolds utilizing biocompatible materials like collagen, gelatin, and fibrinogen, along with PEG, PLA, and PCL.
- 4D printing processes applicable for small-scale structures are μSLA, TPP, laser transfer AM, ink-based micro AM, micro SLS, and BD, with a scale ranging from 0.2 μm for TPP to 80 μm for BD AM. The widely used method for micro 4D printing is μSLA due to the ease of processing the smart material.
- The process, like LIFT, does not depend upon the type of material getting deposited and hence can be used for metals like SMA with high spatial resolution compared to micro SLS.
- Although the SMP and hydrogels that can be processed using TPP are quite low, they can be utilized for indirect micro 4D printing by creating micromolds from the order of a few nanometers to a few micrometers.

- Micro 4D printing has been utilized for various multifaceted applications like microrobotics, biomedical devices, photonics, drug delivery, and microfluidics.
- There are a lot of challenges that 4D microprinting has to overcome before it can be utilized on a large scale. A few of them are materials and their processibility, low resolution, scalability, shape memory characterization at the nanoscale, and cyclic actuation.

As 4D microprinting is in a nascent stage of research, much more efforts from scientific communities and industries have to innovate new methods and related technologies to solve the challenges remaining and find new applications in the field that will greatly benefit from this technology.

Acknowledgments

RRK, AJ, BCGM, & BBS acknowledge the Ministry of Human Resource Development, Government of India, for the graduate fellowship. All authors are grateful to the research scholars and staff of Direct Digital Manufacturing Lab, Department of Mechanical Engineering, & Advanced Photonics Lab, Department of Electronics and Electrical Communication Engineering, Indian Institute of Technology, Kharagpur, for their immense help throughout.

References

1. Gibson, I., Rosen, D.W., Stucker, B., *Additive manufacturing technologies: Rapid prototyping to direct digital manufacturing*, Springer US, Berlin/Heidelberg, Germany, 2010, doi: 10.1007/978-1-4419-1120-9/COVER.
2. Zhang, Y., Jarosinski, W., Jung, Y.-G., Zhang, J., 2 - Additive manufacturing processes and equipment, in: *Additive Manufacturing*, J. Zhang and Y.-G. Jung (Eds.), pp. 39–51, Butterworth-Heinemann, Kidlington, Oxford, United Kingdom, 2018, doi: https://doi.org/10.1016/B978-0-12-812155-9.00002-5.
3. Jena, A., Samal, B.B., Kumar, C.S., Varshney, S.K., Analysis of electro-thermo-mechanical behavior of thin film Ni50-Ti50and Ni40-Ti50-Cu10 shape memory alloys for application in thermal actuators. *Mater. Today Proc.*, 47, 4578–4583, 2021, doi: 10.1016/j.matpr.2021.05.448.
4. Vaezi, M., Seitz, H., Yang, S., A review on 3D micro-additive manufacturing technologies. *Int. J. Adv. Manuf. Technol.*, 67, 5, 1721–1754, 2013, doi: 10.1007/s00170-012-4605-2.

5. Tibbits, S., The emergence of '4D printing,' 2013, Accessed: Jan. 24, 2021. [Online]. Available: https://www.ted.com/talks/skylar_tibbits_the_emergence_of_4d_printing.
6. Samal, B.B., Jena, A., Varshney, S.K., Kumar, C.S., FDM 4D printing: A low-cost approach of shape programming and assessing the shape memory properties using angle measurement methods in hot water actuation testing apparatus. *Mater. Today Proc.*, 2023, doi: 10.1016/j.matpr.2023.02.038.
7. Khalid, M.Y., Arif, Z.U., Ahmed, W., Umer, R., Zolfagharian, A., Bodaghi, M., 4D printing: Technological developments in robotics applications. *Sens. Actuators A Phys.*, 343, 113670, 2022, doi: https://doi.org/10.1016/j.sna.2022.113670.
8. Zhang, Z., Demir, K.G., Gu, G.X., Developments in 4D-printing: A review on current smart materials, technologies, and applications. *Int. J. Smart Nano Mater.*, 10, 3, 205–224, 2019, doi: 10.1080/19475411.2019.1591541.
9. Spiegel, C.A., Hackner, M., Bothe, V.P., Spatz, J.P., Blasco, E., 4D printing of shape memory polymers: From macro to micro. *Adv. Funct. Mater.*, 32, 51, 2110580, 2022, doi: https://doi.org/10.1002/adfm.202110580.
10. Spiegel, C.A. *et al.*, 4D Printing at the microscale. *Adv. Funct. Mater.*, 30, 26, 1907615, 2020, doi: https://doi.org/10.1002/adfm.201907615.
11. Selimis, A., Mironov, V., Farsari, M., Direct laser writing: Principles and materials for scaffold 3D printing. *Microelectron. Eng.*, 132, 83–89, 2015, doi: https://doi.org/10.1016/j.mee.2014.10.001.
12. Lee, K.-S., Yang, D.-Y., Park, S.H., Kim, R.H., Recent developments in the use of two-photon polymerization in precise 2D and 3D microfabrications. *Polym. Adv. Technol.*, 17, 2, 72–82, 2006, doi: https://doi.org/10.1002/pat.664.
13. Bodaghi, M. and Zolfagharian, A., 1 - 4D printing principles and manufacturing, in: *Smart Materials in Additive Manufacturing*, M. Bodaghi and A. Zolfagharian (Eds.), pp. 1–17, Elsevier, Amsterdam, Netherlands, 2022, doi: https://doi.org/10.1016/B978-0-12-824082-3.00014-3.
14. Rajabasadi, F., Schwarz, L., Medina-Sánchez, M., Schmidt, O.G., 3D and 4D lithography of untethered microrobots. *Prog. Mater. Sci.*, 120, 100808, 2021, doi: https://doi.org/10.1016/j.pmatsci.2021.100808.
15. Zolfagharian, A., Kaynak, A., Bodaghi, M., Kouzani, A.Z., Gharaie, S., Nahavandi, S., Control-based 4D printing: Adaptive 4D-printed systems. *Appl. Sci.*, 10, 9, 3020, 2020, doi: 10.3390/app10093020.
16. Sahafnejad-Mohammadi, I., Karamimoghadam, M., Zolfagharian, A., Akrami, M., Bodaghi, M., 4D printing technology in medical engineering: A narrative review. *J. Braz. Soc. Mech. Sci. Eng.*, 44, 6, 233, 2022, doi: 10.1007/s40430-022-03514-x.
17. Khalid, M.Y., Arif, Z.U., Noroozi, R., Zolfagharian, A., Bodaghi, M., 4D printing of shape memory polymer composites: A review on fabrication techniques, applications, and future perspectives. *J. Manuf. Processes*, 81, 759–797, 2022, doi: https://doi.org/10.1016/j.jmapro.2022.07.035.

18. Bodaghi, M., Noroozi, R., Zolfagharian, A., Fotouhi, M., Norouzi, S., 4D printing self-morphing structures. *Mater. (Basel)*, 12, 8, 1353, 2019, doi: 10.3390/ma12081353.
19. Zolfagharian, A., Kaynak, A., Khoo, S.Y., Kouzani, A., Pattern-driven 4D printing. *Sens. Actuators A Phys.*, 274, 231–243, 2018, doi: https://doi.org/10.1016/j.sna.2018.03.034.
20. Samal, B.B., Jena, A., Mishra, D., Controlled shape changing components by using 4D printing technology, in: *10th International Conference on Precision, Meso, Micro and Nano Engineering (COPEN 10), IIT Madras*, no. October, pp. 78–81, 2017, Accessed: Feb. 05, 2021. [Online]. Available: http://www.copen.ac.in/proceedings/copen10/copen/19.revised_manuscript_262.pdf.
21. Samal, B.B., Jena, A., Varshney, S.K., Kumar, C.S., 4D printing: An experimental case study on processing of shape memory polymer by FDM/FFF for nature inspired structures, in: *Advances in Additive Manufacturing Artificial Intelligence, Nature-Inspired, and Biomanufacturing*, A. Kumar, R.K. Mittal, A. Haleem (Eds.), pp. 361–377, Elsevier, Amsterdam, Netherlands, 2023, doi: 10.1016/B978-0-323-91834-3.00019-3.
22. Samal, B.B., Jena, A., Mishra, D., Indian Researchers Develop World's Strongest 4D Printed Parts, Manufactur3D magazine, 2018, Accessed: Jan. 28, 2020. [Online]. Available: https://manufactur3dmag.com/indian-researchers-develop-worlds-strongest-4d-printed-parts/.
23. Khoo, Z.X. *et al.*, 3D printing of smart materials: A review on recent progresses in 4D printing. *Virtual Phys. Prototyp.*, 10, 3, 103–122, 2015, doi: 10.1080/17452759.2015.1097054.
24. Tibbits, S., 4D printing: Multi-material shape change. *Archit. Des.*, 84, 1, 116–121, Jan. 2014, doi: https://doi.org/10.1002/ad.1710.
25. Kuang, X. *et al.*, Advances in 4D printing: Materials and applications. *Adv. Funct. Mater.*, 29, 2, 1805290, 2019, doi: https://doi.org/10.1002/adfm.201805290.
26. Hu, G.F., Damanpack, A.R., Bodaghi, M., Liao, W.H., Increasing dimension of structures by 4D printing shape memory polymers via fused deposition modeling. *Smart Mater. Struct.*, 26, 12, 125023, 2017, doi: 10.1088/1361-665X/aa95ec.
27. Leist, S.K. and Zhou, J., Current status of 4D printing technology and the potential of light-reactive smart materials as 4D printable materials. *Virtual Phys. Prototyp.*, 11, 4, 249–262, 2016, doi: 10.1080/17452759.2016.1198630.
28. Wang, X. *et al.*, Preparation of 4D printed peripheral vascular stent and its degradation behavior under fluid shear stress after deployment. *Biomater. Sci.*, 10, 9, 2302–2314, 2022, doi: 10.1039/D2BM00088A.
29. Zolfagharian, A., Kaynak, A., Kouzani, A., Closed-loop 4D-printed soft robots. *Mater. Des.*, 188, 108411, 2020, doi: https://doi.org/10.1016/j.matdes.2019.108411.
30. Han, D., Lu, Z., Chester, S.A., Lee, H., Micro 3D printing of a temperature-responsive hydrogel using projection micro-stereolithography. *Sci. Rep.*, 8, 1, 1963, 2018, doi: 10.1038/s41598-018-20385-2.

31. Hippler, M. *et al.*, Controlling the shape of 3D microstructures by temperature and light. *Nat. Commun.*, 10, 1, 232, 2019, doi: 10.1038/s41467-018-08175-w.
32. Huang, T.-Y. *et al.*, Four-dimensional micro-building blocks. *Sci. Adv.*, 6, 3, eaav8219, 2020, doi: 10.1126/sciadv.aav8219.
33. Lee, M.R., Phang, I.Y., Cui, Y., Lee, Y.H., Ling, X.Y., Shape-shifting 3D protein microstructures with programmable directionality via quantitative nanoscale stiffness modulation. *Small*, 11, 6, 740–748, 2015, doi: https://doi.org/10.1002/smll.201401343.
34. Chinnusamy, S., Ramasamy, V., Venkatajalapathy, S., Kaliyannan, G.V., Palaniappan, S.K., Experimental investigation on the effect of ceramic coating on the wear resistance of Al 6061 substrate. *J. Mater. Res. Technol.*, 8, 6, 6125–6133, 2019, https://doi.org/10.1016/j.jmrt.2019.10.007.
35. Hu, Y. *et al.*, Botanical-inspired 4D printing of hydrogel at the microscale. *Adv. Funct. Mater.*, 30, 4, 1907377, 2020, doi: https://doi.org/10.1002/adfm.201907377.
36. Dong, M. *et al.*, 3D-printed soft magnetoelectric microswimmers for delivery and differentiation of neuron-like cells. *Adv. Funct. Mater.*, 30, 17, 1910323, 2020, doi: https://doi.org/10.1002/adfm.201910323.
37. Nishiguchi, A., Zhang, H., Schweizerhof, S., Schulte, M.F., Mourran, A., Möller, M., 4D printing of a light-driven soft actuator with programmed printing density. *ACS Appl. Mater. Interfaces*, 12, 10, 12176–12185, Mar. 2020, doi: 10.1021/acsami.0c02781.
38. Carbonell, C. and Braunschweig, A.B., Toward 4D nanoprinting with tip-induced organic surface reactions. *Acc. Chem. Res.*, 50, 2, 190–198, 2017, doi: 10.1021/acs.accounts.6b00307.
39. Marupalli, B.C.G., Adhikary, T., Sahu, B.P., Mitra, R., Aich, S., Effect of annealing temperature on microstructure and mechanical response of sputter deposited Ti-Zr-Mo high temperature shape memory alloy thin films. *Appl. Surf. Sci. Adv.*, 6, 100137, 2021, doi: https://doi.org/10.1016/j.apsadv.2021.100137.
40. Jayanth, D., Sathish Kumar, P., Nayak, G.C., Saravana Kumar, J., Pal, S.K., Rajasekar, R., A review on biodegradable polymeric materials striving towards the attainment of green environment. *J. Polym. Environ.*, 26, 2, 838–865, 2018, https://doi.org/10.1007/s10924-017-0985-6.
41. Zhang, Q., Lei, J., Chen, Y., Wu, Y., Chen, C., Xiao, H., 3D printing of all-glass fiber-optic pressure sensor for high temperature applications. *IEEE Sens. J.*, 19, 23, 11242–11246, 2019, doi: 10.1109/JSEN.2019.2935689.
42. Zhang, Y.S. and Khademhosseini, A., Advances in engineering hydrogels. *Sci. (80-.)*, 356, 6337, eaaf3627, 2017, doi: 10.1126/science.aaf3627.
43. Zhou, X. and Lee, P.S., Three dimensional printed nanogenerators. *EcoMat*, 3, 3, e12098, 2021, doi: https://doi.org/10.1002/eom2.12098.
44. Zheng, X. *et al.*, Multiscale metallic metamaterials. *Nat. Mater.*, 15, 10, 1100–1106, 2016, doi: 10.1038/nmat4694.

45. Körner, C., Additive manufacturing of metallic components by selective electron beam melting — A review. *Int. Mater. Rev.*, 61, 5, 361–377, 2016, doi: 10.1080/09506608.2016.1176289.
46. Reiser, A. *et al.*, Multi-metal electrohydrodynamic redox 3D printing at the submicron scale. *Nat. Commun.*, 10, 1, 1853, 2019, doi: 10.1038/s41467-019-09827-1.
47. Yu, Y., Liu, F., Zhang, R., Liu, J., Suspension 3D printing of liquid metal into self-healing hydrogel. *Adv. Mater. Technol.*, 2, 11, 1700173, 2017, doi: https://doi.org/10.1002/admt.201700173.
48. Carrico, J.D., Traeden, N.W., Aureli, M., Leang, K.K., Fused filament 3D printing of ionic polymer-metal composites (IPMCs). *Smart Mater. Struct.*, 24, 12, 125021, Nov. 2015, doi: 10.1088/0964-1726/24/12/125021.
49. Hwa, L.C., Rajoo, S., Noor, A.M., Ahmad, N., Uday, M.B., Recent advances in 3D printing of porous ceramics: A review. *Curr. Opin. Solid State Mater. Sci.*, 21, 6, 323–347, 2017, doi: https://doi.org/10.1016/j.cossms.2017.08.002.
50. Kotz, F. *et al.*, Three-dimensional printing of transparent fused silica glass. *Nature*, 544, 7650, 337–339, 2017, doi: 10.1038/nature22061.
51. Paunović, N. *et al.*, Digital light 3D printing of customized bioresorbable airway stents with elastomeric properties. *Sci. Adv.*, 7, 6, eabe9499, 2021, doi: 10.1126/sciadv.abe9499.
52. Park, K.D., Wang, X., Lee, J.Y., Park, K.M., Zhang, S., Noh, I., Research trends in biomimetic medical materials for tissue engineering: Commentary. *Biomater. Res.*, 20, 1, 8, 2016, doi: 10.1186/s40824-016-0053-7.
53. Kanakarajan, P., Sundaram, S., Kumaravel, A., Rajasekar, R., Sathish Kumar, P., Acoustic emission testing of surface roughness and wear caused by grinding of ceramic materials. *Mater. Test.*, 57, 4, 337–342, 2015, https://doi.org/10.3139/120.110714.
54. Bertsch, A. and Renaud, P., Chapter 1.2 - Microstereolithography, in: *Three-Dimensional Microfabrication Using Two-Photon Polymerization*, T. Baldacchini (Ed.), pp. 20–44, William Andrew Publishing, Oxford, 2016, doi: https://doi.org/10.1016/B978-0-323-35321-2.00002-9.
55. De Pasquale, G., Additive manufacturing of micro-electro-mechanical systems (MEMS). *Micromachines*, 12, 11, 1374, 2021, doi: 10.3390/mi12111374.
56. Zeng, Y. *et al.*, 3D-printing piezoelectric composite with honeycomb structure for ultrasonic devices. *Micromachines*, 11, 8, 713, 2020, doi: 10.3390/mi11080713.
57. Sugioka, K. and Cheng, Y., Femtosecond laser three-dimensional micro- and nanofabrication. *Appl. Phys. Rev.*, 1, 4, 41303, 2014, doi: 10.1063/1.4904320.
58. Faraji Rad, Z., Prewett, P.D., Davies, G.J., High-resolution two-photon polymerization: The most versatile technique for the fabrication of microneedle arrays. *Microsyst. Nanoeng.*, 7, 1, 71, 2021, doi: 10.1038/s41378-021-00298-3.
59. Paz, V.F. *et al.*, Development of functional sub-100 nm structures with 3D two-photon polymerization technique and optical methods for characterization. *J. Laser Appl.*, 24, 4, 42004, 2012, doi: 10.2351/1.4712151.

60. Serra, P., Duocastella, M., Fernández-Pradas, J.M., Morenza, J.L., 13 - The laser-induced forward transfer technique for microprinting, in: *Advances in Laser Materials Processing*, J. Lawrence, J. Pou, D.K.Y. Low, E. Toyserkani (Eds.), pp. 367–393, Woodhead Publishing, Great Abington, Cambridge, UK, 2010, doi: https://doi.org/10.1533/9781845699819.5.367.
61. Huis in 't Veld, B., Overmeyer, L., Schmidt, M., Wegener, K., Malshe, A., Bartolo, P., Micro additive manufacturing using ultra short laser pulses. *CIRP Ann.*, 64, 2, 701–724, 2015, doi: https://doi.org/10.1016/j.cirp.2015.05.007.
62. Piqué, A., Chapter 1.1 - Laser-based micro-additive manufacturing technologies, in: *Three-Dimensional Microfabrication Using Two-Photon Polymerization*, T. Baldacchini (Ed.), pp. 1–19, William Andrew Publishing, Oxford, 2016, doi: https://doi.org/10.1016/B978-0-323-35321-2.00001-7.
63. Lau, G.-K. and Shrestha, M., Ink-jet printing of micro-electro-mechanical systems (MEMS). *Micromachines*, 8, 6, 194, 2017, doi: 10.3390/mi8060194.
64. Nayak, L., Mohanty, S., Nayak, S.K., Ramadoss, A., A review on inkjet printing of nanoparticle inks for flexible electronics. *J. Mater. Chem. C*, 7, 29, 8771–8795, 2019, doi: 10.1039/C9TC01630A.
65. Lanzetta, M. and Sachs, E., Improved surface finish in 3D printing using bimodal powder distribution. *Rapid Prototyp. J.*, 9, 3, 157–166, Jan. 2003, doi: 10.1108/13552540310477463.
66. Exner, H., Regenfuss, P., Hartwig, L., Kloetzer, S., Ebert, R., Selective laser micro sintering with a novel process, in: *Fourth International Symposium on Laser Precision Microfabrication*, 2003, doi: 10.1117/12.540730.
67. Roy, N.K., Behera, D., Dibua, O.G., Foong, C.S., Cullinan, M.A., A novel microscale selective laser sintering (μ-SLS) process for the fabrication of microelectronic parts. *Microsyst. Nanoeng.*, 5, 1, 64, 2019, doi: 10.1038/s41378-019-0116-8.
68. Matsui, S., Focused-ion-beam deposition for 3-D nanostructure fabrication. *Nucl. Instrum. Methods Phys. Res. Sect. B Beam Interact. Mater. At.*, 257, 1, 758–764, 2007, doi: https://doi.org/10.1016/j.nimb.2007.01.077.
69. Marupalli, B.C.G., Aich, S., Adhikary, T., 2 - Methods for fabrication of NiTi micro- and nanoalloys, in: *Nickel-Titanium Smart Hybrid Materials*, S. Thomas, A. Behera, T.A. Nguyen (Eds.), pp. 9–27, Elsevier, Amsterdam, Netherlands, 2022, doi: https://doi.org/10.1016/B978-0-323-91173-3.00019-5.
70. Askari, M. *et al.*, Additive manufacturing of metamaterials: A review. *Addit. Manuf.*, 36, 101562, 2020, doi: https://doi.org/10.1016/j.addma.2020.101562.
71. Jeong, H.Y., Lee, E., Ha, S., Kim, N., Jun, Y.C., Multistable thermal actuators via multimaterial 4D printing. *Adv. Mater. Technol.*, 4, 3, 1800495, 2019, doi: https://doi.org/10.1002/admt.201800495.
72. Xu, H., Medina-Sánchez, M., Magdanz, V., Schwarz, L., Hebenstreit, F., Schmidt, O.G., Sperm-hybrid micromotor for targeted drug delivery. *ACS Nano*, 12, 1, 327–337, Jan. 2018, doi: 10.1021/acsnano.7b06398.

73. Zhang, B., Lee, H., Byun, D., Electrohydrodynamic jet printed 3D metallic grid: Toward high-performance transparent electrodes. *Adv. Eng. Mater.*, 22, 5, 1901275, 2020, doi: https://doi.org/10.1002/adem.201901275.
74. Zeng, H., Wasylczyk, P., Parmeggiani, C., Martella, D., Burresi, M., Wiersma, D.S., Light-fueled microscopic walkers. *Adv. Mater.*, 27, 26, 3883–3887, 2015, doi: https://doi.org/10.1002/adma.201501446.
75. Tong, K.J., Saad Alshebly, Y., Nafea, M., Development of a 4D-printed PLA microgripper, in: *2021 IEEE 19th Student Conference on Research and Development (SCOReD)*, pp. 207–211, 2021, doi: 10.1109/SCOReD53546.2021.9652719.
76. Kim, S. *et al.*, Fabrication and characterization of magnetic microrobots for three-dimensional cell culture and targeted transportation. *Adv. Mater.*, 25, 41, 5863–5868, 2013, doi: https://doi.org/10.1002/adma.201301484.
77. Schwarz, L., Karnaushenko, D.D., Hebenstreit, F., Naumann, R., Schmidt, O.G., Medina-Sánchez, M., A rotating spiral micromotor for noninvasive zygote transfer. *Adv. Sci.*, 7, 18, 2000843, 2020, doi: https://doi.org/10.1002/advs.202000843.
78. Palaniappan, S.K., Chinnasamy, M., Rathanasamy, R., Pal, S.K., Synthetic binders for polymer division, in: *Green Adhesives: Preparation, Properties and Applications*, pp. 227–272, Wiley Scrivener Publishing LLC, River Street, Hoboken, USA & Beverly, MA, USA, 2020.
79. Vizsnyiczai, G., Frangipane, G., Maggi, C., Saglimbeni, F., Bianchi, S., Di Leonardo, R., Light controlled 3D micromotors powered by bacteria. *Nat. Commun.*, 8, 1, 15974, 2017, doi: 10.1038/ncomms15974.
80. Sun, H.C.M., Liao, P., Wei, T., Zhang, L., Sun, D., Magnetically powered biodegradable microswimmers. *Micromachines*, 11, 4, 404, 2020, doi: 10.3390/mi11040404.
81. Tottori, S., Zhang, L., Qiu, F., Krawczyk, K.K., Franco-Obregón, A., Nelson, B.J., Magnetic helical micromachines: Fabrication, controlled swimming, and cargo transport. *Adv. Mater.*, 24, 6, 811–816, 2012, doi: https://doi.org/10.1002/adma.201103818.
82. Qiu, F. *et al.*, Noncytotoxic artificial bacterial flagella fabricated from biocompatible ORMOCOMP and iron coating. *J. Mater. Chem. B*, 2, 4, 357–362, 2014, doi: 10.1039/C3TB20840K.
83. Lee, S. *et al.*, A needle-type microrobot for targeted drug delivery by affixing to a microtissue. *Adv. Healthc. Mater.*, 9, 7, 1901697, 2020, doi: https://doi.org/10.1002/adhm.201901697.
84. Yang, L. *et al.*, Targeted single-cell therapeutics with magnetic tubular micromotor by one-step exposure of structured femtosecond optical vortices. *Adv. Funct. Mater.*, 29, 45, 1905745, 2019, doi: https://doi.org/10.1002/adfm.201905745.
85. Li, D., Liu, Y., Yang, Y., Shen, Y., A fast and powerful swimming microrobot with a serrated tail enhanced propulsion interface. *Nanoscale*, 10, 42, 19673–19677, 2018, doi: 10.1039/C8NR04907F.

86. Li, D., Liu, C., Yang, Y., Wang, L., Shen, Y., Micro-rocket robot with all-optic actuating and tracking in blood. *Light Sci. Appl.*, 9, 1, 84, 2020, doi: 10.1038/s41377-020-0323-y.
87. Cabanach, P. *et al.*, Zwitterionic 3D-printed non-immunogenic stealth microrobots. *Adv. Mater.*, 32, 42, 2003013, 2020, doi: https://doi.org/10.1002/adma.202003013.
88. Xin, C. *et al.*, Conical hollow microhelices with superior swimming capabilities for targeted cargo delivery. *Adv. Mater.*, 31, 25, 1808226, 2019, doi: https://doi.org/10.1002/adma.201808226.
89. Barbot, A., Decanini, D., Hwang, G., Helical microrobot for force sensing inside microfluidic chip. *Sens. Actuators A Phys.*, 266, 258–272, 2017, doi: https://doi.org/10.1016/j.sna.2017.09.004.
90. Chen, L. *et al.*, Development of direct-laser-printable light-powered nanocomposites. *ACS Appl. Mater. Interfaces*, 11, 21, 19541–19553, 2019, doi: 10.1021/acsami.9b05871.
91. Jin, D., Chen, Q., Huang, T.-Y., Huang, J., Zhang, L., Duan, H., Four-dimensional direct laser writing of reconfigurable compound micromachines. *Mater. Today*, 32, 19–25, 2020, doi: https://doi.org/10.1016/j.mattod.2019.06.002.
92. Ceylan, H., Yasa, I.C., Yasa, O., Tabak, A.F., Giltinan, J., Sitti, M., 3D-printed biodegradable microswimmer for theranostic cargo delivery and release. *ACS Nano*, 13, 3, 3353–3362, Mar. 2019, doi: 10.1021/acsnano.8b09233.
93. Bozuyuk, U., Yasa, O., Yasa, I.C., Ceylan, H., Kizilel, S., Sitti, M., Light-triggered drug release from 3D-printed magnetic chitosan microswimmers. *ACS Nano*, 12, 9, 9617–9625, Sep. 2018, doi: 10.1021/acsnano.8b05997.

15

Novel Technique to Measure Shape Memory Behavior of 4D Material

Nilesh Tiwari[1*] **and Kanif M. Markad**[2]

[1]*Sunrise Industries (India) Limited, Vadodara, India*
[2]*Department of Mechanical Engineering, DVVP COE, Ahmednagar, Maharashtra, India*

Abstract

By using three-dimensional (3D) laser scanning and reverse engineering (RE), it is possible to predict how smart structures made with additive manufacturing will change shape. Through the use of fused deposition modeling (FDM) printing, a temperature-responsive smart material known as polylactic acid (PLA) can be used to create prototypes with a variety of geometric and functional properties. As a first step in validating the estimation of shape memory behavior, rectangular samples are developed, printed, scanned, and regenerated by RE. The results of the shape memory analyses agree very closely with those of the earlier studies, providing support for the strategy that should be used with more complex geometries. The procedure is then applied to the 3D paraboloid surface to generate surfaces for each of the stages of the cycle. Accurate measurements allowed for the successful, quick repair of all surfaces, which would have taken a long time or involved very complex methods with traditional measurement approaches. The paraboloid surfaces' shape recovery and shape fixity correspond to the figures derived for the reference samples. The impact of stresses that are imposed during the programming stage and then released during the recovery stage above PLA's glass transition temperature is also estimated using RE models.

***Keywords*:** Additive manufacturing, shape memory polymer, 3D modeling, 3D laser scanning, 3D shape and deformation measurement, reverse engineering

**Corresponding author*: nileshbwtiwari@gmail.com

R. Rajasekar, C. Moganapriya and P. Sathish Kumar (eds.) Additive Manufacturing with Novel Materials: Processes, Properties and Applications, (459–484) © 2024 Scrivener Publishing LLC

15.1 Introduction

Since its beginnings in the 1980s, additive manufacturing (AM) has drawn a lot of interest because to its incredible flexibility in producing non-geometric prototypes [1]. Additive manufacturing, often known as three-dimensional (3D) printing, encompasses processes including stereolithography, bioprinting, selective laser sintering, fused deposition modeling (FDM), and polyjet printing. These have uses in manufacturing, the creation of unique products, tissue engineering, textiles, sensors, actuators, and other fields [2–7]. With enhanced remote operating capabilities and the potential for artificial intelligence, the current integration of AM and information technology offers greater flexibility [8–10]. Despite recent improvements, it is only able to fabricate static components and has a limited range of materials, including some metals, polylactic acid (PLA), acrylonitrile butadiene styrene (ABS), and polyetheretherketone (PEEK). Its application is also limited by its small bed size and lengthy postprocessing times. Materials that can change form have the potential to be used to create flexible components [11]. In 2013, Tibbit revealed this idea, opening up new avenues for research in AM [12, 13]. Smart materials, often referred to as shape memory materials, have the ability to change their physical characteristics in response to stimuli including temperature, electricity, magnetic, moisture, pH, and light [14, 15]. Complex geometrical constructions created by 3D printing may be flexible thanks to AM made of these materials, which also have the capacity to change form, property, or functionality in response to outside stimuli [16]. The essay also focuses on how such materials' behavior might be estimated.

The extrusion-based methods, commonly used for multiscale and multimaterial 4D printing of hydrogels and polymeric composites, have low production rates, limited structural precision, and exhibit anisotropic mechanical properties in the printed components. polyjet, as an alternative, has restricted resin options and requires expensive equipment. Photopolymerization techniques like stereolithography (SLA), digital light processing (DLP) and continuous liquid interface production (CLIP) are capable of quickly producing high-resolution structures in an efficient manner, but for a limited range of materials and smaller print volumes. Although it is possible to induce prescribed shapeshifting in as-printed parts by adjusting the photopolymerization process, it is still challenging to fabricate multimaterial structures using vat polymerization techniques [17]. Shape memory behavior is a fascinating phenomenon that captures the essence of nature's resilience and adaptability. It refers to a material's ability to return to its original shape after undergoing deformation, without any external assistance. It's almost like the material has a mind of its own and knows how to repair itself, just like the human body does. Imagine a rubber band that has been stretched to its limit. If released, it snaps back to its original shape, as if nothing had happened.

This behavior is especially remarkable in shape memory alloys (SMAs) like nitinol, which can "remember" their original shape even after undergoing extreme deformation. For instance, nitinol wires can be bent into all sorts of shapes, and upon heating, they return to their original form. It's almost like they have a "muscle memory" that allows them to spring back into shape. The beauty of shape memory behavior lies in its versatility. It can be used in a wide range of applications, from medical implants that can adapt to the body's shape to aerospace structures that can withstand extreme temperatures and pressures. Shape recovery is a complicated phenomenon that depends on the cross-linking of molecules that are sensitive to heat, light, magnetic fields, or electricity at the molecular level [18–20]. This remarkable characteristic is due to the reversible nature of their shape-shifting behavior, which is controlled by external stimuli such as heat, light, or an electric field [21]. The mechanism behind SMPs' shape memory properties lies in the material's internal structure, which can be designed to change in response to certain stimuli. In their initial state, SMPs are typically rigid and inflexible. However, when exposed to a stimulus, such as heat, they can undergo a reversible phase transition, changing their internal structure and becoming soft and pliable [22, 23]. This transition can be controlled by varying the temperature at which the SMP undergoes the transformation, as well as the rate and duration of the stimulus. Once the SMP has been deformed into a new shape, it can be "fixed" by cooling it down to a lower temperature, which locks the new shape in place. When the material is heated again, it reverts back to its original shape.

SMPs have a wide range of potential applications in various fields, including biomedical engineering, aerospace, and robotics [24]. For example, in medicine, SMPs could be used to create stents that can be inserted into blood vessels and then expand to their original shape once they reach the desired location. In aerospace, they could be used to create morphing wings that change shape in response to changing flight conditions. In robotics, they could be used to create soft, pliable robots that can adapt to different environments and tasks. In summary, shape memory polymers are a fascinating and versatile class of materials that have the ability to change shape in response to external stimuli. By exploiting this property, researchers and engineers can develop innovative solutions to a wide range of challenges in various fields [25].

Shape fixity and shape recovery are two characteristics that may be used to measure the behavior of shape memory. Shape fixity (R_f), refers to a polymeric material's reversible segments' capacity to hold onto the mechanical deformation that has been applied, whereas shape recovery (R_r) refers to the material's capacity to regain its original shape. In the heat-activated bending test, conventional angle measuring techniques are used to gauge the specimen's shape memory capabilities. Materials including SMA, PLA, ABS, PEEK, polyurethane, and epoxy have different shape memory properties, according to the literature [26, 27]. In this chapter, the unique technique for calculating the shape memory characteristics of a thermally sensitive smart material (PLA) created by FDM is demonstrated.

Through the use of optical technology, 3D laser scanning gathers spatial data from a real specimen in order to produce point clouds depending on the surface's shape. Due to their non-invasive nature and lack of physical specimen alignment, these scanners offer better scanning techniques than conventional coordinate measuring equipment. Reverse engineering (RE), digitization, non-destructive assessment, and quality assurance (QA) are four categories that group various instances of digitizing using 3D laser scanning depending on the postprocessing of point cloud data [28–31]. In this situation, 3D laser scanning has been used for RE of smart material before, during, and during the shape deformation and recovery cycle. In contrast to time-consuming and inaccurate manual processes, RE's computer-aided design (CAD) models are based on point cloud data of actual geometries, providing a fast and precise method of measurement [32]. Relics [33], dentures [34], bones [35], and civil [36] and mechanical constructions [37] are just a few examples of where laser scanning has been used before in the cultural heritage sector for purposes like RE, digitization, non-destructive testing, and quality assurance. The following descriptions of major works are provided for continuity. Applications of the 3D scanning technology were suggested by Reyno *et al.* for measuring surface damage on aircraft structural panels [38]. Dent depths were measured as the difference between a surface fit that approximated the original, undamaged surface and a point cloud depiction of the damaged surface [39]. During on-site aircraft inspection, a non-destructive evaluation approach showed promise for enabling the quick extraction of surface dent measurements [38]. Using data from a terrestrial laser scanner (TLS), Teza *et al.* described a method for automatically identifying mass loss of concrete that involved computing the surface's mean and Gaussian curvatures and piecewise comparing the relevant distributions [40]. Additionally, Xu *et al.* scanned an underground tunnel with TLS technology and used the point cloud data of the tunnel to forecast its real geometry. Based on the expectation–maximization (EM) approach with t-distribution, the best model was examined. This model performed well in terms of automation, time efficiency, and robustness [41]. Measurements of an arch structure with monotonic loads were conducted by Yang *et al.* on the basis of TLS, and they reported a comparison between the original and optimized extraction of point clouds [42]. A drilling tool setup reconstruction utilizing laser triangulation principles was created by Marani *et al.* employing laser scanning [43].

The lack of natural joint specimens with consistent surface morphology for experimental studies [44] prompted Jiang *et al.* to develop a technical method for reproducing natural joint specimens by combining two state-of-the-art techniques: 3D scanning and 3D printing with CAD. The chosen technique involved the use of CAD software and optical scanning equipment to reconstruct a virtual joint specimen that replicated the joint morphology of the natural rock, and then a 3D printer was used to create a physical cast from the CAD model [44]. Dental laser scanning, first developed by Park *et al.* [45], captures the teeth's 3D geometry quickly and precisely. Widespread use of 3D laser scanning techniques is evidence of the field's growing dependence on this technology to improve the accuracy and productivity of a variety of applications. Although it has been used for other purposes, the current literature suggests that it has not been put to use in studying or evaluating shape memory behavior. The conventional method of using heat to bend and restore shape memory polymer is effective for assessing the shape memory properties of flat surfaces in terms of form recovery and shape fixity, but it may not be suitable for 3D surfaces [46]. Therefore, a new method involving 3D laser scanning and RE has been developed to determine the shape memory parameters of a paraboloid surface created using 3D printing with PLA. To confirm the accuracy of the process, comparisons are made between the original CAD model and the model obtained through scanning, and deviation analysis is performed. The deformed specimen and the resulting reshaped structure are then treated using this technology above the glass transition temperature of PLA. This innovative technology has the potential to provide a new avenue for quality control and non-destructive testing of intelligent buildings with thermally induced shape morphing.

15.2 Additive Manufacturing and Methodology

Specimens created by a 3D printer using widely available PLA filament in the shapes of rectangular plates and paraboloid. This section discusses sample preparation and the methods used to estimate shape memory parameters.

15.2.1 Sample Preparation of Rectangular Plates

One of the most popular AM techniques is FDM, which involves melting thermoplastic filament and passing it through a nozzle for layer-by-layer deposition on a heated bed. Due to their accessibility and capacity to print thermoplastic materials that are thermally sensitive, FDM (Omega 3D printers, made by JAK Machinery, India) were used exclusively in the current study. AutoCAD 2020 was used to design CAD models that will be printed using FDM and was then used to convert those CAD models into stereolithography files. To determine the accuracy of the method used to predict the shape memory qualities, rectangular plates were 3D printed and scanned. This technique will be generalized for smart 3D surfaces in the subsections that follow. The PLA filament used in this investigation had a glass transition temperature of 55–60°C [47]. Phases were used to create samples; the first phase involved printing rectangular plates, while the second stage involved printing a paraboloid surface. In order to gauge the shape memory capabilities of PLA, basic rectangular plates with dimensions of 40 mm in length and 10 mm in width were first constructed for the heat induced bending and recovery investigation of the smart material [48]. To comprehend the impact of thickness on shape memory property during the heat activation test, five various thicknesses of 0.8 mm, 1 mm, 1.2 mm, 1.6 mm, and 2 mm were manufactured. Each configuration's ten samples were evaluated to ensure the accuracy and dependability of the findings. After printing PLA plates, a heat activation test that include bending and recovery tests of the cantilever plate in a

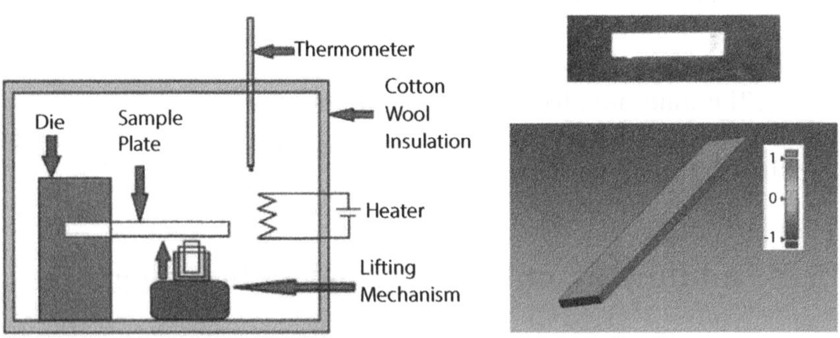

Figure 15.1 Representation of the technique to measure shape memory parameters.

consistent temperature environment were conducted. The temperature of the surroundings was kept at 20°C higher than the PLA's temperature of 80°C while the sample plates were bent at 90° and 180°. As shown in Figure 15.1, a thermal chamber was placed with a connected mechanism to deform the specimen, maintaining homogeneity of condition during shape memory testing.

Shape memory phenomenon has two stages: programming and recovery. During the programming step, the sample is stressed over the material's melting point and then cooled below it while the stress is maintained. The sample would return to its original shape or form during the recovery phase when the temperature was raised above [49]. Using this information, tests were carried out, which include the following steps:

1. A cantilever beam can be created by clamping the sample at one end.
2. The aforesaid temperature increased gradually as the deformation continued.
3. Cooling of the chamber and evaluation of the sample's shape stability.
4. With the increase in temperature above and the computation of form recovery, return to original shape.

15.2.2 Reverse Engineering and Shape Memory Parameters of Rectangular Samples

Reverse engineering (RE) uses point clouds created by scanning to create CAD models of the actual geometry. In comparison to manual procedures for complicated geometries, RE provide a speedier methodology for measuring geometries with higher accuracy and efficiency. Through a 3D laser scanning system of type Picza LPX-600RE 3D Laser Scanner, which has its own internal processing software to provide point cloud data, samples are 3D scanned during each phase of the shape memory cycle. To create CAD models, stereolithography point cloud data are loaded into the RE program Geomagic design X. For each stage of the shape memory cycle, three methods are used to validate the CAD models created by processing them in RE with actual samples. First, samples are measured using a vernier caliper and then models are built in AutoCAD. These models are contrasted with those made with Geomagic Design X and RE from 3D scanning. The deviation results in Figure 15.2 are highlighted in green in the deviation color bar from −1 to +1 mm and are in the range of 0.35 mm. For the sample with dimensions of 40 mm x 10 mm, it can be shown that the thickness

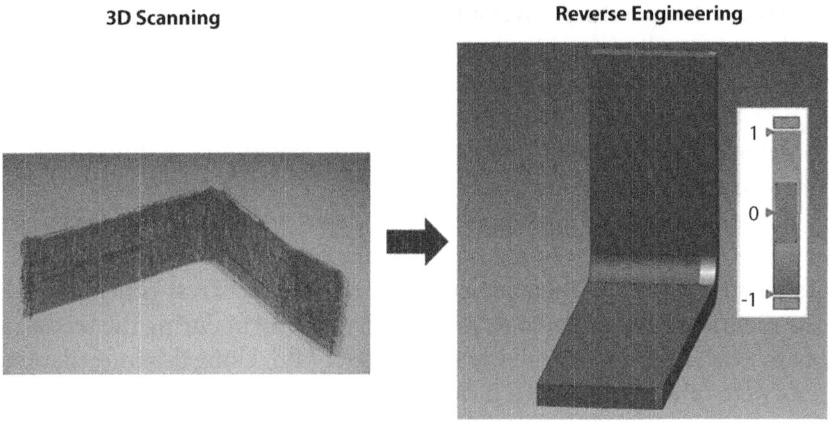

Figure 15.2 Illustration of RE model and CAD model.

difference of 0.35 mm is insignificant. So, it may be considered that a CAD model created with RE is just as accurate as a genuine model. The shape memory cycle testing is conducted by applying a mechanical force to the specimen and measuring its response to the force.

The shape recovery is monitored by measuring the displacement of the specimen at each stage in the cycle. The variation of the shape parameter is then used to calculate the form fixity and shape recovery characteristics of the material. The shape recovery of the specimen is determined by comparing the shape of the specimen before and after the shape memory cycle test. The results from the shape memory cycle testing provide valuable insights into the shape memory behavior of the material. The form fixity and shape recovery characteristics of the material can be calculated from the displacement data. The results can then be used to optimize the material properties for applications such as medical implants, smart structures, and robotics. The technique can also be used to investigate the effects of temperature and mechanical loading on the shape memory behavior of the material. The technique can be used to predict the shape memory behavior of materials under different conditions and to determine the optimum conditions for particular applications. Figure 15.3 summarizes the technique discussed here.

By measuring angles for the specimen using the Rapid-I vision measurement system, across the three stages of shape memory cycles, it is possible to estimate the shape memory properties of the material in terms of shape fixity and shape recovery during the heat activation test. The initial angle of the plate is indicated by the symbol "θ_0" which is taken to be 0° in this investigation. The specimen is bent to the intermediate programming

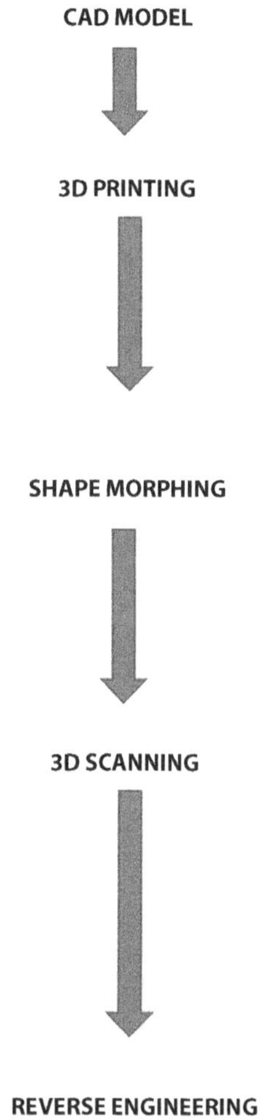

Figure 15.3 Flow of the measurement technique.

angle, θ_p at 90° and 180°," with the introduction of external loads during the bending test of PLA at 80°C. In order to determine the angle, the specimen is cooled to room temperature and tensions are released. The recovery phase is complete, and the recovery angle is measured, when the temperature of the programmed sample is raised to 80°C. To determine the shape fixity and shape recovery of PLA samples, these parameters are substituted in equations 15.1 and 15.2 [48]. To evaluate the process's accuracy and precision so that it may be used for 3D smart surfaces, the values are compared to those RE estimated.

The Rapid-I vision measurement system is used to measure the angles of the specimen in all three stages of the shape memory cycles, which allows us to estimate the shape memory properties of the material, including shape fixity and shape recovery during the heat activation test. In this investigation, the initial angle of the plate is denoted by the symbol "θ_0" which is assumed to be 0°. The specimen is then bent to an intermediate programming angle, θ_p, at 90° and 180°, while an external load is applied. To determine the angle, the specimen is cooled to room temperature and tensions are released. The recovery phase is considered complete when the temperature of the programmed sample is raised to 80°C, and the recovery angle is measured. To determine the shape fixity and shape recovery of PLA samples, these parameters are substituted in equations 15.1 and 15.2 [48]. The accuracy and precision of the process are evaluated by comparing the values to those estimated by RE, which allows us to determine its suitability for 3D smart surfaces.

$$R_f = (\theta_f - \theta_0)/(\theta_p - \theta_0) \tag{15.1}$$

$$R_r = (\theta_f - \theta_r)/(\theta_f - \theta_0) \tag{15.2}$$

15.2.3 Preparation of Samples and RE of Paraboloidal Surfaces

Based on the conclusions made from the analysis of plates, a paraboloid surface is built and 3D printed using PLA to estimate the shape memory characteristics of 3D surface. The use of paraboloid smart structures as antennas and sensors in electronics and space applications makes the analysis profitable to support its ongoing research into the computation of shape recover and form fixity [50, 51]. Based on a parametric equation, an AutoCAD model was created to 3D print a paraboloid surface. For the purpose of maintaining the structure's stiffness while the tests are being conducted, the surface is given a 2-mm thickness. According to FDM compatibility, CAD models are sent in .STL format for 3D printing. To confirm the consistency of the initial CAD model and that produced by RE, the technique of additional 3D scanning to produce point cloud data and development of model by RE is followed, as stated for the plates. Due to the variance in range of 0.35 mm, the findings shown are comparable to the one obtained in Figure 15.2, validating the results for the paraboloid surface. As a result, the approach chosen is used in the steps that follow to record the deformation behavior of smart surfaces throughout the shape memory cycle [52].

To reduce the area and volume of the programmed shape, paraboloid structures are bent in various directions. This may be used in aerospace applications to reduce the amount of area needed for smart antennas and actuators without sacrificing the structure's shape memory behavior [50]. In the programming step, surface deformation is accomplished by raising the temperature in the thermal chamber above PLA. In the current investigation, programmed structures as shown in Figure 15.4—are produced under the impact of arbitrary biaxial stresses of different intensity. Images of paraboloid samples at the programming stage may be found in the first row of Figure 15.4. The relevant CAD models are displayed in the second row with the axis of forces indicated, and the divergence between the point

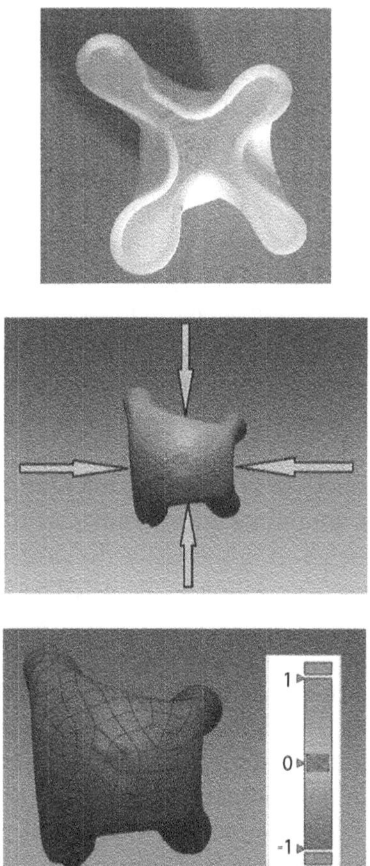

Figure 15.4 Generation of paraboloid surfaces under bi-axial stresses.

cloud data and the associated CAD models is displayed in the third row. Due of the shape memory feature of PLA, when the temperature is raised over a certain point, the structure returns to its original orientation. The sample is 3D scanned at each stage of the cycle to gather point cloud data for RE postprocessing.

15.2.4 RE of Surfaces

At each phase of the form memory cycle, point cloud data from a 3D laser scanning device is fed into Geomagic Design X. The three actual surfaces are shown in Figure 15.6 together with the point clouds created by the scanner and the RE model for each of them throughout the cycle's programming phase. As previously mentioned and also emphasized in Figure 15.6 with color gradient, meshing of the surfaces is controlled in such a way during the patching of RE that deviation is confined within 0.35 mm. The shape memory characteristics, including form fixity and recovery, are estimated using the models produced by RE. These variables would also be dictated by the differences in area and profile along various planes between the original and restored surfaces. In order to demonstrate the accuracy and effectiveness of the process for non-uniform or complicated 3D structures, the findings will be compared to those of regular PLA plates.

Difference in positions during shape memory cycle

The findings of form fixity and shape recovery of paraboloid surfaces are supplemented by the percentage variances between profiles for various orientations throughout various phases of the shape memory cycle. This is done so that the results may be more accurately interpreted. While doing analysis on the RE model, the normal circumferential profile and two side sectional profiles that are mutually perpendicular to one another are the profiles that are evaluated. Along each of the profiles for the individual samples, the percentage of deviation is calculated between the corresponding original orientation and the orientation that was programmed, which is referred to as the programming deviation (σ_p), and the original orientation with the orientation that was recovered, which is referred to as the recovery deviation (σ_r). The radial displacements that are measured from the center of the profile are referred to as 'r_{io}', 'r_{ip}' and 'r_{ir}' accordingly for orientations that are in the original stage, the programming stage, and the recovery stage, where 'i' represents the number of nodes on the profile. Using these values as substitutes in equations 15.3 and 15.4, results in σ_p and σ_r respectively [29].

$$\sigma_p(\%) = \sum_{i=1}^{n} \frac{\left[\frac{|r_{io} - r_{ip}|}{r_{io}}\right]}{n} \tag{15.3}$$

$$\sigma_r(\%) = \sum_{i=1}^{n} \frac{\left[\frac{|r_{io} - r_{ir}|}{r_{io}}\right]}{n} \tag{15.4}$$

Parameters related to the shape memory of paraboloid surfaces

By the process of computing the surface area of the profile that was derived from the RE model, an estimate of the shape memory qualities of the smart surfaces may be obtained. As was said previously, paraboloid sensors and antennae are very important to have when the area that is available for operations is limited, as is the amount of room that is available for storing merchandise. In circumstances like these, the reduction in the circumferential area of the smart paraboloid structures is where the primary attention lies. As a consequence of this motivation, the primary purpose of the current investigation is to compute and investigate the shape memory qualities in terms of the normal circumferential area.

In this section, the formulas of shape fixity and shape recovery in terms of circumferential area will be provided with reference to equations 15.1 and 15.2. At the original, programming, and recovery stages of the experiment, the circumferential area of the specimen is denoted by the letters A_0, A_f and A_r, respectively. At the programming stage of this work, it is difficult to take the sample with a temperature above T_g for scanning, which makes it impossible to determine the sample's circumferential area, A_p. This presents a challenge for the experimentation. Hence, the only area of the sample that is considered for the computation of R_f and R_r is the area of the sample after the cooling in the programming step (A_f), with the assumption that $A_p - A_o \approx A_o$. In light of these factors, the values of R_f and R_r on the circumferential surfaces of a paraboloid may be determined using equations 15.5 and 15.6.

$$R_f = \frac{|A_f - A_0|}{A_0} \tag{15.5}$$

$$R_r = \frac{|A_f - A_r|}{|A_f - A_0|} \tag{15.6}$$

15.3 Analysis

In order to investigate the dependability and effectiveness of 3D laser scanning and RE, several designs are modeled and printed using FDM. This research is done in order to assess the shape memory qualities of rectangular plates and a 3D paraboloid structure built from PLA. The findings of form fixity and shape recovery of cantilever PLA beam are provided and compared with the results that already existed in the earlier research. Following the validation, the approach is then generalized to the paraboloid surfaces in order to develop a mechanism for determining how accurately the smart structure can regain its previous state. In this chapter, a user-defined program called MATLAB is developed to compute and plot all of the parameters that are estimated. Also, the inbuilt applications helped to show curve-fitting and probability distribution in order to improve analysis of the evaluations.

15.3.1 Behavior of Plates With Shape Memory Using Rectangular Shapes

The shape memory cycle is applied to 3D-printed smart PLA samples of varying thickness during heat activation tests conducted at temperatures higher than T_g at 80°C. During this stage of the procedure, the dimensions of the samples are concurrently recorded using both the traditional way of measuring with a vernier scale and the Rapid-I vision measuring system, as well as a methodology that utilizes a mix of 3D laser scanning and RE. The linear dimensions of the specimen are measured with the assistance of a vernier caliper, while the angular measurements are determined with the assistance of a Rapid-I vision measuring system, both of which have an accuracy of ±0.02 mm. Computer-aided design models are built in AutoCAD based on the measurements, and these models are then used to determine the dependability of the chosen process. At this time, samples are being scanned using a 3D laser scanning technology to generate point cloud data. This point cloud data are then uploaded to the Geomagic Design X program so that RE may reconstruct a model of the specimen. This model is contrasted with one that was produced in AutoCAD through the use of more traditional techniques, as seen in Figure 15.1. According to the findings, the difference in dimensions is within the range of ±0.35 mm, that is lower than 0.01% in comparison to the dimensions of the specimen (40×10 mm). Hence, the accuracy of the RE is validated in order to estimate the dimensions of the samples at various stages of the shape memory cycle while they are being subjected to heat activation.

The RE models are mined for the dimensions of the many factors that must be calculated in order to determine form fixity and shape recovery. In equations 15.1 and 15.2, the values are substituted to compute for the properties of PLA samples with thicknesses of 0.8, 1.0, 1.2, 1.6, and 2.0 mm, respectively. Thinner samples are more easily able to unbend and restore their original orientation when compared to thicker samples, and this is true for both the programming angle at 90° and the programming angle at 180°. Figure 15.5 illustrates how the decrease in form recovery that occurs with an increase in thickness is consistent with the findings of Reference [48]. The minimum value is 81% and the maximum value is 91% for a sample with a thickness of 2 mm when bent at angles of 90° and 180°, respectively. The maximum value of R_r is 92% for the sample with 0.8 mm of thickness when bent at angles of 90° and 180°, respectively.

The behavior of thickness with respect to R_r was modeled. The values are rather close to the findings that were discussed in the earlier work [48]. Therefore, the methodology that has been adopted estimates the smartness of material without incorporating human errors that have accumulated during measurement by conventional instruments. This methodology could also be applied to the more complex structure that is made from smart material to verify its shape memory behavior. In addition, all of the

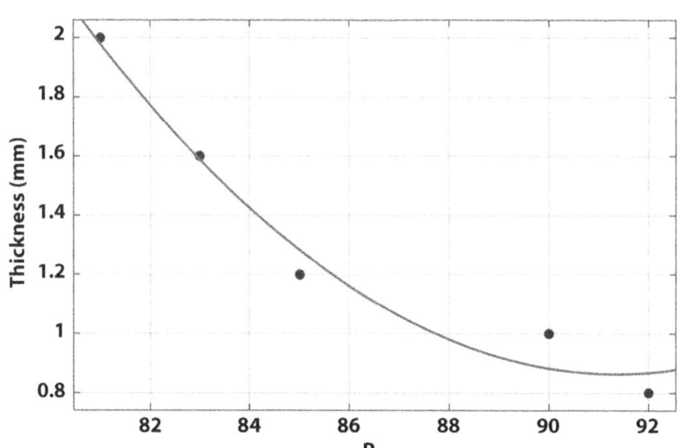

Figure 15.5 Shape recovery of rectangular plates with respect to thickness for θ_p at 90°.

processes that are a part of the method are automatic, with the exception of the picking up and placing of the specimen. This automates the process to the point where 3D scanning and the related RE may be finished in around 15 min. The duration of measurement that is involved in a specimen with simple geometry, such as a rectangle, may be longer compared to conventional measurement techniques; however, the duration of measurement that is involved in a specimen with complex geometry is shorter when this methodology is applied to the specimen. In order to estimate the shape memory parameters of a paraboloid surface that was printed from PLA, a 3D laser scan and RE are used. The interpretations form the basis for these methods.

15.3.2 Behavior of Paraboloid Surfaces With Respect to Shape Memory

As was covered in section 15.2, the introduction of bi-axial forces causes the surfaces of 3D printed paraboloid objects to deform into a variety of intermediate programming orientations. The primary goal in applying the stress is to reduce the specimen's circumferential area. This allows the specimen to occupy less space or pass through a route that is more constrained. The application of various stresses to the surface allows for the estimation of shape memory behavior as well as deviation between the surfaces, which results in the development of three separate surface profiles.

Surfaces Developed through the use of RE

Initial validation was performed to validate the accuracy of 3D laser scanning and RE by measuring the deviation between a CAD model created in AutoCAD based on a paraboloid parametric equation and a model generated by RE from point cloud data. This deviation was measured to validate the accuracy of 3D laser scanning and RE. As was said in the introduction to Figure 15.5, the difference in elevation between the two surfaces is not more than ±0.35 mm. On the basis of this investigation, RE models of all the orientations involved in shape memory cycles for the three surfaces have been built using point cloud data that was imported from a 3D laser scanning system. The comparison between the point cloud data and the corresponding RE models has been compiled in Figure 15.6 to describe the deviation between the orientations of the paraboloid surface during the different phases that make up the shape memory cycle. These phases are original, programming, and recovery.

A color gradient is used to indicate the sample's original data, as well as its programmed data and its recovered data. The deviations serve to gain

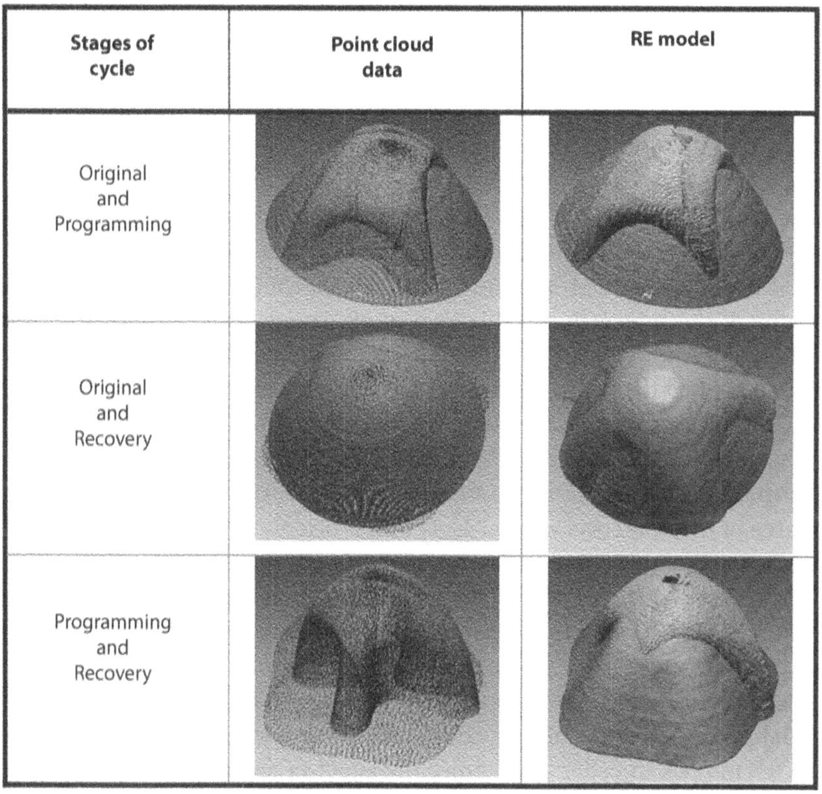

Figure 15.6 Surface comparison between configurations of shape memory cycle.

a clearer understanding of the difference in the surface profile throughout the heat activation test and record the deformation and recovery of the surface during the programming and recovery phases in comparison to the surface's initial state. This representation is necessary for calculating the form fixity and shape recovery of the surface in order to quantify the shape memory behavior of the material.

Deviation estimation

To get an estimate of the difference in area between the two surfaces, first calculate the deviation along each of the various axes. The development of three profiles for each paraboloid specimen is accomplished by extracting trajectories of each orientation at every step of operation for the specimens that are being investigated. One normal circumferential profile and two mutually perpendicular side middle profiles make up the three profiles that are used for each RE model. These profiles are each evenly split into parts by twelve nodes. Estimates are made of the way in which the radial distance varies along each of the three axes. The variance that exists between the original profile and the programming profile is noticeably more than that which exists between the original profile and the recovery profile.

Equations 15.3 and 15.4 are used to get the programming deviation (σ_p) and recovery deviation (σ_r), which are then used to compute the cumulative deviation over each surface. There is a cumulative percentage difference of 60% along the circumferential profile between the original trajectory and the one that was programmed. The equivalent deviation along the side profile of the midsection is 49. As a result, the decrease in area that was obtained in these instances is somewhere in the neighborhood of 45% on average. After then, the difference between the initial trajectory and the recovery trajectory along the circumferential profile was 2.5%, and the difference between the original trajectory and the recovery trajectory along the middle side profile was 4.9%. The construction showed a significant degree of mean recovery towards its initial orientation, which was consistent with the values of recovery ratio for the rectangular plates.

The cumulative deviation illustrates the behavior of the structure, although as shown by the applications, the behavior of various designs of smart structure is typically focused along the usual circumferential axis. This is illustrated by the cumulative deviation. The area of the circumferential profiles is calculated by constructing the contour of the projections for RE models in order to magnify the effect of applied stresses and subsequent recovery of specimens over T_g. This is done in order to maximize the effect of applied stresses. During the programming phase, there is an 80% deviation in the circumferential area, however following the recovery phase, those values are only 10%. So, a substantial reduction in the area has been accomplished in these trials without jeopardizing the structure's capacity for shape memory. This finding would also be validated by the computation of R_f and R_r for these specimens.

***R_f* and *R_r* of paraboloid surface**

When calculating the shape memory parameters R_f and R_r, the normal circumferential area of the RE models is substituted into equations 15.5 and 15.6, respectively. R_f has a value of 80%, and R_r has a value of 88% equivalent to it. The findings are accurate, which verifies the repeatability of the process while also confirming that there is no sign of persistent distortion in the structure as a result of its considerable recovery [53]. It takes roughly 20 minutes to complete the 3D laser scanning of the specimen and the RE necessary to produce the CAD model. This takes into account the total amount of time required to rebuild and compute the surface deviations and shape memory parameters during the various phases of the shape memory cycle. Therefore, the methodology supplements the process to verify the shape memory behavior of 3D surfaces, which could be a cumbersome activity through the convectional measurement technique, particularly in the programming stage, in which complex non-geometric surfaces are encountered. This is because the convectional measurement technique could make the verification of the shape memory behavior of 3D surfaces more difficult.

Distribution of deviation at nodes

As can be seen in Figure 15.7, percentage deviations at various nodes highlight the influence of temperature elevations that are more than T_g over the entirety of the paraboloid structure that is being investigated. This offers the possibility of determining the degree of difference that exists between the profiles at certain nodes throughout all of the profiles. This portrayal highlights magnified representation at all nodes, and it also provides a thorough look in comparison to the outcomes determined in terms of the cumulative deviation or form recovery of the overall smart structures. This also helps to identify the portion of a profile that may not have fully recovered, which might lead to a non-uniform profile during the recovery stage.

The examination of the distribution of deviation across all of the surfaces has been assembled and is shown in Figure 15.8, where a histogram is used to quantify the distribution of the number of nodes with regard to

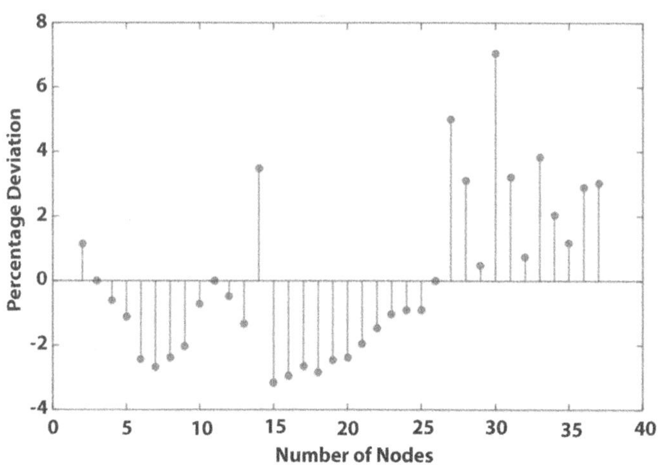

Figure 15.7 Comparison between paraboloid surfaces in recovery stage.

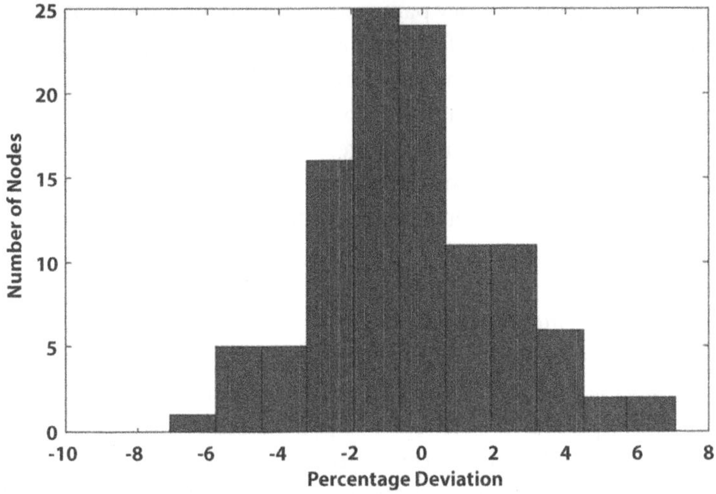

Figure 15.8 Distribution of deviation at nodes in recovery stage with respect to number of nodes.

the percentage deviation. The trajectory that was fitted to the data shows that there is a normal distribution of data within the percentage deviation range of 7 to +7%, which is also in the range recommended by the results of Figure 15.7. A normal probability density is also offered as a means of describing the fluctuation of departure from the mean, which was calculated to be around 0% with a variance of 6.8%. These analyses may be of assistance to the designers of smart components in assisting them to work more carefully on the stresses or materials in order to obtain the desired results with the simpler techniques involving CAD and RE rather than having to deal with the complex computations that are involved with contemporary design methods.

15.4 Conclusions

In the current work, a unique technique for studying the shape memory behavior of thermally actuated smart structures is given. This methodology combines 3D laser scanning with reverse engineering (RE). Validation of the accuracy and precision of the new method was performed on the conventional heat activation test of a rectangular specimen, which resulted in noteworthy findings that were congruent with the existing literature. To investigate the impact of external mechanical stresses at extreme temperatures exceeding T_g and to derive shape memory parameters, the approach is applied to three-dimensional (3D) paraboloid surfaces. The following deductions may be made in light of the research carried out and the analysis that was done for this work:

- The thickness of the printed rectangular specimen has almost little influence on the form's ability to be fixed, but it does significantly govern the shape's ability to recover, as shown by the fact that shape recovery improves as thickness decreases.
- The procedure improves the study and measurement of smart structure during the programming stage, when the surfaces may be deformed to extremely complex and non-linear geometry.
- The redevelopment of smart structure during various stages of the shape memory cycle can be facilitated with high accuracy, precision, and swiftness thanks to the combination of 3D laser scanning and RE. It is possible that standard measuring techniques might have a tough time digesting this, or that more advanced instruments and postprocessing would be required.

As a result, the suggested approach provides a method that is both quick and accurate in estimating the shape memory behavior of intelligent structures. This method also has the capability of being repeated. This procedure may be helpful to design engineers in identifying the anomalies in shape morphing components minutely with the assistance of prevalent design software. If this procedure is not used, design engineers may need to employ advanced software based on the finite element method, which could be avoided by using this procedure.

References

1. Shafranek, R.T., Millik, S.C., Smith, P.T., Lee, C.-U., Boydston, A.J., Nelson, A., Stimuli-responsive materials in additive manufacturing. *Prog. Polym. Sci.*, 93, 36–67, 2019, https://doi.org/10.1016/j.progpolymsci.2019.03.002.
2. Gao, H., Li, J., Zhang, F., Liu, Y., Leng, J., The research status and challenges of shape memory polymer-based flexible electronics. *Mater. Horiz.*, 6, 931–944, 2019, https://doi.org/10.1039/C8MH01070F.

3. Lin, C., Zhang, L., Liu, Y., Liu, L., Leng, J., 4D printing of personalized shape memory polymer vascular stents with negative Poisson's ratio structure: A preliminary study. *Sci. China Technol. Sci.*, 63, 578–588, 2020, https://doi.org/10.1007/s11431-019-1468-2.
4. López-Valdeolivas, M., Liu, D., Broer, D.J., Sánchez-Somolinos, C., 4D printed actuators with soft-robotic functions. *Macromol. Rapid Commun.*, 39, 1700710, 2018, https://doi.org/10.1002/marc.201700710.
5. Sydney Gladman, A., Matsumoto, E.A., Nuzzo, R.G., Mahadevan, L., Lewis, J.A., Biomimetic 4D printing. *Nat. Mater.*, 15, 413–418, 2016, https://doi.org/10.1038/nmat4544.
6. Zhou, Y., Huang, W.M., Kang, S.F., Wu, X.L., Lu, H.B., Fu, J., Cui, H., From 3D to 4D printing: Approaches and typical applications. *J. Mech. Sci. Technol.*, 29, 4281–4288, 2015, https://doi.org/10.1007/s12206-015-0925-0.
7. Lal, A. and Markad, K., Influence of dynamic temperature variation and inplane varying loads over post-buckling and free vibration analysis of sandwich composite beam. *Int. J. Comput. Mater. Sci. Eng.*, 09, 2050012, 2020, https://doi.org/10.1142/S2047684120500128.
8. Jin, Z., Zhang, Z., Gu, G.X., Automated real-time detection and prediction of interlayer imperfections in additive manufacturing processes using artificial intelligence. *Adv. Intell. Syst.*, 2, 1900130, 2020, https://doi.org/10.1002/aisy.201900130.
9. Tiwari, N. and Shaikh, A.A., Hybridization of carbon fiber composites with graphene nanoplatelets to enhance interfacial bonding and thermomechanical properties for shape memory applications. *Polym.-Plast. Technol. Mater.*, 61, 161–175, 2022.
10. Lal, A. and Markad, K., Nonlinear flexural analysis of sandwich beam with multi walled carbon nanotube reinforced composite sheet under thermo-mechanical loading. *Curved Layered Struct.*, 7, 1–16, 2020, https://doi.org/10.1515/cls-2020-0001.
11. Tiwari, N. and Shaikh, A.A., Effect of the amine and carboxyl functionalised graphene on the thermomechanical and interfacial properties of the shape memory polymer nanocomposites. *Polym. Bull.*, 1–19, 2022.
12. Tibbits, S., 4D printing: Multi-material shape change, architectural design. *Arch. Des.*, 84, 116–121, 2014, https://doi.org/10.1002/ad.1710.
13. Tiwari, N., Gagare, S.W., Shaikh, A.A., Shape recovery analysis of the additive manufactured 3D smart surfaces through reverse engineering. *Prog. Addit. Manuf.*, 6, 281–295, 2021.
14. Zhang, Z., Demir, K.G., Gu, G.X., Developments in 4D-printing: A review on current smart materials, technologies, and applications. *Int. J. Smart Nano Mater.*, 10, 205–224, 2019, https://doi.org/10.1080/19475411.2019.1591541.
15. Markad, K. and Lal, A., Synthesis of the multiphase shape memory hybrid composites hybridized with functionalized MWCNT to improve mechanical and interfacial properties. *Polym.-Plast. Technol. Mater.*, 61, 650–664, 2022, https://doi.org/10.1080/25740881.2021.2006709.

16. Tiwari, N. and Shaikh, A.A., Influence of carboxyl-functionalized graphene nanoplatelets on the thermomechanical and morphological behavior of shape memory nanocomposites. *Compos.: Mech. Comput. Appl.: Int. J.*, 13, 75–85, 2022.
17. Falahati, M., Ahmadvand, P., Safaee, S., Chang, Y.-C., Lyu, Z., Chen, R., Li, L., Lin, Y., Smart polymers and nanocomposites for 3D and 4D printing. *Mater. Today*, 40, 215–245, 2020, https://doi.org/10.1016/j.mattod.2020.06.001.
18. Li, F., Scarpa, F., Lan, X., Liu, L., Liu, Y., Leng, J., Bending shape recovery of unidirectional carbon fiber reinforced epoxy-based shape memory polymer composites. *Compos. Part A: Appl. Sci. Manuf.*, 116, 169–179, 2019, https://doi.org/10.1016/j.compositesa.2018.10.037.
19. Tiwari, N. and Shaikh, A.A., Effect of size and surface area of graphene nanoplatelets on the thermomechanical and interfacial properties of shape memory multiscale composites. *Polym.-Plast. Technol. Mater.*, 61, 12, 1–13, 2022.
20. Lal, A. and Markad, K., Probabilistic-based nonlinear progressive failure analysis of piezoelectric laminated composite shell panels in hygrothermal environment. *J. Aerosp. Eng.*, 34, 04021099, 2021, https://doi.org/10.1061/(ASCE)AS.1943-5525.0001345.
21. Tiwari, N. and Shaikh, A., Effect of graphene nanoplatelets on the thermomechanical behaviour of smart polymer nanocomposites, in: *Aerospace and Associated Technology*, pp. 300–304, Routledge, U.K., 2022.
22. Lal, A. and Markad, K., Thermal post buckling analysis of smart SMA hybrid sandwich composite plate. *Polym. Polym. Compos.*, 29, S344–S360, 2021, https://doi.org/10.1177/09673911211001276.
23. Lal, A. and Markad, K., Static and dynamic nonlinear stability analyses of hybrid sandwich composite beams under variable in-plane loads. *J. Mech. Sci. Technol.*, 35, 3895–3908, 2021, https://doi.org/10.1007/s12206-021-0803-x.
24. Anjum, A., Shaikh, A.A., Tiwari, N., Experimental investigations of channel profile and surface roughness on PMMA substrate for microfluidic devices with mathematical modelling. *Optik*, 261, 169154, 2022.
25. Markad, K. and Lal, A., Experimental investigation of shape memory polymer hybrid nanocomposites modified by carbon fiber reinforced multi-walled carbon nanotube (MWCNT). *Mater. Res. Express*, 8, 105015, 2021, https://doi.org/10.1088/2053-1591/ac2fcc.
26. Kuang, X., Roach, D.J., Wu, J., Hamel, C.M., Ding, Z., Wang, T., Dunn, M.L., Qi, H.J., Advances in 4D printing: Materials and applications. *Adv. Funct. Mater.*, 29, 1805290, 2019, https://doi.org/10.1002/adfm.201805290.
27. Tiwari, N. and Shaikh, A.A., Micro buckling of carbon fiber in triple shape memory polymer composites under bending in glass transition regions. *Mater. Today: Proc.*, 44, 4744–4748, 2021.
28. Hoła, J., Sadowski, Ł., Reiner, J., Stach, S., Usefulness of 3D surface roughness parameters for nondestructive evaluation of pull-off adhesion of concrete layers. *Constr. Build. Mater.*, 84, 111–120, 2015, https://doi.org/10.1016/j.conbuildmat.2015.03.014.

29. Koch, C., Georgieva, K., Kasireddy, V., Akinci, B., Fieguth, P., A review on computer vision based defect detection and condition assessment of concrete and asphalt civil infrastructure. *Adv. Eng. Inf.*, 29, 196–210, 2015, https://doi.org/10.1016/j.aei.2015.01.008.
30. Sansoni, G., Trebeschi, M., Docchio, F., State-of-the-art and applications of 3D imaging sensors in industry, Cultural Heritage, Medicine, and Criminal Investigation. *Sensors (Basel)*, 9, 568–601, 2009, https://doi.org/10.3390/s90100568.
31. Tarabini, M., Giberti, H., Giancola, S., Sgrenzaroli, M., Sala, R., Cheli, F., A moving 3D laser scanner for automated underbridge inspection. *Machines*, 5, 32, 2017, https://doi.org/10.3390/machines5040032.
32. Kuş, A., Implementation of 3D optical scanning technology for automotive applications. *Sensors (Basel)*, 9, 1967–1979, 2009, https://doi.org/10.3390/s90301967.
33. Zhao, H., Diao, X., Jiang, H., Zhao, Z., The 3D measurement techniques for ancient architecture and historical relics, in: *Second International Conference on Photonics and Optical Engineering, International Society for Optics and Photonics*, p. 102562L, 2017, https://doi.org/10.1117/12.2257460.
34. Gerasimov, V.A., Kostrin, D.K., Selivanov, L.M., Simon, V.A., Uhov, A.A., Prototype of the intraoral dental scanner based on a laser scanning system. *AIP Conf. Proc.*, 2089, 020010, 2019, https://doi.org/10.1063/1.5095739.
35. DeVries, N.A., Gassman, E.E., Kallemeyn, N.A., Shivanna, K.H., Magnotta, V.A., Grosland, N.M., Validation of phalanx bone three-dimensional surface segmentation from computed tomography images using laser scanning. *Skeletal Radiol.*, 37, 35–42, 2008, https://doi.org/10.1007/s00256-007-0386-3.
36. Jiao, Q., Jiang, H., Li, Q., Building earthquake damage analysis using terrestrial laser scanning data. *Adv. Civ. Eng.*, 2019, e8308104, 2019, https://doi.org/10.1155/2019/8308104.
37. Jo, Y.H. and Hong, S., Three-dimensional digital documentation of cultural heritage site based on the convergence of terrestrial laser scanning and unmanned aerial vehicle photogrammetry. *ISPRS Int. J. Geo-Inf.*, 8, 53, 2019, https://doi.org/10.3390/ijgi8020053.
38. Reyno, T., Marsden, C., Wowk, D., Surface damage evaluation of honeycomb sandwich aircraft panels using 3D scanning technology. *NDT & E Int.*, 97, 11–19, 2018, https://doi.org/10.1016/j.ndteint.2018.03.007.
39. Anjum, A., Shaikh, A.A., Tiwari, N., Comparative assessment of the developed algorithm with the soft computing algorithm for the laser machined depth. *Infrared Phys. Technol.*, 129, 104545, 2023.
40. Teza, G., Galgaro, A., Moro, F., Contactless recognition of concrete surface damage from laser scanning and curvature computation. *NDT & E Int.*, 42, 240–249, 2009, https://doi.org/10.1016/j.ndteint.2008.10.009.
41. Xu, X., Kargoll, B., Bureick, J., Yang, H., Alkhatib, H., Neumann, I., TLS-based profile model analysis of major composite structures with robust B-spline

method. *Compos. Struct.*, 184, 814–820, 2018, https://doi.org/10.1016/j.compstruct.2017.10.057.
42. Yang, H., Omidalizarandi, M., Xu, X., Neumann, I., Terrestrial laser scanning technology for deformation monitoring and surface modeling of arch structures. *Compos. Struct.*, 169, 173–179, 2017, https://doi.org/10.1016/j.compstruct.2016.10.095.
43. Jiang, Q., Feng, X., Gong, Y., Song, L., Ran, S., Cui, J., Reverse modelling of natural rock joints using 3D scanning and 3D printing. *Comput. Geotechn.*, 73, 210–220, 2016, https://doi.org/10.1016/j.compgeo.2015.11.020.
44. Marani, R., Nitti, M., Cicirelli, G., D'Orazio, T., Stella, E., High-resolution laser scanning for three-dimensional inspection of drilling tools. *Adv. Mech. Eng.*, 5, 620786, 2013, https://doi.org/10.1155/2013/620786.
45. Park, S., Kang, H.C., Lee, J., Shin, J., Shin, Y.G., An enhanced method for registration of dental surfaces partially scanned by a 3D dental laser scanning. *Comput. Methods Programs Biomed.*, 118, 11–22, 2015, https://doi.org/10.1016/j.cmpb.2014.09.007.
46. Tiwari, N. and Shaikh, A.A., Micro-buckling of carbon fibers in shape memory polymer composites under bending in the glass transition temperature region. *Curved Layered Struct.*, 8, 96–108, 2021.
47. Song, Y., Li, Y., Song, W., Yee, K., Lee, K.-Y., Tagarielli, V.L., Measurements of the mechanical response of unidirectional 3D-printed PLA. *Mater. Des.*, 123, 154–164, 2017, https://doi.org/10.1016/j.matdes.2017.03.051.
48. Leist, S.K., Gao, D., Chiou, R., Zhou, J., Investigating the shape memory properties of 4D printed polylactic acid (PLA) and the concept of 4D printing onto nylon fabrics for the creation of smart textiles. *Virtual Phys. Prototyping*, 12, 290–300, 2017, https://doi.org/10.1080/17452759.2017.1341815.
49. Anjum, A., Shaikh, A.A., Tiwari, N., Experimental investigations and modeling for multi-pass laser micro-milling by soft computing-physics informed machine learning on PMMA sheet using CO2 laser. *Opt. Laser Technol.*, 158, 108922, 2023.
50. Kalra, S., Bhattacharya, B., Munjal, B.S., Design of shape memory alloy actuated intelligent parabolic antenna for space applications. *Smart Mater. Struct.*, 26, 095015, 2017, https://doi.org/10.1088/1361-665X/aa7468.
51. Tiwari, N. and Shaikh, A.A., Flexural analysis of thermally actuated fiber reinforced shape memory polymer composite. *Adv. Mater. Res.*, 8, 337–359, 2019.
52. Tiwari, N. and Shaikh, A.A., Buckling and vibration analysis of shape memory laminated composite beams under axially heterogeneous in-plane loads in the glass transition temperature region. *SN Appl. Sci.*, 3, 1–15, 2021.
53. Tiwari, N., Shaikh, A.A., Malek, N.I., Modification of the multiphase shape memory composites with functionalized graphene nanoplatelets: Enhancement of thermomechanical and interfacial properties. *Mater. Today Chem.*, 24, 100826, 2022.

16

Additive Manufacturing for Building and Constructions: Overview, Applications and Challenges

Akesh B. Kakarla*, **Ing Kong and Vipulkumar Ishvarbhai Patel**

School of Computing, Engineering and Mathematical Sciences, La Trobe University, Bendigo, VIC, Australia

Abstract

Additive manufacturing (AM), more commonly called three-dimensional (3D) printing, is a rising technology in the civil construction industry. The technique develops the 3D structures rapidly with little personnel, minimal resources, and reduces waste production. However, 3D printing is less explored in building and construction than in aerospace, automobile, and medical applications. This chapter provides an argument for the potential role of AM in building and construction applications. Furthermore, the chapter also provides insights into existing and novel materials and discusses ideal material properties for AM in building structures. The research progress and future of digitalizing construction techniques are systematically reported, and the conclusions will be helpful for designers and civil engineers to recognise the AM implications for minimising cost and time.

Keywords: Additive manufacturing, 3D printing, construction materials, structures

16.1 Introduction

The building and construction (B&C) industry is one of the most significant manufacturing domains in the world, consuming substantial volumes of resources. The global construction industry economy currently spends approximately 13% of the gross domestic product (GDP) value and

**Corresponding author*: a.kakarla@latrobe.edu.au

R. Rajasekar, C. Moganapriya and P. Sathish Kumar (eds.) Additive Manufacturing with Novel Materials: Processes, Properties and Applications, (485–508) © 2024 Scrivener Publishing LLC

is expected to grow by over 13.5% in 2030 [1]. Over the past decade, there have been substantial advancements in the manufacturing industries, and companies constantly incorporated advanced methods to enhance productivity [2, 3]. Hence, labor productivity in the construction industry has increased. However, the industry has recently faced significant challenges in productivity and inadequate performance due to a decline in labor productivity. A study published by the McKinsey firm stated that annual productivity in the construction industry rose to a median of 1% over the past twenty years [4]. Similarly, Nasir *et al.* [5] estimated the labor productivity of 20 countries from Europe, Asia, and America in B&C industries. The studies stated that the United States displayed a -0.84% productivity rate compared to other advanced nations such as the Singapore and the United Kingdom [5]. Furthermore, the industry consumes a significant quantity of resources, which causes resource depletion. For instance, an investigation by Koltz *et al.* [6] stated that approximately 36% of energy, 30% of natural materials, and 12% of water have been used globally in the construction industry. Consequently, construction companies are looking into novel materials to decrease environmental stresses and create climate-neutral buildings [7–9].

Traditional building approaches are simple techniques requiring only two-dimensional drawings, molds and skilled workers to construct the structures [9, 10]. On the other hand, the work in the construction field is labor-intensive, resulting in several problems [10–12]. For instance, a worker operating vibration damping equipment is prone to entire body tremor, loud noise and postural stress that can cause long-term illnesses and injuries [11, 12]. Moreover, according to Man *et al.* [13], a higher death rate and risk of accidents were stated in the B&C industries than in other manufacturing industries. Furthermore, poor communication between structural designers and construction workers resulted in flawed structures and casualties. In addition, the Fourth Industrial Revolution (4IR, or Industry 4.0) [14, 15], which promoted digitization and sustainable manufacturing tasks, significantly impacted the conventional construction industry practices. Accordingly, building companies and researchers are looking for advanced techniques to address the challenges of lower labor productivity, and the 4IR aspects. Furthermore, Alaloul *et al.* [16] and Jazzar *et al.* [17] stated that implementing construction 4.0 technologies can aid the industry in various key aspects, including economics, management, and supply chain areas. From that point of view, additive manufacturing (AM) (3D printing), provides a new set of methodologies for civil engineers and construction industries that could be used by the industry in the direction of digitization and increased productivity.

Over the past few decades, AM has been used in different manufacturing sectors for product development, such as the automobile [18], aerospace [19], healthcare [20], and food industries [21]. The technique was first introduced in 1984 by Charles Hull [22] and described as printing material layer by layer to produce a 3D structure from machine-readable computer-aided design (CAD) files or 3D models [22]. The technique is an innovative approach that provides different opportunities in product development and increased efficiency for the manufacturing industries. The progress of 3D printing became attractive due to the degree of freedom in optimization of topology, reduced usage of materials and energy, reduced dependence on labor, and higher work safety. In addition, the technique assists in addressing the main challenge of Industry 4.0: sustainability issues in the manufacturing sectors [23, 24]. Considering its varied advantages, 3D printing has become a promising technology in the B&C industry. However, it is still in its infancy; studies have yet to prove that utilizing the technique in B&C industries could save time and cost [25]. In addition, the uncertainty in project management objectives involving external and internal risks (which include socioeconomic, environmental, technical, design material, and liability risks) must be addressed before AM is fully applied in B&C industries. Therefore, researchers have investigated various aspects to direct B&C industries toward utilizing AM as mainframe construction technology.

This chapter discusses the current progress and applications of 3D printing in B&C industries. The following section discusses the types of 3D printing techniques used in construction applications. Consequently, distinct types of construction materials, properties, applications, and challenges of B&C in 3D printing are reviewed in this chapter's later sections. Finally, the chapter concludes with a brief overview of AM's adoption into construction applications for future developments.

16.2 Techniques of AM Used in B&C

Three-dimensional (3D) printing techniques in B&C have been investigated since the 1990s [22, 26]. A wide range of manufacturing techniques have been reported and continuously enhanced in the 3D printing of stone-like substances without molds [22, 27]. Recent developments and studies of 3D printing in the civil engineering sector are most commonly based on binder jetting [28] and material deposition [29]. In general, both techniques build up complex designs by adding materials in layer format one over the other. Initially, a CAD file with clear geometry is created using

3D design software, which is later sliced and converted into machine readable files and then printed with a designated material to 3D structures [28].

16.2.1 Binder Jetting

Emanual Sachs developed binder jetting in 1989 at the Massachusetts Institute of Technology, USA [28, 30]. The process is mainly used for powder-based materials. The method comprises of depositing the powdered material in layers (bottom-up) and systematically depositing a binder onto each layer to bond the deposited layers together (Figure 16.1) [31]. The particle size of the materials utilized in the printing process plays a significant role in the process. Accordingly, the techniques required to have uniform particle sizes and shapes to distribute the powder while printing and to consistently avoid geometrical inaccuracies. Studies have shown that particle diameter ranges between 5 and 100 μm could be appropriate for the binder jetting [32]. Moreover, the studies also showed that large particles could be used for a quick printing process. Shakor *et al.* [33] demonstrated the gypsum and cement motor's printability using the binder jetting technique. The analysis showed that the dimensional accuracy of the printed models was equal to the designed CAD model. Furthermore, the wettability and contact angle of the binder and motor spread higher than commercial powder.

Similarly, Na *et al.* [31] studied calcium sulfoaluminate cement's printability using binder jetting. The printing of cement powder was conducted with a combination of sodium silicate and gypsum as binder for interfacial bonding of the layers. The outcomes indicated that the compressive strength of the produced models was 7 MPa but had cracks in a few printed

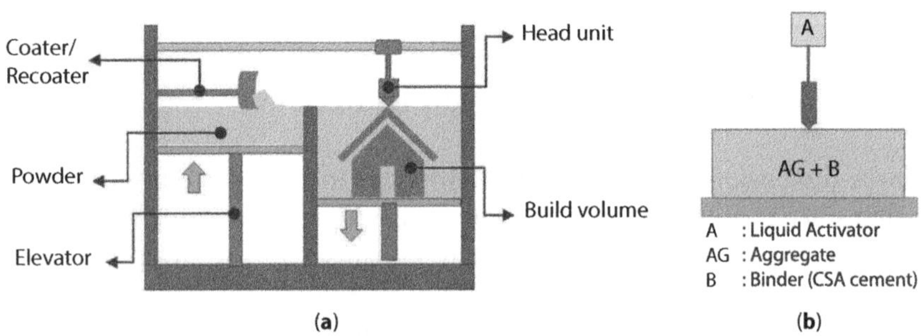

Figure 16.1 (a) Illustration of binder jetting concept, (b) insight view of materials deposition using binder jet printer. Reproduced from Ref [31], Copyrights, MDPI, Materials.

models that needed further curing and post-treatment methods. Salari *et al.* [34] reported the utilization of magnesium oxychloride cement for 3D printing concrete structures using binder jetting. In addition, the team also investigated the parameters influencing the printing. The outcomes indicated that printing was heavily influenced by particle size, quantity of binder and thickness of layer. Despite the advancement of the 3D printing technique, there are some limitations in utilizing the binder jetting process, such as time of printing and material composition, and appropriate binders or liquid activators. Therefore, further investigations are required to be conducted on varied materials to demonstrate the possibilities of this technique for 3D printing cementitious or concrete materials in construction applications.

16.2.2 Material Extrusion

Materials extrusion is the most popular AM technique. The technique commonly extrudes or deposits the materials in layer upon layer through a nozzle. Material extrusion method usually known as fused deposition Modeling, marketed by Stratasys and created by Crump [35–37]. In the context of the B&C industry, the extrusion process involves pumping the cementitious materials from a nozzle in a defined path, either in a mold with a cross-section or freeform printing. Large-scale construction using AM techniques could often be conducted using the material extrusion method because it can be processed widely, including conventional [38, 39], fiber-reinforced cementitious [40], and composite materials [41]. For instance, a research group from Loughborough University reported on a concrete extrusion development based on a gantry system that can extrude concrete with a 1-cm diameter [42]. Ji *et al.* [43] presented the assembly of a material extrusion printer for the B&C industry. The printing system consisted of a reciprocating plugging power system, two feed bins, concrete performance adjustment, and a mixed system (Figure 16.2). In terms of equipment, it was reported that printing was uninterrupted because the technique was modified by replacing the pressure systems with a vibrating technique that yields the compaction of the material. Moreover, the concrete paste extruded using the technique produced 3D structures with good printability and workability [43]. Yu *et al.* [44] demonstrated building aggregate-bed with cement paste binder using extrusion-based concrete printing. Furthermore, a modified extrusion-based printer was developed to perform the studies. The findings suggested that the concrete extrusion printers and mechanical properties are higher in the vertical direction compared to lateral path of printing [44]. However, there are significant

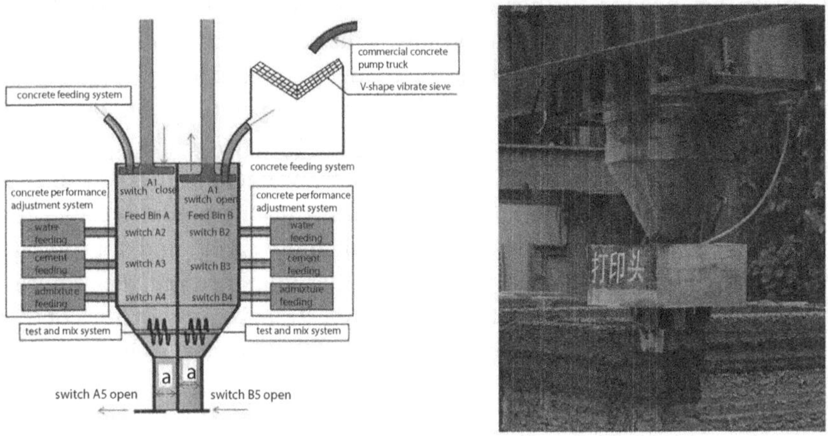

(a) (b)

Figure 16.2 (a) Design overview of the multi-nozzle extrusion print head (b) onsite printing of the developed design [43].

limitations in the extrusion of construction materials causing the blocking of the print head due to longer lines of transporting the mixed concrete to the print head [44].

Material extrusion and binder jetting are commonly studied and applied to print construction materials in off-site and on-site applications. At the same time, other techniques such as contour crafting [45], selective binder activation and selective paste instruction are also used for producing a structure using liquid activator [42], aggregates [42, 46], and cement pastes [42, 47, 48]. Moreover, the techniques advanced the concept of utilizing AM in B&C industries to address Industry 4.0 digitalization and eco-friendly approaches. Nonetheless, further studies are necessary to enhance the concrete or cementitious material printing technology by optimising the printing controls and regimens.

16.3 Construction Materials for AM

High-performance cement-based materials that can provide equivalent properties to conventional building techniques are needed for AM techniques. However, with the transformation of B&C's current industry practices to AM, there is a need to identify novel construction materials for printing. The printable materials combine various materials such as sand,

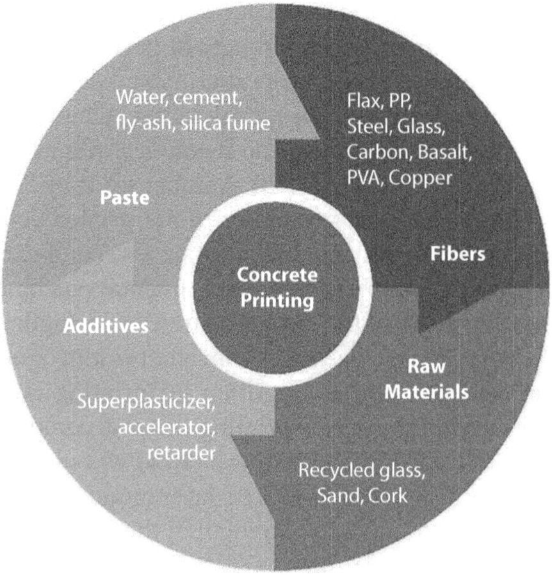

Figure 16.3 Types of 3D printable materials in B&C. Reproduced from Ref [49]. Copyrights, Elsevier.

aggregates, cement, polymers, fibers, fly ash, and other chemical agents (Figure 16.3) [49]. Using AM paved a new way to produce novel materials for B&C, including natural and recycled materials. In this context, the novel materials offer enormous potential to enable sustainability and circular economy advancements.

16.3.1 Cementitious-Based Materials

Cementitious-based materials are widely investigated in AM for construction applications. Moreover, cement is the initial construction element and plays a vital role in building structures [50]. Cement possesses unique chemical and mechanical properties that are beneficial for structures [50, 51]. It is economical and plentiful compared to other materials. In addition, cement is a successful material in traditional construction techniques and can be used for building simple houses to complex structures [50, 51]. Cement-based materials include a wide range of materials, including cement pastes and mortars (concrete, aggregates, sand) [50, 52]. Additionally, new types of cementitious-based materials have been investigated by researchers that satisfy performance and application requirements due to the need for novel materials in civil engineering structures [25]. Perrot *et al.* [53]

presented 3D-printed cylindrical structures with a diameter of 60 mm and a height of 35 mm. The mechanical strength of the materials was predicted using theoretically and validated with experimental data. The analysis displayed that the vertical stress on the models based on building rate and not exceeding critical stress. Moreover, the theoretical results were in agreement with experimental outcomes [53]. Another study by Pierre *et al.* [54] demonstrated the mechanical properties of 3D-printed cement paste in relation to penetration depth. The investigation was conducted with a theoretical approach and compared with experimental outcomes. It showed that the penetration depth of the cement shifted the strength of the printed structure [54]. A systematic design of concrete printing and optimal mix proportions of cement-based composition for printing was reported by Le *et al.* [46]. A high-performance printing material composed of cement (70%), silica fume (1.2 Kg/m^3), and fly ash (20%) was successfully printed with a 9-mm nozzle at constant flow. A 61-layer model was produced using the cement paste. The compressive strength of the produced cementitious material was 110 MPa after 28 days.

Recent studies on printing structures using cement-based materials investigated the rheology properties. The investigations were conducted primarily to understand the flowability of pastes through the pump heads of the 3D printers. Chen *et al.* [55] evaluated the deformation of a 3D-printed model produced with cement paste altered with metakaolin. The experimental analysis stated that structure deformation declined from 7.69% to 4.87% with the adequate tuning of the rheological parameters of cement paste. Similarly, Jin *et al.* [56] described cementitious materials' rheological, mechanical, and structural stability properties. Their results showed the mortar's optimal printability and compressive strength, producing a pentagram of 16 layers without deformation. Mazhoud *et al.* [57] showed the pumpability, shape stability, and rheological and permeability properties of 3D-printed cement samples. The samples were 3D-printed both in air and underwater. The investigation noted that underwater 3D-printed samples displayed lower compressive strength with an increase in the print speed [57]. However, the elastic modulus displayed an increasing trend.

Experimental studies on cement-based materials demonstrated adequate extrudability and buildability, essential for 3D printing. However, detailed structural properties, particularly regarding the workability of the printing cementitious-based materials, need to be addressed. Consequently, continued efforts to develop novel cementitious-based materials for AM techniques can pave the way for more commercial utilization of 3D printing in construction applications.

16.3.2 Polymer and Polymer Composites

Polymer-based concrete materials have gained popularity in the B&C industry due to their excellent characteristics such as waterproof, heat insulation, wear resistance, recyclability, light weight, and good acoustic properties [58]. Many types of polymers have been utilized in developing polymer composites-based construction materials such as vinyl, polyester, resins, and furnace. For instance, Attabadi *et al.* [59] established that polyester resin combined with lightweight clay aggregates displayed high performance with higher strength in destructive tests. Similarly, Ferdous *et al.* demonstrated [60] the epoxy polymer concrete for civil structures building applications. The investigations established that the ideal resin-to-filler ratio was 70:30 to develop a layered material with good mechanical strength. Further, polymers and polymer-based composites can be examined for AM in construction since they are commonly used for 3D printing products in aerospace, automobile, and medical applications. From that perspective, Furet *et al.* [61] demonstrated the construction of polymer foam and concrete walls using 3D printing. Batiprint3D™ was used to build the structure, providing a promising solution for producing complex shapes such as roofs and ceilings. A seminal investigation by Krčma *et al.* [62] pioneered the use of concrete mixed with glass sand and polypropylene for 3D printing. The extruded material's mechanical performance was significantly similar to casted models. Moreover, the 3D-printed models displayed a better degree of porosity. In other studies, Mele *et al.* [63] studied the associations relating to procedure requirements and the geometry of the extruded stands of clay paste-reinforced scrap polymer powder. The analysis discovered that the quality of the extrusion differs with nozzle acceleration, printing speed, and material composition. In addition, higher polymer content results in shape variations and increased impact resistance contrasted to pure clay. Consequently, the fabrication of structures with polymers and polymer composites has been gaining more attention. For instance, ETH Zurich developed a polymer-based 3D mesh structure with a class of robot that can build a large-scale construction [64].

16.3.3 Sustainable Materials

In the new global economy, sustainability in the construction sector has become a central issue [9]. In this context, the materials used for AM should be chosen carefully. The selection of materials must be based on their degree of environmental impact and innovative eco-friendly solutions. With the incorporation of AM in B&C industries, the reuse and recycling of construction and demolition waste could steer effective enforcement of sustainability in the B&C industry. For that reason, researchers and industries are also exploring the incorporation of waste by-products generated by consumers or industries in producing construction materials.

Geopolymers have been researched during the past few decades as a sustainable construction material that evolved significantly in the construction industry [65, 66]. Considered an alternative to Portland cement, geopolymer mortars and geopolymer cements decrease the carbon footprint compared to conventional cement [67]. Furthermore, geopolymer materials possess outstanding physiochemical properties beneficial for construction applications. A schematic illustration of the production of geopolymer is shown in Figure 16.4 [67]. Al-Qutaifi *et al.* [65] produced a geopolymer columns in a special timber mold using 3D printing. The results established that geopolymers as build-up materials in layered methods had higher flexural strength and good durability. In recent studies, Tran *et al.* [68] demonstrated that 3D-printed geopolymer concrete consists of a printable region of slump between 15 and 30 mm, and the range of slump flow was 210–230 mm. The study also stated that sufficient borax content (2–4%) could improve the open time but decreases the compressive strength by over 4%.

Fly ash and recycled waste products generated from the recycling of blast furnaces from steel plants, plastic waste and cement or aggregates waste from demolitions are commonly used to investigate the structural behaviours in civil engineering applications [69]. For instance, Lin *et al.* [70] explored the compressive properties of the mortar produced from waste materials for construction 3D printing. The mortar was prepared using fly ash and biochar, reinforced with acrylonitrile butadiene styrene, and 3D-printed

Figure 16.4 Overview of geopolymer production. Reproduced from Ref [67]. Copyrights, Elsevier.

a building component. Mortars reinforced with polymer provided similar ductility and compressive strength compared to pure mortars. In another study, the printability of mortar composed of recycled sand was reported by Zou *et al.* [71]. Results show that recycled sand can meet reasonable printability requirements and provide an extensive printability window with higher strength compared to natural sand. According to the analysis, using recycled sand to address the shortage of river sand can be beneficial and provides promising applications for 3D printing in B&C applications.

For the task of developing the construction structures using 3D printing techniques, several other polymers such as carbon fiber reinforcement [72], glass fiber reinforcement [73], flax fibers [74], polypropylene, and polyvinyl alcohol [75] were also investigated. Additives are also incorporated to enhance the printability properties of the materials [49]. Building and construction also aims to produce lightweight and functionally graded structures to encourage sustainability and minimize carbon footprint and energy consumption [76]. Drawing on this concept, the Keating and Oxman group from MIT developed lightweight 3D-printed concrete structures by combining aluminium powder, lime, and concrete [77]. Other similar studies performed by Aghaei-Meibodi *et al.* and Rippman *et al.* [78] also proposed methods to produce lightweight structures using 3D printing techniques and novel materials compositions. The significant advances in the materials domain and 3D printing machines will potentially allow one to customize material properties according to the structural functions. However, a detailed assessment of the cost of materials and optimum combinations of materials formulations are needed to make AM more feasible than conventional methods for construction applications.

16.4 Properties of the Printable Construction Materials

Most printable construction materials need to hold four fundamental properties. Those are extrudability, printability, buildability, and open time. However, the characteristics will differ based on the composition of material, formulation technique, and deposition procedures. The procedures may also affect the strength of materials and hardening properties. Lee *et al.* [46] first introduced the analysis technique for 3D-printed concrete. Their qualitative analysis is related to mix designs and the properties required for high-performance mortar for concrete printing. Moreover, the compressive strengths of the deposited or extruded material should be equivalent to cast products, with a range between 80% to 100% [46].

16.4.1 Extrudability

Extrudability of the materials is often related to the competence to transport the mixed mortar or concrete through a nozzle with a constant flow. Gravity is essential for transporting the material from a nozzle to the printing surface during extrusion. Moreover, after deposition, the force of gravity tends to deform the layers related to the materials' strength and rheological properties. Roussel and Cussot [79] and Kaushik *et al.* [80] developed a slump flow theory to understand the material strength and gravitationally induced forces. Similarly, Jacobsen *et al.* [81] worked with mortar and concrete flow conditions in different pipes to understand the properties of extruded concrete. Mu *et al.* [82] theoretically and experimentally demonstrated the extrudability results of cement mixed with fibers. The results established that the viscosity of the cementitious material plays a vital role in extruding the material. In another study by Lu *et al.* [46], the extrudability of concrete was analyzed with 9-mm broad cement strands printed from a 9-mm nozzle. The analysis was conducted to build a free-form structure without blockage or strand fracture. From the analysis, a mix of 60% sand and 40% binder showed optimum extrudability [46]. Despite the advancements in testing the extrudability of 3D printing construction materials, a suitable method is still to be determined.

16.4.2 Printability

Printing the material using the depositing method needs to follow a pre-defined path. The path positions the material layer by layer to realize the designed model. During this process, the length of material extrusion and print speed play a significant role in obtaining an efficient structure (Figure 16.5). Its success also depends on the layer cycle time that could impact the interlaying bond strength between the overlaying layers. Lim *et al.* [83] reported that a circular nozzle with 9-mm operational speeds of 300 mm/s could produce a substantial structure. Zou *et al.* [71] demonstrated the printability of recycled sand by depositing the material in the space of 3.0 m × 3.0 m × 4.0 m in X, Y, and Z directions. The printability of printed strands was found to be within the nozzle's widths. Moreover, the analysis highlighted that the fluidity of the printing mortars ranges from 168 mm to 201 mm. In another study, Kazemian *et al.* [84] demonstrated the print quality of concrete. The outcomes established that the mixture could be considered a good printing mixture based on three aspects: (a) the printed surfaces are free from defects with continuous flow, (b) layer

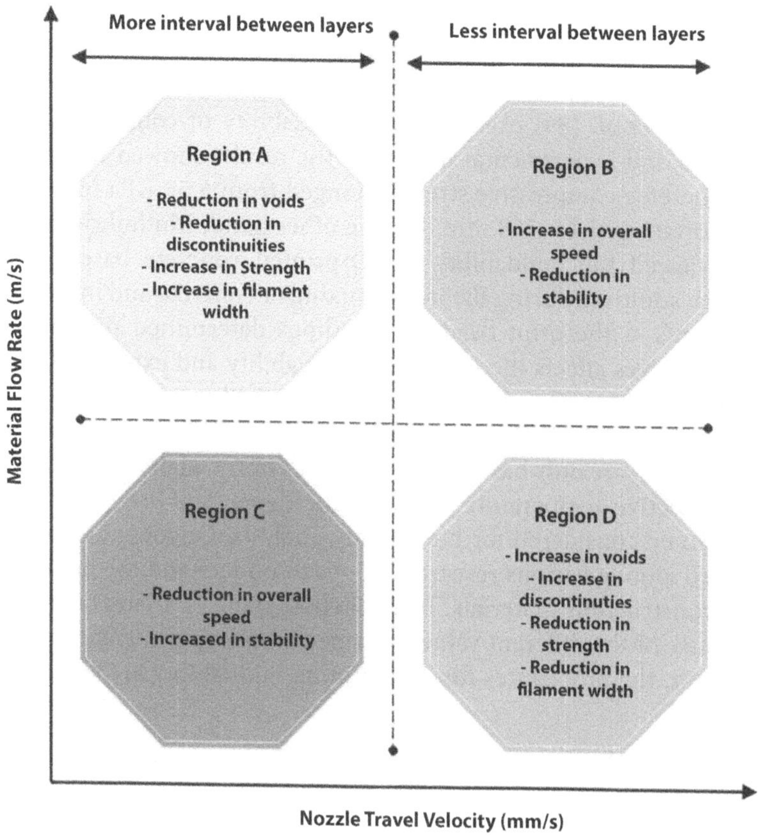

Figure 16.5 Influences of print speed and material flow on printability. Reproduced from Ref [85]. Copyrights, Elsevier.

edges required to be squared instead of rounded, and (c) the printed layer must satisfy the geometric consistency.

16.4.3 Buildability

Additive manufacturing techniques do not require mold that support the builds that are different from traditional construction procedures. As a result, the builds can collapse during the development process due to the deformation of the layers or the load on top layers. Because of this, the buildability of the materials is an essential property to consider while assessing the suitability of the material in 3D printing construction models [86]. However, the buildability refers to various processes and properties, including material failure, loss of stability, yield strength, material hardening, and rigidity of the produced structures [43, 86–88]. The literature has recommended various approaches to theoretically and experimentally evaluate buildability.

Di Carlo *et al.* [89] examined the buildability of concrete based on numerical and experimental methods. The results showed that the layered concrete's compressive strength ranged from 5.52 KPa to 88.3 KPa for a mortar aged 11–288 min. On the other hand, Muthukrishan *et al.* [90] reviewed the buildability of 3D-printed concrete based on two methods: additives during the initial mixing of concrete and interference adjustments at the print head. The findings determined that incorporating additives affects the material's pumpability and extrudability. The modifications at the print head increased the yield strength before extrusion but did not impact the extrudability. Evaluation of current buildability models are only based on the strength of the materials [91, 92]. In this perspective, temperature [93] and time-dependent [94, 95] functions must also be considered for buildability analysis. Ongoing assumptions are tested among various research groups to understand the buildability of the construction materials. Nevertheless, there is a lack of unanimity because of the different values obtained from experimental methods. As a result, the procedures for determining buildability are still in their infancy.

16.4.4 Open Time

Open time is another aspect of evaluating the material properties in AM technique. In general, the Vicat apparatus is used to assess open time [46]. However, the technique is proposed to measure the first and last setting time. Therefore, the approach is unreliable in analyzing the open time of fresh concretes in printing. Because of this, researchers have been exploring the use of the slump test for examining the open time. For instance, Kim *et al.* [96] reported open time using a slump loss test of cement pastes for up to 90 min. In contrast, Alhozaimy *et al.* [97] used the standard cone apparatus to test the combination of concrete and limestone powder to evaluate the slump loss at 15, 30, 60, 90, 120, and 150 min, respectively. The evaluation was conducted to determine the length of time that cement paste can be worked. In contrast, Le *et al.* [46] derived workability time using the printable concrete's shear strength analysis. Open-time analysis of 3D printing construction materials can provide the appropriate period for materials to be formulated and utilized, enabling continuous extrusion without any defects, or clogging of the nozzles.

16.5 Applications of AM in B&C Industries

Although 3D printing in construction is still in its early stages, promising structures have already been developed using AM methods [98].

16.5.1 3D Printing of Building

With advancements in the building sectors, Dutch architects constructed the first 3D-printed house in 2014 [99]. Sand and cement were used to print the building [99]. Later, WinSun Company in China printed a five-storey building [100]. In Dubai, fiber reinforced gypsum and cement was used to 3D-print an office building within 48 h [49]. In 2022, a Melbourne-based company built an entire house using 3D printing to solve the housing shortage in Australia [101]. Fortex Australia used the concrete-based printer to build houses up to 12-m wide, three storeys tall, and 50-m deep [101].

16.5.2 3D Printing of Bridges

Large-scale 3D printing of high-performance structures is minor compared to buildings developed using AM techniques. However, bridges in various parts of the world have been constructed using 3D printing processes. For example, a company in Spain constructed a pedestrian bridge with dimensions of 12-m long and 1.75-m wide using D-shape AM technology [98, 102]. Similarly, in the Netherlands, a concrete bridge was 3D printed by BAM Infra. A freestanding concrete footbridge was developed using low-carbon ECOPact concrete ink that can be easily disassembled and recycled [98]. The footbridge was a collaborative project between ETH Zurich, Zaha Hadid Architects, and incremental3d [103]. Similarly, a 3D-printed concrete staircase with a thickness of 5 mm was created in Scotland [104]. With further advancements based on these commercialized projects, a cutting-edge AM technique could produce a sustainable approach to improve current environmental issues.

16.6 Challenges and Future Developments of AM in B&C Applications

In recent years, AM integration in B&C applications has paved the way for new possibilities for the digital manufacturing of large-scale structures [105]. With the incorporation of AM, workplace safety has increased, and manual work has decreased drastically. In contrast, traditional construction techniques often involve a high level of physical work by laborers and are often unsafe, resulting in worker fatalities [2, 105, 106]. AM techniques can be utilized in building structures in harsh environments such as war zones and natural disaster-affected areas [83, 105]. Further, AM can minimize material usage in developing large-scale structures and reduce labor costs. Those factors are beneficial for increasing production performance. Although current research studies and commercial applications suggest a potential usage of AM in construction, multiple challenges must be addressed to integrate AM in B&C industries fully.

In terms of equipment, more technologies need to be established employing on-site resources, particularly for concrete printing. For instance, the development of multi-nozzle printer heads that can deposit the multimaterials. Novel materials must be evaluated using a broad range of reinforcements such as fiber, steel, and meshes [29, 83, 105]. Furthermore, they also need to deliver structural safety and stability. Currently, there are limited design guidelines for 3D printing in B&C applications. Consequently, standard guidelines for design and operations are required for safe work operations using AM in large-scale developments [8, 49, 83]. Large, randomized controlled trials could make more definitive determinations of a wide range of materials for construction applications [27, 85]. To ensure consistent and reliable structural applications, more information on quality controls and procedures must be developed specifically for on-site applications.

16.7 Conclusions

The rapid developments of AM in construction applications increase construction productivity, cost reduction, and efficient utilization of resources. Thus far, few projects have been completed using 3D printing for commercial building applications. However, in-depth understanding of construction materials and rapid development of AM in B&C could focus on standardizing materials and structural designs. Therefore, the construction industry can adopt AM to facilitate developments in a fast-changing world. This chapter highlighted the incorporation of AM techniques in the B&C industry and the importance of novel materials formulation and their ideal properties. In summarizing the knowledge as presented, with the adoption of AM in construction applications, the B&C industries can also promote Industry 4.0 attributes and sustainable development.

References

1. Robinson, G., J.L., T.W., Future of construction, Available online: https://www.oxfordeconomics.com/resource/future-of-construction/.
2. Paolini, A., Kollmannsberger, S., Rank, E., Additive manufacturing in construction: A review on processes, applications, and digital planning methods. *Addit. Manuf.*, *30*, 100894, **2019**, doi: 10.1016/j.addma.2019.100894.
3. Labonnote, N., Rønnquist, A., Manum, B., Rüther, P., Additive construction: State-of-the-art, challenges and opportunities. *Autom. Constr.*, *72*, 347–366, **2016**, doi: 10.1016/j.autcon.2016.08.026.
4. Barbosa, F., Woetzel, J., Mischke, J., Ribeirinho, M.J., Sridhar, M., Parsons, M., N.B., S.B., Reinventing construction through a productivity revolution. 2017.
5. Nasir, H., Ahmed, H., Haas, C., Goodrum, P.M., An analysis of construction productivity differences between Canada and the United States. *Constr. Manage. Econ.*, *32*, 595–607, **2014**, doi: 10.1080/01446193.2013.848995.
6. Klotz, L.E., Horman, M., Bodenschatz, M., A lean modeling protocol for evaluating green project delivery. *Lean Constr. Inst.*, *3*, 1–18, **2007**.
7. Wu, P., Wang, J., Wang, X., A critical review of the use of 3-D printing in the construction industry. *Autom. Constr.*, *68*, 21–31, **2016**, doi: 10.1016/j.autcon.2016.04.005.
8. De Schutter, G., Lesage, K., Mechtcherine, V., Nerella, V.N., Habert, G., Agusti-Juan, I., Vision of 3D printing with concrete — Technical, economic and environmental potentials. *Cem. Concr. Res.*, *112*, 25–36, **2018**, doi: 10.1016/j.cemconres.2018.06.001.
9. Lima, L., Trindade, E., Alencar, L., Alencar, M., Silva, L., Sustainability in the construction industry: A systematic review of the literature. *J. Cleaner Prod.*, *289*, 125730, **2021**, doi: 10.1016/j.jclepro.2020.125730.
10. Abu-Ennab, L., Dixit, M.K., Birgisson, B., Pradeep Kumar, P., Comparative life cycle assessment of large-scale 3D printing utilizing kaolinite-based calcium sulfoaluminate cement concrete and conventional construction. *Clean. Environ. Syst.*, *5*, 100078, **2022**, doi: 10.1016/j.cesys.2022.100078.
11. Kittusamy, N.K. and Buchholz, B., Whole-body vibration and postural stress among operators of construction equipment: A literature review. *J. Saf. Res.*, *35*, 255–261, **2004**, doi: 10.1016/j.jsr.2004.03.014.
12. Hinze, J.W. and Teizer, J., Visibility-related fatalities related to construction equipment. *Saf. Sci.*, *49*, 709–718, **2011**, doi: 10.1016/j.ssci.2011.01.007.
13. Man, S.S., Chan, A.H.S., Alabdulkarim, S., Quantification of risk perception: Development and validation of the construction worker risk perception (CoWoRP) scale. *J. Saf. Res.*, *71*, 25–39, **2019**, doi: 10.1016/j.jsr.2019.09.009.
14. Nowotarski, P. and Paslawski, J., Industry 4.0 concept introduction into construction SMEs. *IOP Conf. Ser. Mater. Sci. Eng.*, *245*, 52043, **2017**, doi: 10.1088/1757-899X/245/5/052043.

15. Turner, C.J., Oyekan, J., Stergioulas, L., Griffin, D., Utilizing industry 4.0 on the construction site: Challenges and opportunities. *IEEE Trans. Industr. Inform.*, *17*, 746–756, **2021**, doi: 10.1109/TII.2020.3002197.
16. Alaloul, W.S., Liew, M.S., Zawawi, N.A.W.A., Kennedy, I.B., Industrial revolution 4.0 in the construction industry: Challenges and opportunities for stakeholders. *Ain Shams Eng. J.*, *11*, 225–230, **2020**, doi: 10.1016/j.asej.2019.08.010.
17. El Jazzar, M., Schranz, C., Urban, H., Nassereddine, H., Integrating construction 4.0 technologies: A four-layer implementation plan. *Front. Built Environ.*, *7*, 1–14, **2021**, doi: 10.3389/fbuil.2021.671408.
18. Jhinkwan, A., Kalsi, S., Pankaj, An overview on 3D metal printing technology in automobile industry, p. 020028, 2023.
19. Kantaros, A., Ganetsos, T., Piromalis, D., 3D and 4D printing as integrated manufacturing methods of industry 4.0. *Am. J. Eng. Appl. Sci.*, *16*, 12–22, **2023**, doi: 10.3844/ajeassp.2023.12.22.
20. Serrano, D.R., Kara, A., Yuste, I., Luciano, F.C., Ongoren, B., Anaya, B.J., Molina, G., Diez, L., Ramirez, B.I., Ramirez, I.O. *et al.*, 3D printing technologies in personalized medicine, nanomedicines, and biopharmaceuticals. *Pharmaceutics*, *15*, 313, **2023**, doi: 10.3390/pharmaceutics15020313.
21. Li, X., Fan, L., Liu, Y., Li, J., New insights into food O/W emulsion gels: Strategies of reinforcing mechanical properties and outlook of being applied to food 3D printing. *Crit. Rev. Food Sci. Nutr.*, *63*, 1564–1586, **2023**, doi: 10.1080/10408398.2021.1965953.
22. El-Sayegh, S., Romdhane, L., Manjikian, S., A critical review of 3D printing in construction: Benefits, challenges, and risks. *Arch. Civ. Mech. Eng.*, *20*, 34, **2020**, doi: 10.1007/s43452-020-00038-w.
23. Rossi, E., Di Nicolantonio, M., Barcarolo, P., Lagatta, J., Sustainable 3D printing: Design opportunities and research perspectives, in: *Proceedings of the Advances in Additive Manufacturing, Modeling Systems and 3D Prototyping*, M. Di Nicolantonio, E. Rossi, T. Alexander, (Eds.), pp. 3–15, Springer International Publishing, Cham, 2020.
24. Dey, D., Srinivas, D., Panda, B., Suraneni, P., Sitharam, T.G., Use of industrial waste materials for 3D printing of sustainable concrete: A review. *J. Cleaner Prod.*, *340*, 130749, **2022**, doi: 10.1016/j.jclepro.2022.130749.
25. Cao, X., Yu, S., Cui, H., Li, Z., 3D printing devices and reinforcing techniques for extruded cement-based materials: A review. *Buildings*, *12*, 453, **2022**, doi: 10.3390/buildings12040453.
26. Bos, F., Wolfs, R., Ahmed, Z., Salet, T., Additive manufacturing of concrete in construction: Potentials and challenges of 3D concrete printing. *Virtual Phys. Prototyp.*, *11*, 209–225, **2016**, doi: 10.1080/17452759.2016.1209867.
27. Xiao, J., Ji, G., Zhang, Y., Ma, G., Mechtcherine, V., Pan, J., Wang, L., Ding, T., Duan, Z., Du, S., Large-scale 3D printing concrete technology: Current status and future opportunities. *Cem. Concr. Compos.*, *122*, 104115, **2021**, doi: 10.1016/j.cemconcomp.2021.104115.

28. Shakor, P., Chu, S.H., Puzatova, A., Dini, E., Review of binder jetting 3D printing in the construction industry. *Prog. Addit. Manuf.*, *7*, 643–669, **2022**, doi: 10.1007/s40964-021-00252-9.
29. Mechtcherine, V., Bos, F.P., Perrot, A., da Silva, W.R.L., Nerella, V.N., Fataei, S., Wolfs, R.J.M., Sonebi, M., Roussel, N., Extrusion-based additive manufacturing with cement-based materials – production steps, processes, and their underlying physics: A review. *Cem. Concr. Res.*, *132*, 106037, **2020**, doi: https://doi.org/10.1016/j.cemconres.2020.106037.
30. Rajasekar, R., Sumanta, S., Joong, H.L., Ashok, K.D., Mahalakshmi, S., Sathish Kumar, P., Santhosh, S., Carbon-based multi-layered films for electronic application: A review. *J. Electron. Mater.*, 50, 4, 1845–1892, **2021**, doi: 10.1007/s11664-020-08724-4.
31. Na, O., Kim, K., Lee, H., Lee, H., Printability and setting time of CSA cement with Na2SiO3 and gypsum for binder jetting 3D printing. *Mater. (Basel)*, *14*, 2811, **2021**, doi: 10.3390/ma14112811.
32. Bai, Y., Wagner, G., Williams, C.B., Effect of particle size distribution on powder packing and sintering in binder jetting additive manufacturing of metals. *J. Manuf. Sci. Eng.*, *139*, 081019, **2017**, doi: 10.1115/1.4036640.
33. Shakor, P., Nejadi, S., Paul, G., Sanjayan, J., Dimensional accuracy, flowability, wettability, and porosity in inkjet 3DP for gypsum and cement mortar materials. *Autom. Constr.*, *110*, 102964, **2020**, doi: 10.1016/j.autcon.2019.102964.
34. Salari, F., Bosetti, P., Sglavo, V.M., Binder jetting 3D printing of magnesium oxychloride cement-based materials: Parametric analysis of manufacturing factors. *J. Manuf. Mater. Process.*, *6*, 86, **2022**, doi: 10.3390/jmmp6040086.
35. Gibson, I., Rosen, D., Stucker, B., *Additive manufacturing technologies*, Springer New York, New York, NY, 2015, ISBN 978-1-4939-2112-6.
36. Delgado Camacho, D., Clayton, P., O'Brien, W.J., Seepersad, C., Juenger, M., Ferron, R., Salamone, S., Applications of additive manufacturing in the construction industry – a forward-looking review. *Autom. Constr.*, *89*, 110–119, **2018**, doi: 10.1016/j.autcon.2017.12.031.
37. Perrot, A., Rangeard, D., Nerella, V.N., Mechtcherine, V., Extrusion of cement-based materials - an overview. *RILEM Tech. Lett.*, *3*, 91–97, **2019**, doi: 10.21809/rilemtechlett.2018.75.
38. Mohan, M.K., Rahul, A.V., De Schutter, G., Van Tittelboom, K., Extrusion-based concrete 3D printing from a material perspective: A state-of-the-art review. *Cem. Concr. Compos.*, *115*, 103855, **2021**, doi: 10.1016/j.cemconcomp.2020.103855.
39. Kruger, J., du Plessis, A., van Zijl, G., An Investigation into the porosity of extrusion-based 3D printed concrete. *Addit. Manuf.*, *37*, 101740, **2021**, doi: 10.1016/j.addma.2020.101740.
40. Ding, T., Xiao, J., Zou, S., Zhou, X., Anisotropic behavior in bending of 3D printed concrete reinforced with fibers. *Compos. Struct.*, *254*, 112808, **2020**, doi: 10.1016/j.compstruct.2020.112808.

41. Kim, H. and Lee, S., Printability and physical properties of iron slag powder composites using material extrusion-based 3D printing. *J. Iron Steel Res. Int.*, *28*, 111–121, **2021**, doi: 10.1007/s42243-020-00475-0.
42. Buswell, R.A., Leal de Silva, W.R., Jones, S.Z., Dirrenberger, J., 3D printing using concrete extrusion: A roadmap for research. *Cem. Concr. Res.*, *112*, 37–49, **2018**, doi: 10.1016/j.cemconres.2018.05.006.
43. Ji, G., Ding, T., Xiao, J., Du, S., Li, J., Duan, Z., A 3D printed ready-mixed concrete power distribution substation: Materials and construction technology. *Mater. (Basel)*, *12*, 1540, **2019**, doi: 10.3390/ma12091540.
44. Yu, S., Du, H., Sanjayan, J., Aggregate-bed 3D concrete printing with cement paste binder. *Cem. Concr. Res.*, *136*, 106169, **2020**, doi: 10.1016/j.cemconres.2020.106169.
45. Khoshnevis, B., Automated construction by contour crafting—Related robotics and information technologies. *Autom. Constr.*, *13*, 5–19, **2004**, doi: 10.1016/j.autcon.2003.08.012.
46. Le, T.T., Austin, S.A., Lim, S., Buswell, R.A., Gibb, A.G.F., Thorpe, T., Mix design and fresh properties for high-performance printing concrete. *Mater. Struct.*, *45*, 1221–1232, **2012**, doi: 10.1617/s11527-012-9828-z.
47. Panda, B., Lim, J.H., Noor Mohamed, N.A., Paul, S.C., Tay, Y.W.D., Tan, M.J., Automation of robotic concrete printing using feedback control system, July 1 2017.
48. Mahesh Kumar, K.V., Krishnamurthy, K., Sathish Kumar, P., Sahoo, S., Uddin, E., Pal, S.K., Rajasekar, R., Research updates on graphene oxide based polymeric nanocomposites. *Polym. Compos.*, 35, 2297–2310, **2014**, doi: 10.1002/pc.22899.
49. Craveiro, F., Duarte, J.P., Bartolo, H., Bartolo, P.J., Additive manufacturing as an enabling technology for digital construction: A perspective on construction 4.0. *Autom. Constr.*, *103*, 251–267, **2019**, doi: 10.1016/j.autcon.2019.03.011.
50. Robayo-Salazar, R., Mejía de Gutiérrez, R., Villaquirán-Caicedo, M.A., Delvasto Arjona, S., 3D printing with cementitious materials: Challenges and opportunities for the construction sector. *Autom. Constr.*, *146*, 104693, **2023**, doi: 10.1016/j.autcon.2022.104693.
51. Lu, B., Weng, Y., Li, M., Qian, Y., Leong, K.F., Tan, M.J., Qian, S., A systematical review of 3D printable cementitious materials. *Constr. Build. Mater.*, *207*, 477–490, **2019**, doi: 10.1016/j.conbuildmat.2019.02.144.
52. Navaratnam, S., Tushar, Q., Jahan, I., Zhang, G., Environmental sustainability of industrial waste-based cementitious materials: A review, experimental investigation and life-cycle assessment. *Sustainability*, *15*, 1873, **2023**, doi: 10.3390/su15031873.
53. Perrot, A., Rangeard, D., Pierre, A., Structural built-up of cement-based materials used for 3D-printing extrusion techniques. *Mater. Struct.*, *49*, 1213–1220, **2016**, doi: 10.1617/s11527-015-0571-0.

54. Pierre, A., Weger, D., Perrot, A., Lowke, D., Penetration of cement pastes into sand packings during 3D printing: Analytical and experimental study. *Mater. Struct.*, *51*, 22, **2018**, doi: 10.1617/s11527-018-1148-5.
55. Chen, M., Yang, L., Zheng, Y., Huang, Y., Li, L., Zhao, P., Wang, S., Lu, L., Cheng, X., Yield stress and thixotropy control of 3D-printed calcium sulfoaluminate cement composites with metakaolin related to structural build-up. *Constr. Build. Mater.*, *252*, 119090, **2020**, doi: 10.1016/j.conbuildmat.2020.119090.
56. Jin, Y., Xu, J., Li, Y., Zhao, Z., Chen, M., Lu, L., Li, L., Rheological properties, shape stability and compressive strength of 3D printed colored cement composites modified by needle-like pigment. *Addit. Manuf.*, *57*, 102965, **2022**, doi: 10.1016/j.addma.2022.102965.
57. Mazhoud, B., Perrot, A., Picandet, V., Rangeard, D., Courteille, E., Underwater 3D printing of cement-based mortar. *Constr. Build. Mater.*, *214*, 458–467, **2019**, doi: 10.1016/j.conbuildmat.2019.04.134.
58. Shen, J., Liang, J., Lin, X., Lin, H., Yu, J., Yang, Z., Recent progress in polymer-based building materials. *Int. J. Polym. Sci.*, *2020*, 1–15, **2020**, doi: 10.1155/2020/8838160.
59. Sanaei Ataabadi, H., Zare, A., Rahmani, H., Sedaghatdoost, A., Mirzaei, E., Lightweight dense polymer concrete exposed to chemical condition and various temperatures: An experimental investigation. *J. Build. Eng.*, *34*, 101878, **2021**, doi: 10.1016/j.jobe.2020.101878.
60. Ferdous, W., Manalo, A., Wong, H.S., Abousnina, R., AlAjarmeh, O.S., Zhuge, Y., Schubel, P., Optimal design for epoxy polymer concrete based on mechanical properties and durability aspects. *Constr. Build. Mater.*, *232*, 117229, **2020**, doi: 10.1016/j.conbuildmat.2019.117229.
61. Furet, B., Poullain, P., Garnier, S., 3D printing for construction based on a complex wall of polymer-foam and concrete. *Addit. Manuf.*, *28*, 58–64, **2019**, doi: 10.1016/j.addma.2019.04.002.
62. Krčma, M., Škaroupka, D., Vosynek, P., Zikmund, T., Kaiser, J., Palousek, D., Use of polymer concrete for large-scale 3D printing. *Rapid Prototyp. J.*, *27*, 465–474, **2021**, doi: 10.1108/RPJ-12-2019-0316.
63. Mele, M., Campana, G., Pisaneschi, G., De Martino, L., Ricciarelli, M., Dimensional accuracy and impact resistance of 3D printed clay reinforced with scrap polymer powder. *Rapid Prototyp. J.*, 1510–1522, **2023**, doi: 10.1108/RPJ-10-2022-0361.
64. Giftthaler, M., Sandy, T., Dörfler, K., Brooks, I., Buckingham, M., Rey, G., Kohler, M., Gramazio, F., Buchli, J., Mobile robotic fabrication at 1:1 scale: The *in situ* fabricator. *Construct. Rob.*, *1*, 3–14, **2017**, doi: 10.1007/s41693-017-0003-5.
65. Al-Qutaifi, S., Nazari, A., Bagheri, A., Mechanical properties of layered geopolymer structures applicable in concrete 3D-printing. *Constr. Build. Mater.*, *176*, 690–699, **2018**, doi: 10.1016/j.conbuildmat.2018.04.195.

66. Imtiaz, L., Rehman, S.K.U., Ali Memon, S., Khizar Khan, M., Faisal Javed, M., A review of recent developments and advances in eco-friendly geopolymer concrete. *Appl. Sci.*, *10*, 7838, **2020**, doi: 10.3390/app10217838.
67. Hassan, A., Arif, M., Shariq, M., Use of geopolymer concrete for a cleaner and sustainable environment – a review of mechanical properties and microstructure. *J. Cleaner Prod.*, *223*, 704–728, **2019**, doi: 10.1016/j.jclepro.2019.03.051.
68. Tran, M.V., Vu, T.H., Nguyen, T.H.Y., Simplified assessment for one-part 3D-printable geopolymer concrete based on slump and slump flow measurements. *Case Stud. Constr. Mater.*, *18*, e01889, **2023**, doi: 10.1016/j.cscm.2023.e01889.
69. Sivakumar Babu, G.L. and Raja Jaladurgam, M.E., Strength and deformation characteristics of fly ash mixed with randomly distributed plastic waste. *J. Mater. Civ. Eng.*, *26*, 458–467, **2014**, doi: 10.1061/(ASCE)MT.1943-5533.0001014.
70. Lin, A., Kiat Tan, Y., Wang, C.-H., Wei Kua, H., Taylor, H., Utilization of waste materials in a novel mortar–polymer laminar composite to be applied in construction 3D-printing. *Compos. Struct.*, *253*, 112764, **2020**, doi: 10.1016/j.compstruct.2020.112764.
71. Zou, S., Xiao, J., Ding, T., Duan, Z., Zhang, Q., Printability and advantages of 3D printing mortar with 100% recycled sand. *Constr. Build. Mater.*, *273*, 121699, **2021**, doi: 10.1016/j.conbuildmat.2020.121699.
72. Ferrari, V.J., de Hanai, J.B., de Souza, R.A., Flexural strengthening of reinforcement concrete beams using high performance fiber reinforcement cement-based composite (HPFRCC) and carbon fiber reinforced polymers (CFRP). *Constr. Build. Mater.*, *48*, 485–498, **2013**, doi: 10.1016/j.conbuildmat.2013.07.026.
73. Ahmad, J., González-Lezcano, R.A., Majdi, A., Ben Kahla, N., Deifalla, A.F., El-Shorbagy, M.A., Glass fibers reinforced concrete: Overview on mechanical, durability and microstructure analysis. *Mater. (Basel)*, *15*, 5111, **2022**, doi: 10.3390/ma15155111.
74. Yuhazri, M.Y., Zulfikar, A.J., Ginting, A., Fiber reinforced polymer composite as a strengthening of concrete structures: A review. *IOP Conf. Ser. Mater. Sci. Eng.*, *1003*, 12135, **2020**, doi: 10.1088/1757-899X/1003/1/012135.
75. Ohama, Y., Recent progress in concrete-polymer composites. *Adv. Cem. Based Mater.*, *5*, 31–40, **1997**, doi: 10.1016/S1065-7355(96)00005-3.
76. Kumaran, S.T., Ko, T.J., Kumar, S.S., Varol, T., *Materials for lightweight constructions*, CRC Press, Boca Raton, 2022, ISBN 9781003252108.
77. Pajonk, A., Prieto, A., Blum, U., Knaack, U., Multi-material additive manufacturing in architecture and construction: A review. *J. Build. Eng.*, *45*, 103603, **2022**, doi: 10.1016/j.jobe.2021.103603.
78. Rippmann, M., Liew, A., Van Mele, T., Block, P., Design, fabrication and testing of discrete 3D sand-printed floor prototypes. *Mater. Today Commun.*, *15*, 254–259, **2018**, doi: 10.1016/j.mtcomm.2018.03.005.
79. Gosselin, C., Duballet, R., Roux, P., Gaudillière, N., Dirrenberger, J., Morel, P., Large-scale 3D printing of ultra-high performance concrete – a new

processing route for architects and builders. *Mater. Des.*, *100*, 102–109, **2016**, doi: 10.1016/j.matdes.2016.03.097.

80. Kaushik, S., Sonebi, M., Amato, G., Perrot, A., Das, U.K., Influence of nanoclay on the fresh and rheological behaviour of 3D printing mortar. *Mater. Today Proc.*, *58*, 1063–1068, **2022**, doi: 10.1016/j.matpr.2022.01.108.
81. Jacobsen, S., Haugan, L., Hammer, T.A., Kalogiannidis, E., Flow conditions of fresh mortar and concrete in different pipes. *Cem. Concr. Res.*, *39*, 997–1006, **2009**, doi: 10.1016/j.cemconres.2009.07.005.
82. Mu, B., Li, Z., Chui, S.N.C., Peng, J., Cementitious composite manufactured by extrusion technique. *Cem. Concr. Res.*, *29*, 237–240, **1999**, doi: 10.1016/S0008-8846(98)00097-0.
83. Lim, S., Buswell, R.A., Le, T.T., Austin, S.A., Gibb, A.G.F., Thorpe, T., Developments in construction-scale additive manufacturing processes. *Autom. Constr.*, *21*, 262–268, **2012**, doi: 10.1016/j.autcon.2011.06.010.
84. Kazemian, A., Yuan, X., Cochran, E., Khoshnevis, B., Cementitious materials for construction-scale 3D printing: Laboratory testing of fresh printing mixture. *Constr. Build. Mater.*, *145*, 639–647, **2017**, doi: 10.1016/j.conbuildmat.2017.04.015.
85. Hamidi, F. and Aslani, F., Additive manufacturing of cementitious composites: Materials, methods, potentials, and challenges. *Constr. Build. Mater.*, *218*, 582–609, **2019**, doi: 10.1016/j.conbuildmat.2019.05.140.
86. Mechtcherine, V., Nerella, V.N., Will, F., Näther, M., Otto, J., Krause, M., Large-scale digital concrete construction – CONPrint3D concept for on-site, monolithic 3D-printing. *Autom. Constr.*, *107*, 102933, **2019**, doi: 10.1016/j.autcon.2019.102933.
87. Salet, T.A.M., Ahmed, Z.Y., Bos, F.P., Laagland, H.L.M., Design of a 3D printed concrete bridge by testing. *Virtual Phys. Prototyp.*, *13*, 222–236, **2018**, doi: 10.1080/17452759.2018.1476064.
88. Wolfs, R.J.M., Bos, F.P., Salet, T.A.M., Early age mechanical behaviour of 3D printed concrete: Numerical modelling and experimental testing. *Cem. Concr. Res.*, *106*, 103–116, **2018**, doi: 10.1016/j.cemconres.2018.02.001.
89. Di Carlo, T., Khoshnevis, B., Carlson, A., Experimental and numerical techniques to characterize structural properties of fresh concrete, in: *Proceedings of the Volume 9: Mechanics of Solids, Structures and Fluids*, November 2013, American Society of Mechanical Engineers.
90. Muthukrishnan, S., Ramakrishnan, S., Sanjayan, J., Technologies for improving buildability in 3D concrete printing. *Cem. Concr. Compos.*, *122*, 104144, **2021**, doi: 10.1016/j.cemconcomp.2021.104144.
91. Wolfs, R.J.M., Bos, F.P., Salet, T.A.M., Triaxial compression testing on early age concrete for numerical analysis of 3D concrete printing. *Cem. Concr. Compos.*, *104*, 103344, **2019**, doi: 10.1016/j.cemconcomp.2019.103344.
92. Roussel, N., Rheological requirements for printable concretes. *Cem. Concr. Res.*, *112*, 76–85, **2018**, doi: 10.1016/j.cemconres.2018.04.005.

93. Tramontin Souza, M., Simão, L., Guzi de Moraes, E., Senff, L., de Castro Pessôa, J.R., Ribeiro, M.J., Novaes de Oliveira, A.P., Role of temperature in 3D printed geopolymers: Evaluating rheology and buildability. *Mater. Lett.*, *293*, 129680, **2021**, doi: 10.1016/j.matlet.2021.129680.
94. Esposito, L., Menna, C., Asprone, D., Rossino, C., Marchi, M., An experimental testing procedure to assess the buildability performance of 3D printed concrete elements, in: *Second RILEM International Conference on Concrete and Digital Fabrication*. Bos, F., Lucas, S., Wolfs, R., Salet, T. (eds), RILEM Bookseries, vol. 28. pp. 235–245, Springer, Cham, DC, 2020.
95. Zhou, W., McGee, W., Zhu, H., Gökçe, H.S., Li, V.C., Time-dependent fresh properties characterization of 3D printing engineered cementitious composites (3DP-ECC): On the evaluation of buildability. *Cem. Concr. Compos.*, *133*, 104704, **2022**, doi: 10.1016/j.cemconcomp.2022.104704.
96. Kim, B.-G., Jiang, S.P., Aïtcin, P.-C., Slump improvement mechanism of alkalies in PNS superplasticized cement pastes. *Mater. Struct.*, *33*, 363–369, **2000**, doi: 10.1007/BF02479645.
97. Alhozaimy, A.M., Effect of absorption of limestone aggregates on strength and slump loss of concrete. *Cem. Concr. Compos.*, *31*, 470–473, **2009**, doi: 10.1016/j.cemconcomp.2009.04.010.
98. Siddika, A., Al Mamun, M.A., Ferdous, W., Saha, A.K., Alyousef, R., 3D-printed concrete: Applications, performance, and challenges. *J. Sustain. Cem. Mater.*, *9*, 127–164, **2020**, doi: 10.1080/21650373.2019.1705199.
99. Macguire, E., Dutch architects to build world's first 3D printed house.
100. Winsun, Available online: https://www.futureofconstruction.org/case/winsun/.
101. Meacham, S., Melbourne company 3D printing entire homes to solve housing crisis, Available online: https://www.9news.com.au/national/melbourne-company-fortex-3d-printing-entire-homes/710a97e6-7343-43d3-953f-0de832d59f32.
102. Buchanan, C. and Gardner, L., Metal 3D printing in construction: A review of methods, research, applications, opportunities and challenges. *Eng. Struct.*, *180*, 332–348, **2019**, doi: 10.1016/j.engstruct.2018.11.045.
103. Amber, A., 3D-printed bridge "combines ancient wisdoms and modern technology, Available online: https://www.dezeen.com/2023/03/09/striatus-3d-printed-concrete-bridge-zaha-hadid-architects-shajay-bhooshan-holcim-video/.
104. S, A., A bridge incorporating a 3D printed concrete staircase has been inaugurated in Scotland, available online: https://www.3dnatives.com/en/bridge-incorporating-3d-printed-staircase-inaugurated-in-scotland-310320236/.
105. Khan, M.S., Sanchez, F., Zhou, H., 3-D printing of concrete: Beyond horizons. *Cem. Concr. Res.*, *133*, 106070, **2020**, doi: 10.1016/j.cemconres.2020.106070.
106. Krimi, I., Lafhaj, Z., Ducoulombier, L., Prospective study on the integration of additive manufacturing to building industry—Case of a French construction company. *Addit. Manuf.*, *16*, 107–114, **2017**, doi: 10.1016/j.addma.2017.04.002.

Index

Printed and bound by CPI Group (UK) Ltd, Croydon, CR0 4YY

17/09/2026

14964279-0008